ORTHO'S®
HOME
GARDENER'S
PROBLEM SOLVER

Meredith® Books
Des Moines, Iowa

Ortho® Books
An imprint of Meredith® Books

Ortho's Home Gardener's Problem Solver
Editor: Michael McKinley
Contributing Editor: Cheryl Smith
Contributing Writer: L. Patricia Kite
Art Director: Tom Wegner
Assistant Art Director: Harijs Priekulis
Copy Chief: Catherine Hamrick
Copy and Production Editor: Terri Fredrickson
Book Production Managers: Pam Kvitne,
    Marjorie J. Schenkelberg
Contributing Proofreaders: Fran Gardner, Barbara J. Stokes
Contributing Map Illustrator: Jana Fothergill
Indexer: Ellen Davenport
Electronic Production Coordinator: Paula Forest
Editorial and Design Assistants: Kathleen Stevens,
    Karen Schirm

Meredith® Books
Editor in Chief: James D. Blume
Design Director: Matt Strelecki
Managing Editor: Gregory H. Kayko
Executive Ortho Editor: Larry Erickson

Director, Retail Sales and Marketing: Terry Unsworth
Director, Sales, Special Markets: Rita McMullen
Director, Sales, Premiums: Michael A. Peterson
Director, Sales, Retail: Tom Wierzbicki
Director, Book Marketing: Brad Elmitt
Director, Operations: George A. Susral
Director, Production: Douglas M. Johnston

Vice President, General Manager: Jamie L. Martin

Meredith Publishing Group
President, Publishing Group: Christopher M. Little
Vice President, Finance & Administration: Max Runciman

Meredith Corporation
Chairman and Chief Executive Officer: William T. Kerr
Chairman of the Executive Committee: E.T. Meredith III

This book is based on *The Ortho Problem Solver,*
Fifth Edition
Editor: Michael D. Smith
Contributing Writers: Clare A. Binko, Rick Bond, Gene
    Joyner, Wayne S. Moore, Robert D. Raabe, Deni Stein,
    Bernadine Strik, Lauren B. Swezey
Contributing Copy Editor: Barbara Fuller
Contributing Illustrators: Deborah Cowder, Ellen Blonder

Thanks to
Spectrum Communication Services, Inc.

Meredith® Books gratefully acknowledges the following
authorities for their expertise and assistance in expanding
the problem range maps to include Canada:

Dr. Alain Asselin, Agronomist and Professor,
Faculty of Agronomical and Food Sciences,
Université Laval, Quebec, Canada

Adam Brown, Biologist,
Department of Biology
Université Laval, Quebec, Canada

Louise Dumouchel, Entomologist,
Canadian Food Inspection Agency,
Plant Health Risk Assessment,
Agriculture Canada, Nepean, Ontario

Larry Hodgson, project coordinator,
Horticom Inc.,
Sainte-Foy, Quebec, Canada

Dr. Jeremy McNeil, Entomologist,
Professor, Department of Biology,
Université Laval, Quebec, Canada

Susanne Roy, assistant project coordinator,
Horticom Inc.,
Sainte-Foy, Quebec, Canada

All of us at Ortho® Books are dedicated to providing you
with the information and ideas you need to enhance your
home and garden. We welcome your comments and
suggestions about this book. Write to us at:
    Meredith Corporation
    Ortho Books
    1716 Locust St.
    Des Moines, IA 50309–3023

If you would like more information on other Ortho
products, call 800-225-2883 or visit us at www.ortho.com

Note to the Readers: Due to differing conditions, tools,
and individual skills, Meredith Corporation assumes no
responsibility for any damages, injuries suffered, or losses
incurred as a result of following the information published
in this book. Before beginning any project, review the
instructions carefully, and if any doubts or questions remain,
consult local experts or authorities. Because codes and
regulations vary greatly, you always should check with
authorities to ensure that your project complies with all
applicable local codes and regulations. Always read and
observe all of the safety precautions provided by
manufacturers of any tools, equipment, or supplies,
and follow all accepted safety procedures. Always read
pesticide labels carefully and follow label instructions and
warnings.

Contents

Consultants

This has been a vast project, and we have received help from hundreds of people—members of County Extension offices, horticulturalists, nursery professionals, and knowledgeable amateur gardeners. The following names are those of major consultants—people who spent many hours checking manuscripts for accuracy and patiently answering our questions about fine points of obscure garden problems. The accuracy and validity of this book are due to the careful work of these people.

Dr. Jan Abernathie
Plant Pathologist
Chesapeake, VA

Dr. J. Ole Becker
R&D Consultant, Plant Pathology
Riverside, CA

Dr. James Beutel
Extension Pomologist
University of California, Davis

Dr. Darrel R. Bienz
Professor of Horticulture
Washington State University

Dr. Eugene Brady
Professor of Entomology
University of Georgia

Dr. Jerome Brezner
Department of Environmental and Forest Biology
State University of New York, Syracuse

Bartow H. Bridges, Jr.
Landscape Architect and Horticulturalist
Virginia Beach, VA

Dr. Jack Butler
Department of Horticulture
Colorado State University

Dr. Ralph S. Byther
Extension Plant Pathologist
Washington State University

Dr. Robert Carrow
Agronomy Department
University of Georgia Experimental Station, Griffin

William E. Chaney
Cooperative Extension
Gilroy, CA

Eric Clough
Landscape Architect
Winlaw, BC, Canada

Sharon J. Collman
County Extension Agent
Seattle, WA

Dr. Samuel D. Cotner
Extension Horticulturalist
Texas A&M University

Dr. G. Douglas Crater
Extension Horticulturalist, Floriculture
University of Georgia

Dr. T. E. Crocker
Professor of Fruit Crops
University of Florida

Dr. J. A. Crozier
Plant Pathologist
Novato, CA

Dr. R. Michael Davis
Cooperative Extension Specialist, Plant Pathology
University of California, Davis

Dr. Spencer H. Davis, Jr.
Horticultural Consultant
North Brunswick, NJ

Eric R. Day
Manager, Insect Identification Lab
Virginia Polytechnic Institute and State University

Dr. August A. De Hertogh
Professor of Horticultural Science
North Carolina State University

Dr. James F. Dill
Extension Professor of Entomology
University of Maine, Orono

Dr. Clyde Elmore
Extension Weed Scientist
University of California, Davis

Dr. Thomas E. Eltzroth
California Polytechnic University
San Luis Obispo, CA

Barbara H. Emerson
Senior Product Specialist, Union Carbide Agr. Products Company
Research Triangle Park, NC

Dr. James R. Feucht
Extension Professor, Department of Horticulture
Colorado State University

Dr. Ralph Garren, Jr.
Small Fruit Specialist, Cooperative Extension Service
Oregon State University

Greg Giusti
Cooperative Extension
Kelseyville, CA

Cathy Haas
Instructor, Ornamental Horticulture
Monterey Peninsula College, CA

Dr. Frank A. Hale
Associate Professor, Entomology and Plant Pathology
University of Tennessee

Mary Ann Hansen
Instructor and Plant Clinic Manager
Virginia Tech

Dr. Ali Harvandi
Turfgrass Scientist
Oakland, CA

Duane Hatch
Cooperative Extension
Utah State University

Dr. Sammy Helmers
Area Extension Horticulturalist
Stevenville, TX

Kermit Hildahl
Cooperative Extension
University of Missouri

Larry Hodgson
Horticultural Consultant
Quebec City, Quebec, Canada

Gerald J. Holmes
Assistant Professor of Plant Pathology
North Carolina State University

Everett E. Janne
Landscape Horticulturalist, Texas Agricultural Extension Service
College Station, TX

Dr. Alan L. Jones
Professor of Plant Pathology
Michigan State University

Dr. Ron Jones
Department of Plant Pathology
North Carolina State University

Gene Joyner
Palm Beach County Extension
West Palm Beach, FL

L. Patricia Kite
Entomology Writer
Newark, CA

Caroline Klass
Senior Extension Associate, Department of Entomology
Cornell University

Steven T. Koike
Plant Pathologist
Salinas, CA

Dr. Tony Koski
Associate Professor, Horticulture, Extension Turfgrass Specialist
Colorado State University

Charles A. McClurg
Extension Vegetable Specialist
University of Maryland

Frederick McGourty
Mary Ann McGourty
Hillside Gardens
Norfolk, CT

S. Mansour
Extension Service
Oregon State University

Dr. Charles Marr
Extension Horticulturalist
Kansas City University

Bernard Moore
Extension Plant Diagnostician (Retired)
Oregon State University

Dr. Stephen T. Nameth
Associate Professor of Plant Pathology
Ohio State University

Lester P. Nichols
Professor Emeritus of Plant Pathology
The Pennsylvania State University

Dr. Norman F. Oebker
Vegetables Specialist, Cooperative
  Extension Service
University of Arizona

Dr. Howard Ohr
Plant Pathologist, Cooperative Extension
  Service
University of California, Riverside

Dr. Albert O. Paulus
Plant Pathologist
University of California, Riverside

Jay W. Pscheidt
Extension Plant Pathology Specialist
Oregon State University

Dr. Robert D. Raabe
Extension Plant Pathologist
University of California, Berkeley

David Robson
Extension Educator, Horticulture
University of Illinois

Dr. Charles Sacamano
Professor of Horticulture
University of Arizona

Dr. Donald L. Schuder
Professor Emeritus, Department of
  Entomology
Purdue University

Dr. Arden Sherf
Professor, Department of Pathology
Cornell University

Dr. David J. Shetlar
Associate Professor of Landscape
  Entomology
Ohio State University

Dr. Gary Simone
Extension Plant Pathologist
University of Florida

Arthur Slater
Environmental Health and Safety
University of California, Berkeley

Dr. Kenneth Sorensen
Extension Entomologist
North Carolina State University

Dr. Walter Stevenson
Associate Professor of Plant Pathology
University of Wisconsin

Dr. Steven Still
Department of Horticulture
Ohio State University

Dr. Bernadine Strik
Professor of Horticulture, Extension Berry
  Crops Specialist
Oregon State University

Dr. O. Clifton Taylor
Statewide Air Pollution Research Center
University of California, Riverside

William Titus
County Coordinator, Cooperative
  Extension Service
Plainview, NY

Dr. John Tomkins
Associate Professor of Pomology
Cornell University

Carl A. Totemeier
Director, Old Westbury Gardens
Old Westbury, NY

Marian Van Atta
Editor, Living Off the Land
Melbourne, FL

John White
County Extension Agent, Horticulture
El Paso, TX

Dr. Gayle Worf
Extension Plant Pathologist
University of Wisconsin

SPECIAL CONSULTANTS

Many other gardening professionals and gifted amateurs have shared their experience and wisdom with us. The people on the following list are specialists for particular problems. They have often been able to supply answers when nobody else could, and we are deeply indebted to them for their contributions.

Dr. Maynard Cummings
Extension Wildlife Specialist
University of California, Davis

Don Egger
President, Cebeco Lilies
Aurora, OR

Harold E. Greer
President, Greer Gardens Nursery
Eugene, OR

Phil Horne
Mosley Nurseries
Lake Worth, FL

Dr. Lloyd A. Lider
Professor of Viticulture
University of California, Davis

Dr. Wayne S. Moore
Entomologist
Berkeley, CA

Joseph R. Onwinski
Turfgrass Consultant
Lake Worth, FL

John Pehrson
Extension Agent
Parlier, CA

Warren G. Roberts
Arboretum Superintendent
University of California, Davis

Donald Rosedale
Cooperative Extension Service
University of California, Riverside

Dr. Terrel P. Salmon
Extension Wildlife Specialist
University of California, Davis

Ross R. Sanborn
University of California Farm Adviser
Contra Costa County, CA

Joseph Savaage
Entomologist, Cooperative Extension
  Service
Cornell University

Arthur Slater
Senior Environmental Health and
  Safety Technologist
University of California, Berkeley

Richard Tassan
Staff Research Associate
University of California, Berkeley

Introduction: Solving Plant Problems

Ortho's *Home Gardener's Problem Solver* is designed to help you diagnose a problem with a plant, then provide potential solutions.

Diagnosing plant problems requires careful observation of plants and their environment. The key to accurate diagnosis is knowing how to look for clues to a problem and what types of clues to look for. The checklist on pages 8 and 9 gives a step-by-step procedure for gathering clues and diagnosing a problem. It will help you to develop a case history, eliminate unlikely explanations for sources of the problem, and find the real cause.

HOW TO OBSERVE

Begin your observations by examining the plant from a distance. Note its general condition. Is the entire plant affected, or only a few stems, branches, or leaves? If the entire plant shows symptoms, the cause will probably be found on the trunk, on the roots, or in the soil.

Look for patterns and relationships with other plants. Is the problem confined to the sunny side? Is only the young growth affected? Are many sick plants in one spot?

Locate a part of the plant that shows symptoms and take a closer look. Mottled or discolored leaves may indicate an insect or disease problem. A 5- to 15-power hand lens will allow you to see insects or symptoms not easily visible to the naked eye.

If the initial inspection does not reveal any obvious reason for the symptoms, developing a case history for the plant may lead you to a less conspicuous cause of the problem. What has the weather been like recently? Has the temperature been fluctuating drastically? What kind of winter was it? An unusually dry, cold winter can cause dieback that may not become apparent until new growth begins on trees and shrubs in the spring.

Study the recent care of the plant. Has the plant been watered or fertilized regularly? All plants require fertilizer; without regular feeding, their leaves turn yellow, and growth is poor.

When observing the area around the plant, be aware of changes in the environment. Construction around established plants can damage them, although symptoms of decline may not appear for several years. Drastic changes in light may cause problems that appear many days later, as when a houseplant is moved

from a sunny window to a dark corner or when a fruit tree is pruned heavily.

You may have to dig into the plant or the soil to find the cause of a problem. If a stem has a hole in it, cut into it. Or slice a piece of bark off a wilting branch to determine whether the wood is discolored or healthy.

The only way to learn about the roots of a sick plant is to dig up a small plant or to carefully dig a hole to examine the roots of a large plant. The key to a root problem may be in the soil. Investigate the drainage, probe the soil with an auger to determine the soil depth and the type of soil, or test the pH of the soil with a pH kit. Look at all sides of the question, and explore each clue.

Particular types of plants are susceptible to typical problems at certain times of the year. For instance, cherries are usually plagued with fruit flies in the spring when the fruit is ripening, and snapdragons are likely to be infected with rust in spring and summer when the temperature is warm and moisture is present.

PUTTING IT ALL TOGETHER

After studying the ailing plant or plants, read the introduction to the pertinent chapter in this book. The introduction may contain information that relates directly to your problem or give you a tip about where to look next. Then look through the general problem headings at the tops of the pages of the problem-solution chapters. Find the heading that applies to your problem, then

look for your specific plant and see if its problem is listed. If you know the name of your plant, the problem it has, or the insect that is bothering it, you can look up the name in the index at the back of the book. For uncommon problems, you may want to refer to *The Ortho Problem Solver*—a professional edition found in many nurseries and in garden and home improvement centers—but these pages will discuss most of the problems you will encounter.

When you're reading about a plant problem in *Ortho's Home Gardener's Problem Solver*, read carefully. Every word and phrase is important for understanding the nature of the problem. "May" means that the symptom develops only sometimes. And certain phrases offer you clues about the problem, such as the time of year to expect it ("in spring to midsummer…") and where to look for the symptom ("…on the undersides of leaves").

Unfortunately, plants frequently develop more than one problem at a time. Plants have natural defenses against diseases and insects. But when one problem weakens a plant, the plant's defenses are lowered and other problems are able to infect it. For instance, borers are often responsible for a tree's decline, but borers are seldom a serious problem on healthy trees. A borer problem may indicate another problem, such as a recent severe winter or root rot.

HOW TO USE THIS BOOK

All gardeners have problems with pests or diseases from time to time. *Ortho's Home Gardener's Problem Solver* was created to help you solve those problems.

In a straightforward way, it will help you discover what kind of problem you have and what's causing it, and it will tell you more about the problem—such as how serious it can get if you don't do anything. Then it will tell you how to solve the problem.

The photographs at the top of the page are arranged so that similar symptoms are grouped together. Select the picture that looks most like your problem. The small map under the photograph shows how likely the problem is to affect your part of the country. If your region is colored red, the problem is commonplace or severe. If it is colored yellow, the problem is occasional or moderate. If it is uncolored, the problem is nonexistent or minor.

A word of explanation is needed about the solutions we offer. The solution section of each problem assumes that you have seen the problem at the time when the symptoms first become obvious. Each solution begins by telling you what you can do immediately to alleviate the problem. Then it tells you what changes you can make in the environment or in your gardening practices to prevent the problem from returning. In many cases, a chemical spray is recommended as an immediate solution and a cultural change or the planting of a resistant variety as a long-range solution.

We offer several solutions for most of the problems in this book. We tell you, for example, that you can protect your sycamore tree from anthracnose by spraying it with a fungicide in the spring. If your sycamore is 7 feet high, you might choose to spray it the following spring. If it is 40 feet high, however, you will find that hiring an arborist to spray it will be expensive. From our description of the problem, you know that anthracnose seldom does permanent harm to the tree, so you may choose to do nothing.

When Ortho has products to treat any problem, we identify them by name. If Ortho does not make a product to solve a particular problem, we recommend chemical solutions by their generic names—the common name of the active ingredient. For example, we might suggest that you apply a product "containing" a particular active ingredient. This wording alerts you to the fact that you may not find a product by that name but must study the active ingredients listed on product labels. Ask your retailer to help you select an appropriate product, or see pages 334 and 335.

When we recommend an Ortho product, we know that it will do the job for which we recommend it and that it will not harm your plant if you use it according to label directions. But when choosing products by a generic name—even if these are Ortho products—read the label carefully. Although all *malathion* is the same, all products containing it are not. Even though we know *malathion* will solve a particular problem and tell you so, some products that contain *malathion* may be manufactured for a different purpose and may injure your plant. Be sure that the plant you wish to spray is listed on the product label. Always read pesticide labels carefully and follow label directions to the letter.

This book is based in large part on research done for *The Ortho Problem Solver* and its revisions. *The Ortho Problem Solver* is a professional reference tool for solving plant problems. We have drawn on that research to create this book to help you, as a home gardener, solve the problems you are most likely to encounter. The pages of both problem solvers present the experience of many experts, most of them members of cooperative extension services of various states. These men and women have shared the most current and practical information available. If you follow their advice in terms of immediate solutions and long-term prevention, you will approach the realization of every gardener's dream: to garden in such a way that you have no problems at all, but are free to enjoy fully the beauty and bounty of your garden.

Lawns

DEAD PATCHES (continued)

Sod webworm

Damage. Inset: Sod webworm (2x life size).

Problem: From mid-May to October the grass turns brown in patches the size of a saucer in the hottest and driest areas of the lawn. These areas may expand to form large irregular patches. Grass blades are chewed off at the soil level. Silky white tubes are found nestled in the root area. Inside are light brown or gray worms with black spots, from ¼ to ¾ inch long.

Analysis: Several different moths with similar habits are called *sod webworms* or *lawn moths*. These night-flying moths drop eggs into the grass as they fly. The eggs hatch into worms that feed on grass blades at night or on cloudy, rainy days. In the daytime the worm hides in white silky tubes in the soil. Sometimes an entire lawn is killed in a few days.

Solution: Control sod webworms with ORTHO Bug-B-Gon Ready-Spray, ORTHO Dursban® Lawn & Garden Insect Control (granules), or Scotts Lawn Insect Control when large numbers of moths are noticed at dusk or at the first sign of damage. First rake out all the dead grass and mow the lawn. Water thoroughly before spraying. For best results, apply the insecticide in the late afternoon or evening, when the worms are most active. Do not cut or water the lawn for 1 to 3 days afterward. To avoid recurring damage, treat the lawn again every 2 months beginning in late spring or early summer. Damaged lawns may recover rapidly if the insects are controlled early.

Chapters are usually divided into two sections: "Problems Common to Many" and "Problems of Individual Plants"

Group of similar symptoms

The problem name

A photograph (sometimes with an inset) depicting a typical symptom or organism

A range map of the United States and southern Canada accompanies each problem. In areas that are red, the problem is severe or commonplace. In areas that are yellow, the problem is secondary or occasional. In areas that are white, the problem is rare or nonexistent.

Problem section describes the symptom or symptoms.

Analysis section describes the organisms or cultural conditions causing the problem, including life cycles, natural processes, typical progress of the problem, and seriousness.

Solution section provides short-term and long-term techniques to mitigate or cure the problem.

Checklist for Diagnosis

Use this checklist to develop a case history for the problem and to identify symptoms that will lead to an accurate diagnosis. Answer each question that pertains to your plant carefully and thoroughly. When looking for symptoms and answering questions about the condition of the plant, begin with the leaves, flowers, or fruit (unless it is apparent that the problem is elsewhere), because they are the easiest to examine. Once you've eliminated those possibilities, move down the plant to the stems or branches and trunk. Inspect the roots after rejecting all other possibilities.

WHAT TO LOOK FOR

Kind of plant
- ☐ What type of plant is it?
- ☐ Does it prefer moist or dry conditions?
- ☐ Can it tolerate cold, or does it grow best in a warm climate?
- ☐ Does the plant grow best in acid or alkaline soil?

Age
- ☐ Is the plant young and tender, or is it old and in a state of decline?

Size
- ☐ Is the plant abnormally small for its age?
- ☐ How much has the plant grown in the last few years?
- ☐ Is the size of the trunk or stem in proportion to the number of branches?

Time in present site
- ☐ Was the plant recently transplanted?
- ☐ Has it had time to become established, or are its roots still in the original rootball?
- ☐ Is the plant much older than the housing development or nearby buildings?

Symptom development
- ☐ When were the symptoms first noticed?
- ☐ Have symptoms been developing for a long time, or did they appear suddenly?

Condition of plant
- ☐ Is the entire plant affected, is the problem found on only one side of the plant, or are there symptoms throughout the plant?
- ☐ What parts of the plant are affected?
- ☐ Are all the leaves affected, or only the leaves on a few branches?
- ☐ Are the leaves abnormal in size, color, shape, or texture?
- ☐ Do the flowers and fruit show symptoms?
- ☐ Do you see abnormal growth, discoloration, or injuries on the branches, stems, or trunk?
- ☐ Is there anything wrapped around and girdling the plant or nailed into the wood?
- ☐ Does the trunk have a normal flare at the base, or is it constricted or entering the ground straight like a pole?
- ☐ If the entire plant is affected, what do the roots look like? Are they white and healthy, or are they discolored? Use a trowel to dig around the roots of large plants; pull up small, sick plants to investigate the roots; and examine roots of container plants by removing the container. Is the bark brown and decayed?
- ☐ Have the roots remained in the soil ball, or have they grown into the surrounding soil?
- ☐ Are insects on the plant, or is there evidence of insects, such as holes, droppings, sap, or sawdustlike material?
- ☐ Has the problem appeared in past years?

Location of property
- ☐ Is the property near a large body of fresh or salt water?
- ☐ Is the property located downwind from a factory, or is it in a large polluted urban area?
- ☐ Is the property part of a new housing development that was built on landfill?

Location of plant

☐ Is the plant growing next to a building? If so, is the location sunny or shady? Is the wall of the building light in color? How intense is the reflected light?

☐ Has there been any construction, trenching, or grade change nearby within the past several years?

☐ Have there been any natural disturbances?

☐ How close is the plant to a road? Is the road de-iced with salt in the winter?

☐ Is it growing over or near a gas, water, or sewer line or next to power lines?

☐ Is the ground sloping, or is it level?

Relationship to other plants

☐ Are there large shade trees overhead?

☐ Is the plant growing in a lawn or ground cover?

☐ Are nearby plants also affected? Do the same species show similar symptoms? Are unrelated plants affected? How close are other affected plants?

Weather

☐ Have weather conditions been unusual recently (cold, hot, dry, wet, windy, snowy, and so on) or in the past few years?

Microclimates

☐ What are the weather conditions in the immediate vicinity of the plant?

☐ Is the plant growing under something that blocks rainfall or sprinkler water?

☐ How windy is the location?

☐ How much light is the plant receiving? Is it the optimum amount for the type of plant?

Soil conditions

☐ What kind of soil is the plant growing in? Is it predominantly clay, sand, or loam?

☐ How deep is the soil? Is a layer of rock or hardpan beneath the topsoil?

☐ What is the pH of the soil?

☐ Does the soil drain well, or does the water remain on the surface after a heavy rain or irrigation? Does the soil have a sour smell?

☐ Is the soil hard and compacted?

☐ Has the soil eroded away from the roots?

Soil coverings

☐ Does asphalt, concrete, or another solid surface cover the soil around the plant? How close is it to the base of the plant? How long has it been there?

☐ Has the soil surface been mulched or covered with crushed rock?

☐ Was the mulch obtained from a reputable dealer?

☐ Are weeds or grass growing around the base of the plant? How thickly?

Recent care

☐ Has the plant or surrounding plants been fertilized or watered recently?

☐ If fertilizer was used, was it applied according to label directions?

☐ Has the plant or the area been treated with fungicide or insecticide?

☐ Was the treatment for this problem or another one?

☐ Was the pesticide registered for use on the plant? (Is the plant listed on the product label?)

☐ Was the pesticide applied according to label directions?

☐ Did rain wash off the spray immediately after it was applied?

☐ Did you repeat the spray if the label suggested it?

☐ Have weed killers or lawn weed-and-feed fertilizers been used in the area in the past year? How close?

☐ Did you spray on a windy day?

☐ Has the plant been pruned heavily, exposing previously shaded areas tol sun?

☐ Were stumps left after pruning, or was the bark damaged during pruning?

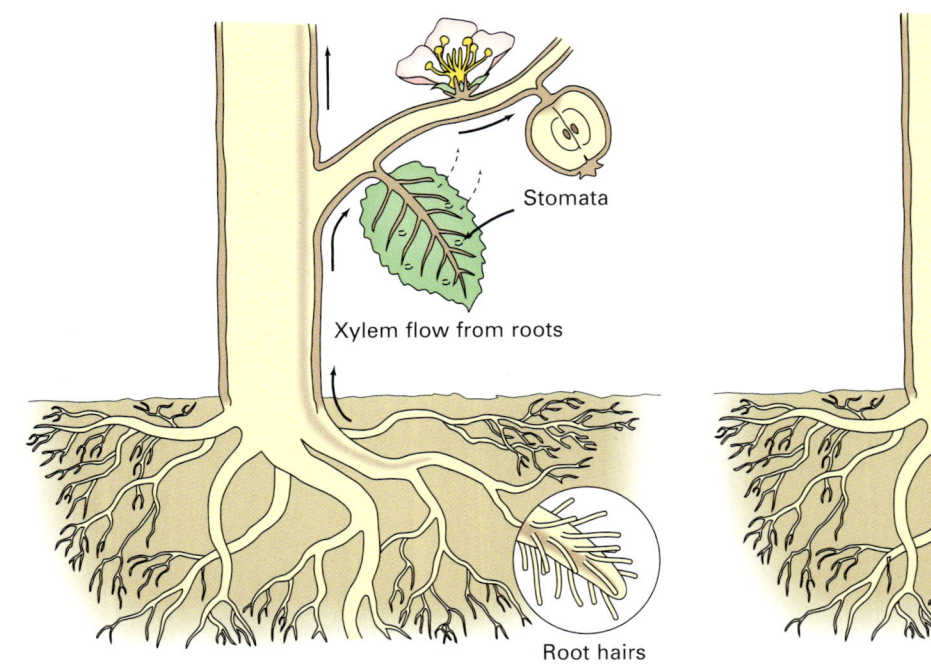

Transport (Xylem)

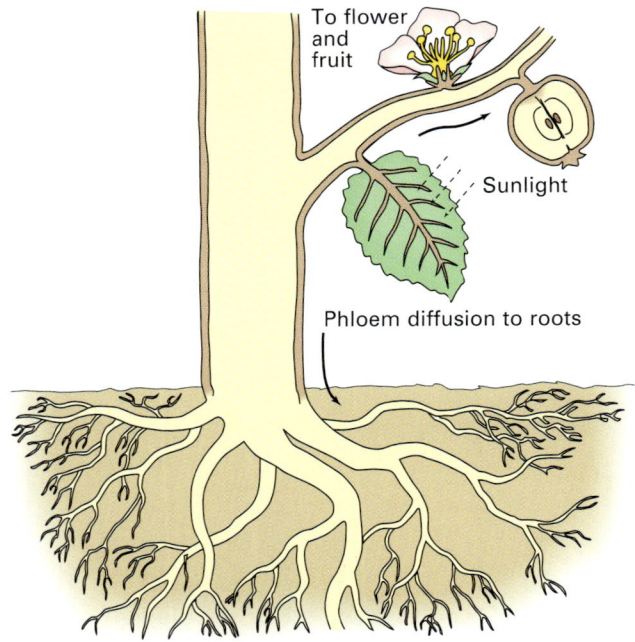

Transport (Phloem)

TRANSPORT

A plant's transport system does many of the same jobs that a human's circulatory system does, but it is not called *circulation* because it does not go in a circle, as the human bloodstream does. Plants have two transport systems: one carrying water and dissolved minerals from the roots to the top of the plant and the other carrying sugars and other manufactured material from the leaves to all other parts of the plant.

Xylem flow: The part of a plant's transport system that carries water upward flows through a system of microscopic tubes called the *xylem* (pronounced ZYE-lem). In woody plants, the xylem is the outermost layer of wood. This wood, which is just under the bark, contains the xylem tubes. These tubes extend into the leaves and through the leaf veins into every part of the leaf. The xylem tubes end in the leaf tissue. Water leaves the xylem in the leaf tissue and evaporates through microscopic pores called *stomata* in the leaf surface.

Transpiration: Evaporation out of the stomata is called *transpiration*. Transpiration not only provides the power that moves water and nutrients into the top of the plant, but also cools the leaf, just as evaporation cools a wet towel. Although the principles are different, you can think of transpiration as pulling water from the roots to the leaves just as soda is sucked through a drinking straw. Because plants can transport only nutrients that are dissolved in water, insoluble material is not available to them.

Three main factors reduce xylem flow: dry soil, a sick root system, or a plugged or cut xylem system. If the xylem flow is reduced below a certain point, the leaves receive fewer nutrients than they need to sustain good health and begin to show symptoms of nutrient deficiencies, usually by turning pale green or yellow, with the veins remaining green. If the xylem flow is reduced when the water demand is high—during hot weather—the leaves may wilt or scorch. This happens because the leaves lose water faster than it can be replaced.

Stomata control: Plants control transpiration by opening and closing their stomata. The stomata are shut at night. They open in the morning but may shut down partially if the light is dim. Stomata also close if a leaf runs out of water. This usually happens before any sign of wilting appears. If the weather is hot when the leaf runs out of water and if the leaf is in the sun, the leaf is likely to overheat and burn. This is called *sunburn* or *scorch,* depending on the pattern of the burning.

Phloem flow: The part of the transport system that carries sugars and other manufactured material from the leaves to other parts of the plant is called the *phloem* (FLO-em). This system of tubes lies inside the bark. If you peel off some bark in the spring, the phloem is visible as the white part of the bark. Flow in the phloem is by diffusion and is much slower than xylem flow. Sugars and other material manufactured in the leaves diffuse through the phloem to growing shoots, flowers, fruit, and roots. Because the roots are so large, most of the phloem flow in the trunk of a tree is downward.

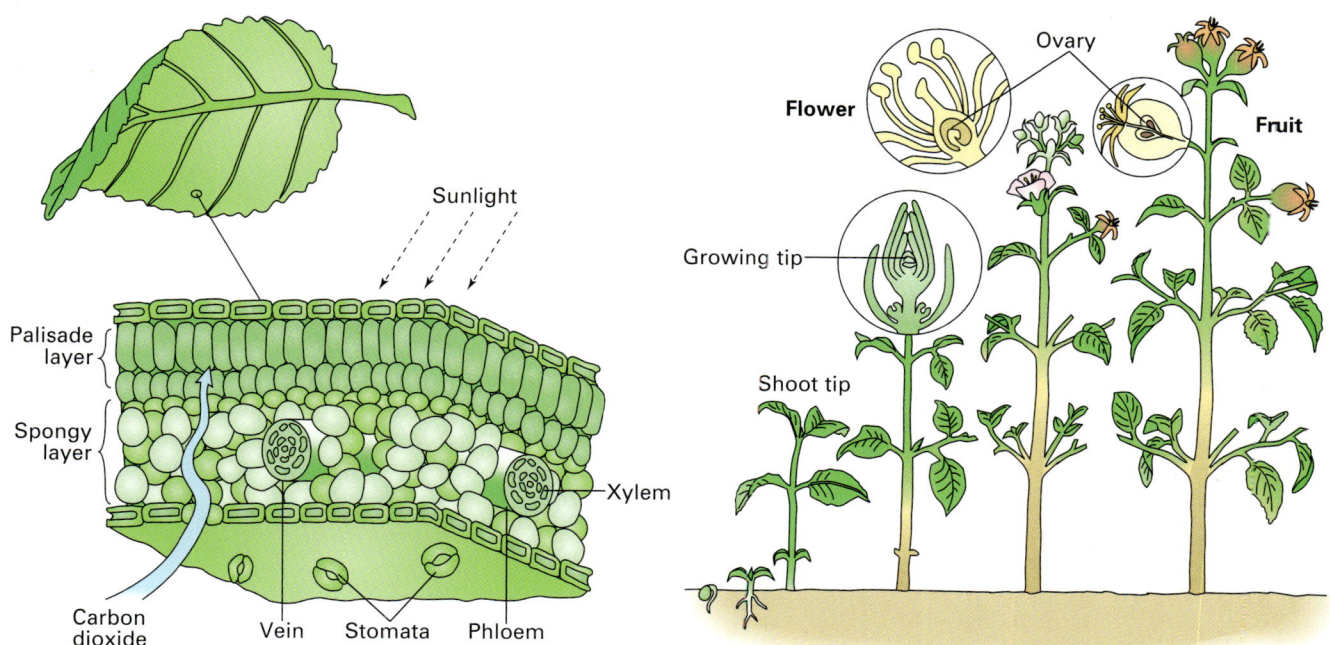

Sunlight

Palisade layer

Spongy layer

Carbon dioxide

Vein   Stomata   Phloem

Xylem

Flower   Ovary   Fruit

Growing tip

Shoot tip

PHOTOSYNTHESIS

Photosynthesis is a chemical reaction that locks up the sun's energy in the form of a chemical so that it can be used to support life. This reaction takes carbon dioxide from the air and combines it with water from the soil to make sugars. Photosynthesis takes place in the green parts of plants. The green pigment is called *chlorophyll*. A plant's chlorophyll collects and focuses the energy in light to make sugar. Most photosynthesis takes place in leaves in a layer of cells, called the *palisade layer*, near the upper surface of the leaf. Carbon dioxide enters the leaf through the stomata in the lower surface and enters the palisade cells, where it is combined with water to make sugar. As a by-product of this reaction, oxygen is released to the air. The sugars are then transported through the phloem to all parts of the plant, where they are burned to release their energy. All animals and humans depend on both the sugars and the oxygen produced by plants, so all life on earth depends on photosynthesis.

Light: Light is the source of energy that combines carbon dioxide and water to make sugar. The more light a plant receives, the more sugar it makes, the faster it grows, and the more flowers and fruit it produces. In dim light, a plant makes barely enough sugar to maintain its life, but in bright light, it makes a surplus that it uses for growth and reproduction.

Water stress: Carbon dioxide enters the leaves through the stomata. Because most plants can't store carbon dioxide, photosynthesis can take place only when the stomata are open. If a plant does not have enough water, it closes its stomata to avoid losing water it can't afford to lose. When the stomata are closed, however, photosynthesis stops. One of the first effects of water stress is that photosynthesis—and growth—stops.

TOP GROWTH

Unlike human growth, which takes place in all parts of the body, plant growth occurs in only a few places, called the *growing points*. The growing points in a plant are in the tips of the root and shoots and, in woody plants, just under the bark.

Tip growth: The growing points in the tips of shoots and roots are composed of tiny bunches of cells that divide repeatedly, building the organs of the plant but always remaining at the tip of the new growth. The new plant parts are tiny while they are in the growing point, but they are complete, with all the cells they'll ever have. As the growing point moves beyond them, the new parts fill with water and swell until they reach their full size. No new cell division takes place while they are expanding, however.

Buds: The growing point moves ahead, but a bud is left at the base of each leaf. A bud is a growing point that is, for the moment, dormant. On some plants, such as tomatoes, these axillary buds begin growing as soon as they are formed. On others, such as apple trees, the buds remain dormant until the following spring. Then they all begin growing at once. Some buds never open by themselves, but pinching off the growing point at the end of the branch they are on makes them open and begin to grow.

Stem growth: The other place a woody plant grows is under the bark. A sheet of cells, called the *cambium*, lies between the wood and the bark. These cells divide repeatedly just as those in the shoot and root tips do. As they divide, they produce xylem cells toward the center of the stem and phloem cells toward the outside. When growth is fastest, in the spring, large xylem cells are made; as growth slows in the summer, smaller cells are made. The difference in the size of these cells makes up the annual rings visible in most wood.

Flowers and fruit: Flowers are produced by the growing point just as the leaves and stems are. Flowers begin as buds, which in some plants open immediately and in others wait until the following spring. At the base of each flower is an ovary, an organ that will contain the seed and, in some plants, become the fruit.

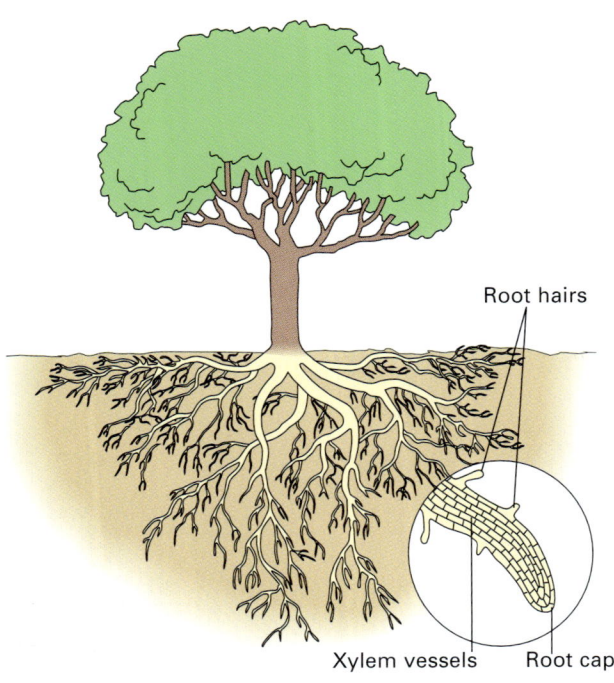

Root hairs

Xylem vessels   Root cap

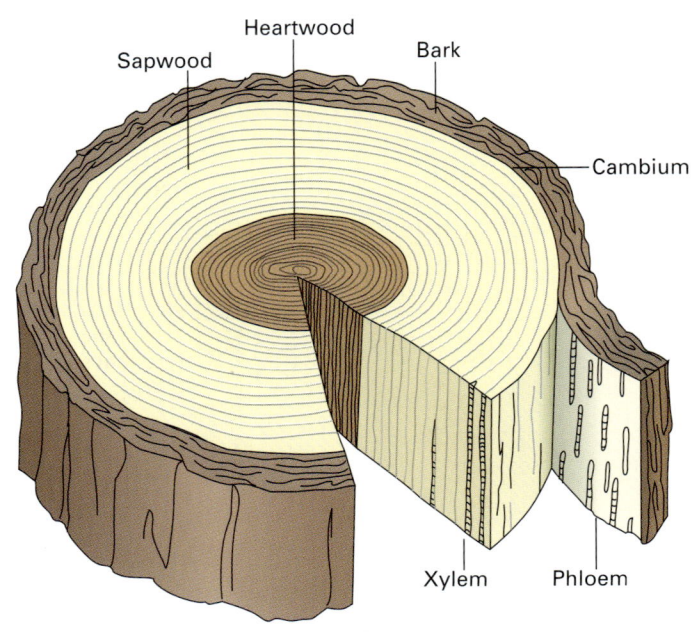

Sapwood   Heartwood   Bark

Cambium

Xylem   Phloem

ROOT GROWTH

Plants live in two distinctly different environments: the air and the soil. The root system of a plant "mines" the soil for water and nutrients, the raw materials used by the plant. These raw materials are transported through the stems to the leaves, where they are combined with carbon dioxide to make sugars and complex chemicals. These processed chemicals are then transported through the stem back to the roots, where they are used for growth.

Root tips: As the growing point at the end of the root moves through the soil, it leaves behind a soft, white, threadlike root tip. Downy root hairs grow from the tip. Most water and nutrient absorption is through the root tips. As the root ages, it becomes yellow and then brown, and the root hairs disintegrate. By the time the root turns brown, it no longer absorbs much water from the soil.

The root system: Roots need three things for vigorous growth: oxygen, water, and nutrients. Of these three, oxygen is most frequently in short supply. Because plants cannot transport oxygen, each part of the plant must absorb it directly from the air. Oxygen diffuses through the spaces between soil particles to reach the roots. Where roots find an abundance of these three necessities, they proliferate.

The root system of a plant is often thought of as extending through the soil about as far as the top growth does above the ground. But plants vary greatly in the extent of their root systems. Trees may have roots that extend dozens or even hundreds of feet beyond their top growth, especially if the soil is frequently dry. Also, the root system is not symmetrical but is more dense on the side that more frequently receives food or water.

Although some plants have deep root systems, garden plants get most of their nutrients from the top foot of soil. For this reason, plants are sensitive to the condition of the soil surface. If the soil is paved over or becomes compacted, the roots near the surface receive less oxygen and water and may die.

TRUNK GROWTH

Growth of tree trunks and expansion of the stems of woody plants take place in the cambium, a layer of dividing cells just under the bark. As each cell in the cambium divides, one of the two new cells becomes either a xylem cell or a phloem cell, and the other remains a cambium cell and divides again.

Wood: As the newest xylem cells expand, they push the cambium layer a little farther away from the center of the tree. Each new cell lives for only a year or two; then it becomes plugged with the detritus of life and ceases to function. By this time, the cell is deep in the wood of the tree and is heavily packed with a material called *lignin*, the material that gives stiffness and rigidity to wood.

Bark: The phloem cells that form in the cambium are pushed to the outside of the tree by their expansion and the expansion of the new xylem cells. Like the xylem cells, phloem cells transport nutrients for a couple of years. Then they die and dry out to become the bark of the tree. The green bark on young stems is composed of a living layer of cells called the *epidermis*. As the twig ages and expands, dead phloem cells replace this living bark.

Girdling: If the phloem and xylem are cut through in a ring around the trunk of a tree, the flow of water to the top of the tree is stopped, the leaves wilt, and the top of the plant dies. The roots may die, or they may resprout and grow a new top. If only the bark is cut through—by root weevils, for instance—water can still reach the top of the tree, and wilting does not occur. But the flow of sugars and nutrients from the leaves to the roots has been stopped, and the roots slowly starve. As they cease functioning, the top also starves, and inevitably the tree dies.

Bark wounds: Any wound in the bark of a tree trunk interrupts the flow of water and nutrients through the tree. If a spot is repeatedly wounded, as frequently happens when a lawn tree is hit with a lawn mower, the growth of the plant is slowed and becomes stunted.

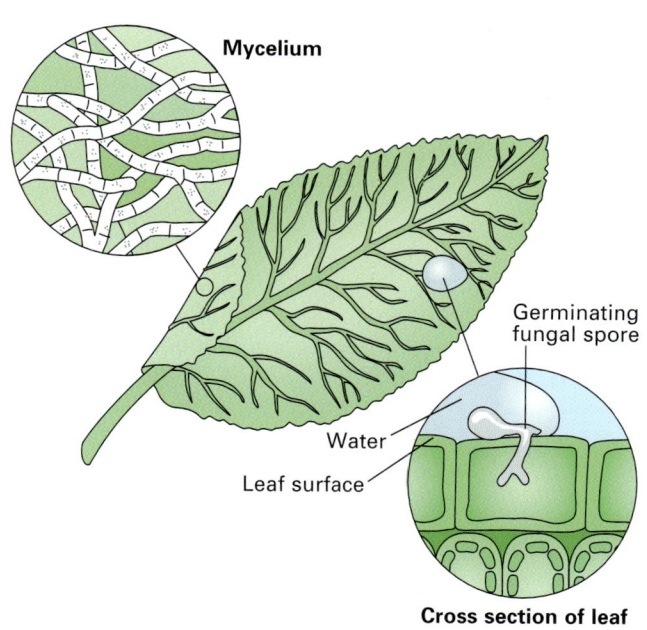

Mycelium

Germinating fungal spore

Water

Leaf surface

Cross section of leaf

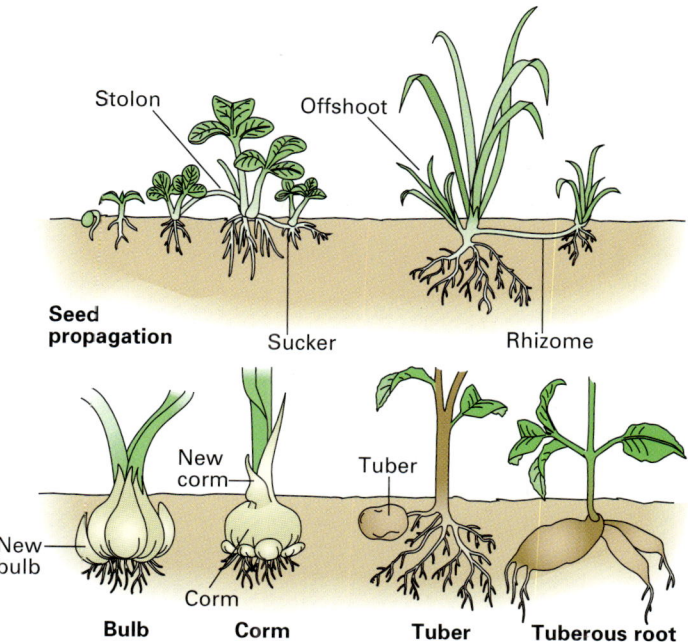

Stolon

Offshoot

Seed propagation

Sucker

Rhizome

New corm

Tuber

New bulb

Corm

Bulb   Corm   Tuber   Tuberous root

FUNGI

Fungi are primitive plants that do not contain chlorophyll and cannot make their own sugars as green plants do. Their structure is simple; most fungi are composed of a network of threads, which are sometimes bundled into cords. These threads and cords are called the *mycelium* (my-SEE-lee-um) and make up the body of the fungus. They are most easily seen as the fragile white threads that form under rotting leaves. Most fungi decompose dead plant matter into simpler chemicals. They are responsible for most of the decay of plant material in nature. But some fungi cause plant disease by invading living plants. Fungus diseases spread in several ways, but most commonly through the soil, by splashing water, or by the wind.

Soil-spread fungi: Fungi that live in the soil and attack plant roots are most commonly spread when soil or infected plants are moved from one place to another. Soil-borne fungi are often brought to the garden initially on the roots of a transplant, then spread from one end of the garden to the other by irrigation water or cultivation.

Water-spread fungi: Other fungi depend on splashing water to spread their spores. Many of the fungi that cause leaf spots on trees and shrubs spend the winter as spores on the bark or on fallen leaves and are then washed to new leaves in the spring by splashing rain.

Wind-spread fungi: The third common way fungi spread is by dustlike spores that are carried by the wind to other plants. Many of these spores live for only a few hours or days after they are released into the air, so they must soon find suitable conditions for germination. For most such fungi, a "suitable condition" is a drop of water on a leaf surface. A fungus spore takes 4 to 8 hours to germinate and penetrate a leaf. If the drop of water in which it is growing dries out during this time, the spore dies. One notable exception, powdery mildew, does not need a drop of water to germinate; its spores can germinate on a dry leaf. Even powdery mildew is more active in wet conditions, however.

HOW PLANTS SPREAD

Plants have two ways of propagating themselves: by seed and by new plants grown from parts of old ones.

Seed propagation: Most plants make seeds, and some make thousands of them. Annual plants (those that live for only one growing season) depend on seeds to continue their species. They often put all their energy into ripening a crop of seeds. Annual weeds, such as crabgrass, usually make thousands of seeds per plant. It is important to kill those weeds before they make seeds, or the same job will need to be repeated the following year.

Offshoots: All other methods of propagation involve some form of growing new plants from parts of old ones. Offshoots are new plants that arise from the base of the old one. If allowed to continue, this will eventually lead to a dense bunch of plants. Offshoots can usually be broken or cut off and planted to make new plants.

Rhizomes and stolons: These horizontal stems form new plants at a distance from the mother plant. Rhizomes travel underground and make new plants where they near the surface. Stolons travel aboveground and form new plants where they come in contact with the ground.

Suckers: The word *sucker* has two different but similar meanings. It refers to a stem that arises from the base of the trunk. It also refers to a plant that arises from the roots of the parent plant, often at some distance from the trunk. Plants that sucker freely, if left to their own devices, will form a thicket in a couple of years. The type of sucker that arises from a root is a whole plant and can be removed and planted elsewhere.

Bulbs: Plants form a wide variety of underground food storage organs. Most of these organs remain dormant for part of the year and then, because of the large amount of food stored in them, burst into vigorous growth. If the storage organ is formed from fleshy leaves, it is called a *bulb*. If it is formed from a vertical stem, it is called a *corm*. If it is formed from a rhizome, it is called a *tuber*. If it is formed from a root, it is called a *tuberous root*. When dormant, these organs can be moved to new locations, or they can be stored before being replanted.

13

Glossary of Horticultural Terms

Abscission: A natural dropping of leaves, flowers, and other plant parts.

Acaricide: A chemical that kills spider mites and other types of mites. Also called a miticide.

Acclimate: To adjust to a change in the environment.

Acid reaction: A property of fertilizer that makes soil more acid. This reaction may be slow and may be apparent only after continued fertilizer use. The strength of the acid reaction is expressed on the fertilizer label in a phrase such as "Potential acidity 400 lb calcium carbonate equivalent per ton." This means that 400 pounds of lime (calcium carbonate) would be necessary to neutralize the acidity produced by a ton of the fertilizer.

Acid soil: Soil with a pH below 7.0. Acid soils can cause problems when their pH is below 5.5.

Adventitious: Forming in unusual locations—for example, roots growing from leaves or aboveground stems.

Aerate: To increase the amount of air space in the soil by tilling or otherwise loosening the soil.

Algae: Simple plants without visible structure that grow in wet locations. Some types of algae form a slippery black or green scum on wet soils, plants, walkways, and other surfaces.

Alkaline soil: Soil with a pH above 7.0. Alkaline soils slow the growth of many plants when their pH is above 8.0.

Annual plant: A plant that grows, flowers, produces seeds or fruit, and dies in 1 year or less. Many herbaceous flowers and vegetables are annual plants.

Antidesiccant: A chemical that slows transpiration from leaves, used when roots have been damaged by transplanting or when water is not available for other reasons. Some antidesiccants coat the leaf with a thin plastic film; others chemically close the stomata. Also called an antitranspirant.

Axil: The location on a stem between the upper surface of a leaf or leafstalk and the stem from which it is growing.

Axillary bud: A bud that forms in a leaf axil.

Bactericide: A chemical that kills bacteria.

Balled-and-burlapped plant: A tree or shrub that is dug out of the ground with the intact soil ball surrounding the roots; the soil ball is then wrapped in burlap or plastic.

Bare-root plant: A tree or shrub that is dug out of the ground and sold with its roots bare of soil. Roses and fruit trees are commonly sold in this manner. Bare-root plants are available in the winter.

Biennial plant: A plant that grows, flowers, produces seeds or fruit, and dies in 2 years. Some herbaceous flowers and vegetables are biennial. Most biennial plants produce foliage the first year and bloom the second year.

Bolting: The rapid development of flowers and seed heads in vegetables. Hot weather, drought, or lack of nutrients may stimulate premature bolting.

Bract: A modified leaf that is sometimes brightly colored, resembling a petal.

Bud: A condensed shoot consisting mainly of undeveloped tissue. Buds are often covered with scales. They develop into leaves, flowers, or both.

Burl: A mass of bud tissue that grows from the side of a tree. Burls are healthy wood and are natural, not to be confused with galls.

Callus: A mass of cells, often barklike in appearance, that forms over wounded plant tissue.

Cambium: A thin ring of tissue within the stem, branch, and trunk; continually forms nutrient- and water-conducting vessels.

Canker: A discolored lesion that forms in a stem, branch, or trunk as a result of infection. Cankers are often sunken and may exude a thick sap.

Castings, worm: Soil that has passed through an earthworm. The earthworm's digestive tract breaks down organic matter into simpler forms that are more readily taken up by plants.

Chilling requirement: The need of some plants to have several weeks below a certain temperature (usually around 45°F) in order for their flower buds to open. Bulbs, trees, and perennial flowers native to cold-winter regions often have chilling requirements. Also called vernalization.

Chlorophyll: The green plant pigment that is necessary for photosynthesis.

Chlorosis: The yellowing of foliage due to a loss or breakdown of chlorophyll. Chlorosis may result from disease or infestation, poor growing conditions, or lack of nutrients. Also called yellowing.

Cold frame: A protective structure that uses the sun's energy to provide heat for plants. Plants may be grown in cold frames early in the spring before all danger of freezing is past.

Complete fertilizer: A fertilizer containing nitrogen, phosphorus, and potassium, the three nutrients most commonly deficient in plants.

Compost: Partially decomposed organic matter used to amend soil. Compost is often made from grass clippings, leaves, and manure.

Conifer: A woody tree or shrub that produces cones. Common conifers are pines, firs, spruces, junipers, redwoods, and hemlocks.

Conk: A mushroomlike fruiting body of any of several different kinds of tree-decaying fungi. They are platelike growths from the trunk of the tree.

Corm: A short, solid, enlarged, underground stem from which roots grow. Corms are food-storage organs, similar to bulbs and tubers. They contain one bud that will produce a new plant.

Crown: (1) The part of a plant where the stem or trunk enters the ground, or the base of a grass plant. (2) The top or leafy portion of a tree.

Cultivar: Short for "CULTIvated VARiety." A plant variety bred by humans and not found in nature.

Dead-heading: The removal of old blossoms to encourage continued bloom or to improve the appearance of a plant.

Deciduous: Shedding all leaves annually, usually in the fall.

Defoliation: Leaf drop that often results from infection, insect infestation, or adverse environmental conditions.

Desiccation: Dehydration or loss of water.

Dormancy: A state of rest and reduced metabolic activity in which plant tissues remain alive but do not grow.

Dormant oil: Oil sprayed on deciduous trees or shrubs while they are dormant. Dormant oils are used to kill overwintering insects or insect eggs on plant bark. For plants with green leaves, summer oil should be used instead. See also Horticultural oil.

Edema: Watery blisters or swellings that form on many herbaceous plants. These swellings may burst open, forming rust-colored lesions.

Espalier: To train a plant (usually a tree or vine) along a railing or trellis so that the branches grow flat against the rail or trellis that supports them.

Evergreen: A plant that retains all or most of its foliage throughout the year.

Fasciation: An abnormal fusion of stems, leaves, or flowers, or the production of distorted growth.

Fertilizer: A substance that contains plant nutrients. Fertilizer may be liquid or dry and may be formulated in many different ways.

Flag: A tuft of dead foliage in an otherwise-healthy tree. Flags are visible from a distance and are used to determine the presence of several insects and diseases.

Force: To cause a plant to bloom outside its normal bloom season by manipulating periods of light and dark. Blooming mums bought any season except fall are forced.

Formulation: The form in which a compound may be produced. For example, a pesticide may be powdered, liquid, or granular, or it may be in an oil solution.

Frass: Sawdustlike insect excrement.

Frond: The large, flat leaf of a fern or palm.

Fruiting body: A fungal structure that produces spores.

Fungicide: A chemical that kills fungi or prevents them from infecting healthy plant tissue.

Gall: An abnormal growth that forms on a plant root, shoot, or leaf. Galls often result from infection or insect infestation.

Germination: The sprouting of seeds.

Girdle: To encircle plant roots, stems, trunks, or branches, constricting the plant part or reducing water and nutrient flow through the girdled plant part.

Graft: To unite a stem or bud of one plant with a stem or root of another plant.

Grub: A thick, soft insect larva (usually a beetle larva).

Gummosis: The oozing of plant sap, often from a plant wound or canker. Gummosis may occur as a result of infection or insect infestation.

Hardening off: The process of plant adjustment to cold temperatures.

Hardiness: The ability of a plant to withstand cold temperatures.

Hardpan: A layer of rocklike soil that sometimes forms a foot or two under the surface. Hardpan is most common in arid regions and is formed by deposits of chemicals that cement soil particles together.

Heartwood: The inner core of wood inside a woody stem or trunk.

Herbaceous: Being mainly soft and succulent, with little or no woody tissue.

Herbicide: A chemical that kills or retards plant growth. Herbicides may kill the entire plant, or they may kill only the aboveground plant parts, leaving the roots alive. Also called a weed killer.

Honeydew: A syrupy substance secreted by many insects of the aphid order. Much of the sugar in the sap the insects drink passes through their bodies undigested and coats leaves and objects under them.

Horticultural oil: Petroleum oil formulated to be mixed with water and sprayed on plants to kill insects. Because the oil kills by smothering, rather than by poisoning, insects do not develop resistance to it. A less-refined version called dormant oil is used only on dormant trees and shrubs. A more highly refined version called summer oil or superior oil can be sprayed on green leaves.

Host: An organism that is parasitized by another organism, such as a plant that is infected by a fungus or infested by an aphid.

Humidity: The amount of water vapor (moisture) in the air.

Humus: The relatively stable end product of the decomposition of organic matter in soil. Humus is a complex mixture of chemicals that supplies plant nutrients to and improves the structure of soil.

Hybrid: The offspring of two distinct plant species; a plant obtained by crossing two or more different species, subspecies, or varieties of plants. Hybrids are often made to produce a plant that has the best qualities of each parent.

Immature: An early growth phase of a plant differentiated from later growth by distinctly different leaf shapes, habits of growth, or other characteristics. Also called juvenile.

Immune: Not susceptible to a disease or insect.

Infiltration, soil: The process by which water moves into the soil.

Inflorescence: The flowering part of a plant, often referring to groups of flowers.

Insect growth regulator: An insecticide that has a hormonelike effect on insects, usually interfering with molting.

Insecticide: A chemical that kills insects.

Insoluble: Not soluble. See also Solubility.

Instar: A stage in the life of insects and mites. Each instar ends with molting the old skin.

Internode: The section of stem between two nodes.

Interveinal: Between the leaf veins. Interveinal yellowing, or chlorosis, refers to a discoloration occurring between the leaf veins.

Juvenile: An early growth phase of a plant differentiated from later growth by distinctly different leaf shapes, habits of growth, or other characteristics. Also called immature.

Larva: An immature stage through which some types of insects must pass before developing into adults. Caterpillars are the larvae of moths and butterflies, and grubs are the larvae of beetles. Larvae are typically wormlike in appearance.

Lateral bud: A bud formed along the side of a stem or branch rather than at the end.

Leach: To remove salts and soluble minerals from the soil by flushing the soil with water.

Leader: The main stems or trunk of a tree or shrub from which side stems or branches are produced.

Leaf margins: The edges of a leaf. Variations in the shapes of leaf margins are used to help identify many plants and differentiate among them.

Leaf scar: A tiny scar left on a twig or stem after a leaf or leafstalk (petiole) drops off.

Leafstalk: A stalk that attaches a leaf to a stem. Also called a petiole.

Leggy: Having leaves spaced too far apart along the stem.

Lesion: A wound, discoloration, or scar caused by disease or injury.

Loam: Medium-textured garden soil containing a balance of sandy and clay soils and organic matter.

Macronutrient: A major nutrient required by plants for normal growth. Most plants need macronutrients, such as nitrogen, phosphorus, and potassium, in large quantities.

Maggot: A legless grub; the larva of a member of the fly family.

Metamorphosis: The changes in body shape undergone by many insects as they develop from eggs into adults.

Microclimate: The environment immediately surrounding a plant; very localized climate conditions. Many different microclimates may occur at the same time in different areas of a garden.

Micronutrient: A minor nutrient required by plants for normal growth. Most plants need micronutrients such as iron, zinc, and manganese in small quantities.

Mite: A tiny animal related to spiders. Many mites feed on plants.

Miticide: A chemical that kills spider mites and other types of mites. Also called an acaricide.

Mulch: A layer of organic or inorganic material on the soil surface. Mulches help to moderate the temperature of the soil surface, reduce loss of moisture from the soil, suppress weed growth, and reduce runoff.

Mycelia: Microscopic fungal strands that form the major part of a fungal growth.

Nematode: A microscopic worm that lives in the soil and feeds on plant roots. Some nematodes feed on plant stems and leaves.

Node: The part of a stem where leaves and buds are attached.

Nodule: A small knob on a plant root. Nematodes, nitrogen-fixing bacteria, and other organisms may cause nodules.

Nonsystemic: Not systemic. See also Systemic.

Nymph: An immature stage through which some types of insects must pass before developing into adults. Nymphs usually resemble the adult form, but they lack wings and cannot reproduce.

Organic matter: A substance derived from plant or animal material.

Overfertilized: Fertilized too much. See also Fertilizer.

Overwinter: To survive the winter season. An "overwintering form" of an insect is the stage of growth, such as an egg, in which the insect spends the winter.

Ovipositor: An insect's egg-laying organ. In many insects, it is a drill-like or knife-like organ that can deposit eggs within a leaf or even deep under the bark of a tree.

Ozone: A common air pollutant that may cause plant injury.

Palisade cells: Columnar cells located in a layer just beneath the upper surface of a leaf.

PAN (peroxyacetyl nitrate): A common air pollutant that may cause plant injury.

Panicle: A complex compound inflorescence. The flowers are often drooping.

Parasite: An organism that obtains its food from another living organism. A parasite lives on or in its host.

Pathogen: An organism (such as a fungus, bacterium, or virus) capable of causing a disease.

Peat: Partially degraded vegetable matter found in marshy areas. Peat is commonly used as a soil amendment.

Perennial plant: A plant that lives for more than 2 years, often for many years. Most woody plants and many herbaceous plants are perennials.

Permanent wilting point: The point of soil dryness at which plants can no longer obtain water from the soil. Once plants have reached the permanent wilting point, they do not recover even if supplied with water.

Pesticide: A chemical used to kill an organism considered a pest.

Petiole: A stalk that attaches a leaf to a stem. Also called a leafstalk.

pH: A measure of the acidity or alkalinity of a substance; a measure of the relative concentration of hydrogen ions and hydroxyl ions.

Phloem: The nutrient-conducting vessels found throughout a plant. Phloem vessels transport nutrients produced in the foliage down through the stems, branches, or trunk to the roots.

Photosynthesis: The process by which plants use the sun's light to produce food (carbohydrates).

Phylum: The largest grouping of the animal kingdom. Mollusks, mammals, and birds are all phyla.

Plant disease: Any condition that impairs the normal functioning and metabolism of a plant. A plant disease may be caused by a fungus, bacterium, or virus or by an environmental factor such as sunburn or lack of nutrients.

Predaceous: Living by preying on others.

Propagation: The means of reproducing plants, such as by seeds, cuttings, budding, or grafting.

Protectant: A chemical that protects a plant from infection or insect infestation.

Pupa: An immature resting stage through which some types of insects must pass before becoming adults.

Pustule: A colored bump or blister caused by a disease organism.

Resistance, chemical: The ability of insects or pathogens to tolerate chemicals meant to control them; acquired through mutation and selection due to frequent exposure to the poison.

Resistance, plant: Not likely to be damaged by a particular problem. For example, a drought-resistant plant withstands more drought than other plants. When discussing plant diseases, a resistant plant is one that has defenses against the pathogen, so is not infected. See also Tolerant.

Resting structure: A dormant phase of a fungus in which it is able to survive for months or years until conditions for growth are again present.

Rhizome: An underground stem from which roots grow. Rhizomes function as storage organs and may be divided to produce new plants. Also known as rootstock.

Root zone: The volume of soil that contains the roots of a plant.

Rootstock: (1) The roots and crown or the roots, crown, and trunk of a plant upon which another plant is grafted. (2) The crown and roots of some types of perennial herbaceous plants, also known as rhizomes.

Rosette: A plant growth pattern in which the leaves form a flat, crowded ring close to the ground.

Runners: Aboveground trailing stems that form roots at their nodes when they make contact with moist soil. See *stolon*.

Sapwood: The outer cylinder of wood in a trunk, between the heartwood and the bark.

Saturated soil: Soil that is so wet that all of the air pores in the soil are filled with water.

Scalp: To mow a lawn close to the ground. This damages the grass and may kill it. Scalping usually happens when the ground surface is uneven or when a mower bounces in dense turf.

Sclerotia: A compact mass of fungal strands (mycelia) that functions as a resting stage for a fungus. Sclerotia are usually brown or black, about the size of a pea, and can withstand adverse conditions. A genus of fungus is called *Sclerotium* after the sclerotia it produces.

Seed piece: A small potato tuber or piece of a tuber planted to grow more potatoes.

Semidormant: Partially dormant. See also Dormant.

Semiparasitic: Partially parasitic. See also Parasite.

Shot hole: Small round holes in leaves that look as if they were made by shotgun pellets; often caused by leaf spot disease, in which the diseased tissue drops out.

Slow-release fertilizer: A fertilizer that releases its nutrients into the soil slowly and evenly over a long period of time.

Soil heaving: The expansion and contraction of soil during periods of freezing and thawing. Plant roots may be sheared off, or plants may be lifted out of the ground during soil heaving.

Soil penetrant: A substance that changes the surface tension of water or another liquid, causing it to wet a repellent surface more thoroughly. Dish soap is an effective soil penetrant.

Solubility: The degree to which a compound will dissolve in water. Compounds with high solubility dissolve in water more readily than compounds with low solubility.

Soluble fertilizer: A fertilizer that dissolves easily in water and is immediately available for plant use.

Spike: A flower stalk with flowers directly attached to the stalk, without stems.

Spikelet: (1) A small, single spike of flowers that forms part of a compound inflorescence. (2) The flower of grasses.

Spore: A microscopic structure produced by fungi, mosses, and ferns that can germinate to form a new plant or a different stage of the same plant.

Spreader-sticker: Spray additive with two functions. It weakens the surface tension of water, keeping it from beading on waxy or fuzzy plant surfaces, and also glues the active ingredient to the leaf, making it more resistant to washing off.

Spur: A short lateral branch bearing buds that will develop into flowers and then into fruit.

Stolon: A horizontal stem above the soil surface that gives rise to new plants. Also called a *runner*.

Stomata: Tiny pores located mainly on the underside of leaves. Oxygen, carbon dioxide, water vapor, and other gases move in and out of the leaf through these pores.

Stone fruit: A tree fruit that contains a large single seed, such as a peach, plum, or cherry.

Succulent: (adj) Full of water. (n) A plant that stores water in its tissues, such as cactus.

Sucker: A shoot or stem that grows from an underground plant part.

Summer oil: Highly refined petroleum oil mixed with an emulsifier to be applied in water. Used to kill soft-bodied insects and smother eggs on actively growing plants. Also called superior oil. See also Horticultural oil.

Sunscald: Damage to leaves, bark, or fruit caused by the heat of the sun.

Surfactant: A substance added to a spray to increase its wetting and spreading properties. Also called a wetting agent.

Systemic: Taken into the plant and spread throughout it. A systemic plant disease, such as a virus, spreads through the plant rather than remaining localized. A systemic pesticide or herbicide is transported throughout the plant.

Systemic insecticide: A pesticide that is absorbed into part or all of the plant tissue.

Taproot: An undivided main root that penetrates deeply into the ground.

Tender plant: A plant that cannot tolerate freezing temperatures.

Terminal bud: A bud at the end of a stem or branch.

Thatch: A layer of dead grass stems on top of the soil. Thatch more than ½ inch deep can be harmful to the lawn.

Tilth: The structure of soil, especially regarding its suitability for plant growth. Soil in good tilth has a soft, loose structure.

Tolerant: A plant that, when infected by a disease, shows few symptoms. A plant tolerant of a disease is infected but not damaged severely, but can transmit the disease to other plants. A plant resistant to a disease does not acquire the disease, so does not transmit it to other plants. See also Resistant.

Toxin: A poisonous substance produced by a plant or an animal.

Translocation: The movement of a compound from one location in a plant to another.

Transpiration: The evaporation of water from plant tissue. Transpiration occurs mainly through the stomata in the leaves.

Tuber: An underground storage and reproductive organ derived from stem tissue; bears dormant buds called "eyes."

Underfertilized: Not fertilized enough. See also Fertilizer.

Vascular system: The system of tissues (phloem and xylem) that conducts nutrients and water throughout the plant.

Vein clearing: A lightening or total loss of color in leaf veins. This often results from infection or nutrient deficiency.

Vernalization: A cooling period required by many plants in order to germinate, grow, or flower properly. Also called a chilling requirement.

Vigor: The health of a plant. A vigorous plant grows rapidly and produces healthy, normal amounts of foliage and flowers. A nonvigorous plant grows slowly, if at all, and produces stunted, sparse growth.

Weed killer: A chemical that kills or retards plant growth. Weed killers may kill the entire plant, or they may kill only the aboveground plant parts, leaving the roots alive. Also called an herbicide.

Wetting agent: A substance that changes the surface tension of water or another liquid, causing it to wet a repellent surface more thoroughly. Wetting agents are often used with pesticides sprayed on waxy or fuzzy foliage. Also called a soil penetrant or surfactant.

Wing pad: An incompletely developed wing, often present on immature stages (instars) of winged insects.

Witches'-broom: A dense cluster of twigs that looks like a witch's broom. Any of several different insects and diseases may cause witches'-broom.

Xylem: The water-conducting vessels found throughout a plant. Xylem vessels transport water and minerals from the roots upward through the plant.

Yellowing: The yellowing of foliage due to a loss or breakdown of chlorophyll. Yellowing may result from disease or insect infestation, poor growing conditions, or lack of nutrients. Also called chlorosis.

Houseplants

Houseplants are an appealing and inexpensive decor for any room. From left to right: heartleaf philodendron, spathiphyllum, cyclamen, African violet, spotted dumb cane, and English ivy topiary.

When a houseplant stops blooming or drops its leaves, the home gardener may be tempted to discard the plant rather than seek out the problem. Bud failure or leaf drop may be due to low temperature, poor soil, drafts, lack of fertilizer, too much or too little water, or too much or too little light. All these problems may be correctable. Saving the plant is entirely possible.

PURCHASING HOUSEPLANTS

When you go to a nursery or garden center to buy a houseplant, you want a specimen that is appropriate for the conditions you can provide. And, of course, you want a healthy plant that is pest-free.

Appropriate Plants: The most important factor in choosing an appropriate species of plant is the amount of light it will receive in your home. To learn how to evaluate indoor lighting, read the section called Providing Light, which appears later in this chapter. Note whether the proposed growing area receives bright, medium, or low light. Take a houseplant reference book with you when you shop, or make sure your nursery has references available on-site. Look up your intended purchase to see whether the site can provide the light the plant needs. Flowering plants and cacti need the most sunlight. Pothos, cast-iron plants, and some ivies grow slowly but well in indirect light. Prima donnas such as orchids have specific light requirements in order to bloom; make sure your site can meet their needs before selecting such plants.

Healthy Plants: How can you tell if a plant is healthy? The leaves of a healthy plant are green unless they are naturally variegated or multicolored, as are some pothos, Chinese evergreens, zebra plants, and others. Unhealthy leaves may have tips or edges that look burnt, brown spots, or a yellowish cast. Unhealthy leaves may appear crumpled or tend to droop. Readily apparent leaf problems can be due to powdery mildew, aphids, whiteflies, spider mites, or other insects. Inspect leaf undersides and leaf-to-stem junctions for signs of disease or insects. Of the plants that appear healthy, select the most compact and fully leafed.

When buying a flowering plant for indoor use, look for a specimen with ample buds as well as flowers. Minimal buds on a plant usually mean it has passed the peak of blooming; it will be another year before it blooms copiously again. A plant with many buds will be colorful throughout the current season. If you find a sturdy, well-budded plant with some flowers, give it a gentle shake. If many flowers drop off, the plant has been subjected to severe stress. Select a healthier specimen.

After bringing your purchase home, set it off by itself for about a week. Even though you did not see any insect pests, the plant may harbor microscopic insect eggs. Check the plant carefully after the quarantine period. If you see even a few insects, treat the new plant with insecticide before placing it near any other plant.

PURCHASING THE PERFECT POT

Among the many choices available for indoor plant potting are unglazed clay, plastic, and glazed ceramic in designs to match every decor.

A clay pot is especially appropriate on a porch or in a rustic atmosphere. Since moisture evaporates quickly through clay pot sides, use clay pots as containers for plants such as succulents, which tolerate dryness. If you place other types of plants in clay pots, they will need more moisture than normal. Since water tends to seep through clay pot bottoms, place a nonporous saucer underneath to prevent rug or counter stains.

Plastic pots are lightweight and often used for hanging plants. Plastic pots hold water longer than clay pots, so be careful not to overwater. Many plastic pots are sold with removable saucers.

Glazed ceramic pots are as effective in water control as plastic pots. However, many ceramic pots do not have drainage holes, a deficiency that can cause overwatering. Place a plastic pot with drainage saucer inside the glazed pot.

PROVIDING SOIL

In nature, plant roots can spread out to seek nutrients. In a pot, what's there is what the plant gets. If vitamins and minerals are lacking, the plant fails to thrive. Nutrients are as important to the plant as adequate light and sufficient moisture.

Most houseplants can thrive in all-purpose potting soil. Fussier plants, such as African violets, may grow better in a commercial potting soil formulated especially for the species. Other types of commercial soils are formulated for specific situations—for example, terrariums. Soilless growing mediums are also available.

You can make your own potting mix, using varying proportions of garden soil, sand, peat moss, vermiculite, and leaf mold. The gardener who uses homemade potting

A sunny kitchen window with a waterproof, tiled surface is an ideal spot for houseplants.

Houseplant potting soil that has been allowed to completely dry out can be difficult to rewet. Submerge the entire pot in a pan of tepid water for half an hour, then drain.

mix, however, runs the risk of bringing in insect pests and disease organisms. Commercial potting soil is inexpensive, convenient, and free of pests and disease.

PROVIDING WATER

Water causes more plant problems than any other single factor. These problems include overwatering, underwatering, and using inappropriate water techniques or tainted water.

Too Much Water: A water overdose without adequate drainage rots roots slowly but steadily, causing plant death. One sign of overwatering is green moss that grows on the surface of the soil. Plant symptoms include lower leaf wilting, faded leaf colors, and poor growth. The lower portion of the plant's main stem, right above the soil line, may darken. Roots are brown and mushy.

If damage has not totally destroyed the roots, rescue attempts can include removing standing water, trimming brown roots, and repotting the plant in well-draining soil. An alternative is to take cuttings from healthy stems and start over.

Too Little Water: Many gardeners worry so much about overwatering that they underwater. A water-stressed plant conserves moisture by slowing or stopping new growth. Without adequate water, green leaves turn dull green or yellow. Drooping

occurs. If buds are present, they may fall off.

When to water? Poke your finger about ½ inch into the potting soil. It should feel moist but not wet. If it feels dry or barely moist, water immediately and thoroughly. Ensure adequate drainage, or standing water will turn the underwatering problem into a case of overwatering.

Watering Techniques and Tainted Water: Water most houseplants from the top. Within an hour, pour off excess water from the saucer underneath the pot.

Most drinkable tap water is adequate for plants. Use it at room temperature. The sodium in some types of artificially softened water can prove a problem, however, if used consistently over a long period. Some tap water may also contain salts, which accumulate quickly in plant containers. Salt damage is evidenced by brown leaftips or edges. Damage occurs on older leaves first, and affected leaves eventually die. The plant may also be stunted, with brittle leaves that curl downward. An accumulation or overdose of fertilizer salts causes similar damage. If you see symptoms of salt accumulation, flush the plant thoroughly with water (for instructions, see page 22). If salts are built up on the rim or at the soil line in the pot, repot the plant in fresh soil to dilute salt levels.

Some gardeners collect rainwater for indoor plant use. Rainwater may carry pollutants, depending on where you live.

In a pot, pollutants may accumulate quickly and deter plant growth.

PROVIDING LIGHT

The secret to providing appropriate light is to match the plant to the site.

Site Evaluation: How do you evaluate light? During prime light time, place a sheet of white paper on the table or sill where a plant will reside. Hold your hand about 12 inches above the paper. If a clearly defined dark shadow results, the site receives bright light. If a muted but clearly definable shadow results, the light is medium. If your hand shadow is barely visible, the amount of light is low. Make sure any plants you purchase can prosper in the lighting conditions you can provide.

Symptoms of Inappropriate Light: Inadequate lighting produces a leggy, weak plant that may suddenly drop its leaves. Growth slows. The lower leaves turn a lighter green, and the plant does not flower. Plants lean toward the light source; rotating the plants regularly prevents uneven growth.

An African violet that does not bloom is probably not receiving adequate light. These plants require about 12 hours of bright light every day.

Although most complaints are about insufficient light, some rooms are too sunny. Dry patches on leaves may be symptoms of sunburn. If the site gets hot enough, buds

Supplemental light fixtures are fairly basic. The main differences are in the bulbs. Choose supplemental lighting according to your plants' needs and your budget.

and flowers drop off and the entire plant may wilt. To prevent further damage, try moving the plant away from the window; shading the window with filmy curtains; or replacing the plant with a heat-tolerant species, such as a cactus.

Cacti are the plants of choice for sunny locations. Watering them once a week will do, except when you want them to flower. Do not allow cacti to dry out totally during the flowering season.

Artificial Light: In sites that receive little sun, artificial light may be the only answer. Fluorescent bulbs and incandescent bulbs provide different types of light. Cool-white fluorescent bulbs give off little heat. They do not bake the moisture out of plants, even if placed within 4 inches of the foliage. With adequate fluorescent lighting, you need no outdoor light; you could grow plants in a closet if you provided a fan for air circulation.

To help support plant growth with incandescent light, you must use a bulb of at least 100 watts. Such a bulb produces a lot of heat; keep incandescent light at least 2 feet above plant tops to keep from burning the foliage and baking the soil.

PROVIDING FERTILIZER

Outdoors, soil is constantly improved with leaf mold, earthworm castings, decaying plants, and animal droppings.

Indoors, once a plant has used up the nutrients in the pot, there's nowhere for it to get more unless the gardener adds some type of fertilizer.

Symptoms of Nutrient Shortage: Plants quickly reflect a nutrient shortage. A nitrogen shortage shows up as yellowing leaves and poor growth. If a plant has leaves a darker green than normal, poor growth, and leaf stems with a purplish tinge, a shortage of phosphorus is probably the cause. A potassium shortage appears as yellowing leaves with brown tips and edges. A lack of iron appears as the yellowing of older leaves, on the bottom portions of stems. This yellowing starts at leaf edges and progresses inward.

Fertilizer Selection and Application: Many types of indoor plant fertilizer are available. The numbers on the container describe the relative proportions

of nitrogen (N), phosphorus (P), and potassium (K). The designation "20-20-20" means the fertilizer contains equal portions of each element. A 5-10-5 mixture is higher in phosphorus than in nitrogen or potassium.

Nitrogen helps make healthy green leaves. Phosphorus encourages a strong root system as well as luxuriant flowers. Potassium aids in disease resistance, promotes plant vigor, increases bloom, and strengthens stems. In addition, plants need trace elements, such as iron, which is essential in minute amounts for chlorophyll production and enzyme functioning. Plants grown in synthetic mediums, such as sand or vermiculite, need a dose of one-third strength fertilizer with each watering, because they contain no soil nutrients.

A little bit of fertilizer may be fine, but a lot of fertilizer is almost always too much. Extra fertilizer accumulates in soil, causing tip burn or browning. Too much nitrogen causes rapid growth at the expense of plant vigor. The plant becomes large and spindly, does not set flowers, and is prone to insect invasion.

If you have applied too much fertilizer, take action quickly. Repot the plant in fresh soil or rinse and drain the current soil to wash out fertilizer residue. When applying fertilizer, always follow label instructions.

GROWING OUTDOOR PLANTS INDOORS

Outdoor potted plants purchased for indoor bloom—such as chrysanthemums, freesias, hydrangeas, hyacinths, and narcissus—may be subject to rapid bud withering if kept in areas with continual hot, dry air. Avoid placing these plants in extremely sunny kitchen windows or near microwaves, ovens, and heating vents.

Given appropriate light, sufficient water, and ample air circulation, outdoor plants will bloom indoors for several weeks. If the soil is allowed to become dry during blooming time, future flowers may not form, even though foliage may recuperate.

When bulbs and other basically outdoor plants cease flowering, they probably will not do so again indoors. Plant them outdoors in appropriate surroundings,

Easter cactus

however, and they will bloom normally.

Holiday plants, such as poinsettias and Christmas cacti, are reared under controlled conditions to produce blooming during specific seasons. Gardeners often expect them to bloom again next year at the same time. This will probably not happen unless light is strictly regulated.

Extended periods of light encourage foliage rather than flowers. For poinsettias and Christmas cacti to reflower, you must provide total uninterrupted darkness during evening and night hours. After poinsettias bloom the first time, cut back stems to 8 inches and repot the plants in fresh houseplant soil. Beginning in October, cover both poinsettias and Christmas cacti with a large carton from sundown to sunup. Remove the carton every morning. Do this until flowers appear.

Gift plants, such as azaleas, may lose their flowers quickly indoors if placed in hot direct sun. Do not let them dry out. Place them away from drafts, and keep them moist and cool. Getting azaleas to bloom indoors a second time is extremely difficult. Even with the best care, it may take several years. After the plants bloom the first time, some gardeners place them outdoors in appropriate surroundings, where the azaleas may do well. Others keep azaleas as foliage plants indoors.

Other gift plants that generally do not bloom again indoors are cineraria and cyclamen. Try placing them outdoors in good soil and growing conditions, and after the first or second spring they may surprise you with flowers.

Too little water

Dry poinsettia.

Problem: Leaves are small, and plants fail to grow well and may be stunted. Plant parts or whole plants wilt; leaves may yellow and drop off. Margins of leaves or the tips of leaves of narrow-leafed plants may dry and become brittle but still retain a dull green color. Bleached areas may occur between the veins. Such tissues may die and remain bleached or may turn tan or brown. Plants may die.

Analysis: Plants need water in order to grow. Besides making up most of the plant tissue, water is also the medium that carries nutrients into the plant, so a plant that is frequently short of water is also short of nutrients. Water also cools the leaves as it evaporates from them. If a leaf has no water to evaporate, it may overheat in the sun and burn. If plants wilt and then are given water, sometimes the margins or tips of the leaves will have completely wilted and will not recover. If this occurs, the margins or tips will die and become dry and brittle but will retain a dull green color.

Solution: Water plants immediately and thoroughly. If the soil is completely dry, soak the entire pot in water for a couple of hours. The rim of the container should be submerged. Or add a soil penetrant (available in garden centers) to the irrigation water. Water again when the soil just below the surface is barely moist, applying enough water so that some water drains from the bottom. In large containers (more than 10 inches wide and 10 inches high), water when the soil 1 to 2 inches below the surface is barely moist.

Too much water or poor drainage

Overwatering damage to Schefflera.

Problem: Plants fail to grow and may wilt. Leaves lose their glossiness and may become light green or yellow. When the plant is lifted from its container and its rootball is examined, the roots are brown and soft and do not have white tips. The soil in the bottom of the pot may be very wet and have a foul odor. Plants may die.

Analysis: If the soil is kept too wet, the air spaces are filled with water, and the roots are weakened and may die. Although plants need water to live, the roots also need air. Weakened plants are more susceptible to root-rot fungi, which wet soils favor. Plants with diseased roots do not absorb as much water as they did when they were healthy, so the soil remains wet. If roots are damaged or diseased, they cannot pick up water and nutrients needed for plant growth.

Solution: Discard severely wilted plants and those without white root tips. Repot into a smaller pot until roots regrow. Clean the pot before reuse. Do not water less severely affected plants until the soil is barely moist. Prevent the problem by using a light soil with good drainage.

Salt damage

Salt damage to Spathiphyllum.

Problem: The leaf margins of plants with broad leaves or the leaf tips of plants with long, narrow leaves turn dark brown and die. This browning occurs on the older leaves first, but when the condition is severe, new leaves may also be affected. Plants may be stunted, with brittle leaves curling downward. On some plants, the older leaves may yellow and die.

Analysis: Salt damage is a common problem on container-grown plants. The roots pick up soluble salts, which accumulate in the margins and tips of the leaves. When concentrations become high enough, the tissues are killed. Salts can accumulate from water or from the use of fertilizers, or they may be present in the soil used in potting. Salts accumulate faster and do more harm if plants are not watered thoroughly. Water that is high in lime does not cause as much salt damage as water that is high in other salts.

Solution: Leach excess salts from the soil by flushing with water. Water thoroughly at least three times, letting the water drain from the pot each time. This is most easily done if the pot is placed in the bathtub, in a basin, or outside in the shade to drain. If a saucer is used to catch the water, empty the saucer 30 minutes after each watering. If the plant is too large to lift, empty the saucer with a turkey baster. Never let a plant stand in drainage water.

Sunburn

Sunburn on Dieffenbachia.

Problem: Dead tan or brown patches may develop on leaves that are exposed to direct sunlight, or leaf tissue may lighten or turn gray. In some cases, the plant remains green but growth is stunted. Damage is most severe when the plant is allowed to dry out.

Analysis: Sunburn or bleaching occurs when a plant is exposed to more intense sunlight than it can tolerate. Plants vary widely in their ability to tolerate direct sunlight. Some plants can tolerate full sunlight; others burn or bleach if exposed to any direct sun. Bleaching occurs when light and heat break down chlorophyll, causing the damaged leaf tissue to lighten or turn gray. On more sensitive plants, or when light and heat increase in intensity, damage is more severe and plant tissues die. Sometimes tissue inside the leaf is damaged but outer symptoms do not develop. Instead, growth is stunted. Plants are more susceptible to bleaching and sunburn when they are allowed to dry out, because the normal cooling effect of water evaporating from leaves is reduced. Plants grown in low-light conditions burn very easily if suddenly moved to a sunny location.

Solution: Move plants that cannot tolerate direct sunlight to a shaded spot. Or cut down light intensity by closing the curtains when a plant is exposed to direct sun. Prune off badly damaged leaves or trim away damaged leaf areas to improve the plant's appearance. Keep plants well watered.

Insufficient light

Leaf drop on zebra plant (Calathea zebrina).

Problem: Plants fail to grow well. Leaves may be lighter green and smaller than normal. Lobes and splits that are normal in leaves may fail to develop. Lower leaves may yellow and drop. On some plants, leaves at first are abnormally large and thinner than normal, then are smaller than normal. Stems and leafstalks may elongate and be spindly and weak. Plants grow toward a light source. Flowering plants fail to produce flowers, and plants with colorful foliage become pale. Variegated plants may lose their variegation and become green.

Analysis: Plants use light as a source of energy and grow slowly in light that is too dim for their needs. If most available light comes from one direction, stems and leaves bend in that direction. If the light is much too dim, plants grow poorly. Although foliage plants generally need less light than plants grown for their flowers or fruit, plants with colorful foliage need abundant light.

Solution: Move the plant gradually to a brighter location. The brightest spots in most homes are the sills of windows that face east, west, or south or locations as close to these windows as possible. To avoid sunburn on sensitive plants, close lightweight curtains when the sun shines directly on the plant. If enough natural light is not available, supply extra light. Move plants from one location to another until you find a place where they grow well.

Nitrogen deficiency

Nitrogen-deficient Swedish ivy plant.

Problem: The oldest leaves—usually the lower leaves—turn yellow and may drop. Yellowing starts at the leaf margins and progresses inward without producing a distinct pattern. The yellowing may progress upward until only the newest leaves remain green. Growth is slow, new leaves are small, and the whole plant may be stunted.

Analysis: Plants use the nutrient nitrogen in many ways, including for the production of chlorophyll. Nitrogen is used in large amounts. When there is not enough of the nutrient for an entire plant, nitrogen is taken from older leaves for use in new growth. Nitrogen is easily leached from soil during regular watering. Of all the plant nutrients, it is the one most likely to be lacking in soil.

Solution: Fertilize houseplants with Miracle-Gro Plant Food. Add the fertilizer at regular intervals, as recommended on the label.

Spider mites

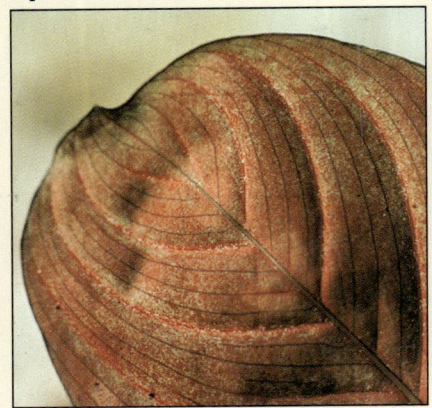

Spider mite damage to prayer plant.

Problem: Leaves are stippled, yellowing, and dirty. Leaves may dry out and drop. There may be webbing over flower buds, between leaves, on the growing points of shoots, or on the lower surfaces of leaves. To determine if a plant is infested with mites, examine the bottoms of the leaves with a hand lens. Or hold a sheet of white paper underneath an affected leaf and tap the leaf sharply. Minute specks the size of pepper grains will drop to the paper and begin to crawl around. The pests are easily seen against the white background.

Analysis: Spider mites, related to spiders, are major pests of many houseplants. They cause damage by sucking sap from the undersides of leaves. As a result of their feeding, chlorophyll disappears, producing the stippled appearance. Spider mite webbing traps cast-off skins and debris, making the plant messy. Under warm, dry conditions, mites can build up to tremendous numbers.

Solution: Isolate infested plants from others. Add ORTHO RosePride Systemic Rose & Flower Care to the potting soil according to label instructions. It also helps to take plants outside or into a shower and wash the mites off the leaves with a strong spray of water. For additional control, take plants outside and spray with ORTHO Isotox Insect Killer. Keep air humid to help prevent infestation and proliferation. Avoid introducing mites into the house with newly-purchased plants, or when bringing houseplants in after a summer on the patio; inspect foliage carefully and treat plants for mite infestation before carrying them inside.

Scales

Scales on Dracaena.
Inset: Immature scale, 50× life size.

Problem: Nodes, stems, and leaves are covered with white, cottony, cushionlike masses; brown, crusty bumps; or clusters of somewhat flattened reddish, gray, or brown scaly bumps. The bumps can be scraped or picked off easily. They don't move when touched. Leaves turn yellow and may drop. A shiny or sticky material may cover the leaves. Mold may be growing on the sticky substance.

Analysis: Several different types of scale insects attack houseplants. Some types can infest many different plants. Scales hatch from eggs. The young, called *crawlers*, are small (about ¹⁄₁₀ inch) and soft-bodied and move about on the plant and to other plants. After moving about for a short time, they insert their mouthparts into the plant, withdrawing the sap. Some develop a soft covering over their bodies, others a hard covering. Some species of scales are unable to digest fully all the sugar in the plant sap, and they excrete the excess in a fluid called *honeydew*, which may cover the leaves or drop onto surfaces below.

Solution: Isolate infested plants as soon as scales are discovered. Remove as many scales as possible with a cloth or toothbrush dipped in soapy water. Spray plants with an insecticide containing *pyrethrins* or a horticultural oil labeled for use indoors, or take the plant outside and spray with ORTHO Orthene Systemic Insect Control. To kill newly hatched eggs, repeat the treatment weekly for 4 weeks. Avoid bringing scale crawlers into the house; when bringing houseplants in after a summer on the patio, inspect foliage carefully and treat plants for infestation before carrying them inside.

Aphids

Aphids on oleander (life size).

Problem: Leaves are curling, discolored, and reduced in size. A shiny or sticky substance may coat the leaves. Leaves may become littered with cast-off insect skins. Tiny (¹⁄₈-inch) nonwinged, green, soft-bodied insects cluster on buds, young stems, and leaves. Winged insects are occasionally seen at the beginning of infestation.

Analysis: Aphids do little damage in small numbers. They are extremely prolific, however, and populations can rapidly build up to damaging numbers on houseplants. Damage occurs when the aphid sucks the juices from the leaves. Aphids are unable to digest fully all the sugar in the plant sap, and they excrete the excess in a fluid called *honeydew*, which often drops onto the leaves below. Any furniture below the plant may be coated with honeydew.

Solution: Spray plants with Miracle-Gro Bug Spray, or take the plant outside and spray with ORTHO Orthene Systemic Insect Control. Inspect new plants, and keep them isolated if there are any signs of infestation. Avoid working in the outdoor garden and then on indoor plants without washing up and changing in between.

Mealybugs

Citrus mealybugs on coleus (2× life size).

Long-tailed mealybugs (4× life size).

Problem: White cottony or waxy insects are on the undersides of leaves, on stems, and particularly in crotches or where leaves are attached. The insects tend to congregate, giving a cottony appearance. Cottony masses that contain eggs of the insects may also be present. A sticky substance may cover the leaves or drop onto surfaces below the plant. Sooty mold can result from the sticky substance. Infested plants are unsightly, do not grow well, and may die if severely infested.

Analysis: Mealybugs are among the more serious problems of houseplants. There are many different types of mealybugs, and one or more species of these insects attack virtually all houseplants. Male mealybugs, rarely seen, are winged and can fly. Female mealybugs have soft bodies that appear to have segments and are covered with waxy secretions, giving them a cottony appearance. The female may produce live young or may deposit hundreds of yellow to orange eggs in white, cottony egg sacs. The young insects, called *nymphs*, crawl about the same plant or to nearby plants. Males do no damage because they do not feed and are short-lived. Female mealybugs feed by sucking sap from the plant. They take in more than they can use and excrete the excess in a sugary fluid called *honeydew*, which coats the leaves and may drop to surfaces below the plant. This fluid may mar finished furniture.

Solution: Separate infested plants from those not affected. If only a few mealybugs are present, wipe them off with cotton swabs dipped in rubbing alcohol or with a damp cloth. Spray infested plants with an insecticide containing *pyrethrins* that is labeled for use indoors, or take the plant outside and spray with ORTHO Orthene Systemic Insect Control. Repeat applications at intervals of 2 weeks and continue for a little while after mealybugs appear to be under control. The waxy coverings on the insects and egg sacs, and the tendency of the insects to group together, protect them from insecticides. Carefully check all parts of the plant to make sure all insects have been removed. Search for egg sacs under the rims or bottoms of pots, in cracks or on the undersides of shelves, and on brackets or hangers. Wipe off any sacs; they are a constant source of new insects. Discard severely infested plants, and avoid taking cuttings from such plants. Thoroughly clean all surfaces around the growing area with soapy water before starting new plants. Be on a constant vigil for mealybugs, and start control measures immediately if the bugs appear. Inspect new plants thoroughly before bringing them in the house.

Whiteflies

Greenhouse whiteflies (2× life size).

Problem: Tiny white-winged insects feed mainly on the undersides of leaves. Nonflying, scalelike larvae covered with white, waxy powder may also be present on the undersides of leaves. When the plant is touched, insects flutter rapidly around it. Leaves may be mottled and yellowing.

Analysis: Whitefly (*Trialeurodes vaporariorum*) is a common pest of many houseplants. The four-winged adult lays eggs on the undersides of leaves. The larvae are the size of a pinhead, flat, oval, and semitransparent. They feed for about a month before changing to the adult form. Both larval and adult forms suck sap from the leaves. The larvae are more damaging because they feed more heavily. Feeding often transmits viral diseases from plant to plant. Adults and larvae cannot completely digest all the sugar in the sap, and they excrete the excess in a fluid called *honeydew*, which coats the leaves and may drop from the plant. Black, brown, or white fungus mold may grow on the honeydew.

Solution: Remove heavily infested leaves. Vacuum plants to pick up adults. Spray plants with Miracle-Gro Bug Spray, or take the plant outside and spray with ORTHO Orthene Systemic Insect Control. Spray the foliage thoroughly, being sure to cover the upper and lower surfaces of leaves. Treat plants at night when insects are not flying. Spray weekly as long as the problem continues. Whiteflies may also be partially controlled with yellow sticky traps. Inspect new plants on arrival.

CHLOROPHYTUM (Spider Plant)

Dead leaf tips

Tip burn.

Salt damage.

Problem: The tips of leaves turn brown or tan. The damaged area develops slowly along the leaf. Older leaves are most severely affected.

Analysis: Spider plant leaf tips die for several reasons. Frequently the problem has a combination of causes.

Solution: Remove the dead tips by trimming the leaves to a point with a pair of scissors. The numbered solutions below correspond to the numbered items in the analysis.

1. Salt damage: Salts from irrigation water or from fertilizer accumulate in the soil. Excess salts are carried to the leaves and deposited in the tips of pointed leaves such as those of spider plants. When enough salts accumulate there, the leaf tip dies. Salts are taken into the plant more rapidly when the soil is dry.

1. Leach excess salts from the soil by flushing with water. Water the plant at least three times, letting the water drain through each time. This is done most easily in a bathtub, in a laundry sink, or outside. Always water spider plants from the top of the pot. Add enough water each time so that some drains through the pot. Empty the saucer after the pot has finished draining. If the plant is too large to handle easily, use a turkey baster to remove the drainage water. Do not overfertilize. Repot into fresh potting mix if salts have accumulated on the pot.

2. Plant too dry: The leaf tip, being farthest from the roots, is the first part of the leaf to die when the plant does not get enough water.

2. Water the plant regularly.

3. Toxic salts: Some chemicals, usually in the form of soluble salts, are damaging in very small amounts. They accumulate in leaf tips as other salts do, killing the tissue there. The most common of these chemicals are fluoride, chloride, and borate.

3. You cannot do much about traces of toxic chemicals in the water, other than find another source of water. Distilled or deionized water is always free of chemicals.

CODIAEUM (Croton)

Insufficient light

New green leaves.

Problem: New leaves are green instead of brightly colored. Stems may be thin and bend toward a light source. Lower leaves may drop.

Analysis: Plants use light as a source of energy and grow slowly in light that is too dim for their needs. If most available light comes from one direction, stems and leaves bend in that direction. Leaves remain green, without bright colors.

Solution: Move the plant gradually to a brighter location. A lightly curtained sunny window is ideal. If a brighter location is not available, provide supplemental lighting. Crotons may be grown outside in the summer. When moving them from indoors to outdoors, place the plants in light shade for at least a week before putting them in full sun. If you wish to grow a houseplant in a dim location, select a plant that tolerates dim light from the list on page 347.

GROWING GUIDE

Light: Flowering plants need more light than foliage plants. To keep cyclamen plants in flower, place them in a south-facing window.

Soil: Grow cyclamens in a loose, well-drained soil mix.

Fertilizer: Fertilize once a month with Miracle-Gro Plant Food according to directions on the label.

Water: How much: Add enough water to the pot so that a little water drains from the bottom.
How often: Water whenever the top inch of soil is slightly moist. Do not let the soil get dry.

Temperature: Cyclamen will continue to flower when daytime temperatures are warm and nighttime temperatures are cool. In mild climates, put the plant outside at night. In areas with extremely cold winters, put the plant in a cool area but do not allow it to freeze. When days and nights are warm, the leaves will yellow and the plant will stop flowering.

Special Conditions: Do not allow cyclamen to become dry during their growing season. In bright light, plants that are dry will sunburn. Avoid overwatering. Cyclamen are very susceptible to root rot.

High temperatures

Damage to Cyclamen from high temperature.

Problem: Outer leaves turn yellow. Leaves may die and turn brown, and their stems become soft. Plants stop flowering. Plants eventually lose their leaves and go dormant.

Analysis: Although cyclamen are cool-weather plants, they tolerate warm days as long as they have cool nights (below 55°F). Cool temperatures initiate flower buds. Constant high temperatures inhibit flower buds, and plants stop flowering. High temperatures also prevent the plant from growing well, causing leaves to lose their green color and die. Most cyclamen naturally go dormant during the summer.

Solution: Grow cyclamen plants in a cool room with as much light as possible. If a cool room is not available, put them near a window at night. If temperatures are not below freezing, put the plants outside at night. Under alternating temperatures, they will flower for long periods. Keep plants adequately watered and fertilized.
If cyclamen do go dormant, keep them dry and store them in darkness until cool fall temperatures return, then water lightly until growth begins.

Cyclamen mite

Cyclamen mite damage.

Problem: Leaves become curled, wrinkled, and cupped in scattered areas. New leaves may be more severely affected, remain very small, have a bronze discoloration, and be severely misshapen. Flower buds are distorted and may drop or fail to open.

Analysis: Cyclamen mites (*Steneotarsonemus pallidus*), related to spiders, are too small to be seen with the naked eye. These mites attack several houseplants and can be very damaging on cyclamen. Their feeding injures the plant tissues, causing the leaves and flower buds to be malformed and stunted. Cyclamen mites infest new growth most heavily but will crawl to other parts of the plant or to other plants. They reproduce rapidly.

Solution: Spray plants with a miticide containing *hexakis* that is labeled for use indoors, or take the plant outside and spray with ORTHO Isotox Insect Killer Repeat every 2 weeks until new growth is no longer affected. Discard severely infested plants. Houseplants showing cyclamen mite damage should be isolated from other plants until the mites are under control. Nearby plants should be observed closely so that they can be sprayed if symptoms appear. Avoid touching leaves of infested plants and then touching leaves of other plants. Avoid working in the outdoor garden and then on indoor plants without washing up and changing clothes in between.

FERN

GROWING GUIDE

Light: Ferns prefer bright, indirect light, but will usually adapt to moderate light. They will burn in direct sun.

Soil: Use a standard potting mix and add an equal amount of peat moss.

Fertilizer: Fertilize once a month during the growing season (early spring through late summer) with a liquid or soluble plant food.

Water: How much: Add enough water so that 10 percent of the water drains through the pot.
How often: Water when the top inch of soil is still moist, but not wet.

Temperature: Ferns will do well in average house temperatures.

Special Conditions: Ferns are sensitive to salts, and salts must be flushed from the soil with extra water. Do not let ferns stand in drainage water.

Low humidity

Frond dieback caused by low humidity.

Problem: Leaves turn yellow and eventually die. Fronds may die from the tips down. The center parts of the plant are more severely affected than are the outer portions.

Analysis: Most ferns need higher humidity than homes provide. The ferns used as houseplants are adapted to forest floors and creeksides, where the air is usually moist. In the winter, when homes are heated, the air can become as dry as desert air. Air is particularly dry near heater vents or radiators. The problem is made more severe if the soil in which the fern is growing is allowed to dry out.

Solution: Move the fern to a more humid location, such as a well-lit bathroom. Place several plants together, and keep them away from drafts. Misting does not help relieve stress on the fern; it dampens the fronds for only a few minutes at a time. Placing the plant in a tray of gravel in which some water is kept raises the humidity around the plant only if the damp air is not allowed to escape. If air moves freely around the plant, the practice is of little value. A portable humidifier raises the humidity in the immediate vicinity while it is operating. Plant ferns in a potting mix that drains quickly and contains a high proportion of organic material, such as peat moss or ground bark. Never allow the potting mix to dry out.

FICUS (Ornamental Figs)

GROWING GUIDE

Light: Most figs should be grown in the best light available. Fiddle-leaf fig can tolerate somewhat lower light.

Soil: Any standard houseplant mix.

Fertilizer: Fertilize once a month with Miracle-Gro Plant Food. Plants in reduced light should be fertilized once every 2 months.

Water: How much: Add enough water so that some drains out the bottom of the pot.
How often: Water when the soil under the surface is moist but not wet. Ficus must not be allowed to dry out.

Temperature: Keep ficus as warm as possible under house conditions.

Special Conditions: Avoid moving ficus to areas where light levels are different. Abrupt changes in light cause leaf yellowing and dropping on some species. If moving plants to different light intensities, move them a little at a time so that the change in light is gradual.

Leaf drop

Leaf drop.

Yellow leaves.

Viruses

Mottled, blotched leaves caused by virus.

Problem: Leaves of weeping fig drop. The leaf dropping may cause defoliation of many branches or, in severe cases, the entire plant. Dropping leaves may be green and healthy looking or yellow and discolored.

Analysis: Weeping figs may drop their leaves in response to any of the following conditions.

Solution: The numbered solutions below correspond to the numbered items in the analysis.

1. Overwatering: When plants are watered too frequently or soil drainage is poor, the roots are susceptible to root-rotting fungi. Weak and decaying roots cannot provide enough water and nutrients for proper plant growth.

1. Allow the plant to dry out slightly between waterings. The soil just beneath the surface should be moist but not wet when you water. Empty the saucer after the pot has drained. If the pot does not drain well, transplant to a pot with a good drainage hole, and use a light, well-draining soil mix.

2. Underwatering: Weeping figs need constantly moist soil. If plants are not watered frequently enough or if the soil is not thoroughly soaked at each irrigation, the plants respond by dropping their leaves.

2. Check the soil periodically. Water when the soil just below the surface is still moist but is no longer wet.

3. Insufficient light: Weeping figs need bright indirect light or direct sunlight for best growth. They may drop their leaves even in locations that are bright enough for most other foliage plants.

3. Move plants to a location in bright indirect light or in direct sunlight. If the plants have been growing in a dark area, first move them for 2 weeks to a location that receives bright indirect light or only 1 to 2 hours of morning sun; then place them in direct sunlight.

4. Transplant shock: Transplanting always results in some disturbance to the rootball. Weeping figs are likely to drop some leaves even when the disturbance is minimal.

4. Transplant weeping figs carefully, so as not to disturb the rootball. Some leaf drop after transplanting is normal. It will stop after a few weeks if the plant is given proper care.

5. Changes in environment: Drafts and extreme fluctuations in temperature, light levels, and watering are likely to cause leaf drop. When a greenhouse-grown plant is brought into a drier, darker, cooler home environment, it often responds with leaf drop.

5. Avoid drafty areas and sudden environmental changes. Place new plants where conditions are as similar as possible to those in which they were grown. Some leaf drop is normal for a few weeks until the plant becomes acclimated to its new location.

Problem: Leaves are mottled with yellow blotches, or they have yellow streaks or flecks that later turn dark brown or black. Leaves may have partial, complete, or concentric rings of dark-colored tissue. New leaves may be stunted and cupped. Flowers may be mottled and the colors broken. They may be moderately or severely distorted and may have brown flecks or streaks in them. Affected flowers do not last long.

Analysis: Several viruses infect orchids. Some infect many different types of orchids, and others infect only a few. Viruses cause a plant to manufacture new viruses from its protein. In the process, the metabolism of the plant is upset. Different symptoms appear, depending upon the orchid and the virus present. Viruses are easily transmitted on tools used in cutting orchid plants.

Solution: Observe the effects of viruses on the plants. If flowers are badly malformed or discolored, discard the plants. Keep plants that are virus-free separate from those showing virus symptoms, because the viruses are easily transmitted. Disinfect cutting tools when moving from one plant to another by dipping tools in a solution of 1 part chlorine bleach and 9 parts water. Do not purchase plants that show such symptoms.

SAINTPAULIA (African Violet)

GROWING GUIDE

Light: African violets need abundant light to produce flowers. Grow them in bright light, but not in direct sunlight.

Soil: Any standard houseplant mix.

Fertilizer: Fertilize monthly with Miracle-Gro Liquid African Violet Food during the summer or when plants are growing.

Water: How much: Add enough water so that some drains out the bottom of the pot.

How often: Water plants when the soil just under the surface is moist but not wet.

Temperature: Average house temperatures are adequate, but do not leave African violets in rooms where the temperature falls below 50°F at night.

Special Conditions: Avoid getting cold water on the leaves when watering, and don't let the plant sit in drainage water. Do not expose plants to cold drafts.

Insufficient light

Failure to bloom due to insufficient light.

Problem: Although the plant seems healthy, it does not bloom.

Analysis: African violets, like other flowering plants, won't bloom unless they are properly fed and watered. If the plant is a good green color and is growing well but not blooming, however, it is probably not receiving enough light. Plants use light as a source of energy and will not bloom unless they can afford the energy to do so. African violets bloom at lower levels of light than most other plants, but they do require a fairly bright location to bloom well.

Solution: Move the plant gradually to a brighter location. The ideal light level for African violets is as bright as possible without being direct sun. If the light is coming through a window exposed to the sun, the window should be curtained so that the sunlight is not quite bright enough to make shadows. If the light is too bright, the leaves will lose their bright green color and become pale with an orange or yellow cast. If the light is both bright and hot, the leaves will burn. If you don't have a bright enough location in your house, give the plants supplemental light. Use fluorescent fixtures, and place them as close to the top of the plants as possible.

Water spots

Water spots on African violet foliage.

Problem: White to light yellow blotches in various patterns, including circles, occur on the older leaves. Small islands of green may be separated by the discolored areas. Brown spots sometimes appear in the colored areas.

Analysis: Members of the African violet family are very sensitive to rapid temperature changes. Water spots occur most commonly when cold water is splashed on the leaves while the plant is being watered. If this happens in light, chlorophyll is destroyed. In this plant family, all of the chlorophyll in the leaves is found in a single layer of cells near the upper surface. If the chlorophyll in that layer is broken down, the green color disappears, and the color of the underlying leaf tissue is exposed.

Solution: Avoid getting cold water on African violet leaves when watering. Or use water at room temperature, which will not cause spotting if it touches the leaves. Spotted leaves will not recover. Pick them off if they are unsightly.

Powdery mildew

Powdery mildew on African violet flower.

Problem: White or gray powdery patches appear on the leaves, stems, buds, and flowers. Leaves and flowers may be covered with the powdery growth. This material usually appears first on the upper surfaces of the older leaves. The affected plant parts may turn yellow or brown and shrivel up and die.

Analysis: Powdery mildew on African violet is caused by a fungus (*Oidium* species). The powdery patches consist of fungal strands and spores. Air currents carry the spores to healthy leaves and flowers of the same plant and to other African violets. The fungus robs the plant of its nutrients, causing yellowing or browning of the tissues. Dim light, warm days, and cool nights encourage the growth of powdery mildew.

Solution: Spray with a fungicide containing *triforine* (Funginex®) or *thiophanate-methyl*. Remove infected flowers and flower buds and badly infected leaves. Keep plants in bright, indirect light, and away from cold drafts.

Viruses

Virus causing blotches on Tolmiea foliage.

Problem: All leaves have blotches of various shades of yellow and green. Plants showing color blotches fail to grow as rapidly as green plants of the same type. Infected plants are of a variegated variety.

Analysis: Several viruses have been found in variegated piggyback plants. The most common is cucumber mosaic virus. Viruses disrupt the normal functioning of the cells and, as a result, not as much chlorophyll is produced. This causes different shades of green and yellow to appear in the leaves. Aphids and other insects transmit the virus from one plant to another.

Solution: Keep variegated piggyback plants isolated from other plants, because aphids can transmit the viruses to other susceptible plants. Control aphids by spraying plants with Miracle-Gro Bug Spray, or take them outside and spray with ORTHO Orthene Systemic Insect Control. Inspect new plants, and keep them isolated if there are any signs of aphid infestation. Avoid working in the outdoor garden and then on indoor plants without washing up and changing in between.

Dieback

Dieback on wandering Jew.

Problem: Tips of older leaves turn yellow, then die and turn brown. This condition occurs more frequently on long stems than on short stems. The longer the stem, the more tip burn is found.

Analysis: The cause of dieback is unknown. It looks like salt damage (see page 22) but often occurs without an accumulation of salts in the soil. As wandering Jew stems grow long, the plant does not support the old leaves, and they turn yellow and die starting at the tips.

Solution: Remove dead leaves as they appear. Keep pinching the tips of the stems so they do not become too long. This will force new buds to grow farther back on the stems. New leaves will not show this problem. Plants may occasionally need to be cut back severely so that only several inches are left on each stem. After cutting back, reduce the amount of water and fertilizer until the plant is actively growing again.

No other groundcover is as inviting for foot traffic as luxurient turf grass. When installing pavers in a lawn, set them flush with the soil surface to allow for easy mowing with no need for edging.

Replanting the lawn is often the first phase of landscaping for a newly purchased home. Putting in a new lawn involves removing the old one, preparing the soil, then planting seed or installing sod.

A new lawn placed over a still-viable old lawn results in holdover weeds rapidly taking the place of delicate seedlings. Before sodding or seeding, spray with a glyphosate compound to kill any growing vegetation and avoid much hand weeding later.

Soil pH: Before seeding or laying sod, test the soil in the lawn area to determine pH. The pH scale ranges from 0 to 14, with neutral at 7.0. The ideal pH for grass is between 6.5 and 6.8. The pH determines the rate at which nutrients are released from the soil, if these nutrients are already present.

Acidic soil: In soil that is acidic (with a pH of 5.5 or lower), magnesium, phosphorus, and calcium are less available for plant use than in neutral soil. Nitrogen is only partially released in acidic soil because the soil organisms that free it are less active. When soil has a pH of 5.0 or lower, soil organisms cease working altogether and no nitrogen is released. Acidic soil is also called sour soil.

Lawns in overly acidic soil may grow slowly. Leaves may be pale and root development may be poor. Overly acidic soil promotes disease mechanisms. Applying fertilizer may not help because the low pH stops or slows nutrient release. Neutralize acidic soil by applying finely ground dolomitic limestone, working it into the soil.

Alkaline soil: Sometimes called sweet soil, alkaline soil has a pH above 8.0. When pH exceeds 8.0, iron and manganese are no longer available to the grass. The lawn becomes pale or yellow. To correct alkaline soil, spade in iron and sulfate compounds.

Testing for pH: In general, acidic soils are high in organic material and occur in areas with annual rainfall of 30 inches or more. Alkaline soils are high in calcium and are common in areas where rainfall is minimal.

To measure pH precisely, take soil samples to a professional testing service. In some states, soil testing is offered by university extension services. To find a soil-testing service, look under "Soil" in the local telephone directory or ask the local county extension agent for advice.

Take samples from about 15 areas within a prospective lawn site. A soil-sampling tube, available at your local nursery, works best, but the hollow shaft of an old curtain rod will do. Insert the tube in the soil to a depth of 6 inches. If your home is on an established property, mix the various samples together in a freshly cleaned nonmetal container. You should now have about 2 cups of soil. Remove any stones, roots, or debris. If your home is on land subject to recent construction activities, keep the samples separate. This allows identification of areas that need special work. Label all containers, cover them, and keep them dry. In an accompanying letter, tell the service about land slope and other pertinent details.

Alternatively, you can use a home soil-testing kit you have purchased at a nursery. Though this is not as accurate as a professional analysis, it is inexpensive, easy, and usually adequate. When you know the pH of the lawn area, you can correct it, if necessary, when you add organic additives.

Organic Amendments: Although some grasses do well enough in poor soil, none is at its best. Seedlings may fail to thrive or even emerge. Mature grass remains straggly, bare spots emerge, and insect pests can attack the weakened blades.

For the lawn to thrive, you must improve poor soil. Topsoil should be used only when you must raise the land grading. Knowing the origin of topsoil is difficult. The soil may be poor; contain weed seeds, disease organisms, nematodes, or insect pests in larval or egg form; or have a high chemical content. Purchase topsoil carefully.

Organic amendments actually do improve poor soil. Adding ample organic material to clay soil can lessen the problems of runoff and compaction. This causes particles to form small crumbs rather than a sticky stiff configuration. Ample organic matter mixed into sandy soil helps hold moisture and nutrients in the root zone. Five organic amendments are most often used: peat moss, compost, manure, ground bark, and sawdust.

■ **Compost:** From a compost pile or purchased by the bag or truckload at a nursery, compost is any organic material that has begun to decompose. By decomposing, dead plant parts release their nutrients. In addition, compost makes clay or compacted soil crumblier. When soil particles have air spaces between them, water penetrates better, nutrients reach root zones, and drainage is improved. In sandy soil, compost acts like a sponge to hold moisture and nutrients in the root zone. These benefits bode well for delicate grass seedlings, which need all the help they can get.

■ **Manure:** Steer and horse manure, if treated to remove insect eggs and weed seeds, is an effective soil conditioner. It is used by itself or with compost. The amount of nutrients in the manure varies with what the animals have been eating. Manure may have a high salt content, which can prevent seedling germination. Avoid steer or horse manure additives if the soil is already high in salt or is alkaline. Fresh manure cannot be used safely on or around a new or established lawn. Gases given off by the ammonia in fresh manure may severely damage grass and grass roots. If obtained directly from farm or stable, manure must first be composted to destroy undesirable organisms. Other forms of manure—such as chicken, sheep, or swine—are sometimes available; as with steer or horse manure, compost them first.

■ **Peat moss:** Peat moss is best reserved for acid-loving plants such as azaleas. Few lawn grasses do well in soil heavily fortified with peat moss, although it can be used sparingly. By itself, peat moss has an acidic pH of 3.5 to 4.5. If acidic soil is fortified with peat moss, the highly acidic result may not release valuable plant nutrients. Also, peat moss sheds water if allowed to dry out, so a soil high in peat moss may be harmful to drought-sensitive seedlings. If your lawn has a high peat moss component, add a wetting agent to the water. Wetting agents are available at most nurseries.

■ **Ground bark and sawdust:** Although less expensive than compost or soil mix, ground bark and sawdust are wood by-products that add little nutrients. In the soil, sawdust decomposes with the aid of bacteria that use nitrogen, taking nitrogen away from growing grass. Add a 10-10-10 general-purpose plant food while working wood by-products into the soil.

■ **Amendment application:** Regardless of which organic amendment you choose, a sprinkling of it will not solve your soil problems. You must add an amendment layer 1 to 4 inches deep to effect a change in soil structure. Work the new material in thoroughly with a tiller, spade, or rake.

Rocks or large dirt clods can cause problems by altering water flow. Rake them out of the top 2 inches of soil. Make sure the ground is level. Water the soil, then go over it with a garden roller; do this several times to see how the soil settles. Make corrections as needed to avoid puddling and runoff.

When overseeding an existing lawn, prepare the old lawn by mowing it as closely as possible, raking up the clippings, then scratching the soil vigorously with a metal rake. You'll need to sow seed at two to three times the amount recommended on the package.

LAWNS FROM SEED

Successful seed sowing is a matter of seed quality, timing, effective distribution, and proper care.

■ **Seed quality:** Always check the date on the seed container before purchasing grass seed. Buy only seed that is produced for sale in the current year. If the box is a holdover from the previous year, some seeds may have sprouted in the container. After planting, the others may sprout poorly or not at all.

■ **Sowing time:** Planting at the right time helps prevent seedling death. Fall and spring planting give the best results. Fall sowing is preferred by many experienced gardeners. It reduces the problem of heat damage to seedlings. Allow 6 weeks of growing time before the weather turns cold. Seed sown later may not germinate. Fall seeding is usually done no later than mid-September. Estimates are that for every day after this, 10 percent of the seeds are lost, except in areas of continual warm weather.

The benefits of spring lawn planting include ample sunlight. Combined with ample water, the result may be a deep cushiony lawn. However, weed growth is quite active in spring. Even the most dense mature grass cannot crowd out weeds while in the seedling stage.

Summer heat is also hard on seedlings. Most grasses do best between 50°F and 70°F. A successful springtime planting requires careful watering. Make sure the sprinkler

system irrigates the entire planted area. Water it for 20 minutes in the morning; 20 minutes in the evening; and, if the air temperature rises above 80°F, an additional 20 minutes in the afternoon. If you must use a hose to water, provide a fine spray or mist to avoid washing seeds away. In dry weather, keep seedlings moist by covering them with a ¼- to ½-inch-deep sawdust or straw mulch. Do not use peat moss, which promotes water runoff when dry.

■ **Seed distribution:** Use a drop or broadcast spreader to sow seeds. Hand-distribution tends to be irregular, resulting in both bare and overly seeded areas. As soon as seeds are distributed, rake the planting area lightly. Then roll the earth with a light roller. This avoids wind drift by pressing the seeds firmly into the soil. Seeds should be no more than ½ inch deep. Improper planting depth may slow or stop seed germination. Seeds spread on top of thatch, or left on top of soil instead of being pressed into the soil, may dry out without sprouting.

■ **Care of newly planted areas:** Keep foot and animal traffic off a newly seeded lawn to avoid uncovering or disturbing germinating seeds. Insert small brightly colored flags to warn off pedestrians. To keep animals off, put up a temporary barrier.

Improper watering is a common cause of seed failure. Once a seed has germinated and the protective seed coat is broken, adequate moisture is a must. If the soil around the seedling dries out, so does the seedling.

However, overly wet soil with a high nitrogen component encourages the proliferation of fungi that cause damping-off. These fungi are a major cause of seed failure to thrive. The infection may attack seeds before germination. If seeds do germinate, the fungi infect the growing tips of the grass and kill the seedlings before they emerge from the soil. Even after seedling emergence, the fungi remain a problem. Fungi may attack stems and roots at or just below soil level. The lawn seedlings lie on the ground instead of standing upright. Surviving seedlings may have stunted roots, lessening their nutrient retention. If damping-off causes lawn seed failure, improve drainage before reseeding. Do not add nitrogen fertilizer until seedlings have fully sprouted.

Sod Lawns: Gardeners who do not have time to plant seed and monitor seedlings or who want an immediate lawn thick enough to forestall wind-carried weed seeds often prefer to lay sod.

■ **What to install:** Sod comes in strips from 6 to 9 feet long. It should be uniformly green, moist, and ¾ to 1 inch thick. Overly thick sod roots slowly. Overly thin sod dries out too fast, depriving leaves and roots of necessary water. If sod falls apart when handled, reject it. Do not put down sod that is wet, dry, spotted with brown, or yellowing. Sod with any of these characteristics has not been grown under healthy conditions, is old, has been stored improperly, or has been injured in transit.

■ **When to install:** Lay cool-season sod, such as bluegrass or bentgrass, in early spring or in late summer to early fall. Lay warm-season grass sod, such as bermudagrass, bahiagrass, centipedegrass, or St. Augustine-grass, in late spring or early summer.

■ **How to install:** Healthy sod can be expensive and must be carefully handled to avoid ruining it. Prepare the soil before sod installation to avoid problems from incorrect pH, holdover weeds, uneven ground, insects, nematodes, and disease. Install the sod as soon as possible after it has been cut.

Unroll one strip of sod at a time. Lay the strips so that the seams are staggered from row to row, as the mortar lines between bricks in a walkway are staggered. The sod strips should fit snugly against each other, but be careful not to stretch them. Sod tends to shrink, and stretching increases that tendency and results in yellow stripes that run through the lawn. For 2 weeks, you may have to water daily to keep the soil moist but not wet at all times. Pay special attention to pathway and driveway borders; these are the first to dry out and the last to knit with the soil. Test for sod rooting by tugging gently at the corner of a strip. If it resists, the sod has taken hold. A sod lawn can be functional in as little as 2 weeks, although heavy foot traffic should be routed elsewhere initially.

■ **Warning signals:** Symptoms of sod failure include newly laid areas turning yellow, then brown. Instead of meshing with the underlying soil, failing sod rolls up easily, like a carpet. When you roll it, no roots are visible on the bottom of the sod. If your sod exhibits these symptoms, it may have been subjected to heat or water stress either in the field or in storage; sod may be damaged within 2 days of digging and initial rolling if the weather is hot. To try to save failing installed sod, water it diligently. Damaged sod can recover partially with good care.

Sod failure can also occur if the soil dries out, either because of direct water lack or because sod is placed directly on thatch. If sod is placed directly on hard soil, the roots cannot penetrate well. Soil preparation prevents these problems.

Once the sod is well-established, the next issue is when to begin mowing it.

MOWING LAWNS

In new lawns, begin mowing when grass is at least 3 inches high. Before that, seedlings are not well rooted and you may pull out young plants. After the initial mowing, mow to the height recommended for your grass variety.

In mature lawns, you may be blaming yellow lawn patches on insects or fungus when your mower or mowing technique may be the culprit. Like most chores, mowing can be done correctly or incorrectly.

Mower Types: To mow correctly, you must use good-quality equipment appropriate for the job.

■ **Reel mowers:** These power mowers work with a scissorslike motion. They are precise and recommended for grass that must be short and well tended—golf course–type grass. If you have zoysiagrass or hybrid bermudagrass or bentgrass, you must use a reel mower.

The disadvantage of reel mowers is that their cutting does not follow the lay of the land. The results may be slightly longer grass in a sunken area and slightly shorter grass on a rise. Reel mowers are not as effective as rotary mowers in lawns that contain high

One of the best ways to cut your mowing time is to use the largest mower that's practical for your lawn. Mowing a one-acre lawn with a 24-inch mower will save about an hour over mowing the same size lawn with an 18-inch-wide mower.

weeds, high grass, or rough ground. To avoid problems with a reel mower, have it serviced 3 times during the growing season. Signs of dull reel blades are an overly "striped" effect after mowing, untouched grass blades, or lawn areas that appear rough.

■ **Rotary mowers:** These power mowers are easier to maintain than reel mowers. Also, the rotary models are lighter and easier to handle. They are effective on high grass, plant stalks, and tall weeds. Rotary mowers can trim close to trees, walls, fences, and other standing objects. In addition, they chop up lawn leaves effectively. Rotaries work well if you cannot mow your lawn on a precision timetable; however, do not expect that precision look. Rotary mowers give a knifelike cut, fraying and bruising leaftips more than reel mowers. This is particularly true if you neglect to have the blades sharpened often enough.

Dull mower blades shred grass tips, and the result is a lawn with a sickly brown or gray tinge. This is most noticeable in dry weather. Also, shredded tips act as entry points for lawn diseases. Sharpen the blades of a rotary mower after every 6 to 8 mowings. Some gardeners do this themselves. For any major work, hire a professional.

■ **Push mowers:** Small hand-powered push mowers are time-effective for lawns under 2,000 square feet. You do not have to hunt for fuel, check the parts, and so on; you just have to oil the bearings and get the mower sharpened once a year to keep it in top form. But hand-powered mowers do require more muscle to operate than the power varieties.

■ **Electric mowers:** Those with large lawns often consider electric mowers. The major disadvantage of an electric mower is that it remains anchored to a power outlet. This creates a maneuvering problem when you are cutting around trees, for example. However, electrics are quiet and easy to start. Never use a corded electric mower or edger on wet grass.

Frequency of Mowing: Cutting a lawn too often, particularly with the blade set low, exposes the lower portions of grass leaves to bright sun, burning them. If this happens repeatedly, grass reacts by developing shallow roots. Shallow-rooted lawns are particularly prone to disease and weed problems. In addition, poor rooting does not provide enough nutrients to the grass leaves. A lawn with shallow roots may eventually die out.

Mowing too seldom often means taking off too much grass at one time. There is a direct relationship between root depth and lawn height. Long roots are to your advantage. They not only reach out for more lawn nutrients, they help the grass resist drought. If you let grass grow too long and then lop off more than half of it, the roots go into shock. If you do this often enough, your lawn may develop a thin, spotty, or burned look. For best results, never remove more than one-third of the height of the grass.

The table on page 36 gives recommended grass heights according to grass type. If your lawn is a mixture of types, cut to the length recommended for the dominant one. Mow whenever grass is one-third to one-half higher than the height given.

Bentgrass is the finest-bladed, lowest-growing, and highest-maintenance of all the cool-season turfgrasses. It requires frequent mowing, watering, and fertilizing.

CHOOSING A LAWN TYPE

Given all available lawn grasses, you can choose from more than 40 varieties. There are also grass substitutes, such as dichondra, to consider. Basically, grass is categorized as a hardy, or cool-season, variety for cold-winter areas or as a subtropical, or warm-season, variety for areas where frost is rare. Whether you are installing a new lawn or caring for the lawn you already have, knowing the needs of the grass is important. If you are installing a lawn, you must choose a type that is appropriate for the climate and has maintenance requirements you can satisfy. Knowing the variety of an existing lawn helps you provide the care it needs and diagnose problems.

Most lawns are a blend of grasses, which makes them more resistant to disease and infestation than lawns of a single type. The problems that afflict one grass may not affect another.

Bahiagrass: Warm season. Makes a coarse, open lawn. So tall and fast growing that in finer lawns it is thought of as a weed. Spreads by runners. Most varieties are grown for hay; Pensacola and Argentine are lawn varieties. Easy to maintain, drought-resistant, shade tolerant, and needs little fertilizer. Problems are chlorosis (yellowing), dollar spot (see page 49), and mole cricket invasion (see page 53).

Bentgrass: Cool season. Leaves are small, well textured, fine, fairly flat, and upright. Color may be bluish green, medium green, or apple green. Several varieties are available. Spreads by rooting at the joints. If well cared

Type of grass	Recommended height (inches)
Bahiagrass	2–3
Bentgrass	3/8–3/4
Bermudagrass	1/2–1 1/2
Bluegrass	2–3
Centipedegrass	1–2
Dichondra	1/2–1 1/2
Fescue	
Chewing	1–2
Red	2–3
Tall	3–4
Ryegrass	
Annual	1 1/2–2
Perennial	1–2 1/2
St. Augustine grass	1–2 1/2
Zoysiagrass	1/2–1 1/2

RECOMMENDED GRASS HEIGHTS

for, bentgrass presents an exceptionally uniform appearance. Often used on golf courses and putting greens. Offers some shade tolerance. Problems arise when bentgrass is not mowed often enough or cut short enough, and it builds a heavy thatch layer quickly. A few varieties require mowing three times per week. Bentgrass is fussy. It suffers under drought and requires regular, heavy applications of fertilizer. In hot, muggy weather or cool, damp conditions, it is

susceptible to fungal diseases. Bentgrass is not recommended for most home lawns.

Bentgrass becomes a weed when it is accidentally introduced into lawns of other types that do not require such stringent mowing. In bluegrass, fescue, or ryegrass lawns, bentgrass looks matted. In areas between the other grasses, bentgrass may appear dead with long straggly stems. In the spring, it remains brown much longer than the surrounding grasses. It may also take over the lawn, because it grows quickly.

If you do not want the high maintenance of a predominantly bentgrass lawn or do not like its appearance in your chosen lawn type, use a systemic herbicide containing *glyphosate*. The treatment does not affect roots of surrounding trees or shrubs. Bentgrass areas die out within 4 weeks of treatment, and you can sow selected lawn seed within 7 days.

Bermudagrass: Warm season. Regional names include devilgrass, manienie, and wiregrass. Color differs with variety; bermudagrass may be grayish green, bright deep green, deep blue-green, light green, yellow-green, or dark green. Leaf blades range from 1/8 to 1/4 inch wide. Common bermudagrass has bronze seed spikes. Stolons, or stems, creep along the soil surface. Stolons range from 6 to 18 inches long. Bermudagrass requires sun. It roots deeply and is drought-tolerant. In general, bermudagrass lawns are dense. New varieties are medium- to fine-textured.

Bermudagrass is usually pest- and disease-free if well tended. When thick, it resists weed sprouting. Because it provides rapid coverage and withstands a lot of foot traffic, it is used for play areas and athletic fields. In winter, expect bermudagrass to turn brown or straw-colored. Help it stay green longer by applying fertilizer in late fall and removing thatch, which blocks sunlight. Some gardeners dye dormant bermudagrass green in winter. Once bermudagrass is fully dormant (without naturally green stems or leaves) in winter, use contact weed killer to kill a wide range of grassy weeds and broadleaf weeds in the lawn.

Before extensive hybridization, early varieties of bermudagrass were considered a worse lawn pest than crabgrass. Even modern "common" varieties can occasionally become a lawn problem. Bermudagrass spreads rapidly by surface and underground runners. If it gets in nonbermudagrass lawns or flower beds, its deep root system may make it difficult to eradicate. Bermudagrass is sometimes confused with quackgrass (see page 56), a weed, because both spread in the same stolen-creeping manner. See page 43 for control in lawns. Control invasive bermudagrass in ground cover or shrubs with the selective herbicide ORTHO Grass-B-Gon Grass Killer which kills grassy weeds but does not affect most broadleaf plants.

Kentucky Bluegrass: Cool season. Despite its name, Kentucky bluegrass is not a Kentucky native; English settlers brought it to the state in packing hay. There is "common" Kentucky bluegrass and "improved" Kentucky bluegrass. Both spread by rhizomes. Bluegrass is generally dark green and fine-textured. Density is moderate to thick. It is considered the best of all lawn grasses in appearance. The improved version, usually sold as a blend of varieties, has deeper color, higher density, and better heat resistance. Improved Kentucky bluegrass is also more resistant to diseases such as leaf spot (see page 55), stripe smut, and fusarium blight (see page 49).

When considering bluegrass for your lawn, select varieties carefully to avoid problems. Delta, Kenblue, Newport, and Park are susceptible to leaf spot. Delta and Park have problems with chlorosis (yellowing) in alkaline soils. Newport, Fylking, and Park seem susceptible to fusarium blight. Merion does not do at all well in shade and is prone to mildew, stripe smut, and rust. All bluegrass requires ample water; the grass goes dormant and turns brown in even short drought periods, but it does recover if watered well. Bluegrass needs regular applications of a medium to large amount of fertilizer. The grass does not prosper if mowed severely.

Do not confuse perennial Kentucky bluegrass with its pest relative, annual bluegrass (see page 56), which is also called annual speargrass, dwarf speargrass, and walkgrass. Annual bluegrass, which is generally considered a weed, is pale green. In mid- to late spring, white seed heads appear.

Centipedegrass: Warm season. Medium green with a tendency toward yellowing from chlorosis. Needs iron supplements. Centipedegrass has a coarse texture and will grow in some shade. Adaptable to poor soil, it does well in acidic conditions. Is aggressive enough to crowd out weeds. Spreads by runners. Centipedegrass needs little general maintenance, though its shallow root system makes it drought-sensitive. With sufficient

watering, however, a centipedegrass lawn recovers quickly. This hardy grass is resistant to chinch bug attack and rhizoctonia disease. If centipedegrass gets into ornamental plant areas, control it selectively with ORTHO Grass-B-Gon Grass Killer.

Dichondra: Warm season. Dichondra is not a grass, but a ground-hugging broad-leafed plant that can make a lush, bright green carpet. It spreads by reseeding and from underground runners, and stays green throughout the year. Because it needs little mowing, it is used instead of grass for many small to medium-sized areas that are not subject to foot traffic. The plant thrives in heat but tolerates some shade. It needs much water and fertilizer. Weeds are hard to control in dense dichondra.

Dichondra is susceptible to a few diseases, including brown patch and alternaria leaf spot. Dichondra is also susceptible to cutworms, snails, slugs, red spider mites,

gnats, flea beetles, and nematodes. To rid dichondra of nematodes (see page 54), you may have to remove the entire lawn, cultivate deeply, and then replant. Because nematodes can be brought in with flats of dichondra, treat new plants with a soil fumigant before installing them.

Fescue: Cool season. Types of fescue include chewing fescue, red fescue, and tall fescue. Most fine fescue spreads slowly; but tall fescue can spread rapidly. Fescue tends to present a rather stiff and windswept appearance. It remains medium to dark green all year, and makes a rugged lawn. Tall fescue is used for play areas. Fescue can survive in city conditions, including smog. It adapts to dry growing conditions and poor soil and needs little fertilizer. Tall fescue produces bunchy clumps that may be considered a lawn nuisance. Occasionally it does so well it becomes a weed. Red fescue tolerates acidic soil, dry areas, and some

St. Augustinegrass is a robust, fast-growing, warm-season perennial with broad, dark-green blades. It is among the most shade-tolerant of warm-season grasses.

Because it exhibits the best wear tolerance of any cool-season grass, perennial ryegrass has received a lot of attention from turf breeders. The result has been turf-type perennial ryegrass varieties that are fine-bladed, rich green, and resistant to pests and diseases.

shade. In moist fertile soil and in hot climates, it is prone to summer diseases such as red thread. This disease, also called pink patch, affects ryegrass, bluegrass, and bentgrass as well as fescue. Infected grass turns light tan to pink in areas ranging from 2 inches to 3 feet in diameter. Pink webs, almost resembling tangled sewing thread, bind the leaves together. Though seldom fatal, red thread does affect lawn appearance. Try adding nitrogen to the soil as a control.

Ryegrass (Annual): Cool season. Also called Italian ryegrass and common ryegrass, the annual type of this grass is coarse, with leaves far apart. It does not make a tightly knit lawn. Annual ryegrass is often used for quick lawn cover. It needs a lot of water but tolerates some shade. Annual ryegrass does not survive cold winters or extremely hot summers. It must be replanted each year.

Ryegrass (Perennial): Cool season. This grass is somewhat coarse with a waxy shine on the leaves, which are far apart. Perennial ryegrass tolerates some shade and needs only moderate applications of fertilizer and water. Traditionally, its bunchy growth habit did not create the lush lawn look, but selective breeding has developed ryegrass with a fine leaf resembling bluegrass. This improved ryegrass creates a beautiful lawn

that is tough enough to plant in play areas. And it is an easy grass to grow, though it can be hard to mow in summer.

St. Augustinegrass: Warm season. A dark blue-green coarse grass with hard flat stems and flat broad leaves. Grows quickly and tolerates salty soil and shade. St. Augustinegrass is not always durable under heavy traffic, however, and it turns brown in winter. Its coarse texture makes cutting with a power mower a necessity. This type of lawn needs much iron. It tends to creep into flower beds; fortunately, however, it is shallow-rooted and easily removed by hand. Unfortunately, chinch bugs (see page 51) find this grass a favorite food.

Zoysiagrass: Warm season. The several varieties of this grass have wiry blades that may be broad at the base and taper to a point at the tips. Zoysiagrass is dark green and easy to maintain. It requires a moderate amount of water and fertilizer. This grass tolerates heavy foot traffic and is drought- and weed-resistant. Few pests trouble it. This grass is almost winter-hardy. Winter dormancy turns it straw-colored except in mild-winter areas. Zoysiagrass does not turn green again until warm weather. Lawns of this grass are slow to establish. Without regular rotary mowing it becomes difficult to cut. Zoysiagrass tends to build thatch.

GRASS CLIPPINGS

Some say grass clippings are good for a lawn; some say they prevent sunlight from reaching the lawn; and others say they foul the mower. There's truth in all three views. If you mow regularly and the clippings are less than 1 inch long, you can leave them on the lawn without causing a problem. When grass is mowed properly, the clippings take about a day to disappear. They are 90 percent water and dry up to almost nothing; therefore, they cannot pile up or entangle with thatch and impede the mower. Since as much as one-third of a lawn's nitrogen requirements may be supplied by decomposing lawn clippings, they can certainly be beneficial. This is welcome news because many communities now ban yard wastes and grass clippings from normal trash collection, to slow down the rapid filling of available landfills.

But in dry-summer areas or with infrequent mowing, clippings do not decay quickly enough. The clippings can mound, entangling in thatch and making mowing difficult. Excess clippings are not only unsightly, they also shade growing grass underneath. They provide an excellent source of nutrition and humidity for disease fungi, which may soon attack living grass underneath the matting. Some of the new rotary mowers effectively chop clippings small and scatter them so they do not build up into a problem.

Clipping disposal is not difficult if you have an active compost pile. Gather up the clippings, scatter them in the pile with other garden leaves and waste material, and turn the pile regularly to provide air. An alternative is spreading clippings no more than 1 inch thick as a mulch in dahlia, rose, or shrubbery beds.

Clippings can create a problem if you pile them up, however—especially if the pile is near the house. Grass has a high water content. Piled up, no room remains for air movement or water evaporation. Smell-producing bacteria thrive, turning the grass into yellow-brown slime. The pile becomes a breeding place for houseflies, false stable flies, soldier flies, and little houseflies. Turning over just one small section of decaying piled grass may expose as many as 3,000 fly maggots.

WATERING MATURE LAWNS

Even the most drought-tolerant lawns—such as bermudagrass, zoysiagrass, or fescue—cannot live with prolonged lack of water (see page 53). Other grasses suffer from even brief water deprivation. The first sign of a drought problem is the dark bluish green tinge caused partly by leaf folding. The lawn loses its springiness, and footprints remain following any foot traffic. Most lawns go dormant about three days after drought-caused wilting begins.

Many factors determine a proper watering program. These include turf type, soil, climate, temperature, wind velocity, humidity, rain, and maintenance practices. If only the top few inches of soil are regularly watered, roots do not seek water any deeper down. Shallow roots force you to water more often to maintain a green lawn. But frequent watering keeps soil constantly wet, encouraging weeds and disease. Watering should regularly penetrate 6 to 8 inches deep to encourage deep lawn rooting. This enables lawns to go longer between waterings, cutting down on disease potential.

Sprinkler Efficiency: Inefficient sprinkler head placement results in overwatering in some areas, underwatering in others. Test the sprinkler by setting shallow containers of the same size at regular intervals throughout the lawn area. Put some close to the sprinkler heads and some at the farthest reach of the water. Make a diagram of your container layout, then record the amount of water in each container after a normal sprinkling. Over its entire surface, a lawn needs 1 to 2 inches of water per week. Adjust sprinkler heads accordingly.

Soil Type: The type of soil in your lawn affects how much water the grass actually receives. Two soil extremes cause watering problems: clay soil and sandy soil.

■ **Clay soil:** This type of soil is composed of individual mineral particles of less than $\frac{1}{125000}$ inch. Because of their extremely small size, clay particles tend to pack together and become dense. Dense soil slows water penetration.

Water clay soil slowly to avoid water runoff. Many gardeners believe they are giving ample water to lawns growing in clay soils, while runoff is causing a water shortage that results in brown or yellow grass. Clay soil packing and subsequent runoff become

Oscillating-arm sprinklers are designed to apply water over large areas, and are highly adjustible. Individual sprinklers differs; test yours to be certain of its pattern.

even worse if the grass is walked on when wet.

Once wet, clay subsoil holds water for quite a while. This is true even if the topsoil looks dry and cracks. Waterlogged soil decreases air penetration and may lead to many fungal diseases. Because the topsoil is dry, you may mistakenly continue watering.

■ **Sandy soil:** This type of soil provides quick drainage and effective air circulation, but it does not retain moisture well. You may be giving enough water, yet still see lawn yellowing or browning. Water moves rapidly through sand particles, which range in size from $\frac{1}{500}$ inch for fine sand to $\frac{1}{12}$ inch for coarse sand.

Type Determination: To determine the type of soil in your lawn, fill a quart jar about two-thirds full of water. Fill the jar with soil until it is almost full. Add a bit of commercial dispersing agent, such as Calgon, to get best results. Replace the top tightly. Shake the jar vigorously. Now let the soil settle. A sand layer becomes visible in a short time, the heaviest sand settling out first. Clay and silt take hours to settle; fine clay may remain suspended indefinitely. Many soils are a mixture of types, but one type usually predominates.

Effectiveness Tests: If your watering schedule is not producing desired results, purchase a soil-moisture tester, or coring tube. The coring tube takes up a long earth plug and lets you see and feel the deep-down moisture level. A simple test, but not as diagnostic, is poking a long screwdriver into the ground. If it pokes through 6 soil inches easily, the lawn is usually wet enough. Test soil moisture 12 hours after watering. Soil should be moist 6 to 8 inches down.

Changes in Water Availability: Since lawn browning can occur quickly, the response must be prompt. When water availability becomes restricted (see page 53), do not apply fertilizer except during the fall rainy season. Fertilizer promotes growth that is not supported by adequate moisture. Remove all weeds, which compete for water. Do not cut grass as short, and mow less often. But do not let it grow higher than one third over the recommended mowing height. Less frequent mowing may mean clippings will have to be removed.

With restricted watering, a lawn does not look lush. It may develop a spotty, thin appearance. Some gardeners under mandatory water rationing prefer to let a lawn die out altogether. If drought is recurrent, replant with drought-tolerant turf or drought-resistant ground cover.

If a lawn suddenly becomes spongy, the cause may be thatch buildup. Some gardeners confuse thatch with loose grass clippings. Clippings can be raked out.

PREVENTING WEEDS

Weeds produce huge numbers of seeds. The seeds are lightweight and often have built-in travel mechanisms that allow them to stick to fur or be wafted by the wind. In addition, weed seeds travel by means of transported soil, soil amendments, and garden seeds that are not weed-free; birds and other animals; rain; equipment; and the gardener who unknowingly has weed seeds on his or her clothing. Some planted grasses become weeds when they invade nonlawn areas or lawns of different types.

In general, lawn weeds are a sign that growing conditions are not optimal for grass. In nature, the strongest vegetation usually survives in an area for which it is best suited. The correct lawn grass for the area, put in properly and given prime care, can usually preclude newly arriving weeds. Conversely, weeds thrive in an area where ground preparation has been poor, soil unimproved, water minimal, and fertilizer scarce.

In the ongoing lawn weed battle, your primary strategy in lawn weed control is creating optimal growth conditions for planted grass, thereby crowding out weeds. To grow the healthiest grass possible, consider the questions that follow.

■ Is soil acidity slowing grass growth?
■ Does poor drainage affect grass growth adversely?
■ Is soil compaction inhibiting roots?
■ Are the grass types planted best suited to the climate?
■ Is fertilizer application timed to meet the needs of the lawn throughout its seasonal life cycle?
■ Is mowing technique and frequency appropriate for the type of grass?
■ Is watering deep enough?
■ Is intermittent overwatering creating soggy soil?
■ Are trees and shrubs blocking necessary sunlight?
■ Have pest insects, nematodes, or disease weakened the grass?
■ Does foot traffic prevent the grass from thriving?

If your analysis shows that growing conditions are hindering the grass, correct the conditions by following the suggestions in the appropriate sections of this book. Time spent encouraging grass will be time saved from pulling weeds.

In addition, reduce the opportunity for weed seeds to take hold by purchasing weed-free seed and treating the ground with a preventative before planting. During prime lawn-growing season, consider spraying emerging weeds before they send out roots or runners or create seeds. Control measures to prevent weed seed formation, keep new seeds from finding a place to set roots, and quickly eliminate any seedlings that do take hold, form an effective prevention trio of methods.

Weed seeds are extremely durable. Some seeds may germinate quickly upon finding a satisfactory site; others may lie dormant for a year, or even two, before germinating. Repeat weed-control measures each season. If you are thorough, weeds will not invade your lawn or your free time.

IDENTIFYING WEEDS

Correctly identifying lawn weeds allows you to plan effective control measures. Knowing whether your weeds reproduce from roots, rhizomes, stolons, seeds, or a combination allows you to take appropriate action. For example, the dandelion multiplies not only from seeds, but from its long root system. If you leave even a small portion of root in the ground, way down, the dandelion will re-emerge with vigor.

In general, weeds are categorized as annuals or perennials and as warm-season or cool-season plants.

Warm-season annuals: About 80 percent of lawn weeds are annuals. Summer, or warm-season, annual weeds peak at midsummer, when heat slows the growth of competing cool-season grass. The seeds of these plants germinate as soon as soil temperature reaches 60° to 65°F. These seeds are not shade-tolerant, so if perennial lawn grasses are present and growing strongly, warm-season annual weeds have problems germinating because of lack of light. Even tree shade slows germination. But where lawn is thinning, weed seedlings establish rapidly and aggressively crowd out grass.

In warm climates, applying fertilizer early in the season gives lawn grass a growth spurt before annual summer weeds germinate. Letting the lawn grow a little tall before cutting, particularly in spring, provides shade that slows annual weed seedlings. Water carefully. Light, frequent watering helps weeds and discourages grass growth. Water deeply and only when needed.

In cool climates, apply fertilizer and mechanically improve the growing area at the end of summer. Follow this with an application of fertilizer early the next spring. Leave the grass longer throughout spring and summer, letting established lawn smother out emerging weed seedlings.

Common warm-season annual weeds include foxtail, goosegrass, sandbur, and shepherdspurse.

When applying lawn weed killer with a hose-end sprayer, avoid spraying on windy days and follow all label directions carefully.

■ **Foxtail:** Annual foxtail is found throughout the United States and in parts of Canada. Its leaves are 2 to 6 inches long, flat, and ¼ to ½ inch wide; they sometimes appear twisted. The topsides of green foxtail leaves are hairy. Yellow foxtail leaves are smooth. Both thrive in sunny, bare spots. It grows best in damp, well-fertilized soil.

If a lawn is kept mowed, foxtail forms a low mat. If unmowed, hairy flower spikes resembling bristles appear between June and September. Each bristle may be 2 to 4 inches long and resemble a fox's tail. Foxtail is often confused with crabgrass because both grow in clumps. However, foxtail clumps are not as wide as crabgrass clumps. Since foxtail reproduces from seeds only, rather than from reproductive stems or runners, it can be removed with a trowel or by hand. Keep it under control by removing lawn clippings that contain seed heads. Chemical controls include ORTHO Weed-B-Gon Crabgrass Killer, ORTHO Grass-B-Gon Grass Killer, or Roundup® Weed & Grass Killer.

■ **Goosegrass:** This warm-season annual resembles crabgrass but is darker green with a silver center. Its smooth, flat stems form a rosette. Leaf blades are 2 to 10 inches long and ⅓ inch wide. Goosegrass germinates when soil temperature is between 60° and 65°F, several weeks after crabgrass. It multiplies from seeds, expands by spreading, and has an extensive root system. It does not root at stem joints. Seeds are produced on stalks 2 to 6 inches high that appear from July to October. Mature plants die with first frost. Seeds are dormant over winter. It prefers compacted soils with poor drainage and light, frequent watering. To control, treat with ORTHO Grass-B-Gon Grass Killer, ORTHO Weed-B-Gon Crabgrass Killer, ORTHO Kleeraway Grass & Weed Killer, or Roundup® Weed & Grass Killer.

■ **Sandbur:** Also called burgrass and sandburgrass, this annual grassy weed has yellow-green leaf blades that are 2 to 5 inches long and ¼ inch wide. The topsides of weeds may be rough. In mowed lawns, sandbur tends to form low mats. In unmowed lawns, it may reach 2 feet tall. Spiked straw-colored seed burrs ½ inch tall appear from July to September. Sandbur grows best in sandy dry soil. Begin control by improving soil with organic matter and encouraging strong lawn growth. If weeds persist, spot-treat with ORTHO Weed-B-Gon Crabgrass Killer, ORTHO Grass-B-Gon Grass Killer, or Roundup® Weed & Grass Killer.

■ **Shepherdspurse:** Also called shepherd's-bag and lady's-purse, this annual weed may appear throughout the year in warm-winter areas. Its arrow-shaped leaves are toothed or lobed and form a rosette. Tiny white flower clusters appear on stems that can reach up to 18 inches high. Seeds are in triangular pods resembling small sacklike purses. Seeds can remain dormant for several years before germinating in spring. In warm-weather areas, seeds may germinate in fall. Shepherdspurse is not fussy about soil, but it will not grow in shade. Mechanical control consists of hand-pulling. Chemical control consists of a treatment containing MCPP. Treat when plants are actively growing.

Cool-season Annuals: Cool-season annual weeds generally start growth from seeds in late summer or fall. They grow rapidly until the first solid frost, then go into a partial resting phase. With spring, they grow rapidly again, this time setting seeds. They die out in early summer.

Cool-season weeds have the advantage of growing when most grasses are partially or entirely dormant. They thrive in early spring, when desired grasses are just getting started, and in fall, when desired grasses are slowing down for winter. Unrestricted, cool-season annual weeds may ruin a lawn before it gets into full spring growth.

If your lawn has just a few cool-season weeds, eliminate them completely before any flowers or seeds appear. In southern states, plant a winter grass in fall that crowds weeds out. In northern states, leave the grass slightly higher in spring and fall to shade out weed seedlings. Rake up grass clippings if weed seeds are present. Chemical controls are effective when cool-season grasses don't respond to manual measures.

■ **Downy brome:** Downy brome is a slender, upright annual weed that grows throughout the United States except some southeastern areas. It grows to 2 feet tall. Light green coarse, hairy leaves are 2 to 6 inches long. Drooping purple flower clusters appear in spring. Seeds can remain viable in soil for more than two years. Germination is in fall or early spring. Downy brome weeds turn purplish when mature. They prosper in poor sandy or gravelly soil and cool growing conditions. Remove by hand-pulling. Chemical controls include ORTHO Grass-B-Gon Grass Killer and Roundup® Weed & Grass Killer. To keep downy brome from returning, apply Scotts Halts Crabgrass Preventer.

Top: Goosegrass
Bottom: Shepherd's purse

■ **Prostrate knotweed:** Prostrate knotweed grows quite low to the ground with smooth blue-green oval leaves, each about 1 inch long and ¼ inch wide. The leaves attach to wiry stems at visible joints. The stems range from 4 to 24 inches long. Knotweed is an annual that forms mats that can reach 2 feet wide, crowding out lawn grasses. It grows throughout the United States and southern Canada. The weed is usually found in compacted soil. The fastest growth period for prostrate knotweed is from early spring to early fall. Tiny greenish white flowers bloom in clusters at the leaf and stem joints from June to November. Knotweed reproduces from seeds, which are plentiful. Though knotweed cannot get started in healthy, dense turf, it is common in areas of heavy foot traffic. Aeration helps control prostrate knotweed. Pull out young plants. There is no preemergent control. If necessary, treat the lawn with ORTHO Weed-B-Gon Lawn Weed Killer or ORTHO Weed-B-Gon Jet Weeder in early spring.

■ **Black medic:** Other names for black medic include yellow trefoil and black clover. It is sometimes confused with clover. Black medic has three-leaflet cloverlike leaves that are slightly toothed at the tips. Its low-growing stems are slightly hairy. Black medic is common in lawns throughout the United States from May through September.

Ground ivy

Broadleaf plantain

It forms thick mats that crowd out desirable lawn grasses. Small bright yellow flowers bloom in late spring and early summer. In warm-weather areas, black medic can bloom until December. Blooms are followed by black kidney-shaped seed pods. Black medic is an annual that multiplies from seeds only. It is prevalent in nitrogen-deficient lawns. To eliminate small black medic patches, try hand-pulling. Increase lawn nitrogen with fertilizer where appropriate. Treat the lawn with GreenSweep Weed & Feed (20-0-0) or ORTHO Weed-B-Gon Lawn Weed Killer.

■ **Common groundsel:** This weed grows 6 inches to 1½ feet tall, with 4-inch-long toothed leaves of medium green. Yellow flowers 1 inch long appear from April to October. It reproduces by seeds and by stems that may take root at the lower joints. The plant grows best in moist, rich soil. Hand-pull groundsel before it produces seeds. Spot-treat with Roundup® Weed & Grass Killer.

Warm-season grassy perennials: These weeds are more difficult to control than annual weeds. Since they are a different color and texture than lawn grasses, they stand out, possibly ruining a lawn's appearance. If grassy perennial weeds take over, it may be necessary to pull out the present lawn and replace it.

■ **Dallisgrass:** A clumpy rosette-type weed, dallisgrass has coarse-textured leaves. Each leaf is 4 to 10 inches long and ½ inch wide. Stems 2 to 6 inches long emerge from the plant center in a starlike pattern. Dallisgrass reproduces from seeds and underground stems. This perennial has extremely deep roots. It has a tendency to turn brown in the center. Although it is primarily a summer weed in cool areas, it grows all year in mild climates. Growth begins quite early in spring. Dallisgrass grows best in warm weather; low, wet ground; and high-cut lawns. However, once established, it spreads rapidly in low-cut lawns. This

weed is a severe problem in the southern United States. There is no preemergent control. Try draining soil to eliminate dallisgrass. If the weed persists, treat it in early spring or summer with ORTHO Grass-B-Gon Grass Killer, ORTHO Kleeraway Grass & Weed Killer, or Roundup Ready-to-Use Weed & Grass Killer.

■ **Nimblewill:** Also termed nimbleweed and dropseed, nimblewill has smooth, flat, light green or bluish green leaves up to 2 inches long on wiry stems up to 10 inches tall. In spring nimblewill turns green after other grasses. The result is brown patches in infested lawns. Nimblewill stems root at lower nodes as the plants reach outward. It thrives in hot dry areas and in thin turf during drought. For small infestations, dig out patches of the weed. Eliminate it in lawns with ORTHO Grass-B-Gon Grass Killer. Begin control in early spring.

■ **Nutsedge:** Other names for nutsedge are nutgrass, cocosedge, and cocograss. The most common nutsedge forms are yellow or purple. There are annual varieties as well as perennial. Nutsedge has greenish yellow leaves emerging from triangular stems. In flower, nutsedge has umbrellalike clusters topping its stems. Seed heads are yellow-brown. In summer, nutsedge grows more rapidly than lawn grass, is easily seen, and disfigures lawns. It can be extremely difficult to eliminate, because it multiplies from tubers, seeds, and underground stems. The tuber stores nutrients. If any tuber is left in the ground after the rest of the plant is removed, nutsedge can regrow. This weed prefers overly watered soil. Control includes changing lawn-watering techniques or increasing soil drainage. To treat nutsedge chemically, use ORTHO Weed-B-Gon Crabgrass Killer. Begin spraying in early spring and repeat the treatment according to label instructions.

Cool-season Perennial Broadleaf Weeds: The leaves of these weeds can be lance-shaped, arrow-shaped, scalloped, or oval. Basically, they are not grasslike. All make a ragged appearance in lawns. This category includes broadleaf plantain, dandelion (see page 59), mouse-ear chickweed, ground ivy, speedwell, and white clover (see page 57).

■ **Broadleaf plantain:** Also called common plantain, broadleaf plantain has thick, egg-shaped, wavy-edged leaves growing in a ground-hugging rosette. It grows throughout the United States and southern Canada. The gray-green leaves of broadleaf plantain range from 2 to 10 inches long. From May to September seed heads appear in a long cluster from a central upright stem. Plantain multiplies from seeds and from resprouting roots. The weed germinates best in rich, moist, compacted soil. As broadleaf plantains grow, they suffocate surrounding lawn grass. To mechanically control plantain, dig out roots with a trowel. Do not let flowers or seeds form. Aerate the lawn. There is no preemergent control. Spray the lawn with ORTHO Weed-B-Gon Lawn Weed Killer, ORTHO Weed-B-Gon Weed Killer, or ORTHO Weed-B-Gon Jet Weeder in the spring or fall.

■ **Mouse-ear chickweed:** This plant can appear in the finest of well-kept lawns. Not directly related to the annual common chickweed, this broadleaf perennial multiplies rapidly and can crowd out desired grasses. Mouse-ear chickweed has a different color and texture than grass.

The ½-inch-long leaves of the weed are both fleshy and fuzzy. Stems are low and spreading. Small white flowers appear from April through June. The period of fastest growth is in early spring. Mouse-ear chickweed multiplies from seeds and runners, which root easily at the nodes. This prevalent weed thrives in moist, poorly drained soil in sun or shade. Mechanical control consists of keeping runners and stems off the ground so they cannot take root. Mouse-ear chickweed is extremely difficult to eliminate by hand-pulling, since it can resprout from pieces left in the soil. To discourage mouse-ear chickweed, cut the lawn short. Remove clippings containing runners and discard them. There is no preemergent control. Treat the lawn with GreenSweep Weed & Feed (20-0-0) or ORTHO Weed-B-Gon Lawn Weed Killer in early spring or late fall.

■ **Ground ivy:** This plant is also called creeping ivy and gill-over-the-ground. It is a cool-season, perennial weed that grows 3 to 6 inches tall. Bright green leaves are scalloped, round, and about 1 inch wide. Tiny light blue to purple flowers appear from spring through summer. Reproduction is from seeds and creeping stems that root upon soil contact. Ground ivy does well in sun or shade as long as soil is damp. Plants form a dense mat that can completely crowd out lawn grass. There is no preemergent control. Cut grass short and rake to keep runners from touching the ground. Hand-pulling must be thorough because pieces left in the ground can resprout. Treat the lawn with ORTHO Weed-B-Gon Lawn Weed Killer or ORTHO Weed-B-Gon Jet Weeder in the the spring or fall.

■ **Speedwell:** Also called creeping veronica, in a few years this perennial weed can cover an entire lawn. It has bright green, roundish, scallop-edged leaves ½ inch long. Each plant is about 4 inches high. Tiny bluish white flowers bloom on stalks that grow somewhat above leaves. Heart-shaped seedpods form on stems below flowers. Speedwell reproduces from creeping stems that root easily upon touching ground. It is not usually found in well-drained sunny areas that receive fertilizer regularly. It grows best in moist, shady lawn and acidic soil, but it can grow in sunlight if soil remains moist. To discourage speedwell, cut grass short and remove all clippings to avoid stem rooting. Spray with ORTHO Weed-B-Gon Lawn Weed Killer or ORTHO Weed-B-Gon Jet Weeder when the plant is flowering or actively growing.

Cool-season Perennial Grassy or Grasslike Weeds:

■ **Bermudagrass:** This plant can be a valued warm-season turfgrass or a nasty weed in cool season lawns. It is often confused with quackgrass because both spread in the same creeping fashion and form mats. Bermudagrass blades are about ⅛ inch wide. Stems are gray-green, hairy, and 6 to 18 inches long. This grass reproduces from seeds, aboveground stems, and underground stems. Roots may grow several feet deep. Seeds are formed on 3-inch-wide fingerlike segments that grow slightly above stems. From three to seven of these segments grow on each plant. Bermudagrass is a fast-growing perennial that turns brown when temperatures drop below 50°F. It is slow to

turn green in spring, creating brown patches throughout the lawn.

Though drought- and heat-tolerant, this annual does not grow vigorously in shade. In sunny areas, however, bermudagrass may quickly crowd out desirable lawn. Mechanically control bermudagrass by hand-pulling. Remove all roots—any piece of root left can resprout. The best time to begin chemical control is in spring, when shoots are just appearing; use ORTHO Kleeraway Grass & Weed Killer or Roundup Ready-to-Use Weed & Grass Killer according to label instructions. Six weeks later, retreat any new green stems and leaves. Bermudagrass can be spot-treated with ORTHO Kleeraway Grass & Weed Killer or Roundup Ready-to-Use Weed & Grass Killer any time grass is actively growing. After 1 week, mow as close as possible and reseed the area. Multiple treatments may be needed. There is no preemergent treatment. To prevent bermudagrass invasion in cool-season areas, apply a heavy dose of fertilizer in fall. Give the lawn adequate water during the summer.

■ **Zoysiagrass:** Like bermudagrass, zoysiagrass is a valued lawn grass in warm-weather areas but a weed in other regions. In cold areas, this perennial becomes dormant with the first fall frost and remains so until late spring. The result is that small to large irregular patches of brown suddenly develop amidst cool-weather grasses that stay green, such as bluegrass, fescue, or bentgrass. The browning is often attributed to insect invasion, but the zoysiagrass is still perfectly healthy. Its tops have died back, but its roots are merely resting. Though it grows slowly, zoysiagrass is hardy. It often takes hold in a lawn in areas of heavy traffic, then invades the surrounding area. Mechanical control consists of digging up clumps and reseeding. If zoysiagrass is a continuing problem, some gardeners let it take over lawns growing in full sun, then spray it with green dye in winter. Chemical control consists of spot treatment with Roundup Ready-to-Use Weed & Grass Killer. Treat while zoysiagrass is still green; once it turns brown, anything but digging is ineffective.

■ **Wild garlic and wild onion:** Wild garlic is often mistaken for wild onion and vice versa. Though they are similar in habits, they are not the same. Both are often the first growth seen in spring. They resemble grasses but are not. Both grow from small underground bulbs and have a garlicky or oniony odor. Leaves are slender, hollow, and

round, joining together near the plant base. Greenish purple or white flowers appear at leaf height. Germination occurs in spring and fall. Bulblets may appear at leaftips. The bulblets fall to the ground and sprout. Wild garlic and onion spread rapidly from spring to midsummer. They thrive in heavy, wet soil. They are drought and cold hardy. There is no preemergent control. Mowing when wild garlic and wild onion first appear can lessen infestation, but bulbs must be totally removed for full control. Postemergent control is most effective in late fall, when wild garlic or onion is still small and vulnerable. Once these perennials take hold, they can be extremely difficult to eliminate. Treat the lawn with ORTHO Weed-B-Gon Lawn Weed Killer, ORTHO Weed-B-Gon Jet Weeder, or an herbicide containing *dicamba* as the leaves emerge in the spring. Repeat the application in 2 weeks.

■ **Timothygrass:** This blue-green bunching grass thrives in thin, poorly fertilized lawns. Its leaves are broad and pointed. Timothygrass grows best in spring and fall. To control it, dig up all clumps.

■ **Velvetgrass:** This perennial can grow to 4 feet tall in unmowed areas. In mowed areas, the bright green velvety leaves stay flat. The plants root wherever joints touch the soil. Seed heads 2 to 4 inches long appear from July to August. Seeds remain dormant over winter and germinate in spring. Perennial velvetgrass thrives in damp areas with good soil, but it tolerates partial shade. To eliminate velvetgrass, use Roundup Ready-to-Use Weed & Grass Killer at any time, but spraying before seed heads mature will be most effective.

Hand-pulling weeds is easier on the back with long-handled weeders, but dandelions will regenerate if even a tiny piece of root is left in the soil.

USING CHEMICAL CONTROLS

Herbicides are powerful formulations that must be used on the appropriate plants at the proper time. In fact, the time of application defines into which of two major categories a chemical fits. These categories are for preemergent herbicides and postemergent herbicides.

A preemergent control stops sprouting at an early stage. It is most effective when placed in or on soil before weed seedlings poke out of the ground. If applied properly, few, if any, of the targeted weeds emerge. A postemergent control is effective after the weeds have emerged and begun growth.

Many modern preemergent and postemergent controls are quite specific as to what weeds they eliminate. Used improperly, these herbicides can create more problems than the weeds themselves. Desired grass and surrounding plants can be temporarily or permanently harmed.

Read the herbicide label before purchasing the product. Make sure the chemical is appropriate for the plant you want to treat, and see if using the product requires special precautions. Use all controls at the appropriate time. Postemergent controls are not effective during the weed's dormant season. Use them on lawn grasses when soil is moist and weeds are growing strongly. Avoid spraying or dusting on windy days. Wind can waft weed killer to nearby

Most mushrooms do not damage a lawn, but some gardeners find them unsightly. The easiest solution, although it is only temporary, is to break the mushrooms with a rake or mower.

plants. If they are susceptible to the control, they may soon brown or die. Apply chemical controls early in the morning or at dusk, when the air is generally calm. If the herbicide label lists a nearby plant as particularly sensitive to the control, protect the plant with a cardboard or wood barrier.

Lack of patience may result in control overdose. A treatment can take from 3 to 10 days to produce visible results. Do not reapply the product because weed browning does not occur a day or two after spraying or dusting. Herbicide overdoses are dangerous to surrounding grasses.

Weed killers do not gain effectiveness if mixed at a concentration stronger than the instructions recommend. Using too strong a mixture can damage or kill desirable plants.

Do not use a weed killer on newly seeded lawns, even if the control is not supposed to affect grass of the type that is planted. Any type of weed control used around seedling grasses can kill them. Seedlings are far more sensitive to control ingredients than mature plants. If you want to use a postemergent control, do so far in advance of reseeding the grass; this will give the control ingredients time to weaken in the soil. The time required differs for different controls. Read the label of the product you buy to determine a safe time for treatment.

Sometimes desirable plants brown long after the application of a weed control but soon after the application of an insecticide. If the same applicator was used for both, the cause could be herbicide contamination. If even a little bit of an herbicide remains in the applicator, it can affect plants. Wash all applicators thoroughly with water and detergent to remove traces of herbicide. Better yet, use separate applicators for herbicides and insecticides.

SOLVING TREE-RELATED LAWN PROBLEMS

When trees and grass grow in the same area, two problems often develop: surface roots, and excess shade from tree canopies and fallen foliage.

Surface Roots: Because of their marked natural tendency to develop surface roots, some trees are best left out of lawn areas. These include acacia, ailanthus, silver maple, alder, Pacific dogwood, fig, evergreen ash, honey locust, mulberry, sycamore, poplar, elm, sumac, black locust, and willow. If you move into a home where the lawn has severe

surface root problems, consider removing the trees and replacing them with appropriate species. Those include maple; silk tree; smoke tree; hawthorn; Modesto ash; golden rain tree; crape myrtle; magnolia; and flowering cherry, peach, and plum.

Surface rooting is caused by external factors as well as natural tendencies. One of the most common is lawn watering. Tree roots need regular watering that penetrates from 6 inches to 3 feet into the soil—the depth necessary depends on the tree species. Sprinkler systems tend to water shallowly. Tree roots, suffering from drought in deeper, unirrigated soil, move upward instead of downward to get moisture.

Excess standing water around lawn trees is another cause of surface rooting. Water fills the air spaces in the soil. As a result, the only readily available oxygen is near the surface, so the tree roots move upward. If a natural or created basin surrounds the tree, you might want to dig a drainage area. Irrigate less often near the tree base so soil dries out between waterings. When you do irrigate the tree, water it deeply, perhaps using a root irrigator.

Compacted soil can also cause tree roots to move upward, seeking oxygen. To allow deeper oxygen penetration, loosen soil around tree roots. Try not to harm the roots themselves. Severe root injury can permit disease and insect attack.

Fertilizer application practices can also encourage surface rooting. If nutrients remain at the surface, the roots do, too. To prompt downward growth, place fertilizer in 12-inch-deep holes or borings. Space the holes evenly under the full expanse of the leaf canopy.

If surface rooting remains a problem after watering, drainage, soil compaction, and fertilizer application has been adjusted, root pruning may be necessary. Some root pruning can be done without harming the tree. If you prize the tree, consider hiring a professional to do the job.

If surface roots are allowed to remain, you may have to adjust your mowing practices. If you do not want to leave the root area bare, consider placing a ring of bark mulch around the tree or planting a hardy ground cover.

Regardless of tree type and maintenance practices, some lawn upheaval may occur near old or large trees. Their big roots, even at a depth, displace ground. Adding good weed-free soil as a yearly topdressing can improve appearance.

Tree Shade: The fact that your lawn is shaded by trees does not necessarily mean the grass gets too much shade. Trees with naturally sparse canopies, such as birch, can allow enough filtered light to permit shade-tolerant grass to grow. If the lawn thins and turns dark green, however, or if moss and algae take hold, corrective action is needed.

To increase the amount of sun under trees, try trimming back or cutting off all limbs that extend below or grow less than 6 feet from the ground. Thinning a dense tree crown can also help. If the lawn does not improve after the pruning, you may have to replant the area.

In an area that receives less than 2 hours of direct sunlight each day, consider planting a shade-tolerant nongrass species such as vinca; wild ginger; sweet woodruff; mondo grass; winter creeper; ajuga; Japanese spurge; or even ivy, if it can be controlled.

If the area receives at least 2 hours of direct sun daily, a shade-tolerant grass can probably survive. Under deciduous trees in cool-winter areas, put in shade-tolerant lawns in late summer or early fall. This gives the grass time to establish during the tree's leafless period.

Under nondeciduous trees in cool climates, reseed or resod in early spring. Consider creeping, red, or chewing fescue. Some of the bluegrass varieties also tolerate part shade. Where warm-season grasses can grow, replant just after the grass-growing season begins in early spring. Among the shade-tolerant, warm-season choices are zoysiagrass, centipedegrass, St. Augustinegrass, and carpetgrass. Since tree roots interfere with soil preparation, adding a layer of topsoil will help make a healthy seedbed.

Always remove fallen leaves under deciduous trees planted in lawn areas. Fallen foliage shades grass even when light is available. When removing leaves from newly seeded areas, take care not to damage the seedlings or seedbed.

Shady areas call for care in mowing as well. When light is scarce, grass grows a bit taller in its reach for the sun. Close mowing of shaded grass can be harmful since it reduces the productive leaf blade area. Adjust your lawn mower accordingly when moving from turf in sun to turf in tree shade; let the shaded grass grow a little taller than the grass in the sun.

In lawn areas that have been overplanted with trees, regardless of tree type, shading will eventually become a problem.

Overplanting is fairly common in new-home areas where owners want, in a short time, to establish the leafy look of an old neighborhood. In addition to creating too much shade, overplanting impedes air flow. Lack of air circulation discourages grass growth and encourages lawn diseases. If the planting is extremely thick, pruning is not enough; trees must be removed to alleviate the shading problem.

CONTROLLING LAWN PESTS

Gardeners usually think of lawn pests in two categories: insects and larger animals.

Lawn insects: Insect pests do an ample share of lawn destruction. At varying times of the year, chiggers (see page 319), chinch bugs (see page 51), billbugs (see page 52), armyworms, cutworms, European crane flies, grubs (see page 50), sod webworms (see page 50), flea beetles (see page 69), fiery skipper butterfly larvae, fleas (see page 318), fire ants, fruit flies, grasshoppers, greenbugs, leafhoppers, mites, scales, wireworms, and mole crickets (see page 53) can invade in small or large numbers. Healthy lawns can tolerate more insect damage than poorly maintained lawns. Some perennial ryegrasses are not attractive to pests such as armyworms, billbugs, cutworms, and sod webworms. Large populations of any pest insect species usually call for intervention by the gardener. Predators—such as birds, parasitic wasps, *Bacillus thuringiensis* (Bt), ladybugs, and green lacewing larvae—can help the gardener control insect pests without chemicals.

■ **Armyworms:** These pests chew grass blades and stems, causing circular bare patches in lawns. In large numbers, armyworms can chew a lawn to the ground in three days. Found throughout the United States except in the coldest areas, the fall armyworm is one of the worst southern lawn pests. These worms move from lawn to garden, then may return to do more damage. Newly hatched worms are white with black heads. Mature worms are green, tan, or brown with dark or orange back and side stripes. Adult size is 1½ inches long. Parents are 1-inch-wide tan or mottled gray moths. Like the adults, armyworms are most active at night and on overcast days. In daylight, they hide in the soil around grass roots. The first generation, which appears in spring, causes the most damage.

Bacillus thuringiensis is partially effective

as a natural control of larvae. Other controls include ORTHO Diazinon Ultra Insect spray (liquid) or ORTHO Dursban® Lawn & Garden Insect Control (granules).

■ **Cutworms:** The larvae of moths, cutworms feed on grass stems and leaf blades. Cutworms are brown, gray, or nearly gray; there are spotted and striped varieties. A full-sized larva can be up to 2 inches long. Cutworms curl up when touched. Adults are dark 2-inch-wide night-flying moths. Often called miller moths, they are common at night around outdoor lights. Cutworms feed at night. During the day, they hide in the upper soil layers. Some types never emerge, feeding only on grass roots. Their feeding causes 2-inch-wide bare spots in the lawn. A closer look shows grass sheared off at or below ground level. Birds often seek out cutworms as food. *Bacillus thuringiensis* sprays and parasitic wasps help destroy cutworms. Apply chemical controls in late afternoon or early evening; Controls include ORTHO Diazinon Soil & Turf Insect Control, ORTHO Dursban® Lawn & Garden Insect Control, or Scotts Lawn Insect Control.

■ **European crane fly larvae:** These larvae eat grass roots, causing yellow-brown patches in summer dry seasons. Damage often begins at the lawn periphery and moves inward. The brownish wormlike maggots develop a tough skin and are sometimes called leatherjackets. A larva is about 1 inch long. Crane flies are found throughout the United States. Adults look like long-legged mosquitoes. An adult's body size, not including the legs, is about 1 inch long. Crane flies do not sting or do other harm. To ascertain whether damage is caused by crane fly larvae or other lawn pests, water damaged areas thoroughly, then cover them overnight with black plastic. If crane fly larvae are present, they will be lying on the soil surface under the plastic the next morning. Crane fly feeding stops naturally in mid-May. Treatment is most effective in early April; use ORTHO Bug-B-Gon Ready-Spray, ORTHO Diazinon Soil & Turf Insect Control, or Scotts Lawn Insect Control.

■ **Fiery skipper butterfly larvae:** By destroying grass blades, these insects create isolated brown spots in lawns. Initially these dead areas are 1 inch wide, but they can expand to cover larger lawn areas. The adult butterfly is orange, brown, or both. It is usually seen during warm weather, flying over lawns in midday. The larvae are small brownish yellow worms that may be found within grass blades. White cottony masses may appear in the lawn—these are the cocoons of a butterfly parasite. The parasites can sometimes control an infestation.

■ **Fire ants:** Infestations of fire ants are becoming increasingly serious as their territory moves from southern states to other warm-winter areas. Their tunnels and mounds can obstruct mowers, and tunneling can eliminate a lawn. Their bites are painful and, if numerous, can severely injure animals and people. Fire ants are more of a problem in sunny clay soils than other areas. Control fire ants by using ORTHO Ant-Stop Orthene Fire Ant Killer, ORTHO Fire Ant Killer Granules, or various baits.

■ **Fruit fly larvae:** These pests live in young grass stems. They eat and gradually destroy the central shoots. Grass then sends out side shoots. Parents are tiny black flies. The females lay eggs on the grass blades. As many as 10 larvae can live inside a single blade. Control is difficult because larvae are well-protected by the stems. *Diazinon* granules can be partly effective as a control.

■ **Grasshoppers:** These insects can become lawn problems in areas near farmland; they migrate to yards when crop sustenance is insufficient. Dry windy weather encourages grasshopper populations. In large numbers, grasshoppers can eat grass to the base. If the number is small, hand-picking can be effective. This job is easier early in the morning, when grasshoppers move slowly. If grasshopper lawn invasions occur repeatedly, slow down the next infestation with a bran bait containing *Nosema locustae*, a grasshopper disease organism. This may take several years to achieve full effect. Seasonal controls include ORTHO Orthene Systemic Insect Control applied with a lawn sprayer over the entire lawn.

■ **Greenbugs:** Small light green aphids that feed on plant sap, greenbugs usually infest Kentucky bluegrass lawns. Their damage appears as rusty-looking lawn areas. These areas expand as greenbug populations increase. Greenbugs do not do much damage in lawns with enough sun; in shaded areas, they can become pests. To control greenbugs, use ORTHO Orthene Systemic Insect Control or an insecticidal soap.

■ **Leafhoppers:** These ⅛-inch wedge-shaped yellow, green, or gray insects live on most lawns. They hop and fly easily from one leaf blade to another and suck out leaf sap. As a result, individual leaves develop white spots. With large infestations, leafhopper damage is demonstrated by lawn fading. Severe infestations can eradicate an emerging lawn. Leafhoppers are most abundant in warm weather. The appearance of damaged seedlings may mimic drought injury. However, if leafhoppers are present, they are almost surely doing the damage. Small infestations are usually not bothersome to plants. Leafhoppers hop about in groups when disturbed, however, and their presence may annoy gardeners. Use ORTHO Orthene Systemic Insect Control, ORTHO Bug-B-Gon Ready-Spray, ORTHO Dursban® Lawn & Garden Insect Control (granules), or Scotts Lawn Insect Control.

■ **Mites:** Grass turns straw-colored as mites suck sap from the blades. The lawn then becomes brown and sparse. Some gardeners working in areas where mites are about may experience skin irritation. Three types of mites generally infest lawns: bermudagrass mites, which prey on bermudagrass only; clover mites; and winter grain mites, which attack bluegrass, fescue, and bentgrass. Most mites are too small to be seen without a microscope. Under magnification, these 1/30-inch pests vary in color depending on species. They have eight legs and are insect relatives rather than insects. Bermudagrass mites may be seen by shaking an infested plant over a sheet of dark paper. The mites are visible as creamy specks that begin crawling around. Mites thrive in hot, dry weather. Adequate watering keeps populations down. Controls include insecticidal soap, ORTHO Bug-B-Gon Ready-Spray, ORTHO Diazinon soil & Turf Insect Control, or Scotts Lawn Insect Control.

■ **Scales:** Infestations of scales on lawns are difficult to control. Scales are legless insects with hard shells. They are extremely small and look like bumps on leaves or roots. Pearl scales attack the roots of bermudagrass, St. Augustinegrass, and centipedegrass. Bermudagrass scale feeds on bermudagrass stems, giving the plant a moldy appearance. Rhodesgrass scales attack grass crowns, causing blades to wither and die. Control consists of spot-treating with *diazinon* every 7 to 14 days from May 1 through June.

■ **Wireworms:** These brown hard-shelled larvae are the offspring of click beetles. A full-sized larva is 1½ inches long. Wireworms feed in groups on grass roots, causing irregular areas of wilted grass. The larvae are most prevalent in soggy soil. Create an organic control by digging several 3-inch-wide by 3-inch-deep holes in the lawn. Bury a potato in each hole and mark each one by inserting a stake or some other device in the ground. In a few days the potatoes will be filled with feeding wireworms. Remove and destroy the potatoes; do not compost them. Chemical controls include diazinon and chlorpyrifos. In areas of serious wireworm infestation, use *diazinon* or *chlorpyrifos* as a preventive before seeding the lawn. This treatment will help keep wireworms from destroying seedlings.

Other Lawn Pests: Moles, rabbits, and gophers top the list of lawn-destroying animals, although in certain areas armadillos, skunks, crayfish, birds, voles (meadow mice), and field mice can cause significant damage.

■ **Moles:** In digging tunnels that serve as feeding pathways, moles create raised ridges that may eventually crisscross a lawn. Ridges range from 3 to 5 inches wide. Because mole tunneling destroys grass roots, ridge areas brown quickly. Moles build new feeding tunnels constantly and may not use the same one twice. If you look carefully, you may find the entry and exit mounds. These are round, conical, fan-shaped, or irregular in shape. The hole usually has dirt in it but is still visible. The mounds are connected to main runways, which moles use repeatedly. These are 12 to 18 inches underground and not usually visible above ground.

Moles are 6 to 8 inches long with gray to black velvety fur. They have slender hairless snouts and small eyes and ears. Moles' front feet are large with long claws, and they function for digging much like a hoe. Despite the many tunnels, one lawn is usually home to only one mole. Except for breeding season, in early spring, moles tend to live solitary lives. Moles eat insects and earthworms. Before you try any other control, consider using *diazinon* to eliminate pest bugs in the lawn. Without ample food, a resident mole may seek supplies elsewhere. Other controls include trapping, bait, repellents, and fumigation. Bait and repellents are not as effective as other controls. Above- and below-ground mole traps are available. Move tunnel traps daily if a mole is not caught.

■ **Gophers:** These rodents are occasionally seen poking their heads out of newly constructed dirt mounds early in the morning. Gophers are brown, have small eyes and ears, and have conspicuous pouches on both sides of their mouths. Gophers protect their territories. The number of mounds in your lawn may seem to indicate the presence of a gopher colony; however, each lawn usually contains just one gopher. The mounds consist of finely pulverized soil that is quite visible in a green lawn. Each mound may contain a visible hole, or an earth clump may camouflage the hole. Tunnels 6 to 8 inches below the lawn surface connect the mounds. Gophers do not create lawn ridges. The rodents eat roots of grass and other plants, pulling them down into underground burrows. Traps are the most efficient form of control. Both wire traps and box traps are sold at many hardware stores and plant nurseries. Dig down to an active horizontal tunnel, and place two traps in it. Follow instructions that accompany the trap.

■ **Rabbits:** When hungry, rabbits eat almost every type of green plant, including lawn grass. They can be minor pests or major ones, depending on the supply of food in the area. The most effective control is keeping rabbits out. Owning a cat or dog that annoys animal pests is sometimes a solution. Another control is a fence made of 1-inch-wide wire mesh. This should be 2 feet high and extend 6 inches underground to avoid rabbit jumping and tunneling. Keeping jackrabbits out requires an even higher fence. Rabbit repellents are sometimes effective. They repel rabbits by making grass taste unpleasant. Repellents must be reapplied as grass grows or is mowed.

■ **Field mice and voles:** These rodents sometimes take up residence in abandoned gopher and mole burrows or create their own burrows. If they make a lawn their winter residence, they may work under snow cover. Lawn runways become visible when snow disappears. Voles feed on grass during the winter and can cause extensive damage. A cat that annoys these rapidly multiplying pests can be an effective control. Other controls include mousetraps, rattraps, and box traps. If children or pets are in the area, do not use poison bait in the open.

■ **Skunks and armadillos:** In residential areas adjacent to woodland or farmland, skunks and armadillos can cause lawn damage as they dig up grass to feed on insects, particularly grubs. Lawn grubs from an infestation of pest beetles, such as Japanese beetles or June beetles, attract them. Removing their food supply usually causes skunks and armadillos to go elsewhere. If the problem becomes serious, apply a pest-beetle control.

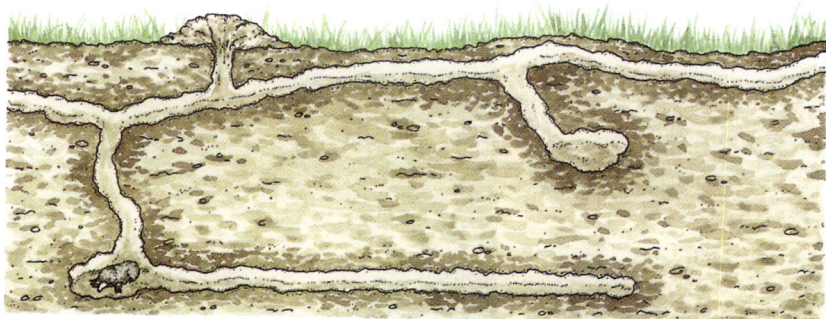

A typical mole tunnel system. The surface runway punctuated with cone-shaped mounds is actually a feeding tunnel that may only be used once. The burrow, or main tunnel system that serves as the animal's living quarters, is usually 12 to 18 inches deep.

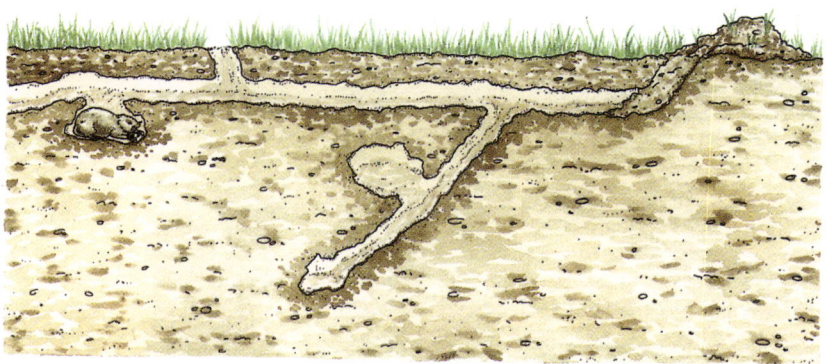

A typical pocket gopher tunnel system. Unlike those of moles, the feeding runways of gophers are not visible from the surface. The large mounds created through excavation are crescent-shaped with an obviously visible plug of earth. In addition to the mounds, each burrow may have several smaller feeding tunnel openings that give the animal access to plants aboveground.

■ **Birds:** Lawns containing earthworms, chinch bugs, sod webworms, cutworms, and caterpillars are a valuable food source for birds. Birds do not cause many problems in high-cut lawns; the insects they eliminate would do much more damage. However, in low-cut lawns, mass feeding may create an abundance of pecking holes that detract from appearance. Since pest bugs are a prime attraction, using ORTHO Diazonon Ultra Insect Spray (liquid) or ORTHO Dursban® Lawn & Garden Insect Control (granules) usually causes the birds to turn elsewhere. Be certain to treat lawns in late afternoon so insecticides are dry by the following day when birds return—freshly applied insecticide may be harmful to birds.

■ **Crayfish:** Also called crawfish and crawdads, crayfish are water-loving creatures that resemble miniature lobsters. Crayfish become a lawn problem only if a lawn is constantly soggy. This can be due to watering practices, poor drainage, or a high water table. Crayfish construct soil mounds around a hole about 1 inch wide. In severe infestations, these mounds must be leveled out to permit mowing. Control consists of correcting drainage problems or easing up on watering. Without hospitable surroundings, crayfish usually go elsewhere. If the growing site is not fully correctable, consider using crayfish bait. Spring treatment is most effective, particularly after a rain.

■ **Snails and slugs:** The silvery trails of snails and slugs wind across lawns and are visible in the morning and on overcast days. These familiar lawn pests hide under ground cover or leaves during the day to avoid sunlight. Both mollusks eat grass. Control snails and slugs by using ORTHO Bug-Geta Snail & Slug Killer. Place the bait in the same areas each time, near the animals' hiding places as well as in a band around the areas you wish to protect.

Lawns

DEAD PATCHES

Dog urine injury

Dog urine spots.

Problem: Circular spots, straw brown in color and 8 to 10 inches in diameter, appear in the lawn. A ring of deep green grass may surround each patch. Other patches may be dark green in color, without any dead areas in them. No spots or webbing appear on the grass blades, and the grass does not mat down. Dogs have been in the area.

Analysis: Dog urine burns grass. The salts in the urine cause varying stages of damage, from slight discoloration to outright death. The nitrogen in the urine may encourage immediately surrounding grass to grow rapidly, resulting in a dark green, vigorously growing ring. Lawns suffer the most damage in hot, dry weather.

Solution: Water the affected areas thoroughly to wash away the urine. This reduces but does not eradicate the brown discoloration. Surrounding grass will eventually fill in the affected areas. For quick repair, spot-sod. If possible, keep dogs off the lawn.

Chemical or fertilizer burns

Burn caused by a fertilizer spill.

Problem: Grass dies and turns yellow in irregular patches or in definite, regular stripes or curves. Grass bordering the areas is a healthy green color. Yellow areas do not spread or enlarge. They appear within 2 to 5 days after fertilization or after a chemical has been spilled on the lawn.

Analysis: Chemicals such as pesticides, fertilizers, gasoline, and hydrated lime may burn the grass if applied improperly or if accidentally spilled. When excessive amounts of these materials contact grass, they cause the blades to dry out and die.

Solution: Prevent or minimize damage by picking up the spilled material, then washing the chemical from the soil immediately. If the substance is water soluble, water the area thoroughly— three to five times longer than usual. If the sub-stance is not soluble in water, such as gasoline or weed oil, flood the area with a solution of dish soap diluted to about the same strength as used for washing dishes. Then water as indicated above. Some substances, such as preemergent herbicides, cannot be washed from the soil. In such a case, replace the top foot of soil in the spill area. Prevent further damage by filling gas tanks, spreaders, and sprayers on an unplanted surface, such as a driveway. Apply chemicals according to the label instructions. Apply fertilizers when the grass blades are dry and the soil is moist. Water thoroughly afterward to dilute the fertilizer and wash it into the soil. Keep drop spreaders closed when stopped or turning.

Fusarium patch

Fusarium patch. Inset: Close-up.

Problem: Yellow-green spots a few inches in diameter form, usually when the snow melts but possibly any time during cool weather from late fall to early spring even if no snow is present. Spots grow and become pinkish white. Affected leaves become matted and turn light tan while patches grow outward from a rusty pink border. A patch can grow up to 1 foot in diameter. Circular patches can join together, causing extensive damage.

Analysis: A plant disease that primarily affects cold-season grasses, fusarium patch (also known as *pink snow mold*) is caused by the fungus *Microdochium nivale*. Active only at cold temperatures (32° to 60°F) when moisture is abundant, fusarium patch is most likely to occur after snow has been on the ground for several months. Prolonged cold weather worsens symptoms; turf quickly recovers if warm weather follows the melting of snow. Prolonged rainy periods in winter also favor this disease. Serious infection leads to crown and root rot.

Solution: Reduce shade in infected areas. Do not apply excessive nitrogen-rich fertilizer in the fall. Overly tall grass is susceptible to fusarium patch, so mow lawns in autumn before snowfall. Reduce thatch buildup. Lightly infected turf usually recovers on its own. For seriously affected areas, apply a fungicide containing *fenarimol*, *mancozeb*, or *triadimefon* according to label directions.

Brown patch

Brown patch, common in warm, humid areas.

Problem: Circular patches of dead grass a few inches to a few feet in diameter appear in the lawn during periods of high humidity and warm temperatures (75° to 85°F). Dark purplish smoky rings sometimes surround brown areas. Filmy white to tan tufts may cover blades in the early morning if the dew is heavy. After 2 to 3 weeks, the center brown grass may recover and turn green, giving brown areas a doughnut shape.

Analysis: Brown patch is caused by a fungus (*Rhizoctonia solani*). It is one of the most prevalent diseases in warm, humid areas, attacking all types of turfgrass. Lush, tender growth from excessive nitrogen fertilization is the most susceptible to attack. Sometimes only the blades are affected and the grass recovers in 2 to 3 weeks. When the infection is severe and warm weather continues, the disease attacks plant crowns and kills the grass.

Solution: Control brown patch with ORTHO Multi-Purpose Fungicide Daconil 2787® Plant Disease Control or Scotts Lawn Fungus Control. Spray when the disease is first noticed and at least three more times at 7- to 10-day intervals. Repeat the treatments as long as warm, humid weather continues. Keep grass as dry as possible to slow down disease spread. Water only in the morning, one or two times per week.

Dollar spot

Dollar spot damage. Inset: Close-up.

Problem: The grass turns light brown to straw-colored in circular areas from the size of a silver dollar to 6 inches in diameter during the warm, wet weather of May to June and September to October. The small dead areas may merge to form large irregular patches. Small, light brown blotches with reddish brown borders appear on the leaf blades. These spots extend across the entire width of the blade. In the early morning before the dew dries, a white cobwebby growth may cover the infected blades.

Analysis: Dollar spot, also called *small brown patch*, is caused by a fungus (*Sclerotinia homeocarpa*). It is most active during mild (60° to 80°F), moist days and cool nights. It attacks many kinds of lawn grasses but is most severe on bentgrass, bermudagrass, and Kentucky bluegrass. Lawns troubled by dollar spot are usually under stress from lack of moisture and nitrogen. An infection seldom causes permanent damage, although the lawn takes several weeks or months to recover. Shoes, hoses, mowers, and other equipment spread the fungus.

Solution: Control dollar spot with ORTHO Multi-Purpose Fungicide Daconil 2787® Plant Disease Control or Scotts Lawn Fungus Control. Make two applications, 7 to 10 days apart, beginning when the disease is first evident. The grass recovers quickly if treated promptly. Keep grass as dry as possible. Water only in the morning, one or two times per week. It is important to maintain proper nutrient levels; applying nitrogen will help the lawn to recover if it has a nitrogen deficiency.

Fusarium blight

Fusarium blight caused by a fungus.

Problem: During hot weather, dead patches form in bluegrass lawns in enlarging rings that may grow up to a foot in diameter. Often weeds invade the center of a dead spot, creating a frog's-eye appearance. Patches may merge to form large dead areas. Each patch begins as a circular grayish green area about 2 inches in diameter. The grass in these patches grows slowly, wilts easily, and dies to a yellow-brown color in hot weather. Eventually weeds invade the entire dead circle.

Analysis: Fusarium blight is caused by a soil-borne fungus (*Fusarium* species), often in combination with one or more other pathogenic fungi. It primarily attacks Kentucky bluegrass and annual bluegrass. The disease begins as a small spot, then grows. As it kills the grass, the dead spot fills in with resistant plants, usually weeds, causing the frog's-eye look. This is a hot-weather disease, being favored by hot, dry, windy conditions.

Solution: Once symptoms are well developed, it's too late to spray for the current season. Irrigate regularly to keep the thatch and soil evenly moist. Avoid heavy nitrogen fertilization. The following spring, treat with Scotts Lawn Fungus Control or a fungicide containing *mancozeb* or *triadimefon* before or as soon as symptoms appear. When replanting, select a resistant variety of bluegrass, or choose ryegrass or fine fescue, or plant a mix of 20 percent (by weight) perennial ryegrass and 80 percent Kentucky bluegrass, which is more resistant than pure bluegrass.

DEAD PATCHES *(continued)*

Sod webworm

Damage. Inset: Sod webworm (2× life size).

Problem: From mid-May to October the grass turns brown in patches the size of a saucer in the hottest and driest areas of the lawn. These areas may expand to form large irregular patches. Grass blades are chewed off at the soil level. Silky white tubes are found nestled in the root area. Inside are light brown or gray worms with black spots, from ¼ to ¾ inch long.

Analysis: Several different moths with similar habits are called *sod webworms* or *lawn moths*. These night-flying moths drop eggs into the grass as they fly. The eggs hatch into worms that feed on grass blades at night or on cloudy, rainy days. In the daytime the worm hides in white silky tubes in the soil. Sometimes an entire lawn is killed in a few days.

Solution: Control sod webworms with ORTHO Bug-B-Gon Ready-Spray, ORTHO Dursban® Lawn & Garden Insect Control (granules), or Scotts Lawn Insect Control when large numbers of moths are noticed at dusk or at the first sign of damage. First rake out all the dead grass and mow the lawn. Water thoroughly before spraying. For best results, apply the insecticide in the late afternoon or evening, when the worms are most active. Do not cut or water the lawn for 1 to 3 days afterward. To avoid recurring damage, treat the lawn again every 2 months beginning in late spring or early summer. Damaged lawns may recover rapidly if the insects are controlled early.

Improper mowing

Scalped spot in lawn.

Problem: Grass has yellow patches or strips after mowing. A few days later, patches may turn brown and die.

Analysis: Grass becomes damaged when too much is removed during mowing. This can happen in two ways.
1. Scalping: Mowing near the soil level, or scalping, occurs when the mower cuts too low in some areas. If the lawn is bumpy, high spots may get scalped. Where the grass has been cut too short, the lower parts of the grass blades are exposed to and burned by sunlight. If scalping damages the base of the grass plant, the grass may die. Otherwise, it will probably recover in a week or so.
2. Insufficient mowings: If the mower is cutting at the proper height but the grass has grown too long between mowings, the lower parts of the grass blades are exposed and burned. Waiting too long to mow seldom kills the grass, but the lawn remains unsightly for a week or two.

Solution: The damaged spots need no special care. The numbered solutions below correspond to the numbered items in the analysis. To prevent further damage:
1. If the lawn surface is uneven, level the high spots. If the lawn is spongy from an accumulation of thatch, power rake or dethatch the lawn. Raise the mower blade to the suggested height.
2. Mow frequently enough so that you don't remove more than half the grass blade. If the grass is very tall when you mow, raise the mower blade to half the grass height and lower it gradually over the next few mowings.

Grubs

Damaged lawn. Inset: White grub (2× life size).

Problem: From August through October the grass appears to wilt and turns brown in large, irregular patches. Brown areas of grass roll up easily like a carpet. Milky white grubs from ⅛ to 1 inch long, with brown heads and three pairs of legs, lie curled in the soil. Birds and animals may be digging in the lawn.

Analysis: Grubs are the larvae of different kinds of beetles, including May and June beetles (also called *white grubs*) and Asiatic, Japanese, and masked chafer beetles. The grubs feed on turf roots and may kill the entire lawn. Some birds and animals relish the grubs and dig up the lawn to feed on them. The adult beetles do not damage the lawn, but they do lay eggs in the soil. May and June beetles lay eggs in the spring and summer. The Asiatic, Japanese, and masked chafer beetles lay eggs in mid- to late summer. The eggs hatch and the grubs feed on roots 1 to 3 inches deep in the soil. In late fall they move deep in the soil to overwinter; they resume feeding in the spring.

Solution: Apply ORTHO Grub-B-Gon when you first notice damage and grubs. This insect growth regulator is effective all season, and can be applied at any time. Or apply Scotts GrubEx from spring through summer to prevent young grubs from damaging your lawn. Water in well after applying. To save areas just beginning to fade, keep the soil moist but not wet.

Pythium blight

Pythium blight on grass seedlings.

Problem: In hot, humid weather from April to October, the grass wilts, shrivels, and turns light brown in irregular spots ½ to 4 inches in diameter. Spots enlarge rapidly, forming streaks 1 foot wide or wider or patches 1 to 10 feet in diameter. Infected blades mat together when walked on. Blades are often meshed together by white cobweblike threads in the early morning before the dew dries. Grass sometimes dies within 24 hours.

Analysis: Pythium blight, also called *grease spot* or *cottony blight*, is caused by a fungus (*Pythium* species). It attacks lawns under stress from heat (85° to 95°F), poorly drained soil, and excessive moisture. Dense, lush grass is the most susceptible. All turfgrasses are affected, ryegrasses the most severely. The fungus spores spread easily in free-flowing water, on lawn mower wheels, and on the soles of shoes. The disease is difficult to control because it spreads so rapidly, killing large areas in hours. Pythium blight is common in the fall on overseeded ryegrass.

Solution: Treat the lawn with a fungicide containing *chloroneb* or *ethazole* as soon as the disease is noticed. Repeat treatments every 5 to 10 days either until the disease stops or until cooler weather resumes. Keep traffic off the diseased area to avoid spreading spores. Don't overwater in hot, humid weather. Severely infected areas often do not recover, so reseed or resod to re-establish the lawn. Treat the new lawn during hot, humid weather. Wait until cool weather to overseed ryegrass.

Chinch bug

Damage. Inset: Chinch bug (6× life size).

Problem: In April through October the grass wilts, turns yellowish brown, dries out, and dies in sunny areas and along sidewalks and driveways. To check for chinch bugs, select a sunny spot on the edge of an affected area where yellow grass borders healthy green grass. Cut out both ends of a tin can. Push one end of the can 2 to 3 inches into the soil. Keep it filled with water for 10 minutes. Black to brown insects with white wings, ⅛ to ¼ inch long, float to the surface in 5 to 10 minutes. Pink to brick red nymphs with a white stripe around the body may also be numerous.

Analysis: Chinch bugs (*Blissus* species) feed on many kinds of lawn grasses, but St. Augustinegrass and zoysiagrass are favorites. Both the adults and the nymphs suck the juices out of the blades. At the same time, they inject a poison that causes the blades to turn brown and die. Heavy infestations may completely kill the lawn. These sun- and heat-loving insects seldom attack shady lawns. They can move across an entire lawn in several days.

Solution: Control chinch bugs with ORTHO Bug-B-Gon Ready-Spray, ORTHO Dursban® Lawn & Garden Insect Control, or Scotts Lawn Insect Control as soon as you see damage. Mow and water the lawn before spraying, applying ½ to 1 inch of water to bring the insects to the surface. To prevent recurring damage from newly hatched nymphs, treat every 2 months until frost. In southern Florida, repeat the applications year-round.

Annual bluegrass dying

Annual bluegrass dying.

Problem: Areas of grass that were once lush and green die and turn straw brown. Grass appears whitish in late spring, and with the onset of hot summer weather, these places become irregular dead patches.

Analysis: Annual bluegrass (*Poa annua*) is one of the most troublesome but least noticed weeds in the lawn. This member of the bluegrass family is lighter green, more shallow rooted, and less drought tolerant than Kentucky bluegrass. As its name suggests, annual bluegrass usually lives for only 1 year, although some strains are perennial. The seed germinates in cool weather from late summer to late fall. Annual bluegrass grows rapidly in the spring, especially if the lawn is fertilized then. Seed heads appear in mid- to late spring at the same height that the grass is cut. The seed heads give the lawn a whitish appearance. When hot, dry weather arrives, the plants turn pale green and die. The seeds fall to the soil and wait for cooler weather to germinate. Annual bluegrass is most serious where the soil is compacted.

Solution: When dead patches appear in hot weather, the annual bluegrass is dead. The lawn is laced with its seeds, however, which will germinate with cooler fall weather. Patch the dead spots. When the weather begins to cool in the fall, treat the lawn with Scotts Halts Crabgrass Preventer to kill the seeds as they germinate. Do not cut the lawn too short. Lawns more than 2½ inches tall have very little annual bluegrass. Aerate the lawn in compacted areas.

DEAD PATCHES (continued)

Crabgrass dying

Crabgrass dying in early fall.

Problem: Brown patches develop in the lawn with the first fall frost. Close examination 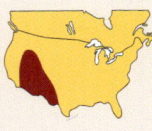 of the dead spots reveals not dead lawn grass but a weed.

Analysis: Crabgrass (*Digitaria* species) is an annual grassy weed. It forms large, flat clumps, smothering lawn grass as it spreads. Crabgrass dies with the first killing frost in the fall or with the onset of cold weather, leaving dead patches in the lawn. Crabgrass sprouts from seeds in early spring.

Solution: In early spring, 2 weeks before the last expected frost, treat the lawn with Scotts Halts Crabgrass Preventer. This preemergent weed killer kills the seed as it germinates. Kill actively growing crabgrass with ORTHO Weed-B-Gon Crabgrass Killer. Maturing plants are harder to kill. Repeat treatments two more times at 4- to 7-day intervals if necessary. To control crabgrass growing in cracks in sidewalks and driveways, use Roundup Fence & Yard Edger. A thick lawn seldom contains much crabgrass.

Billbugs

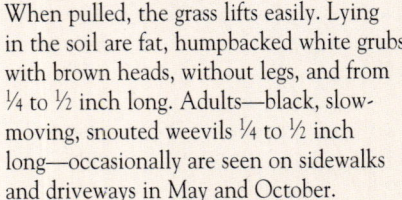

Billbug damage. Inset: Larvae (life size).

Problem: The grass turns brown and dies in expanding patches from mid-June to late August. When pulled, the grass lifts easily. Lying in the soil are fat, humpbacked white grubs with brown heads, without legs, and from ¼ to ½ inch long. Adults—black, slow-moving, snouted weevils ¼ to ½ inch long—occasionally are seen on sidewalks and driveways in May and October.

Analysis: The larvae of billbugs (*Sphenophorus* species) damage lawns by hollowing out the grass stems and chewing off the roots. They can destroy an entire lawn. In May, the adults lay eggs in holes they chew in grass stems. The newly hatched larvae feed inside the stems, hollowing out the stem and crown and leaving fine sandlike excrement. Large larvae feed on roots.

Solution: Control billbugs with ORTHO Diazinon Soil & Turf Insect Control or Scotts Lawn Insect Control. Repeated treatments are not usually necessary unless the billbugs are migrating from neighboring yards. Small damaged areas usually recover if the larvae are killed. Water and fertilize the lawn to stimulate new growth. Reseed or resod large areas. Maintain proper soil moisture and fertility.

Salt damage

Dead patch from salt accumulation.

Problem: Grass slowly dies, especially in the lowest areas of the lawn. A white or dark crust may be present on the soil.

Analysis: Salt damage occurs when salt accumulates in the soil to damaging levels. This can happen in either of two ways: (1) the lawn does not receive enough water from rainfall or irrigation to wash the salts from the soil or (2) the drainage is so poor that water does not pass through the soil. In either case, as water evaporates from the soil and grass blades, salts that were dissolved in the water accumulate near the surface of the soil. In some cases, a white or dark brown crust of salts forms on the soil surface. Salts can originate in the soil, in irrigation water, or in fertilizers.

Solution: The only way to eliminate salt problems is to wash the salts through the soil with water. If the damage is only at a low spot in the lawn, fill in the spot to level the lawn. If the entire lawn drains poorly, improve drainage by aerating. If the soil drains well, increase the amount of water applied at each watering by 50 percent or more, so that excess water will leach salts below the root zone of the grass. Fertilize according to product label instructions.

Mole crickets

Damaged lawn. Inset: Mole cricket (¾ life size).

Problem: Small mounds of soil are scattered on the soil surface. The lawn feels  spongy underfoot. Large areas of grass turn brown and die. To determine if the lawn is infested with mole crickets, make a solution of 2 tablespoons of liquid dishwashing detergent to 2 gallons of water. Drench 4 square feet of turf with the mixture. Mole crickets—greenish gray to brown insects, 1½ inches long, with short front legs and shovellike feet—will come to the surface within 3 minutes.

Analysis: Several species of mole crickets (*Scapteriscus* and *Gryllotalpa* species) attack lawns. They prefer bahiagrass and bermudagrass but also feed on zoysiagrass, St. Augustinegrass, and centipedegrass. They damage lawns by tunneling through the top 1 to 2 inches of soil, loosening it and uprooting plants so that the plants dry out. Mole crickets also feed on grass roots, weakening the plants. They feed at night and may tunnel as many as 10 to 20 feet per night. In the daytime, they return to their underground burrows. Adults migrate from their burrows to new areas twice a year, from March to July and again from November to December.

Solution: In June or July, after the eggs hatch and before the young nymphs cause much damage, treat the lawn with ORTHO Dursban® Lawn & Garden Insect Control (granules), ORTHO Bug-B-Gon Ready-Spray, or Scotts Lawn Insect Control. Mole crickets are not active in dry soil, so mow and water before applying. Do not water for 36 hours after application. If damage continues, treat again in late summer to early fall. Keep the lawn watered to encourage new root growth.

Drought

Lawn damaged by drought.

Problem: Footprints in the lawn make a long-lasting imprint instead of bouncing right back. The grass blades turn a dull bluish green or slate gray color and wilt. In the cool evening, the grass recovers until the sun and heat of the following day make it darken and wilt. Areas begin to thin out. After a few days the lawn begins to look and feel like straw and dies.

Analysis: A lawn suffers from drought damage when water evaporates from the lawn faster than the roots absorb it. Drought damage occurs first in the hottest and driest areas of the lawn—along sidewalks and driveways and on south- or west-facing slopes, south sides of reflecting buildings, and areas with sandy soil. Grass blades don't wilt as broadleaf plants do. They don't droop, but roll or fold up lengthwise.

Solution: Water the lawn immediately. If the grass has turned yellow, the affected areas will require several weeks to recover. If you are not very conscientious about watering your lawn, consider planting a drought-tolerant turfgrass.

Thatch

Thatch between grass and soil surface.

Problem: Grass thins out in sunny or shady areas of the lawn. Weeds invade the sparse areas. Grass may suddenly die in large patches during summer heat and drought. Cut and lift several plugs of grass 2 to 3 inches deep. Look to see if the stringy, feltlike material between the grass and soil surface is thicker than ½ inch.

Analysis: Thatch is a tightly intermingled layer of partially decomposed stems and roots of grass that develops between the actively growing grass and the soil surface. Thatch slows grass growth by restricting the movement of water, air, and nutrients in the soil. Thatch is normal in a lawn, but when it is thicker than ½ inch, the lawn begins to suffer. As the layer accumulates, the grass roots grow in the thatch instead of down into the soil. Thatch accumulation is encouraged by overly vigorous grass growth caused by excessive fertilization and frequent watering.

Solution: To reduce the thatch and increase the lawn's vigor, power rake or dethatch the lawn. Dethatch cool-season grasses in the fall and warm-season grasses in late spring or early summer. Avoid dethatching while new growth is turning green. The machines for the job can be rented, or hire a contractor. Dethatchers, also called *verticutters*, have vertical rotating blades that slice through the turf, cutting out thatch. First, mow the lawn as short as possible. Go over it one to three times with the dethatcher. Take up the debris, fertilize, and water to hasten the lawn's recovery.

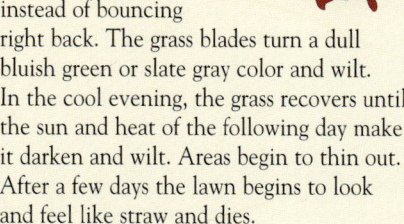

GRASS THIN (continued)

LAWN PALE OR YELLOW

Nematodes

Nematode damage.

Problem: The grass grows slowly, thins out, and turns pale green to yellow. In hot weather the turf may wilt in irregular patterns. Main roots are short with few side roots, or many roots may grow from one point.

Analysis: Nematodes are microscopic worms that live in the soil and feed on grass roots, damaging and stunting them. They are unrelated to earthworms. The damaged roots cannot supply sufficient water and nutrients to the leaf blades, and the grass is stunted or slowly dies. Nematodes are found throughout the country but are most severe in the South. They prefer moist, sandy loam soils. They can move only a few inches each year on their own, but they may be carried long distances by soil, water, tools, or infested plants. Testing roots and soil is the only positive method to confirm the presence of nematodes. Contact your county extension office for sampling instructions and addresses of testing laboratories. Soil and root problems such as poor soil structure, drought stress, nutrient deficiency, and root rots can also produce symptoms similar to those caused by nematodes. Eliminate these problems as causes before sending soil and root samples for testing.

Solution: No chemicals available to homeowners kill nematodes in planted soil. Soil fumigation or solarization can be used to control nematodes before a new lawn is planted, however. To solarize, cultivate and wet an area thoroughly before covering it with clear polyethylene film. Leave the film in place 4 to 6 weeks during the hottest time of the year.

Iron deficiency

Iron deficiency; blades yellow between veins.

Problem: Irregular patches of grass are yellow. Individual blades are yellow between the veins; the veins remain green. If the condition persists, the leaves may become almost white and die back from the tips. In severe cases, the grass is stunted.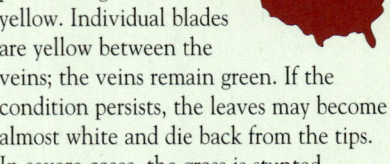

Analysis: Iron deficiency is a common problem in many plants and is usually caused by alkaline soil conditions. In alkaline soil, much of the iron forms insoluble compounds that are unavailable to grass plants. Lack of iron may also be caused by an iron deficiency in the soil, excess phosphorus in the soil, a poor root system, overwatering, or the use of water that contains large amounts of bicarbonate salts. Plants use iron in the formation of chlorophyll in the leaves. When iron is lacking, new growth is yellow. Many turfgrass species—including Kentucky bluegrass, perennial ryegrass, fine fescue, creeping bentgrass, and bermudagrass— are susceptible to iron deficiency.

Solution: For a quick green-up, spray the lawn with a liquid iron supplement. In the future, fertilize with Miracle-Gro All-Purpose Lawn Fertilizer, which contains iron. Lower the alkalinity of the soil by adding ferrous sulfate or ferrous ammonium sulfate. Water the lawn thoroughly after applying one of these amendments. Never add lime to soil in which iron deficiency is a problem. Some turfgrass varieties are resistant to iron deficiency; when replanting, ask for one of these at your nursery.

Nitrogen deficiency

Unfertilized lawn with green clover.

Problem: Grass is pale green to yellow and grows more slowly than usual. If the condition persists, the grass becomes sparse and weeds invade the lawn.

Analysis: Nitrogen is a key element in maintaining a healthy lawn with few insect and disease problems. Clover stays green because it obtains nitrogen from the air, but grasses cannot. It is best to maintain a level of nitrogen in the soil that (1) does not stimulate excessive leaf growth, which would increase the frequency of mowing, (2) does not encourage shoot growth at the expense of root growth, and (3) varies according to the cultural and environmental conditions present. Because heavy rains and watering leach nitrogen from the soil, periodic feedings are necessary throughout the growing season. Acid soil may cause nitrogen to be unavailable to the grass.

Solution: Apply Scotts Turf Builder fertilizer according to the instructions on the label. Properly fertilized lawns are dense and have a nice green color without excessive growth. To prevent burning and to move nutrients into the soil, water thoroughly after application. Grass begins using the nitrogen in the fertilizer within 15 to 24 hours. Recycle the nitrogen by leaving grass clippings on the lawn if they are not extremely long. If the soil is acid (below pH 5.5), liming is necessary for effective nitrogen utilization.

Septoria leaf spot

Septoria leaf spot, common in spring and fall.

Problem: In the spring and fall, the lawn has a gray cast. The tips of the blades are pale 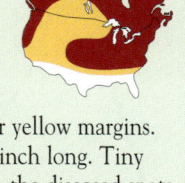 yellow to gray, with red or yellow margins. Pale areas may be ⅛ to 1 inch long. Tiny black dots are scattered in the diseased spots on the blades. From a distance, damage may resemble dull mower injury.

Analysis: Septoria leaf spot, also called *tip burn,* is caused by a fungus (*Septoria* species) that infects most northern grass species and bermudagrass. It is most prevalent in the cool, wet weather of early spring and fall. Lawns that have not been fertilized are most susceptible. The disease usually attacks in the spring, declines during the hot summer months, and returns in the fall. Because the disease infects leaf tips first, frequent mowing removes much of the diseased part of the blades.

Solution: Treat the infected lawn with a fungicide containing *maneb* or *chlorothalonil* (ORTHO Multi-Purpose Fungicide Daconil 2787® Plant Disease Control) as soon as discoloration appears. Repeat the treatment three more times, 7 to 10 days apart, or as long as weather favorable for the disease continues. Keep the lawn healthy and vigorous. Mow the lawn regularly. Because no variety is completely resistant, plant a blend of two or three disease-tolerant varieties.

Rust

Rust; powdery orange spores that rub off.

Problem: Grass turns orange-yellow or reddish brown and begins to thin out. An orange powder that looks like rust coats the grass blades and rubs off on fingers, shoes, and clothing. Reddish brown lesions under the powder do not rub off.

Analysis: Rust is caused by a fungus (*Puccinia* species) that occurs most frequently on Kentucky bluegrass, ryegrass, tall fescue, and zoysiagrass. It is most active during moist, warm weather (70° to 75°F) but can be active all winter in mild winter areas. Heavy dew favors its development. Grasses under stress from nitrogen deficiency and lack of moisture are most susceptible to attack. Rust is also more severe in the shade. The orange powder is composed of millions of microscopic spores that spread easily in the wind. Lawns attacked severely by rust are more likely to suffer winter damage.

Solution: Rust develops slowly, often more slowly than the grass grows. Apply a high-nitrogen fertilizer to maintain rapid growth. Mow frequently, removing the clippings. If the disease is severe, treat with a fungicide containing *triadimefon* or with ORTHO Multi-Purpose Fungicide Daconil 2787® Plant Disease Control. Repeat the application every 7 to 14 days until the lawn improves.

Powdery mildew

Powdery mildew; a whitish gray mold.

Problem: Whitish gray mold develops on the upper surfaces of grass blades during cool  rainy weather. The lawn looks as if it has been dusted with flour. The leaf tissue under the mold turns yellow, then tan or brown. Severely infected plants wither and die.

Analysis: Powdery mildew is caused by a fungus (*Erysiphe graminis*) and occurs when the nights are cool (65° to 70°F) and damp and the days warm and humid. It is most severe on Kentucky bluegrass but also attacks fescues and bermudagrass. Lawns growing in the shade are the most affected. Powdery mildew slows the growth of leaves, roots, and underground stems, causing gradual weakening of the grass and making the grass more susceptible to other problems. Lawns growing rapidly because of excessive nitrogen fertilization are very susceptible to attack from this fungus. The fine white mildew on the blades develops into powdery spores that spread easily in the wind.

Solution: Treat the lawn with a fungicide containing *triadimefon* when the mildew is first seen. A second application 7 to 10 days later may be needed in severe cases. Reduce shade and improve air circulation by pruning surrounding trees and shrubs. Keep the lawn healthy and vigorous.

GRASSLIKE WEEDS

Annual bluegrass

Seed heads. Inset: Annual bluegrass.

Problem: In midspring, abundant seed heads give the grass a whitish appearance. Pale green grassy weeds grow among desirable grasses. They turn yellow and die with the onset of hot weather.

Analysis: Annual bluegrass (*Poa annua*) is one of the most troublesome but least noticed weeds in the lawn. This member of the bluegrass family is lighter green, more shallow rooted, and less drought tolerant than Kentucky bluegrass. As its name suggests, annual bluegrass usually lives for only 1 year, although some strains are perennial. The seed germinates in cool weather from late summer to late fall. Annual bluegrass grows rapidly in the spring, especially if the lawn is fertilized then. Seed heads appear in mid- to late spring at the same height that the grass is cut. The seed heads give the lawn a whitish appearance. When hot, dry weather arrives, the plants turn pale green and die. The seeds fall to the soil and wait for cooler weather to germinate. Annual bluegrass is most serious where the soil is compacted or over-irrigated and where drainage is poor.

Solution: Weed killers are only partially effective in controlling annual bluegrass. Prevent seeds from germinating by applying Scotts Halts Crabgrass Preventer as a preemergent treatment in late summer to early fall. Do not use if you plan to reseed the lawn in the fall. Replace the dead areas in summer with sod. Do not cut the lawn too short. Lawns more than 2½ inches tall have very little annual bluegrass. Aerate the lawn in compacted areas. Space irrigations far enough apart that the ground surface has time to dry.

Quackgrass

Quackgrass, showing underground stems.

Problem: A grassy weed with hollow stems grows in a newly seeded lawn. Wheatlike spikes grow at the tips of the stems. The narrow leaf blades are bluish green and rough on the upper surface. A pair of "claws" occurs at the junction of the blade and the stem. Rings of root hairs grow every ¾ to 1 inch along the underground stems.

Analysis: Quackgrass (*Elytrigia repens*), a cool-season perennial, also called *couchgrass* or *witchgrass,* spreads extensively through the lawn by long white underground stems. It reproduces by seeds and these underground stems. The seeds may lie dormant in the soil for up to 2 years. Quackgrass is found most frequently in fertile, newly seeded lawns. It grows much more rapidly than grass seedlings, often crowding them out.

Solution: Quackgrass is difficult to control in lawns. If the entire lawn is infested with it, the lawn will need to be killed and another planted. If only isolated areas are infested, kill them and replant those spots. The quackgrass must be actively growing before it is sprayed. Let it grow to 4 to 6 inches high, then spray with ORTHO Kleeraway Grass & Weed Killer or Roundup Weed & Grass Killer. If regrowth occurs, repeat the treatment.

Crabgrass

Crabgrass seedhead.

Problem: A grassy weed forms broad, flat clumps in thin areas of the lawn. It grows rapidly through the summer, rooting at the stem joints. The pale green blades are 2 to 5 inches long and ⅓ inch wide. Seed heads 2 to 6 inches tall grow from the center of the plant.

Analysis: Crabgrass (*Digitaria* species) sprouts from seeds in the early spring, growing rapidly and producing seeds all summer until the first killing frost in the fall. Then the plants turn brown and die. The seeds lie dormant over the winter and sprout in the spring. Crabgrass is one of the most common lawn weeds in its area of adaptation. When a lawn begins to thin out from insects, disease, or poor maintenance, crabgrass is one of the first weeds to invade the area.

Solution: Kill actively growing crabgrass with ORTHO Weed-B-Gon Crabgrass Killer. Older plants are harder to kill; repeat the treatment two more times at 4- to 7-day intervals. To kill crabgrass seeds as they germinate, apply Scotts Halts Crabgrass Preventer in late winter or early spring, 2 weeks before the last expected frost (about the time forsythia and dogwood bloom). Keep the lawn healthy and vigorous; crabgrass is usually not a serious problem in a thick, healthy lawn.

Tall fescue

Tall fescue, a cool-season perennial weed.

Problem: Clumps of very coarse, tough grass invade thin areas of the lawn. The medium-dark green blades, each ½ inch wide, are ribbed on the top surface and smooth on the bottom. In the spring and fall, the lower parts of the stems turn reddish purple. The blades tend to shred when mowed.

Analysis: Tall fescue (*Festuca arundinacea*), a cool-season, perennial bunch-type grass, is very durable. It is commonly used on athletic fields because it holds up well under hard wear. Tall fescue makes an attractive turf when grown by itself. When it is seeded with, or invades, bluegrass, bermudagrass, or ryegrass lawns, however, it is considered a weed. It becomes very clumpy and makes an uneven turf. When insects and diseases attack the desirable grasses in the lawn, the tall fescue is usually not affected. It resists diseases and grubs, and sod webworms attack it only if they've eaten everything else. It is also somewhat heat tolerant, and its deep roots help it survive periods of heavy moisture and drought.

Solution: Kill clumps of tall fescue with ORTHO Kleeraway Grass & Weed Killer, ORTHO Grass-B-Gon Grass Killer, or Roundup Weed & Grass Killer while it is actively growing from early summer to early fall. Omit a regular mowing before treating to allow for enough leaf tissue to absorb the chemical. These herbicides will also kill any desirable grasses they contact. One week after spraying, mow the tall fescue and reseed the area. The tall fescue may still be green when it is removed, but the roots will die in 3 to 4 weeks.

Barnyardgrass

Barnyard grass, a warm-season annual weed.

Problem: In summer and fall, a low-growing grassy weed with reddish purple stems 1 to 3 feet long grows in the lawn. The smooth leaves are ¼ to ½ inch wide, with a prominent midrib.

Analysis: Barnyard grass (*Echinochloa crus-galli*), also called *watergrass*, is a warm-season annual weed that is usually found in poorly managed lawns of low fertility. It reproduces by seeds and develops into a plant with a shallow root system. Although the natural growth habit of barnyard grass is upright, when mowed regularly it forms ground-hugging mats.

Solution: Kill mats of actively growing barnyard grass with ORTHO Grass-B-Gon Grass Killer, ORTHO Kleeraway Grass & Weed Killer, or Roundup Weed & Grass Killer. Repeat the treatment two more times, at intervals of 7 to 10 days, until the plants die. It may discolor the turf for 2 to 3 weeks. Improve soil fertility and maintain a dense, healthy lawn. To kill barnyard grass seedlings as they sprout, apply Scotts Halts Crabgrass Preventer in the early spring, 2 weeks before the last expected frost.

Clover

Clover spreads by seed and rooting stems.

Problem: A weed with leaves composed of three round leaflets at the top of a hairy leafstalk, 2 to 4 inches tall, grows in the lawn. The leafstalks sprout from the base of the plant. White or pink-tinged flowers, ½ inch in size, bloom from June to September. They often attract bees.

Analysis: Clover (*Trifolium* species) is a common perennial weed in lawns throughout the United States. Although some people like it in a lawn, others consider it messy, or they don't like the bees attracted to the flowers. Clover reproduces by seeds and aboveground rooting stems. The seeds can live in the soil for 20 years or more. The plant, which has a creeping, prostrate habit, suffocates lawn grasses, resulting in large patches of clover. When buying grass seed, be sure to read the label carefully. Clover seeds are sometimes contained in seed mixtures. Because clover produces its own nitrogen, it thrives in lawns that are low in this plant nutrient.

Solution: Treat the lawn with ORTHO Weed-B-Gon Lawn Weed Killer or GreenSweep Weed & Feed (20–0–0) in the spring and early fall. Repeated treatments are often necessary.

BROADLEAF WEEDS

Oxalis

Oxalis, a perennial weed.

Problem: A weed with pale green leaves divided into three heart-shaped leaflets invades thin areas of the lawn. The leaves are ¼ to ¾ inch wide and are similar to clover. The stems root at the lower joints and are often thinly covered with fine hairs. Small, bright yellow flowers are ½ inch long with five petals. Cucumber-shaped, light green seedpods develop from the fading flowers. Plants may be 4 to 12 inches high with a prostrate-to-erect growth habit.

Analysis: Oxalis (*Oxalis stricta* and *O. corniculata*), also called *yellow woodsorrel* or *creeping woodsorrel*, is a perennial plant that thrives in dry, open places but may also be a problem in moist, well-fertilized lawns. It often invades lawns that are beginning to thin from insect, disease, or maintenance problems. Oxalis reproduces from the seeds formed in the seedpods. When the pods dry, a light touch causes them to explode, shooting their seeds several feet in all directions.

Solution: Control oxalis with ORTHO Weed-B-Gon Lawn Weed Killer, ORTHO Weed-B-Gon Weed Killer, GreenSweep Weed & Feed (20–0–0), or an herbicide containing *triclopyr*. The most effective time to spray is when the weeds are actively growing, in the spring or late summer to fall. Oxalis is not easy to kill; several treatments are usually needed. Keep the lawn healthy and vigorous. A healthy lawn helps smother the oxalis.

Spotted spurge

Spotted spurge produces many seeds.

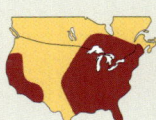

Problem: A low-growing weed with oval pale to dark green leaves ¼ to ¾ inch long appears in the lawn. Each leaf may have a purple spot. The leaves are slightly hairy on the underside and smooth on top. The slender, reddish or pale green stems ooze milky white sap when broken. The sap may irritate the skin. The many stems fan out on the soil surface and over the top of the grass, forming mats up to 2 feet in diameter. Tiny pinkish white flowers bloom and produce seeds from May to October; a plant can begin producing seed when only 2 weeks old. Many of the seeds remain dormant over the winter and sprout the following spring. Some sprout immediately. Plants die with the first fall frost.

Analysis: Spotted spurge (*Euphorbia maculata*), also called *milk purslane* or *prostrate spurge*, invades thin areas of the lawn, smothering the grass. Spurge sprouts from seeds in the spring and dies with the first frost. This weed commonly invades lawns that are dry and infertile, but it can also be found in well-maintained lawns.

Solution: In the late spring or early summer, treat the lawn with GreenSweep Weed & Feed (20–0–0) or ORTHO Weed-B-Gon Weed Killer, or spot-treat with ORTHO Weed-B-Gon Jet Weeder. Keep the lawn well watered to discourage spurge from invading dry areas.

Common chickweed

Common chickweed forms dense mats.

Problem: A weed with small (½-inch-long) teardrop-shaped leaves and starlike white flowers grows in thin spaces in the lawn. A single row of white hairs appears on one side of the stem. The stems root easily at their joints. Common chickweed is most often found in shady or moist areas.

Analysis: Common chickweed (*Stellaria media*) is a weed that grows from seeds that sprout in the fall; the plants live for less than a year. Common chickweed grows primarily in damp, shady areas under trees and shrubs and on the north side of buildings, but it can also occur in dormant warm-season grasses. Common chickweed invades home lawns when they begin to thin out from insects, disease, mechanical damage, or shade. It reproduces by seeds and by the creeping stems that root at their joints wherever they touch the soil. It has a low prostrate growing habit, forming a dense mat that crowds out the grass.

Solution: Treat the lawn with ORTHO Weed-B-Gon Weed Killer, ORTHO Weed-B-Gon Lawn Weed Killer, or GreenSweep Weed & Feed (20–0–0) when it is growing actively in the early spring or late fall. Repeated applications may be necessary. Do not water for 2 days after applying.

Henbit

Henbit spreads by seeds and rooting stems.

Problem: A weed with rounded, toothed leaves, ¾ inch wide, grows in the lawn. The lower leaves are attached to the square (four-sided) upright stems by short leafstalks; upper leaves attach directly to the stems. Stems root easily at lower joints. Lavender, ½-inch flowers appear from March to June and again in September.

Analysis: Henbit (*Lamium amplexicaule*), also known as *dead nettle* or *bee nettle*, is a weed found in lawns and flower and vegetable gardens across the country. It is a winter annual that sprouts from seeds in September and grows rapidly in the fall and the following spring. Henbit also reproduces by stems that root easily wherever the stem joints touch the soil. Henbit most frequently invades thin areas in lawns with rich soil.

Solution: Treat the lawn with GreenSweep Weed & Feed (20–0–0) or ORTHO Weed-B-Gon Weed Killer in early spring when henbit is growing most rapidly. Do not water for 24 hours after treating. A few small plants can be hand-pulled.

Purslane

Purslane, most common in hot dry weather.

Problem: A low-growing weed with reddish brown, thick, succulent stems is found in thin areas or newly seeded lawns. The leaves are thick, fleshy, and wedge shaped. Small yellow flowers sometimes bloom in the leaf and stem joints. Stems root where they touch the soil.

Analysis: Purslane (*Portulaca oleracea*), a summer annual weed that thrives in hot, dry weather, is seldom found in the spring when the lawn is being treated for other weeds. Purslane grows vigorously, forming a thick mat. The small yellow flowers open only in the full sunlight. Purslane primarily invades bare spots in lawns or thin lawns that have not been watered properly. Purslane stores water in its thick, fleshy stems and leaves and therefore survives longer than grass during dry weather.

Solution: Spray the lawn with ORTHO Weed-B-Gon Lawn Weed Killer or ORTHO Weed-B-Gon Jet Weeder when the weed is actively growing. If the lawn has just been reseeded, do not treat until the seedlings have been mowed three times. Wait 3 to 4 weeks before seeding bare areas.

Dandelion

Dandelion; wind can carry the seed for miles.

Problem: From spring to fall, a weed with bright yellow flowers blooms in the lawn. In some southern states it may bloom all winter. Flower stems grow 2 to 10 inches above the plants. The medium-green leaves, 3 to 10 inches long, are deeply lobed along the sides. The plant has a deep, fleshy taproot.

Analysis: Dandelion (*Taraxacum officinale*) is the most common and easily identified perennial weed in the United States. It reproduces by seeds and from shoots that grow from the fleshy taproot. This taproot grows 2 to 3 feet deep in the soil, surviving even the severest of winters. Dandelions are most numerous in full sunlight. In the early spring, new sprouts emerge from the taproot. As the yellow flowers mature and ripen, they form white puff balls containing seeds. The wind carries the seeds for miles. The tops die back in late fall, and the taproot overwinters to start the cycle again in the spring. Dandelions prefer wet soil and are often a sign of overwatering.

Solution: Treat the lawn with ORTHO Weed-B-Gon Lawn Weed Killer or GreenSweep Weed & Feed (20–0–0). Use ORTHO Weed-B-Gon Jet Weeder for spot-treatment. For best results, make two applications, first in the early summer and again in the early fall. Do not water or mow for 2 days afterward. Feed the lawn adequately to keep it dense. Mow frequently enough to keep the flowers from becoming seed heads. Hand-digging is impractical because any piece of root left in the soil will sprout into new plants.

BROADLEAF WEEDS (continued)

Sheep sorrel

Sheep sorrel, common in acid, infertile soils.

Problem: Arrow-shaped leaves, 1 to 4 inches long, with two lobes at the base of each leaf, form a dense rosette. Erect, upright stems grow 4 to 14 inches tall. Two types of flowers appear in midspring; one is reddish green, the other yellowish green.

Analysis: Sheep sorrel (*Rumex acetosella*) a cool-season perennial, is also called *red sorrel* or *sourgrass* because of its sour taste. It grows in dry, sterile, sandy or gravelly soil and is usually an indication of acid soil or low nitrogen fertility, although it will survive in neutral or slightly alkaline soil. Sheep sorrel reproduces by seeds and red underground root stalks. The root system is shallow but extensive and is not easily removed.

Solution: In spring or fall, treat with ORTHO Weed-B-Gon Lawn Weed Killer. Don't mow for 5 days before or 2 days after treating. Sheep sorrel is difficult to control, so several treatments may be necessary. To discourage sheep sorrel, test the soil pH and correct to between 6.0 and 7.0 if necessary. Improve the soil fertility.

Mallow

Mallow, sometimes mistaken for ground ivy.

Problem: A weed with hairy stems, 4 to 12 inches long, spreads over the lawn. The stem tips turn upward. Round, heart-shaped, hairy leaves, ½ to 3 inches wide and slightly lobed along the edges, are attached to the stems by a long leafstalk. White to lilac flowers, 2½ inches in diameter with five petals, bloom singly or in clusters at the leaf and stem junction. Mallow is often mistaken for ground ivy, but the spreading branches do not root when they contact soil as the branches of ground ivy do.

Analysis: Mallow (*Malva* species), also called *cheeseweed*, is found throughout the country in lawns, in fields, and along roadways. It is an annual, or sometimes a biennial, and reproduces by seeds that remain dormant in the soil over winter and germinate in the spring. Seeds may remain fertile in the soil for many years. Mallow has a straight, nearly white taproot that is difficult to pull from the soil. Mallow is commonly found in poorly managed lawns and in soils high in manure content.

Solution: Treat the lawn with ORTHO Weed-B-Gon Lawn Weed Killer or ORTHO Weed-B-Gon Jet Weeder from midspring to early summer. Don't water for 5 days before treating or 2 days after. Maintain a thick, healthy lawn.

Field bindweed

Field bindweed, a troublesome perennial.

Problem: A plant with long twining stems grows across the lawn. The leaves are arrowhead-shaped and up to 2 inches long. White to pink funnel-shaped flowers, about 1 inch across, bloom and produce seed from May to September. The seeds may remain dormant in the soil for many years before sprouting.

Analysis: Field bindweed (*Convolvulus arvensis*), a deep-rooted perennial weed, also known as *wild morning glory*, is found throughout most of the United States in lawns, gardens, and fields and along roadways. It is one of the most troublesome and difficult weeds to eliminate because of its extensive root system. The roots may grow 15 to 20 feet deep. Roots or pieces of roots left behind from hand-pulling or spading easily resprout. Field bindweed, which reproduces by seeds and roots, twines and climbs over shrubs and fences and up into trees. It prefers rich, sandy, or gravelly soil but will grow in almost any garden soil.

Solution: Treat plants from late spring through early summer or from early to late fall with ORTHO Weed-B-Gon Lawn Weed Killer, ORTHO Weed-B-Gon Jet Weeder, or ORTHO Brush-B-Gon Poison Ivy, Poison Oak & Brush Killer. Because of the deep roots, repeated treatments may be necessary. Treat again whenever new growth appears. Or spot-treat the lawn with ORTHO Kleeraway Grass & Weed Killer, and reseed.

MISCELLANEOUS

Mushrooms

Mushrooms.

Problem: Mushrooms sprout up in the lawn after wet weather. They may be growing in circles of dark green grass. When the weather gets colder or the soil dries out, they disappear.

Analysis: Mushrooms, also called *toadstools* or *puffballs*, live on organic matter buried in the soil. The mushroom is the aboveground fruiting or reproductive structure of a fungus that lives on and helps to decay the organic matter. The organic matter may include buried logs, lumber, roots, or stumps. Most mushrooms do not damage the lawn but are unsightly. Mushrooms growing in circles of dark green grass, called *fairy rings*, may make the soil impervious to water, injuring the grass.

Solution: There is no practical or permanent way to eliminate mushrooms. When buried wood is completely decayed, the mushrooms will disappear. The easiest and most practical solution, although it is only temporary, is to break up the mushrooms with a rake or lawn mower.

Algae

Algae, a problem in shady, wet areas.

Problem: A green to black slimy scum covers bare soil and crowns of grass plants. When dry, it becomes crusty, cracks, and peels easily.

Analysis: Algae (*Symploca* and *Oscillatoria* species) are freshwater plants that invade shady, wet areas of the lawn. They injure grass by smothering or shading it as they grow over the crowns of the plants. Invaded areas become slippery. Algae live in compacted soil and soil that is high in nitrogen and organic matter. They need constantly or frequently wet conditions to survive. Organic fertilizers encourage algae, especially in the cool seasons. Algae may be carried from place to place by animals, equipment, people, and birds. Water taken from ponds, lakes, and streams and used for irrigation usually contains algae.

Solution: Patches of algae may be sprayed with a fungicide containing *maneb*, *mancozeb*, or wettable sulfur two times, 1 month apart, in early spring. This is only a temporary solution. Algae will soon return if the conditions promoting them are not corrected. Correct soil compaction and improve drainage by tilling and adding organic matter before planting, or by aerating lawns with the type of aerator that removes cores of soil. Prune nearby trees to reduce shade. Avoid high-nitrogen fertilizers in late fall and winter. Maintain a healthy, vigorous lawn.

Moss

Moss, most common in shade and acid soil.

Problem: Green, velvety, low-growing plants cover bare soil in shady areas of the lawn.

Analysis: Moss invades thin or bare areas of the lawn. It does not grow in a vigorous lawn. Moss is encouraged by poor fertility, poor drainage, compacted soil, shade, and high acidity. Moss plants sprout from spores and fill in bare or thin areas.

Solution: Shortly after mowing, apply Scotts Moss Control Granules while grass is moist. Moss may also be removed by hand or power raking. Reduce shade by pruning nearby trees. Correct soil compaction and improve drainage by tilling and adding organic matter before planting, or by aerating lawns with the type of aerator that removes cores of soil. Test the soil pH, and correct it if necessary. Maintain a healthy, vigorous lawn.

Algerian ivy is an excellent, vigorous ground cover for slopes in mild Pacific Coast areas.

Ground covers shade out weeds and add both texture and color to small or large areas. They can be remarkably drought-resistant. But ground covers, like other lawn and garden plants, require nurturing to stay lush.

CONSIDERING THE SOIL

The first step in establishing ground cover is to ensure that the soil can provide plants with the aeration, nutrients, and pH they need.

Dense soil: Some soils, such as clay, tend to be naturally dense. Clay soil cracks and becomes quite hard when dry. Water may run off rather than sinking in, creating droughtlike conditions.

To tell if you have dense soil, dig a hole 2 feet deep. Fill the hole with water. If the water level drops less than $\frac{1}{10}$ inch per hour, your soil is dense and has a drainage problem.

Correcting dense soil is hard work but necessary. The most permanent solution is to dig the entire planting area to a depth of 6 inches and work in such soil amendments as gypsum and organic matter. This is also a good time to correct soil pH (see the Inappropriate pH section).

Compacted soil: Some soils are naturally compact. Others are compacted by construction work or foot traffic. Whatever the cause, the result can be hardpan and struggling ground cover unless the gardener corrects the situation.

Hardpan: A planting hole in hardpan, even when backfilled with organic amendments, collects water, creating a basin where roots rot. If your ground cover site contains hardpan, you must correct the soil before planting. In mild cases, correction consists of using a soil auger to bore deep holes in the soil. The holes allow air and water to flow downward and ease the way for growing roots. If the planting area is small but the soil is quite compact, use a hand-operated aerating tool to make 3-inch-deep holes about 3 inches apart. The practical approach to breaking up hardpan in large areas is deep plowing—you may have to hire a contractor to do the job. Working gypsum into the soil can also be effective. Talk to a knowledgeable nursery professional or a landscape gardener about the best approach to your situation.

If you have an area subject to constant foot traffic, consider installing paving stones or building a paved pathway. Plant shrubbery to prevent pedestrians from taking shortcuts.

Nutrient shortage: Slow growth is often a sign of a nutrient shortage. Other symptoms are yellowing or pale green leaves that remain small. To correct the soil, apply a general-purpose fertilizer regularly according to label instructions. If leaves turn pale but leaf veins remain normal, supplemental iron may be needed.

In some cases, the soil contains sufficient nutrients but its pH keeps plants from absorbing them. Read the next section to determine if your planting site is too acidic or alkaline.

Inappropriate pH: Kits for testing soil pH are available at nurseries. The kits are easy to use and supply valuable information.
- **Acidic soil:** Plants unsuited to acidic soils (see page 66) grow slowly. Their leaves turn yellow or pale green. To improve acidic soils, apply a ground dolomitic limestone additive formulated especially for gardens.
- **Alkaline soil:** Plants grown in overly alkaline soil develop yellow areas between the veins on new leaves. This yellowing may be due to the lack of iron and manganese, which alkaline soil keeps plants from absorbing. To remedy the problem, make soils more acidic by applying aluminum sulfate, ferrous sulfate, or other sulfur. Use a fertilizer that creates an acidic reaction, such as a fertilizer formulated for azaleas.

Amend ground cover beds with organic matter and fertilizer before planting.

When planting many ground cover plants over a large area, lay out the bed in a staggered, or triangular, pattern.

Supplementary soil: If you decide to add topsoil to your planting site, make sure the new soil is free of weeds and weed seeds. Soil delivered by the truckload from construction sites is particularly suspect. Buy supplementary topsoil from a reputable nursery, and make sure you specify a weed-free product. If it saves you from weed control, clean topsoil is worth the extra cost.

CONSIDERING GROUND COVERS

In selecting a ground cover, consider its growth habit, its ability to crowd out weeds, and its requirements for sun or shade and water.

Invasive ground covers: A ground cover is often selected for its rapid growth habit. But this can turn the ground cover into a nuisance that moves into areas reserved for other plants. Any ground cover marked "Rapid growth, very hardy" might be invasive given the right environment. Included in the possibly invasive list are honeysuckle, some mints, goutweed, sweet woodruff, crown vetch, Indian mock strawberry, gill ivy, Aaron's beard, and dwarf bamboo. If a ground cover becomes too energetic and regular cutting does not deter it, treat it as a weed and kill it by applying a systemic herbicide.

Weed-shading ground covers: Some ground covers eventually grow thick enough to shade the ground below them. This decreases weed germination. Given the best environment, essentially weed-free ground covers include ajuga, chamomile, Irish moss, snow-in-summer, wild strawberry, cinquefoil, lamb's-ears, and vinca.

Sun and shade requirements: A shade-loving ground cover, such as Irish moss, will not grow well where sunlight plays on it most of the day. A sun-loving ground cover, such as chamomile, may die out in shady areas.

Study the planting area before rushing out to buy plants. You must know what the environment can support. If you live in or near an established area, one way to find out what will grow is to walk past thriving gardens in your neighborhood. If a neighbor's planting area faces the same direction yours does, given the same care, the same ground cover will probably do well for you. Do not limit yourself by choosing a ground cover that is already established nearby, however. If you see a thriving ground cover, research its growing requirements. Then find other plants that thrive in similar conditions.

Signs of inadequate sunlight include poor growth, leggy growth, leaf drop, unusually dark green leaves, and insect attack. Increase the sunlight the ground cover receives by trimming adjacent shrubbery or tree limbs. If providing more sunlight is not possible, carefully transfer the affected plants to a more suitable area. Replant with a shade-tolerant species.

Signs of too much sun include faded leaves. In severe situations, leaves may turn yellowish white. Growth slows. Too much sun breaks down plant tissue. Chlorophyll (green leaf pigment), which is necessary for leaf functioning, disappears. Wind and drought make the problem more severe. Place affected plants in a shadier location with sufficient water. Replace them with a more sun-tolerant species.

Resilient ground covers: If your planting site is subject to heavy foot traffic, choose a ground cover that can tolerate it. The list of such plants includes dichondra, chamomile, Korean grass, Irish moss, lippia, and mazus.

Drought- and moisture-tolerant ground covers: If you live in a drought area, consider using a drought-tolerant ground cover.

The selection includes woolly yarrow, coyotebrush, ground cover ceanothus, low-growing California buckwheat, and lippia.

Quite a few ground covers can prosper in damp soil, as long as you provide adequate drainage. These include mint, ajuga, wild ginger, bunchberry, and pachysandra.

PREVENTING WEEDS

Unfortunately, prime conditions for ground cover also encourage weeds. Hand-picking weeds from ground cover is difficult; getting a firm grip on a weed usually means tearing out some ground cover, too. Practical solutions include plastic or fabric sheets that block weeds, weed blocks, and herbicides.

Weed blocks: If it is thick enough to block sunlight, a black plastic sheet spread over the planting area can effectively prevent weeds. Holes cut in the plastic allow the ground cover to grow. In addition to preventing weeds, weed blocks conserve soil moisture. Weed block fabric, a fairly new innovation, is better than black plastic because it allows water, air, and nutrients to pass into the soil but does not allow weed growth.

Weed blocks present a few disadvantages, however. Plastic cracks with age and needs to be replaced. Weeds may sprout through cracked plastic or through the holes intended for ground cover. Though a weed block is an excellent defense against weeds, you may need to use herbicides as well.

Herbicides: One way herbicides are classified is according to the time they are applied.

■ **Preemergent herbicides:** Weed killers of this type are applied before weeds sprout; these herbicides prevent weed germination while allowing desirable plants to grow normally. Preemergent herbicides, which are usually applied in granular form, permeate the top inch or so of soil. When applied under a weed block, these products contribute to highly effective weed control. Check the label to ensure that the product is safe for your ground cover.

Preemergent herbicides do not kill existing weeds, which must be treated separately. The products lose some of their effectiveness if the soil is used as a path or disturbed by feeding birds, digging animals, or cultivation.

■ **Postemergent herbicides:** If weeds emerge, consider postemergent herbicides. Some are systemic, which means they work throughout the plant (including the root). Some can be applied without injury to specific ground covers. Others must be applied to weeds only; they will kill your desirable plants as well as weeds.

Before selecting a postemergent herbicide, read the label instructions carefully. Find out if your ground cover is susceptible to the product and how you need to apply it to protect garden plants. In lush ground cover you may have to use a narrow paintbrush to treat weeds. Use postemergent herbicides on windless days only, when chemicals cannot drift on the breeze. Shield susceptible ground cover to protect it from inadvertent exposure.

WATERING GROUND COVERS

Soil that is too dry or soggy may be due to factors other than too little or too much rain. Soil permeability and irrigation systems affect the amount of water plants actually absorb.

Drought conditions: During droughts, plants begin to yellow. Leaf edges may turn brown. Bare spots may appear. If you have been watering adequately, drought can still

occur because compacted soil is prohibiting water penetration. Rapid watering causes runoff rather than allowing moisture to sink in. Hand-watering is often the culprit— which is not to say that sprinkler systems are perfect. Sprinkler heads may fall out of adjustment and leave some areas dry. To test the coverage, put a few small cans throughout the watering area. All areas should get an equal amount of water,

appropriate to the type of ground cover. If the cans contain different amounts, correct the system.

Overwatered ground covers: Too much water causes root rot. Ground cover may die in spots or fail to thrive. Lower leaves turn yellow, then upper leaves. If too much rain is the cause—not excess watering—install drain tile.

By its third season of growth the pachysandra shown being planted on page 64 has completely filled in to become a lush, solid bed of green.

DISCOLORED LEAVES

Lack of nutrients

Underfertilized Ajuga.

Problem: Leaves are smaller than usual and turn pale green to yellow. Plants grow very little.

Analysis: Fertilizer supplies the nutrients plants need to stay vigorous and healthy; adequate fertilizing may lessen insect and disease problems. The three essential nutrients that plants require in the largest amounts are nitrogen (N), phosphorus (P), and potassium (K). *Nitrogen* gives the leaves their green color and encourages rapid growth. Without nitrogen, plants turn pale green to yellow. *Phosphorus* encourages flowering and fruiting and helps build a strong root system. *Potassium* increases the plant's resistance to disease and aids in overall growth. Most garden fertilizers contain all three nutrients. On the bag are three numbers, called the N–P–K ratio or fertilizer grade, which state the nutrient content. For example, a fertilizer bag reading 5–10–10 has, by weight, 5 percent nitrogen, 10 percent phosphate (phosphorus), and 10 percent potash (potassium). Many fertilizers also supply trace, or minor, nutrients that plants need in minute amounts, which may be lacking in the soil.

Solution: Fertilize your ground cover at regular intervals with Scotts All-Purpose Plant Food. Follow label instructions for amounts and timing of fertilizing. Water well after fertilizing to dilute the fertilizer and wash it into the soil.

Iron deficiency

Iron deficiency in Pachysandra.

Problem: Leaves turn pale green or yellow. The newest leaves (those at the tips of the stems) are most severely affected. Except in extreme cases, the veins of affected leaves remain green. In extreme cases, the newest leaves are small and all-white or yellow. Older leaves may remain green.

Analysis: Plants frequently suffer from deficiencies of iron and other minor nutrients such as manganese and zinc, elements essential to normal plant growth and development. Deficiencies can occur when one or more of these elements is depleted in the soil. Often minor nutrients are present in the soil, but alkaline (pH higher than 7.0) or wet soil conditions causes them to form compounds that the plant cannot use. An alkaline condition can result from overliming or from lime leached from concrete or brick mortar. Regions where soil is derived from limestone or where rainfall is low usually have alkaline soils.

Solution: Spray the foliage with a chelated iron fertilizer, and apply the fertilizer to the soil around the plants to correct the deficiency of minor nutrients. Check the soil pH. Lower the pH of alkaline soil by treating it with ferrous sulfate or aluminum sulfate and watering it in well. Maintain an acid pH by fertilizing with Miracid Plant Food.

Acid soil

Leaf yellowing on vinca caused by acid soil.

Problem: Plants grow slowly and leaves turn pale green to yellow. Plants don't improve much when fertilized. A soil test shows a pH of below 6.0.

Analysis: Soils with a pH of less than 6.0 are common in areas of heavy rainfall. Heavy rains leach lime from the soil, making it more acid. The amounts and types of nutrients available to plants are limited in acid soils. Below a pH of 5.5, the availability of nitrogen, phosphorus, potassium, calcium, magnesium, and other nutrients decreases. These nutrients are essential for healthy plants. Plants vary in their tolerance for acid soils. Most plants grow best with a soil pH between 6.0 and 7.5.

Solution: Test your soil pH with an inexpensive test kit available in garden centers. Many county extension offices also test soil pH. To make a soil less acid, apply lime, following the directions given with the soil testing kit. Soil acidity corrections are temporary. Add lime to your soil every 1 to 2 years if you live in an acid-soil area.

Scorch

Winter burn on English ivy.

Powdery mildew

Powdery mildew on Euonymus.

Greenhouse whiteflies

Greenhouse whiteflies (20× life size).

Problem: Leaf tips and edges are brown and dead. Leaves may fall.

Analysis: Leaf scorch may be caused by any of several conditions.
1. Heat scorch: In hot weather, water evaporates rapidly from the leaves. If the roots can't absorb and convey water fast enough to replenish this loss, the leaves turn brown and wither. This condition often occurs when shade-loving plants receive too much sun.
2. Winter burn occurs on plants growing in full sun. On a clear winter day, the sun heats the leaf surface, increasing the need for water. If the ground is frozen, or if it has been a dry fall and winter, the roots can't get enough water.
3. Salt injury results from excess salts in the soil. These salts can come from irrigation water, salts used to melt snow and ice, or fertilizers. This condition is worse in poorly drained soils, where salts can't easily be leached.

Solution: Follow these guidelines to prevent scorching. The numbered solutions below correspond to the numbered items in the analysis.
1. Keep ground covers well watered during hot weather. For a list of sun- and shade-tolerant ground covers, see page 350.
2. To prevent winter burn, be sure the soil is moist before the ground freezes. Provide shade during clear, cold weather.
3. Leach the salts from the soil with very heavy waterings. If your irrigation water is salty, leach regularly to keep salt from accumulating in the soil. When you fertilize, apply only the amounts recommended on the label, and water thoroughly afterward.

Problem: Grayish white powdery patches partially or entirely cover leaves and stems, primarily the upper surfaces of leaves. Leaves die and may drop off.

Analysis: Powdery mildew, a common plant disease, is caused by several fungi that thrive in both humid and dry weather. The powdery patches consist of fungal strands and spores. The spores are spread by the wind to healthy plants. The fungus saps plant nutrients, causing yellowing and sometimes death of the leaf. A severe infection may kill the plant. Because powdery mildew attacks many different kinds of plants, the fungus from a diseased plant may infect other types of plants in the garden. See page 343 for a list of powdery mildews and the plants they attack. Under favorable conditions, powdery mildew can spread through a ground cover in a matter of days or weeks.

Solution: Spray with ORTHO RosePride Funginex Rose & Shrub Disease Control or ORTHO RosePride Orthenex Insect & Disease Control. Make sure that your plant is listed on the product label. These fungicides do not kill the fungus on leaves that are already diseased. They do, however, protect healthy leaves by killing the mildew spores as they germinate. Follow label directions regarding frequency of application. Remove infected leaves and debris from the garden.

Problem: Tiny, white, winged insects $\frac{1}{16}$ inch long feed mainly on the undersides of leaves.

Nonflying, scalelike larvae covered with white, waxy powder may also be present on the undersides of leaves. When the plant is touched, insects flutter rapidly around it. Leaves may be mottled and yellow. In warm-winter areas, black mold may cover the leaves.

Analysis: Greenhouse whitefly (*Trialeurodes vaporariorum*) is a common insect pest of many garden and greenhouse plants. The adult lays eggs on the undersides of leaves. The larvae are the size of a pinhead and look quite different from the adult. The larvae are flat, oval shaped, and semitransparent, with white, waxy filaments radiating from the body. They feed for about a month before changing into adults. Both larval and adult forms suck sap from the leaves. The larvae are more damaging because they feed more heavily. In warm-winter areas, the insect can be active year-round, with eggs, larvae, and adults present at the same time. The whitefly does not survive freezing winters. Spring reinfestations in freezing-winter areas come from migrating whiteflies and infested plants placed in the garden.

Solution: Control whiteflies by spraying with ORTHO Orthene Systemic Insect Control or with an insecticide containing *diazinon or malathion*. Make sure that your plant is listed on the product label. Treat every 7 to 10 days as necessary. Spray the foliage thoroughly, covering both the upper and lower surfaces of the leaves. Whiteflies may also be partially controlled with yellow sticky traps.

INSECTS *(continued)*

Spider mites

Spider mite (50× life size).

Problem: Leaves are stippled, yellowing, and dirty, and may dry out and drop. There may be webbing over flower buds, between leaves, or on the lower surfaces of leaves. To determine if a plant is infested with mites, examine the bottoms of the leaves with a hand lens. Or hold a sheet of white paper underneath an affected branch and tap the branch sharply. Minute specks the size of pepper grains will drop to the paper and begin to crawl. The pests are easily seen against the white background.

Analysis: Spider mites, related to spiders, are major pests of many garden and greenhouse plants. They cause damage by sucking sap from the undersides of leaves. As a result of their feeding, the plant's green leaf pigment disappears, producing the stippled appearance. Spider mite webbing traps cast-off skins and debris, making the plant messy. Some spider mites are active throughout the growing season, but most thrive in hot, dry weather (70°F and up). By midsummer, they may have built up to tremendous numbers. Other spider mites are most prolific in cooler weather. They feed and reproduce primarily during spring and, in some cases, fall. By the onset of hot weather (70°F and up), these mites have caused their maximum damage.

Solution: Mites can be difficult to control because their eggs are resistant to most chemical sprays. Treat infested plants with ORTHO Isotox Insect Killer or ORTHO Diazinon Ultra Insect Spray when mites first appear. Make sure your plant is listed on the product label. Repeat the treatment two more times 7 to 10 days apart. Continue treatments if mites reappear.

Aphids

Aphids on ivy leaf (2× life size).

Problem: Leaves are curled, distorted, and yellow. A shiny, sticky substance may coat the leaves. Tiny (⅛-inch) pale green to black soft-bodied insects cluster under leaves and stems. Ants may be present. If infestation continues, plants may become stunted.

Analysis: Aphids (also called *plant lice*) do little damage in small numbers. They are extremely prolific, however, and populations can rapidly build up to damaging numbers during the growing season. Damage occurs when the aphid sucks the juices from the leaves of the ground cover. The aphid is unable to digest fully all the sugar in the plant sap, and it excretes the excess in a fluid called *honeydew*. Ants feed on honeydew and are often present where there is aphid infestation. Large numbers of aphids may cause little damage to a plant, or just a few aphids on a plant may cause severe distortion and stunting. Certain aphids also transmit plant diseases.

Solution: Aphids are usually easy to control if they are not protected by tightly curled leaves, galls, or cottony material. Spray with ORTHO Orthene Systemic Insect Control, ORTHO Malathion 50 Plus Insect Spray, or an insecticidal soap as soon as the insects appear. Repeat the spray if the plant becomes reinfested. Make sure that your plant is listed on the product label. Aphids may continually reinfest the garden from other plants neaby. Inspect your plants regularly for aphids.

Snails and slugs

Snail and slug damage to ivy.

Problem: Stems and leaves may be sheared off and eaten. Silvery trails wind around on the plants and soil nearby. Snails or slugs may be seen moving around or feeding on the plants, especially at night. Inspect the garden for them at night by flashlight.

Analysis: Snails and slugs are mollusks and are related to clams, oysters, and other shellfish. They feed on a wide variety of garden plants. Like other mollusks, snails and slugs need to be moist all the time. For this reason, they avoid direct sun and dry spots and hide during the day in damp places, such as under flowerpots or in thick ground cover. They emerge at night or on cloudy days to feed. Snails and slugs are similar, except that the snail has a hard shell into which it withdraws when disturbed. Slugs lay white eggs encased in a slimy mass in protected places. Snails bury their eggs in the soil, also in a slimy mass. The young look like miniature versions of their parents.

Solution: Scatter ORTHO Bug-Geta Snail & Slug Killer in bands around the areas you wish to protect. Also scatter the bait in areas where snails or slugs might be hiding, such as in dense ground cover, weedy areas, compost piles, or pot storage areas. Before spreading the bait, wet down the areas to be treated to encourage snail and slug activity that night. Repeat the application every 2 weeks for as long as snails and slugs are active. Remove scrap lumber and other debris near the ground cover that create cool, damp breeding places.

AJUGA

Crown rot

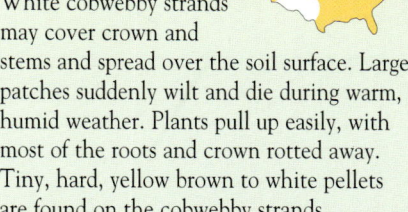

Crown rot.

Problem: Lower leaves turn yellow. White cobwebby strands may cover crown and stems and spread over the soil surface. Large patches suddenly wilt and die during warm, humid weather. Plants pull up easily, with most of the roots and crown rotted away. Tiny, hard, yellow brown to white pellets are found on the cobwebby strands.

Analysis: Crown rot is caused by a fungus (*Sclerotium* species) that occurs mostly in wet, poorly drained soil. The fungus can be a serious problem on other herbaceous perennials besides ajuga. The fungus enters the plant through the roots and crown and spreads into the stem, rotting it and causing the plant to wilt and die. In mild infections, new growth sometimes sprouts from buds that have not been killed. The tiny fungal pellets found in the soil survive winters and other unfavorable conditions to infect other plants. The fungus most frequently enters the garden originally in infested soil or plants and lives for years in the soil.

Solution: Remove and destroy all infected plants. Do not replant in the area until the fungus-infested soil has been either removed or drenched with a fungicide containing PCNB. To prevent crown rot, plant ajuga in well-drained soil.

DICHONDRA

Flea beetles

Flea beetle damage. Inset: Flea beetle (life size).

Problem: From May to October, dichondra leaves turn brown, first along the edges of the lawn, then progressing toward the center. Small round holes like shot holes are chewed in the upper surfaces of the leaves. To determine if this is an insect problem, spread a white handkerchief on the border between a damaged area and a healthy area. Black insects, about $\frac{1}{25}$ inch long, hop onto the white cloth, where they are easily seen.

Analysis: The adult flea beetle is the most damaging pest of dichondra but does not feed on grass lawns. Although it hops like a flea, it is a true beetle. The adult spends the winter in garden trash and weeds, emerging with the warm spring weather to begin feeding and laying eggs. Damage is spotty at first because the beetles are so small. In a short time, however, they can destroy a lawn.

Solution: Spray infested dichondra with ORTHO Diazinon Ultra Insect Spray or ORTHO Dursban® Lawn Insect Spray at the first sign of damage. Water the area first; then apply the insecticide according to the directions on the label. Repeat once a month throughout the growing season. A healthy planting of dichondra is more resistant to attack from flea beetles than an unhealthy one and recovers more quickly if it does become infested. Fertilize with Miracle-Gro Plant Food once a month from March to September.

FRAGARIA
(Wild Strawberry)

Leaf spot

Leaf spot is caused by several different fungi.

Problem: Spots and blotches appear on the leaves. The spots may be yellow, red, tan, gray, or brown and range in size from barely visible to $\frac{1}{4}$ inch in diameter. Several spots may join to form blotches. Leaves often turn yellow, die, and fall off. Leaf spotting is most severe in warm, humid weather. In damp conditions, a fine gray mold sometimes covers the infected leaf tissue.

Analysis: Several different fungi cause leaf spots on wild strawberry. Some of these eventually kill the plant. Others merely spot the leaves and are unsightly but not harmful. The fungi are spread from plant to plant by splashing water, wind, and contaminated tools. They survive the winter on diseased plant debris not cleaned out of the garden. Most of the leaf spot fungi do their greatest damage in humid conditions and mild weather (temperatures between 50° and 85°F).

Solution: Spray the infected planting with a fungicide containing mancozeb every 3 to 10 days. Because leaf spot fungi are most active during warm, humid weather, spray more frequently during those weather conditions. These fungicides protect the new, healthy foliage but do not kill the fungus on leaves that are already infected. Remove infected leaves and debris from the garden. Avoid overhead watering, particularly late in the day.

HEDERA (Ivy)

Scales

Brown soft scale (2× life size).

Problem: Raised tan to reddish black crusty or waxy bumps cover the stems or the undersides of leaves. The bumps can be picked off; the undersides are usually soft. Leaves may turn yellow and fall. In some cases, a sticky substance coats the leaves. A black, sooty mold often grows on the sticky material. Ants may be present.

Analysis: Scales are insects that spend the winter on the trunk and twigs of the plants. They lay eggs in spring and in early summer (late spring in the South); the young scales, called *crawlers*, settle on leaves and twigs. The small (1/10-inch), soft-bodied young feed by inserting their mouthparts and sucking sap from the plant. The legs usually atrophy, and a hard crusty or waxy shell develops over the body. Mature female scales lay their eggs underneath their shell. Some species of scales that infest English ivy are unable to digest all the sugar in the plant sap, and they excrete the excess in a fluid called *honeydew*. A black, sooty mold fungus may develop on the honeydew. Ants feed on the sticky substance and are often present where scales cluster. An uncontrolled infestation of scales may kill the vines after 1 or 2 seasons.

Solution: Spray with an insecticide containing *malathion* in early summer (late spring in the South) when the young are active and in the crawler stage. Early the following spring, before new growth begins, spray the vines with a dormant oil spray to control overwintering insects.

MESEMBRYANTHEMUM (Ice Plant)

Root rot

Root rot. Inset: Close-up view of roots.

Problem: Leaves turn yellow, starting with the older, lower leaves and progressing to the younger ones. Plants grow very little. When they are pulled up, the roots appear black, soft, and rotted. The soil is very moist.

Analysis: Root rot is caused by several species of fungi that are present in moist soils. These fungi normally do little damage, but they can cause root rot in waterlogged soils. Waterlogged soil may result from overwatering or from poor drainage. Infection causes the roots to decay, resulting in wilting, yellowing leaves and the death of the plant.

Solution: Ice plant is drought-tolerant and needs little watering. Remove and discard severely infected plants. Allow the soil around the plants to dry out between waterings until the spread of the disease is halted, using the following steps. Spray leaves with an antitranspirant (available in many nurseries) and provide shade to reduce drought stress while drying the soil. Further reduce drought stress by placing clear plastic tents over the plants to maintain high humidity. Use tents only in the shade. Begin watering again when the plant shows signs of drought stress, such as heavy wilting or yellowing and dropping of leaves. To avoid future root rot problems, improve the drainage by tilling the soil and adding organic matter before replanting.

PACHYSANDRA (Japanese Spurge)

Leaf and stem blight

Leaf and stem blight; irregular blotches.

Problem: Irregular brown and black blotches appear on the leaves and spread to cover most of the leaf. Leaves are soft and water-soaked. Areas on the stem turn black, soft, and sunken. In wet weather, pinkish spore masses appear along the stem and on leaf spots.

Analysis: Leaf and stem blight is caused by a fungus (*Volutella pachysandrae*). Plants are more susceptible to attack from this fungus if they are weak from spider mites or winter injury, if they are too crowded, if they are planted in the full sun or in waterlogged soil, or if the area stays too moist because of tree leaves falling into the bed. The fungus is most active in rainy, cool spring weather and spreads from plant to plant in splashing water and on contaminated tools. The spores survive the winter on infected stems and leaves left in the garden. If not controlled, this blight can kill an entire planting of pachysandra in 1 or 2 seasons.

Solution: Pull out and discard all diseased plants. Although this may seem like an endless job, doing so will give you much better control of this stubborn plant disease. Then spray the cleaned bed with ORTHO Multi-Purpose Fungicide Daconil 2787® Plant Disease Control at intervals of 1 week, beginning when the disease is discovered or when new growth starts in the spring. Thin and shear the beds periodically to improve air circulation. Rake out leaves in the fall to reduce the chance of the disease occurring in the spring. Keep pachysandra healthy by controlling scale and reducing winter burn. Do not overwater.

VINCA (Periwinkle)

Root and stem rot

Root and stem rot on Vinca minor.

Problem: Shoot tips wilt and die. There are no black specks on the affected stems. Plants 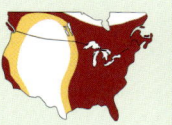 pull up easily, with most of the roots and lower stem soft and rotted away. Individual plants are affected first, and within several weeks, entire clumps wilt and die.

Analysis: Root and stem rot is caused by a fungus (*Pellicularia filamentosa*). It is the most serious disease affecting periwinkle. It occurs mostly in heavy, poorly drained soil and during periods of wet weather. It can be found throughout the growing season in most periwinkle plantings. The fungus enters the plant through the roots and crown and rots the cells, causing the plant to wilt and die. The fungus persists indefinitely in the soil and spreads from plant to plant on contaminated tools and in splashing water.

Solution: Remove and destroy badly infected plants. Allow the soil to dry between waterings until the spread of the disease is halted, using the following steps. Spray leaves with an antitranspirant (available in many nurseries) and provide shade to reduce drought stress while drying the soil. Further reduce drought stress by placing clear plastic tents over the plants to maintain high humidity. Use tents only in the shade. Begin watering again when the plant shows signs of drought stress, such as heavy wilting or yellowing and dropping of leaves. If the area is replanted, improve the drainage by tilling the soil and adding organic matter. To prevent recurrence of the disease, allow the soil to dry between waterings until it is barely moist.

Gray mold

Gray mold, a common fungal disease.

Problem: Brown or black spots appear on the edges of the leaves and spread inward,  sometimes covering the entire leaf. Flowers may be discolored or spotted. As the disease progresses, a fuzzy brown or grayish mold forms on the infected tissue during cool, wet weather.

Analysis: Gray mold (also called *botrytis blight*) is a widespread plant disease caused by a fungus (*Botrytis cinerea*) that is found on most dead plant tissue. The fungus initially attacks foliage and flowers that are weak or dead, causing spotting and mold. The fuzzy mold that develops is composed of fungus strands and millions of microscopic spores. Once gray mold has become established on plant debris and weak or dying leaves and flowers, it can invade healthy plant tissue. The fungus is spread by splashing water or by infected pieces of plant tissue contacting healthy tissue. Cool temperatures and high humidity favor gray mold growth. Rain and overhead watering enhance the spread of the fungus. Infection is more of a problem in spring and fall, when temperatures are lower. In mild-winter areas where freezing is rare, gray mold can be a year-round problem.

Solution: Remove and discard all fading flowers and diseased leaves. Treat plants with a fungicide containing *captan*. For best control, add a spreader-sticker to the spray.

Canker and dieback

Dieback on Vinca minor.

Problem: Shoot tips wilt, turn brown, and die. The infection progresses down the  stem to the soil surface, killing the plant. Affected stems turn black. Tiny black specks are often seen on the diseased stems. Dark brown spots sometimes develop on leaves, which then die and fall. Individual plants are affected first, but within several weeks entire clumps wilt and die.

Analysis: Canker and dieback are caused by two fungi (*Phomopsis livella* and *Phoma exigua*) that attack periwinkle in the spring soon after the new growth begins. This disease is most prevalent during rainy seasons. The fungal spores spread from plant to plant on splashing water and contaminated tools. This disease can be devastating, killing an entire planting in a few weeks.

Solution: Spray infected plants every 7 days with a fungicide containing copper or *mancozeb*. Remove badly infected plants. Avoid overhead watering. To reduce the chances of new infection, water early in the day, rather than late afternoon or evening, so the leaves have time to dry out before nightfall.

Lavish flower borders are easy to grow when you choose plants appropriate to your site. In this sunny northeastern garden, daisies, daylilies, Asiatic lilies, delphiniums, feverfew, and baby's breath transform a modest house into a lavish wonderland.

Flowers can turn a drab yard into a showplace. Indoors, a bouquet of cut flowers—whether a formal arrangement in a porcelain vase or a bunch of black-eyed Susans in a jar—seems to bring sunlight and a touch of elegance into the house. Home gardeners turn to annuals, perennials, biennials, and bulbs to provide a supply of blooms. Each type of plant requires a different approach in terms of planning, planting, and care.

PICKING PLANTS TO SUIT THE JOB

Buying plants calls for research. What climate does each species need? What requirements does it have for sunlight, drainage, or fertilizer? (Before purchasing plants and fertilizer, read the section called Assessing and Improving Soil, which appears later in this chapter.) Buying plants that flower takes special attention in terms of learning the blooming patterns and life cycle of each species.

Climate: Passion flowers are from South America, a climate that stays warm most of the year. Logically, a warm climate is what this showy vine needs to be at its best. Peonies hail from China and range into northern Siberia; hot climates do not always promote their flowering. Agapanthus and pelargoniums originally hail from South Africa, so while they will take some drought, they are also accustomed to warmth. Though a plant may be available for purchase in your area, it may not be well suited to grow there.

Research growing conditions carefully, keeping your microclimate in mind.

Time of purchase: Outdoor plants are seasonal. Each may flower for a week, or a month, but then enter a foliage-only phase, or may die back altogether. If you buy a plant in full bloom, it may be at the close of its blooming season. Buy flowering plants at the beginning of their blooming season.

Time of flowering: Plan ahead for color areas by checking the full flowering period of each plant in your growing zone. Go beyond learning whether the species you are considering is an annual, perennial, or biennial; ascertain the habits of this particular variety.

Some plant species, such as pinks and bellflowers, have biennial, perennial and annual varieties. Labels don't always identify which a plant is. If you unknowingly plant an annual variety, assuming it's perennial, you may think you have a plant problem when the plant dies after a single blooming period. To prevent such surprises, bring a reference book with you to the nursery or garden center, or consult one there. Make sure the habits of the variety you choose suit the plans you have for your garden.

If you are buying through a catalog, be aware that the catalog description may not jibe with what will happen in your neighborhood. A catalog may state clearly that a certain plant is a perennial, but in your garden it may turn out to be an annual. Blooming time given in the catalog is usually dictated by weather where the catalog originates, and may be far from what you experience in your part of the country. Bacteria and viruses found in local soil can markedly change flower color, so the red tulips you bought may come up streaked. Talk with your neighbors and gardening professionals to learn how specific plants behave in your area.

Containerized camouflage: If you have a highly visible site that is homely unless kept camouflaged by colorful plants, select containerized flowering plants. Chrysanthemums, daylilies, and many other vivid plants thrive this way. The containers can be hidden with a natural mulch or redwood chips, or you can use large decorative pots. When the plants' blooming period is over, move the containers elsewhere and replace them with another set of full-flowered containerized plants.

Buying seeds: Packets of seeds can provide an ample rainbow of beauty for a small amount of money. Patient gardeners may find planting seeds more satisfying than installing transplants, and seed packets often offer a greater variety of plant species than local nurseries. Take care when buying seeds, however. Buy only from a reputable dealer and always check the package date to be sure the seeds are fresh. Stale seeds don't germinate as well as fresh ones. In addition, the information given on the packet is often sparse and fairly generic. Read up on the varieties you have chosen to determine appropriate sowing and transplanting times, and proper garden conditions.

PLANTING AND GROWING ANNUALS

Annual plants—such as pansies, marigolds, impatiens, or phlox—live and die in a year or less. If annuals get through this period, grow, flower, and produce seed, their disappearance is a natural phenomenon, not a problem. You have wisely satisfied their requirements for direct sun, ample water, effective drainage, and soil of at least average quality.

If you want flowers in your garden on a continuous basis throughout the season, plant annuals. Healthy annuals may bloom for months, one flower following another, from the moment of first bud to the first severe cold spell. Gardeners living in an area with no frost or only occasional frost can plant a series of annuals to provide flowers almost all year long.

Annuals do have disadvantages, however. You might not have abundant flower production unless you prepare soil beforehand, preferably a month in advance. This gives added ingredients—such as compost, manure, or fertilizers—a chance to blend and mellow. The roots of most annuals tend not to reach out as much as other plants, so they generally must draw on the nutrients they find where they are placed. If the required vitamins and minerals are not present, an annual might continue

Impatiens is North America's most popular annual for shade.

Many long-lived and easy-care perennials make effective landscape plants. Above: 'Goldsturm' black-eyed Susan, Siberian iris, 'Autumn Joy' sedum, and garden phlox make a dramatic statement when massed in generous numbers. Below: two ornamental grasses (Pennisetum alopecuroides and Miscanthus sinensis), variegated yucca, and asters provide a late-season backbone for annuals and vegetables next to a deck.

blooming for a time after purchase, drawing on nutrients already within the leaves. However, flowers will become smaller and smaller and finally cease altogether. As a result of their relatively limited root systems, annuals are susceptible to transplant shock.

If you are planning on bouquets of summer annuals, do not wait for summer weather to buy the plants. Summer may seem prime for planting, but it is often too hot for newly transplanted annuals. The summer sun encourages top growth before the annuals' meager root systems can support it. The result may be stunted plants and minimal flowering. Solve the problem by planting summer-flowering annuals when spring weather is still a bit on the cool side but not cold enough to risk frost damage.

PLANTING AND GROWING PERENNIALS

Perennials—such as begonias, bergenias, herbaceous peonies, and asters—bloom for a limited time per year, but they make up for it by having a longer life. Unlike shrubs, which are woody plants, perennials are nonwoody, or soft-stemmed, plants.

Perennials usually bloom once a season. This can be for a week or a month. Some perennials completely die back and disappear after flowering, then emerge the following year. Their disappearance can be a problem if you forget to mark their planting site. Resting perennials often get chopped up when enthusiastic gardeners think they have empty space and begin planting

something new. Avoid this problem by marking perennial plant sites with small labeled stakes. Or, if you live in a warm area, buy perennials that stay green the entire year for a leafy backdrop in winter.

You may see the phrase "Perennial treated as an annual" on plant labels or in books. The phrase refers to perennials such as wax begonia, snapdragon, and coleus, which may not survive cold winters. Gardeners who live in areas where winters are harsh should treat such a plant as an annual and plan to replace it each year.

Division: Some perennials, such as chrysanthemums, require division every few years, when the plant grows into a crowded unattractive clump. Other signs that you need to divide perennials include extra-tall growth, weak stems, and few blooms. Divide perennials in the spring. Expose the rootball and divide its segments to create separate plants. Transplant the newly formed perennials to a site where they do not have to compete for light and nutrients. To increase the number of flowers, pinch plants back after transplanting. Do this every few weeks until the plants become bushy and full.

Transplant shock: To ease transplant shock in divided or new perennials, work when the weather is cool, in early mornings or late evenings. Try not to damage roots. Each root is valuable in water acquisition. After transplanting, nip back about one third of the old growth. Water well, using a vitamin booster with a liquid plant food. Despite the best of care, you may still see some symptoms of transplant shock, including leaf drop, flower drop, and wilting.

PLANTING AND GROWING BIENNIALS

Biennial plants have two-year lives. They start their life cycle one year, go dormant in winter, then bloom and complete their lives the following year. Many gardeners think something is wrong with plants like foxgloves, hollyhocks, and Canterbury bells because the first year after being set out they grow but do not flower. Waiting for these plants to flower takes patience. If you want blooming biennials on a continual basis, plant some new ones each year to flower the following year, being careful not to disturb those planted previously.

PLANTING AND GROWING BULBS

Bulbs should be as easy to grow as annuals, biennials, and perennials, but some gardeners seem to have no luck with them. The most frequent problems are caused by unhealthy bulbs, poor drainage, and improper care.

Bulb selection: Some problems are inherent with bulb purchase. With bulbs, you tend to get what you pay for. In many cases, bargain bulbs are "bargains" for a reason. They may be undersized, diseased, poorly stored, inaccurately labeled, improperly matured, or of inferior stock.

Examine bulbs carefully before purchase. Do a light squeeze test. Healthy bulbs, regardless of type, are quite firm and feel heavy for their size. They are free of deep, dark areas; cuts; or soft spots. No mold grows on the outside covering.

Bulb diseases: The bulb holds nutrients much as a storage tank does. The nutrients enable the plant to send up flowers in spring or early summer. Viruses and fungi can interfere with the scenario, however.

■ **Viruses:** If soil- or insect-borne viruses enter the bulb, they move quickly throughout the entire plant, affecting every stage of growth. The many types of viruses can only be seen with an electron microscope. Virus damage is highly visible, however, and accounts for much plant damage.

■ **Fungi:** Root and stem rots are common causes of failure to thrive. They are caused by fungi, which are attracted by standing water. Plant fungi invade and plug the nutrient channels. The infection remains long after standing water has finally disappeared. Sometimes plant descriptions can help you avoid bulb infection. Be wary of bulbs whose labels say "Needs well-drained soil," "Keep on dry side," or "Requires good aeration." Standing water suffocates plant roots of all kinds by filling necessary air spaces in the soil with water. Roots cannot absorb water and nutrients unless oxygen is present. A plant top can wilt from drought while its roots are standing in water.

Do not confuse standing water with ample water. If soil is permeable to air and water, the danger of overwatering is slight. Standing water, however, offers an open invitation to fungi.

Bulbs in containers need excellent drainage; use a fast-draining potting mix, and be sure pots have adequate drainage holes. In cold-winter areas, plant bulbs in containers in the fall. Keep the pots moist over winter in a cool place protected from freezing (such as an attached garage). In mild-winter areas, purchase pre-chilled bulbs or pre-chill them yourself in the vegetable bin of a refrigerator for 10 weeks prior to planting in pots. After planting, place the pots in the coolest part of your garden until shoots emerge in spring.

Hole preparation: In poorly drained areas, improve drainage and minimize the chance of disease by digging planting holes at least 1½ to 2 feet deep. Place a layer of stones and gravel at the bottom of the holes, and slant the excavation so that it drains water away from the site. If you are placing a great number of bulbs in the same poorly drained area, install drain tiles to keep excess water from accumulating.

Proper soil preparation encourages bulb longevity and increase. If you have clay or some other heavy soil, dig planting holes 12 inches deep. Plant the bulbs, then backfill the holes with commercial potting soil.

Bulb foliage: The health of the bulb through the dormant season is related to the longevity of the foliage. Do not cut back the leaves of a bulb plant until after the foliage dies back naturally. Cutting leaves before they wither or cutting off many leaves when you cut flowers can impede the storage of nutrients and cause bulbs to decline.

SOIL TEXTURE

Good soil texture that provides roots with aeration as well as retention of water and nutrients is important for healthy plants. As a rough check on your garden soil, try to squeeze a moist handful into a ball. Soil high in sand (large, irregularly shaped soil particles) will crumble and not hold its shape well. Sandy soil offers roots lots of air but little water and nutrient retention.

Soil high in clay (extremely small, platelike particles that adhere closely together) will form a solid, sticky mass that doesn't easily break apart. Clay soil retains water and nutrients, but little air. Good garden loam, such as the soil shown above, is roughly even in its percentage of silt (medium-size soil particles), clay, and sand. It will hold together when squeezed, yet break apart easily when prodded by the fingers. Especially when high in organic matter, good garden loam retains water and nutrients well, yet also provides the air that roots need.

More precise than the "squeeze test" is the "settling test," shown at right. Take enough samples from the top 6 inches of soil at different garden locations to add up to about 3 cups. Let the samples dry, then pulverize them with a rolling pin. Fill a quart jar two-thirds full of water, stir in 1 teaspoon of low-suds dishwasher detergent or a water softener, add the soil, cover, and shake vigorously. After two minutes, the sand will have settled to the bottom of the jar. After several hours, the silt will have settled into a layer on top of the sand. After a week, most of the clay will have settled into a layer on top of the silt. Measure the three levels to determine the relative proportions of sand, silt, and clay in your soil.

How do you know the pH of the soil you have? Generally, acidic soils are found in areas of high rainfall; alkaline soils are found in desert areas or areas of low rainfall. However, every region contains microclimates, so the general principle may not provide an accurate assessment. To get a precise pH reading, take soil samples to a professional testing service. The local county extension office can tell you about such services in your area. Or try looking under "Soil" in the telephone directory. The testing service will give instructions about taking the samples. Most services ask you to take samples from different garden areas by digging to a depth of 6 inches and taking a thin soil slice from the edge of each hole. Then you mix the samples together in a clean nonmetal container. You should now have about 2 cups of soil. Remove any stones, roots, or debris. Cover the container, and keep it dry until analysis.

An alternative to professional testing is using a home soil-testing kit, which you can buy at a nursery or garden center. A home test is not as accurate as a professional analysis, but a kit is inexpensive, easy to use, and usually adequate.

■ **Acidic soil:** Plants placed in acidic soil often grow slowly and have pale green or yellow leaves. Roots are few and small. Overly acidic soil promotes disease. Adding fertilizer does not help the plants, because low pH slows nutrient release.

To correct acidic soil, apply finely ground dolomitic limestone. This not only raises soil pH, but supplies needed calcium and magnesium. The heavier the soil, the more limestone you need. Liming to increase alkalinity is especially helpful for clay soils, making them more friable, or crumbly. Liming is not permanent. You will have to recheck pH every 2 years and reapply limestone as necessary.

If you have acidic soil and are unable to correct it, consider planting rhododendrons, azaleas, camellias, or other acid-loving plants. These plants do not have to depend on soil organisms to release nitrogen from soil. A beneficial fungus on the plants' roots can convert soil nitrate into usable nitrate. If you put acid-loving plants in nonacidic soil, sufficient iron may not be released for their needs. Yellow areas may appear between the veins of new leaves, affecting plant appearance. Leaves weakened in this way cannot support normal flowering.

ASSESSING AND IMPROVING SOIL

In terms of soil, annuals, perennials, biennials, and bulbs need a growing site that provides an appropriate pH and essential nutrients.

Soil pH: The pH of soil reflects the concentration of hydrogen ions in it; this determines how acidic or alkaline the soil is. pH is measured on a scale from 0.0 to 14.0. A soil that measures pH 7.0 is neutral; it is neither acidic nor alkaline.

In terms of plant growth, acidity or alkalinity is important because it determines how quickly the soil can release nutrients—or whether the nutrients can be released from the soil at all. For example, in acidic soil—that with a pH of 5.2, say—magnesium, phosphorus, and calcium are less available for plant use than they would be in neutral soil. Nitrogen is released only partially because soil organisms that create it are less active in acidic soil. When soil has a pH of 5.0 or lower, these organisms often cease working altogether so that no nitrogen is released. Conversely, at a pH below 5.0, manganese and aluminum may become available in harmful amounts.

■ **Alkaline soil:** Soil with a pH over 7.0 is alkaline. Some plants thrive at this level. When pH reaches 8.0, however, the soil does not release certain nutrients. Iron and manganese are no longer available to plants. Older leaves remain green, but new leaves have yellow areas between their veins. Flowering plants do not generally tolerate alkaline soil. If you live in a highly alkaline region and want to put in annuals, perennials, or bulbs, consider using raised beds or containers in the garden. Fill these with a commercial soil mix.

To correct mildly alkaline soil, apply acidic peat or an acidic mulch made of pine needles. Use a fertilizer that promotes acid-loving plants, which will create reactions that lower soil pH. Ordinary powdered sulfur corrects alkalinity, but you must strictly regulate the rate of application according to soil needs. Use aluminum sulfate instead of powdered sulfur; the combination product is less prone to application error. For the first treatment, use 2 pounds of aluminum sulfate per 100 square feet. Applying lime-sulfur spray is also helpful.

Soil nutrients and fertilizers:
Buying the wrong type of fertilizer may cause as many flowering plant problems as overfertilizing. The "big three" fertilizer elements are nitrogen, phosphorus, and potassium. Other elements, present in trace amounts, are also important to plant growth.

■ **Nitrogen:** This element is a vital component of plant protein. Symptoms of nitrogen shortage include slow growth and yellowing leaves. Too much nitrogen forces lush foliage as opposed to structural growth. The plant becomes vulnerable to weather variables, attacks by insects, and diseases.

■ **Phosphorus:** Only with phosphorus can plants create stiff stems to hold leaves and flowers up to sunlight and pollination. A shortage of phosphorus slows root growth, flowering, and seed production. Leaves turn purplish or become dark gray-green. Symptoms appear in older leaves first. With severe phosphorus deficiencies, plant flowering is minimal.

■ **Potassium:** In North American soils potassium tends to be plentiful but in a form difficult for plants to use. Potassium is necessary for photosynthesis; without photosynthesis, plants starve. In addition, potassium is essential for stiff stems on flowering plants and for the formation of bulbs and tubers. Symptoms of potassium

Annuals, perennials, ornamental grasses and flowering shrubs combine in a mixed border with abundant, long-lasting color.

shortage include mottled yellow or pale green mature leaves with scorched edges. Flower yield is minimal.

■ **Other soil nutrients:** Fertilizer labels often list elements in addition to nitrogen, phosphorus, and potassium. These include iron, calcium, magnesium, zinc, manganese, and sulfur. Though plants use them in minuscule amounts, these trace elements have significant growth effects. Of the secondary nutrients, iron is the one usually in shortest natural supply. Iron deficiency causes yellowing plant leaves, though leaf veins generally remain green. The newest leaves are the most severely affected. Applying an iron-containing fertilizer to the soil corrects yellowing. If necessary, also spray foliage and soil with an iron solution.

■ **Fertilizer selection:** To correct nitrogen, phosphorus, or potassium imbalance, purchase a fertilizer that meets your immediate and long-term needs. Fertilizer labels look more complex than they really are. The numbers on the container tell you, in relative terms, how much of each of the three major elements is present. Nitrogen (N) is always listed first, followed by phosphorus (P) and potassium (K). If the label says "20-20-20," the fertilizer contains equal portions of nitrogen, phosphorus or phosphate and potassium or potash. A 5-10-5 mixture is higher in phosphorus than in nitrogen or potassium.

PREVENTING AND SOLVING PROBLEMS

Cultural techniques such as disbudding, pinching back, and staking can prevent many growth problems. When growth is healthy, plants are in the best position to fend off external threats, such as frost and insects.

Cultural techniques for problem prevention: Plants—like people—have a finite energy supply. By using cultural techniques to direct a plant's energy toward healthy growth, you can save the energy you would otherwise have to expend on treating disease or replacing damaged plants.

■ **Flower removal:** Make certain poor flowering isn't caused by old flowers left on the stems. In many plants, especially roses, flower production slows when older blossoms use energy to create seed. To encourage continual flowering, cut off flowers past their prime.

■ **Disbudding:** If your plants produce many flowers but the flowers are not large enough to be attractive in cut displays, try the technique of disbudding. Disbudding involves removing some flower buds while they are about the size of a bean. With fewer

Cosmos respond to the removal of spent flowers by producing more abundant blooms. Many other flowers respond well to deadheading, including Gaillardia, Coreopsis, Pelargonium, Dianthus, Tagetes, Petunia, Calendula, Salvia, Leucanthemum, Antirrhinum, Stokesia, Achillea, and Zinnia.

buds to develop, the plant has more energy to use in creating larger blooms from the remaining buds. Disbudding is a common practice on peonies, dahlias, carnations, chrysanthemums, and roses.

■ **Pinching back:** Snapdragons, chrysanthemums, dahlias, fuchsias, blood leaf, salvias, and garden geraniums tend toward ranginess, though the problem can strike just about any plant. To encourage sturdy, bushy growth, pinch back the stems by using your thumb and forefinger to nip them off just under the flower buds. Pinching back does delay blooming, but the long-term result is a stronger, more attractive plant. Snapdragons do well with just one pinching. Chrysanthemums may need up to four.

■ **Stem support:** Long-stemmed flowering plants may fall over because of stem or blossom weight. Grabbing a handful of stems and tying them to a pole is not the answer to this problem. If the plants have just a few slim stems, such as delphiniums or carnations, loosely tie each stem to a slender stake. If the plants are bushy, provide support by surrounding them with metal hoops that are made for the purpose. Be sure to anchor the hoops securely in the ground. Whether using hoops or stakes, always maintain the natural growth habit of the plants.

Problems in the environment: Encroaching shade, burning sun, sudden frosts, and hungry insects threaten flowering plants. With a little help from you, your flowers can survive, thrive, and reward you with an array of blossoms.

■ **Insufficient sunlight:** Flowering plants need plenty of sun. Though they may continue to grow in less than optimum light conditions, they may not flower or flowering may be sparse. In some cases, just enlarging the open space around the plants may give them the boost of sunlight they need to prosper. Sometimes the remedy is to cut back overhanging branches that shade the garden. (Before cutting, make sure the limbs are not critical to the survival of the tree or shrub. Read about pruning techniques or consult an arborist to prevent damage.) In other cases, the only course is to move the struggling plants to a sunnier spot. Though transplanting is always a risk, continuing shade could mean certain failure.

■ **Sunburn:** Though plenty of sunlight encourages flowering plants, too much sunlight can damage them. Symptoms of sunburn include faded and bleached leaves or leaves that turn pale and yellowish. Foliage may become brittle. To prevent sunburn, temporarily shade the plants with some sort of loose cover. Make sure they get all the water they need during periods of sunburn danger.

■ **Frost:** Whenever it occurs, frost is a danger to plants because it can damage plant cells. Symptoms of frost damage include blackened leaf parts and failure to thrive.

Some perennials and biennials must survive frost during the dormant season. Protect them with a covering of mulch. Mulch vulnerable plants before the first hard frost; the layer of insulation protects them from low temperatures. Mulch hardy perennials and biennials after the ground freezes. Because the plants are hardy, they are likely to begin spring growth after the first significant thaw. The first thaw is often a false spring, and the frost that follows may damage or kill delicate shoots. By mulching hardy perennials after the ground freezes, you insulate them from weather fluctuations. In this case, the mulch keeps the plants cold until sprouting is safe.

Evergreen branches, wood chips, straw, pine needles, and corn stalks are effective mulches because they allow air circulation. Avoid using fallen leaves as mulch unless the leaves are well composted. Uncomposted leaves tend to form a thick soggy mass that compacts and smothers most plants.

A 3-inch layer of mulch usually provides the needed protection. The layer should be loose so air can penetrate to the soil. Through the winter, some of the mulch decomposes. By spring, the soil is clear enough for plants to sprout but still sufficiently covered to get some protection from drying and weed encroachment.

■ **Pests:** Among the hungry horde of insects that plague flowering plants are thrips, leafminers, leafhoppers, aphids, ants, spittlebugs, mealybugs, bulb flies, cutworms, and a multiplicity of caterpillars. Correct diagnosis is 90 percent of cure. Consult the problem-solving section of this book for photographs and a detailed discussion of insect pests.

Some problems are easily recognizable. Look closely and you can see budworms of varying ages hiding in your decimated petunia blossoms. Cut the stem of a dying hollyhock or aster and 1-inch-long yellow or pale pink corn borers are suddenly out in the open. The 1½-inch-long green cabbageworm can eliminate your entire nasturtium bed and then move over to your alyssum, carnations, and geraniums. Hungry caterpillars can defoliate and deflower an entire flower bed. Woolly bears like cannas, dahlias, violets, petunias, hollyhocks, and verbena, among many other perennials and annuals. Caterpillar parents are moths and pretty butterflies. The sight of parents flying about each spring should make you alert to potential caterpillar problems in the flower garden.

Earwigs are familiar to gardeners in all but the coldest parts of the country. Normally they are beneficial scavengers, eating waste materials such as decaying fruit and plant litter. But sometimes they multiply and become pests, feeding on flowers and seedlings. Since earwigs like to hide in dark, damp places, try trapping them in rolled-up newspapers and overturned flowerpots. Each morning shake the catch out into a pail of soapy water. Continue trapping until no new earwigs appear. If that does not work, spray with ORTHO Diazinon Ultra Insect Spray or dust with ORTHO Ant-Stop Ant Killer Dust.

Grasshoppers emerge only during certain weather conditions, but their appearance can signify disaster to flowering plants. They often move into planted areas from nearby fields in dry weather. Seedlings disappear first. Then grasshoppers begin chewing large holes around leaf edges. They often begin work in areas near weed patches. Begin your control program as soon as you see grasshoppers, because the first few act as scouts for the rest. Treat with ORTHO Bug-Geta Plus Snail, Slug & Insect Killer, or spray with an insecticide containering *diazinon* or *malathion*. Repeat at weekly intervals if plants become reinfested.

Like insect pests, animal pests can endanger the flower garden. Bulbs sometimes seem particularly vulnerable to attack by animal pests.

Mice use mole runs to get at bulbs underground. Tulip, lily, and crocus bulbs are particular favorites. Bulbs wilt rapidly after being gnawed, and they may be completely destroyed. Adding a liberal load of stone chips or pebbles to the soil when planting discourages mice. Another solution is encircling the entire bulb bed with 12-inch-high fine-mesh netting. Bury this vertically so that only about 3 inches extend above the surface. Do not mulch the bulb bed in winter until the ground is frozen. If depredations continue, plant bulbs that mice do not like as well, such as narcissus.

Deer often seek out spring bulbs. Since keeping high-jumping deer out of a garden is extremely difficult, one solution is to plant more than you need so you have enough to share. Keep deer favorites close to the house, where the wild animals may be more reluctant to venture. Place mothballs around the perimeter of the yard if children or pets are not nearby. The mothballs may act as a deer repellent. On yard outskirts, plant narcissus. Deer normally will not touch this bulb, and you can choose from many beautiful varieties.

Rabbits can be a garden menace. They eat almost any growing thing they can reach. Rabbits devour all parts of crocus plants and can finish a tulip bed off while it is still in bud form. For emergency protection, apply a thick layer of blood meal around bulb plants. Cover early spring bulbs with evergreen limbs from your holiday tree—a good recycling project. Leave the green covering over crocus bulbs until the spring sprouts leaf out. Keep tulips covered until stems are 12 inches high. Though dogs and cats can cause other garden problems, they do help keep rabbits away.

Behind every successful flower garden lie regular watering, fertilizing, weeding, and pest control. This well-tended border of daylilies, sage, Asiatic lilies, loosestrife, Shasta daisy, baby's breath, coreopsis, astilbe, feverfew, and rudbeckia shows what is possible.

79

POOR FLOWERING

Transplant shock

Transplant shock on cineraria.

Problem: Recently transplanted flowers drop their flower buds before they open. Often, blossoms and leaves also drop prematurely. The plant may wilt during the hot part of the day even if the soil is moist.

Analysis: Transplant shock can occur even under ideal conditions, as many plants drop some of their buds, flowers, and leaves when transplanted. Bud and leaf drop result from root damage that occurs during transplanting. Tiny hairlike rootlets that grow at the periphery of the root system absorb most of the water the plant uses. When these rootlets are damaged during transplanting, less water is supplied to the foliage and flowers. Flower buds, flowers, and leaves fall off, and the plant wilts. The more the roots are damaged during transplanting, the greater the leaf and bud drop. Because plants lose water rapidly during hot, dry, windy periods, transplanting at these times will cause plants to undergo greater shock. They will not recover as quickly. As the root system regenerates, new flower buds will form.

Solution: Transplant when the weather is cool, in the early morning, in the late afternoon, and on cloudy days. Whenever possible, transplant small plants rather than large ones. When transplanting, disturb the soil around the roots as little as possible. Preserve as much of the root system as possible. If the roots have been disturbed or if the plant is large or old, pinch off about one-third of the growth to reduce the amount of foliage needing water.

Thrips

Thrips damage to Gladiolus.

Problem: Flower buds turn brown and die before they open. Silvery white streaks are often seen on the leaves. Flowers that have opened are often streaked and distorted. If the flower buds are peeled open, tiny ($\frac{1}{20}$-inch) insects resembling brown or straw-colored wood slivers can be seen moving around at the base of the petals.

Analysis: Several species of thrips, a common insect pest, attack garden flowers. Thrips are found in protected parts of the plant, such as the insides of flower buds, where they feed by rasping the soft plant tissue, then sucking the released plant sap. The injured tissue dies and turns white or brown, causing the characteristic streaking of the leaves and flowers. Because thrips migrate long distances on wind currents, they can quickly infest widespread areas. In cold climates, thrips feed and reproduce from spring until fall. With the onset of freezing weather, they find sheltered areas such as grass clumps and hibernate through the winter. In warm-weather climates, thrips feed and reproduce all year. These pests reach their population peak in late spring to midsummer. They are especially troublesome during prolonged dry spells.

Solution:
Thrips cannot be eliminated completely, but they can be kept under control. Spray infested plants with ORTHO Orthene Systemic Insect Control or ORTHO Isotox Insect Killer. Spray two or three times at intervals of 10 days as soon as damage is noticed. Make sure that your plant is listed on the product label. Repeat the spray if reinfestation occurs. Pick off and destroy old, infested flowers.

Many small flowers

Chrysanthemums that have not been disbudded.

Problem: The plant produces many small flowers rather than a few large, showy flowers. The leaves are healthy.

Analysis: Large, showy flowers will not be produced on certain plants, such as chrysanthemums, dahlias, and carnations, unless most of the flower buds are removed. Generally, these plants produce long stems with a terminal flower bud at the end and secondary flower buds at the base of each leaf on the flower stem. The plant has only a limited amount of nutrients with which to nourish each bud. If the plant has many buds, each bud will receive only a small amount of nutrients and will develop into a small flower.

Solution: Pinch off the side flower buds as soon as they are large enough to be handled. The earlier they are removed, the larger the terminal flower will be.

SLOW GROWTH

Poor growth

Zinnias planted too late.

Phosphorus-deficient columbine.

Insufficient light

Geraniums in low light.

Problem: The plant fails to grow, or it grows very slowly. No signs of insects or diseases are present.

Analysis: A plant might grow slowly for any of many reasons.

Solution: The numbered solutions below correspond to the numbered items in the analysis.

1. Improper planting time: Many plants require warm temperatures and long hours of sunlight to grow well. If transplants that need warm weather are set out too early in the spring or too late in the fall, when temperatures are cool, they will not grow.

1. Check page 351 for a list of common flowers and their growing seasons. As the weather warms up, the plants will start to grow.

2. Unseasonable cool spell: If the weather is unseasonably cool or cloudy, most plants—even those adapted to cool temperatures—will slow down their growth rate.

2. Plants will start to grow again when unseasonable cool spells have passed. If the weather is especially cloudy and moist, check for signs of disease. Fungal and bacterial infections are especially troublesome during periods of moist weather.

3. Natural dormancy: Many perennials and all bulbs undergo a period of no growth, which usually occurs soon after they have flowered. Although the plant may seem to be inactive during this period, it is actually developing roots, bulbs, or rhizomes for the following year's growth. The foliage of many perennials and bulbs eventually dies back completely, and the plant becomes entirely dormant.

3. Check the list of common flowers on page 351 to see if your plant is a bulb or a perennial with a natural dormancy period.

4. Phosphorus deficiency: Phosphorus is a plant nutrient essential to normal plant growth and development. Many garden soils are deficient in phosphorus. When plants do not receive enough phosphorus, they usually grow very slowly or stop growing altogether. Sometimes their foliage also turns dark green, or it may redden slightly.

4. For a quick response, spray the leaves with Miracle-Gro Bloom Booster. Fertilize the plants with the same fertilizer, which is high in phosphate.

Problem: The plant is located in a shaded area and grows slowly or not at all. Growth is weak and leggy, and flowering is poor. The leaves may be dark green and larger than normal. The oldest leaves may drop off. There are no signs of disease or insect pests.

Analysis: Any plant that receives less light than it requires cannot produce as much food as it needs. It grows slowly and is weak and leggy. Plants contain chlorophyll, which uses sunlight to produce energy. This energy is used to make food for plant growth and development. Plants differ in the amount of light they need to grow properly. Some plants need many hours of direct sunlight daily, while others thrive in shaded locations.

Solution: Look up your plant in the alphabetical section that begins on page 93 to determine its light requirement. If your plant is not receiving enough light, transplant it to a sunnier location.

WILTING PLANTS

Root problems

Wilting snapdragons.

Wilting pincushion flowers.

Lack of water

Leaf-edge damage caused by wilting.

Problem: The plant is wilting. The leaves are discolored to yellow or brown and may be dying. The soil may be moist or dry.

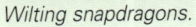

Analysis: These symptoms are caused by one of several root problems.

1. Stem and root rot: Many fungi and bacteria decay plant roots and stems. In addition to leaf wilting and discoloration, spots and lesions frequently form on the leaves and stems. The infected tissue may be soft and rotted, and the plant pulls out of the ground easily to reveal rotted roots. Most of the disease-causing organisms thrive in wet soil.

2. Fertilizer burn: Excessive amounts of fertilizer can cause plants to wilt. The leaves wilt and become dull and brown. Later they become dark brown or black and dry up. When too much fertilizer is applied and not watered in well, a concentrated solution of fertilizer salts is formed in the soil, making it difficult for plants to absorb the water they need and even drawing water out of them. A high concentration can cause roots to die and kill the entire plant.

3. Nematodes: These microscopic worms (not related to earthworms) live in the soil and feed on plant roots. While feeding, they inject a toxin into the roots. The result is that roots can't supply adequate water and nutrients to the aboveground plant parts, so the plant slowly dies. Infested plants are weak; are slow growing; often turn bronze or yellowish; and wilt on hot, dry days, even when the soil is wet. If you pull the plant up, you see stunted roots that are often dark and stubby and may have nodules on them.

Solution: The numbered solutions below refer to the numbered items in the analysis.

1. Look up your plant in the alphabetical section beginning on page 93 to determine which root and stem rot diseases may affect it. Treat accordingly. Allow soil to dry while minimizing water needs of the plants. Provide shade, spray with antitranspirant, and place clear plastic tents over plants in shade to maintain humidity. Resume watering when plants show signs of stress.

2. Dilute the fertilizer in the soil and leach it below the root zone by watering the soil heavily. Soak the affected area thoroughly with plain water, let it drain, then soak again. Repeat three or four times. Cut off dead plant parts. Follow directions carefully when fertilizing.

3. If you have a chronic problem with wilting, yellowing plants that slowly die and you've eliminated all other possibilities, test for nematodes. Testing roots and soil is the only positive method for confirming the presence of these pests. Contact your county extension office for sampling instructions, addresses of testing laboratories, and control procedures for your area.

Problem: The plant wilts often, and the soil is frequently or always dry. The leaves turn brown, shrivel, and may be crisp.

Analysis: The most common cause of plant wilting is dry soil. Plant roots take up water, which moves up into the stems and leaves and evaporates into the air through microscopic breathing pores in the surface of the leaf. Water pressure in the plant cells keeps the cell walls rigid and prevents the plant from collapsing. When the soil is dry, the roots are unable to furnish the leaves and stems with water, the water pressure in the cells drops, and the plant wilts. Most plants will recover if they have not wilted severely. Frequent or severe wilting, however, will curb a plant's growth and eventually kill it.

Solution: Water the plant thoroughly, applying enough water to wet the soil to the bottom of the root zone. If the soil is crusted or compacted, cultivate the soil around the plant before watering. To help conserve soil moisture, apply a mulch around the plant, or incorporate peat moss or other organic matter into the soil. Do not allow the plant to wilt between waterings. Look up your plant in the alphabetical section beginning on page 93 to determine its moisture requirement.

Heat or acute root damage

Transplant shock on begonias.

Poor drainage on impatiens.

Lack of nitrogen

Nitrogen-deficient impatiens.

Problem: The plant is wilting, but the foliage looks healthy. No signs of disease or insects are present. The soil is moist.

Analysis: If a plant wilts in moist soil but the leaves look healthy, one of the following situations has probably occurred recently.

Solution: The numbered solutions below correspond to the numbered items in the analysis.

1. Intense heat or wind: During hot, windy periods, plants may wilt even though the soil is wet. Wind and heat cause water to evaporate quickly from the leaves. The roots can't take in water as fast as it is lost.

1. As long as the soil is kept moist during periods of intense heat and wind, the plants will probably recover without harm when the temperature drops or the wind dies down. Shading the plants and sprinkling them with water to cool off the foliage and reduce the rate of evaporation from the leaves may hasten recovery.

2. Transplant shock: Plants frequently wilt soon after being transplanted. Although called transplant shock, this is not the same condition as shock in a human being, but refers to wilt resulting from injured roots. Roots are usually broken or injured to some degree during transplanting. Damaged roots are unable to supply the plant with enough water, even when the soil is wet. As the root system restores itself, its water-absorbing capacity increases. Unless its roots are severely injured, the plant will soon recover.

2. Preserve the root system as much as possible when transplanting. Keep as much of the soil around the roots as possible. Transplant when the weather is cool, in the early morning, in the late afternoon, or on cloudy days. If the roots have been disturbed, or if the plant is large and old, prune off about one-third of the growth. If possible, transplant when the plant is dormant.

3. Rodents: The roots, underground stems, and bulbs of many plants are often disturbed or fed upon by rodents, including pocket gophers and field mice. Root, bulb, and stem damage can result in rapid wilting and sometimes in death of the plant.

3. Rodents may be trapped or baited. For details about rodent control, see page 47.

4. Mechanical injury: Cultivating, digging, hoeing, thinning, weeding, and any other kind of activity that damages roots can cause wilting.

4. Prevent mechanical injury to plants by working very carefully around them. Cultivate as shallowly as possible.

Problem: Leaves turn pale green, then yellow, beginning with the older leaves. Growth is slowed. Older leaves may drop. New leaves are small. Severely affected plants may die.

Analysis: Garden soils are frequently deficient in nitrogen, the most important nutrient for plant growth. Nitrogen is essential in the formation of plant protein, chlorophyll (green leaf pigment), and many other compounds. When a plant becomes deficient in nitrogen, it breaks down proteins and chlorophyll in the oldest leaves to recover nitrogen to be recycled for new growth. This loss of chlorophyll causes the older leaves to turn yellow. A continuing shortage of nitrogen results in overall yellowing. Because nitrogen is leached from the soil more readily than are other plant nutrients, and because it is needed in larger quantities than are other nutrients, it must be added to most garden soils and for all flowers. Nitrogen leaches from sandy soil more readily than from clay soil, and it leaches more quickly when rainfall or irrigation is heavy.

Solution: Fertilize with Miracle-Gro Plant Food. Repeat applications according to directions. Fertilize more frequently in sandy soils or where rainfall is heavy. Check that plants are not suffering from saturated soil or root rot, as these plants may exhibit symptoms of nitrogen deficiency.

DISCOLORED OR SPOTTED LEAVES (continued)

Leaves discolored

Iris leaves discolored by lack of water.

Dried-out crocus.

Iron deficiency

Iron-deficient Pelargonium.

Problem: Leaves turn pale green to yellow. The plant may be stunted. In many cases, leaf edges turn brown and crisp, and some leaves shrivel and die. No signs of disease or pests are present. Leaves do not discolor from the base of the plant upward as they do with nitrogen deficiency. For information on nitrogen deficiency, see page 83.

Analysis: Leaves discolor for several reasons.

Solution: The numbered solutions below correspond to the numbered items in the analysis.

1. Frequent water stress: Plants require at least a minimal supply of water to remain healthy and grow properly. When they are allowed to dry out once or twice, they usually survive. Plants that suffer from frequent drought stress, however, undergo changes in their metabolism that result in leaf discoloration, stunting, and lack of growth. If the soil is allowed to dry out completely, the plant will die.

1. Do not let plants wilt between waterings. Consult the alphabetical section beginning on page 93 for the moisture needs of your plant. Provide plants with adequate water.

2. Salt buildup in the soil: Leaf discoloration and browning occur when excess salts dissolved in the soil water are taken up into the plant and accumulate in the leaf tissue. Soil salts build up to damaging levels in soils that are not flushed occasionally. Salt buildup commonly occurs in arid regions.

2. Flush out soil salts periodically by watering deeply and thoroughly.

3. Sunburn: Shade-loving plants placed in a sunny location will develop discolored leaves. Sunburned leaves often have a whitish or yellow bleached appearance. Leaves not directly exposed to the sun usually remain green and uninjured.

3. Check to see whether your plant is adapted to sun or to shade. Consult the alphabetical section beginning on page 93 for the light needs of your plant. Transplant shade-loving plants to a shaded location.

Problem: Leaves turn pale green or yellow. The newest leaves (those at the tips of the stems) are most severely affected. Except in extreme cases, the veins of affected leaves remain green. In extreme cases, the newest leaves are small and all-white or yellow. Older leaves may remain green.

Analysis: Plants frequently suffer from deficiencies of iron and other minor nutrients such as manganese and zinc, elements essential to normal plant growth and development. Deficiencies can occur when one or more of these elements are depleted in the soil. Often these minor nutrients are present in the soil, but alkaline soils with a pH of 7.5 or higher or wet soil conditions cause them to form compounds that cannot be used by the plant. Alkalinity can result from overliming or from lime leached from cement, concrete, or mortar.

Solution: Spray the foliage with a chelated iron fertilizer, and apply the fertilizer to the soil around the plants to correct the deficiency of minor nutrients. Check the soil pH. Lower the pH of the soil before planting by using aluminum sulfate, ferrous sulfate, or soil sulfur. Maintain an acid-to-neutral pH by fertilizing with Miracid Plant Food.

Leaf and flower spots

Fungal leaf spot on impatiens.

Leaf spot on peony.

Powdery mildew

Powdery mildew, a fungal disease.

Problem: Spots and blotches appear on the leaves and flowers.

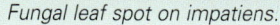

Analysis: Several disease and environmental factors contribute to spotting and blotching.

1. Fungal leaf spots: Spots caused by fungi are usually small and circular and may be found on all of the leaves. The spots range in size from barely visible to an inch in diameter. They may be yellow, red, tan, gray, brown, or black. Often the leaves are yellow and dying. Infection is usually most severe during moist, mild weather (50° to 85°F).

2. Bacterial leaf spots: Spots caused by bacteria are usually tiny and angular in shape. They are usually dark and are sometimes accompanied by rotting and oozing from infected areas. Bacterial spots may be found on all parts of the plant and are most often favored by warm, moist conditions.

3. Sunburn: Shade-loving plants placed in a sunny location will develop spots and blotches. Sun-loving plants may also develop sunburn symptoms if they are allowed to dry out or if they are recently transplanted from greenhouses. Initially, sunburned leaves develop a whitish or yellowish bleached appearance. Large, dark blotches form on the damaged tissue.

4. Spray damage: Spotting of foliage and flowers may also be caused by insecticide, fungicide, and herbicide spray damage. Sprays may drift in from other areas.

Solution: The numbered solutions below refer to the numbered items in the analysis.

1. Picking off the diseased leaves may give adequate control. Clean up plant debris, especially during the winter. If plants are severely infected, spray them with a fungicide containing *chlorothalonil* (Daconil 2787®) or *mancozeb*. Make sure that your plant is listed on the fungicide label before spraying. These sprays are protectants, not controls. Old leaves will remain diseased. Water early in the day so foliage can dry thoroughly.

2. If practical, pick off and destroy spotted leaves. If the plant is severely infected, discard it. Clean up plant debris. Avoid overhead watering. Disinfect pruning tools.

3. Pick off the injured leaves and plant parts. Check to see whether your plant is adapted to sun or shade. Transplant shade-loving plants to a shaded location, or provide shade. Provide plants with adequate water, especially during hot, sunny, or windy days.

4. Once damage has occurred, you cannot do anything about it. Read and follow directions carefully when spraying. Avoid spraying on windy days when the spray can drift. If spray drifts onto the wrong plant, rinse off the leaves immediately with water.

Problem: Grayish white powdery spots and patches cover the leaves and stems, often primarily the upper surfaces of leaves. Infected leaves eventually turn yellow.

Analysis: Powdery mildew is caused by several closely related fungi that thrive in both humid and dry weather. The powdery patches consist of fungal strands and spores. The spores are spread by the wind to healthy plants. The fungus saps plant nutrients, causing the leaves to turn yellow and sometimes to die. A severe infection may kill the plant. Since some powdery mildews attack many different kinds of plants, the fungus from a diseased plant may infect other types of plants. Under conditions that favor it, powdery mildew can spread through a closely spaced planting in a matter of days. In the late summer and fall, the fungus forms small, black, spore-producing bodies, which are dormant during the winter but which can infect more plants the following spring. Powdery mildew is generally most severe in the late summer and under humid conditions.

Solution: Look up your plant in the alphabetical listing beginning on page 93 to determine which fungicide should be used. Spray at regular intervals of 10 to 12 days or as often as necessary to protect new growth. Remove and destroy severely infected plants. Where practical, pick off diseased leaves. Clean up and destroy plant debris. Plant in sites that have good air circulation and receive early morning sun.

DISCOLORED OR SPOTTED LEAVES (continued)

Rust

Pelargonium rust.

Problem: Yellow or orange spots appear on the upper surfaces of leaves. Yellowish orange, rust, or chocolate-colored pustules of spores develop on the undersides of leaves. Infected leaves usually dry up and die. The plant may be stunted.

Analysis: Rust is caused by any of several related fungi. Most rust fungi spend the winter as spores on living plant tissue and, in some cases, in plant debris. Some rust fungi also infect various weeds and woody trees and shrubs during part of their life cycle. Flower infection usually starts in the early spring as soon as conditions are favorable for plant growth. Splashing water and wind spread the spores to healthy plants. Some rust fungi cannot infect the flower host unless the foliage is wet for 6 to 8 hours. Rust is favored by moist weather, cool nights, and warm days.

Solution: Several fungicides are used to control rust. Look up your specific plant in the alphabetical listing beginning on page 93 to determine which fungicide to use. Spray plants thoroughly at the first sign of disease, covering both the upper and lower surfaces of the leaves. Some plants are so susceptible to rust that you may need to spray at weekly intervals throughout the summer. Fungicides will only protect uninfected tissues; they will not cure diseased leaves. To allow wet foliage to dry out more quickly, water in the morning rather than in the late afternoon or evening. Remove and destroy all infected plants in the fall to prevent them from infecting new plantings. Plant rust-resistant varieties, if available.

Spider mites

Spider mite damage to columbine.

Problem: Leaves are stippled, yellowish, bronze, or reddish and are often dirty. There may be webbing over flower buds, between leaves, or on the lower surfaces of leaves. To determine if a plant is infested with mites, examine the bottoms of the leaves with a hand lens. Or hold a sheet of white paper underneath an affected leaf and tap the leaf sharply. Minute specks the size of pepper grains will drop to the paper and begin to crawl around. These pests are easily seen against the white background.

Analysis: Spider mites, a common pest of many garden and greenhouse plants, are related to spiders. They cause damage by sucking sap from the undersides of leaves. As a result of their feeding, the plant's green leaf pigment disappears, producing the stippled, discolored appearance. Spider mites are active throughout the growing season but thrive in hot, dry weather (70°F and up). By midsummer, they may have built up to tremendous numbers. During cold weather, spider mites hibernate in the soil, on weeds and plants retaining foliage, and on tree bark.

Solution: Spray infested plants with ORTHO Isotox Insect Killer or ORTHO RosePride Orthenex Insect & Disease Control. Make sure that your plant is listed on the product label. Spray plants thoroughly, being sure to cover both the upper and lower surfaces of the leaves. Repeat the spray at least two more times at intervals of 5 to 7 days. If the plant can tolerate heavy watering, heavy irrigation of the foliage can help to reduce the number of spider mites.

Leafminers

Leafminer damage to Dahlia.

Problem: Light-colored irregular trails wind through the leaves. Blotches may eventually appear on infested leaves. Some of the trails and blotches are filled with black matter. Severely infested leaves may dry up and die.

Analysis: Most insects that cause this type of damage belong to the family of leafmining flies. The tiny adult flies lay their eggs on the undersides of leaves. The maggots that hatch from these eggs penetrate the leaf and live between the upper and lower surfaces. They feed on the inner leaf tissue, creating winding trails and blotches. Their dark excrement may dot or partially fill sections of the trails. Generally, the larvae emerge from the leaves to pupate. Leafminers are present continually from spring until fall. The last generation of maggots pupates in the soil or plant debris through the winter and emerges as adult flies the following spring.

Solution: Spray infested plants with ORTHO Orthene Systemic Insect Control. Pick off and destroy infested leaves. Remove and destroy all plant remains in the fall.

Viruses

Virus-damaged Gladiolus.

Problem: Leaves may be mottled yellow-green or may be uniformly yellowing. In some cases, the foliage develops yellow rings, or the veins may turn yellow. Flowers and leaves may be smaller than normal and distorted. The plant is usually stunted, and flowering is generally poor.

Analysis: Several different plant viruses infect flowering plants. These viruses include mosaics, yellows, and ring spots. The severity of viral infections depends on the plant and on the strain of virus. In some cases, symptoms of infection may not show up unless several viruses are present in the plant at the same time. Viral infections do not generally kill the plant but may greatly reduce its overall vigor and beauty. Many viruses are spread by aphids, which feed on diseased plants and transfer the virus to healthy plants. Some viruses can be transferred to healthy plants on hands and equipment contaminated by plant sap. Viruses usually persist in the plant indefinitely.

Solution: No chemicals control or eliminate plant virus diseases. Remove and destroy weak, infected, and stunted plants. Wash your hands thoroughly and disinfect pruning shears after working on infected plants by dipping the shears after each cut into a solution of 1 part chlorine bleach to 9 parts water. Purchase only healthy plants. Keep the aphid population under control. Remove weeds that may attract and harbor aphids.

Air pollution

Nasturtiums damaged by ozone.

Problem: The upper surfaces of leaves may be bleached, with white flecks or reddish brown spots. Sometimes the leaves are distorted. Older leaves are affected more than younger ones.

Analysis: Some gases released into the atmosphere from cars and factories damage plants. The most common type of pollution is *smog*. Air pollution damage is most commonly a problem in urban areas, but it also occurs in rural areas where gardens are located downwind from factories. Some plants are severely affected and may even die. Flower production is reduced on pollution-damaged plants. The three most common pollutants are ozone, PAN (peroxyacetyl nitrate), and sulfur dioxide. Many environmental factors affect a plant's susceptibility to air pollution, including temperature, air movement, light intensity, and soil and air moisture.

Solution: Air pollution injury is usually a localized problem. Check with your neighbors to see if the same kinds of plants in their gardens have been affected the same way. Because injury from air pollutants is similar in appearance to injury from nutrient deficiencies, insects, diseases, and mites, these problems should be eliminated as causes before the damage is attributed to air pollution. Nothing can be done about air pollutants.

Aphids

Aphids on pepper (4× life size).

Problem: Leaves are curled, distorted, and yellowing. Often the flowers are malformed. Tiny (⅛-inch) yellow, green, or dark-colored soft-bodied insects are clustered on leaves, stems, and flowers. A shiny, sticky substance may coat the leaves. Ants are sometimes present.

Analysis: Aphids do little damage in small numbers. They are extremely prolific, however, and populations can rapidly build up to damaging numbers during the growing season. Damage occurs when the aphid sucks the juices from the leaves and buds. The aphid is unable to completely digest all the sugar in the sap, and it excretes the excess in a fluid called *honeydew*, which often drops onto the leaves below. A sooty mold fungus may develop on the honeydew, causing the leaves to appear black and dirty. Ants feed on this sticky substance and are often present where there is an aphid infestation. In warm areas, aphids are active year-round. In cooler climates, where winter temperatures drop below freezing, the adults cannot survive. Eggs the aphids lay in the fall on tree bark, old leaves, and plant debris, however, can survive the winter and cause reinfestation in the spring. Aphids transmit viral plant diseases such as mosaics.

Solution: Spray with an insecticide containing *acephate* (Orthene®), *diazinon*, or *malathion*. Make sure that your plant is listed on the product label. Clean up plant debris in the fall.

INSECTS (continued)

Ants

Ants on Euphorbia *(½ life size).*

Problem: Ants crawl on the plants and soil. In many cases, these plants are infested with aphids, scales, mealybugs, or whiteflies.

Analysis: Ants, familiar to gardeners throughout the country, do not directly damage plants. Ants may be present for any of several reasons. Many ants feed on honeydew, a sweet, sticky substance excreted by several species of insects, including aphids, scales, mealybugs, and whiteflies. Ants are attracted to plants infested with these pests. In order to ensure an ample supply of honeydew, ants may actually carry aphids to healthy plants. Aphid infestations are frequently spread in this manner. Ants may also feed on flower seeds and nectar. Although they do not feed on healthy plants, they may eat decayed or rotted plant tissue. Ants generally live in underground colonies or nests. Certain species may form colonies in trees or in building foundations.

Solution: Destroy ant nests by treating anthills with ORTHO Diazinon Granules or by spraying the nest and surrounding soil with ORTHO Diazinon Ultra Insect Spray. Control aphids, scales, mealybugs, and whiteflies by spraying the plants with an insecticide containing *malathion* or *diazinon*. Make sure that your plant is listed on the product label. Avoid spraying when flowers and bees are present. Ants can be kept off individual plants with sticky barriers.

Spittlebugs

Spittlebug froth.

Problem: Masses of white, frothy foam are clustered between the leaves and stems. If the froth is removed, small, green, soft-bodied insects can be seen feeding on the plant tissue. The plant may be stunted.

Analysis: Spittlebugs, also known as *froghoppers*, appear in the spring. Spittlebug eggs, laid in the fall, survive the winter to hatch when the weather warms in the spring. The young spittlebugs, called *nymphs*, produce a foamy froth that protects them from sun and predators. This froth envelops the nymphs completely while they suck sap from the tender stems and leaves. Adult spittlebugs are not as damaging as the nymphs. The adults are ¼ inch long, pale yellow to dark brown, and winged. They hop or fly away quickly when disturbed. Spittlebugs seldom harm plants, but if infestation is very heavy, plant growth may be stunted. The presence of spittlebugs is usually objectionable only for cosmetic reasons.

Solution: Wash spittlebugs from plants with a garden hose. If plants are heavily infested, spray with an insecticide containing *malathion*, *acephate* (Orthene®), or *methoxychlor*. Make sure that the plant is listed on the product label. Repeated treatments are rarely necessary.

Caterpillars

Caterpillar (life size).

Problem: Holes appear in the leaves and buds. Leaves, buds, and flowers may be entirely sheared off. Caterpillars are feeding on the plants.

Analysis: Many species of these moth or butterfly larvae feed on garden plants. Usually the adult moths or butterflies begin laying eggs on garden plants with the onset of warm spring weather. The larvae that emerge from these eggs feed on the leaves, flowers, and buds for 2 to 6 weeks, depending on weather conditions and species. Mature caterpillars pupate in cocoons either buried in the soil or attached to leaves, tree bark, or buildings. Some caterpillar species have only one generation per year. With these species, all the caterpillars hatch, grow, and pupate at the same time. Other species have numerous generations, so caterpillars of various sizes may be present throughout the growing season. The last generation of caterpillars in the fall survives the winter as pupae. The adult moths and butterflies emerge the following spring. Many caterpillars feed only at night and are not obvious during the day. Look at night with a flashlight.

Solution: Spray infested plants with an insecticide containing *acephate* (Orthene®), *carbaryl*, or *diazinon* or with the bacterial insecticide *Bacillus thuringiensis* (Bt). Make sure that your plant is listed on the product label. Avoid spraying when bees are present. Small numbers of caterpillars may be removed by hand.

Beetles

Cucumber beetle feeding on zinnia (½ life size).

Problem: Insects are chewing holes in the leaves and flowers. Their hard wing covers are folded across their backs, meeting in a straight line down the center of the back. They are frequently shiny and brightly colored.

Analysis: Many species of beetles infest flowers. In the spring or summer, beetles fly to garden plants and feed on flowers, buds, and leaves. Punctured flower buds usually fail to open, and fully open flowers are often eaten. Because many beetles feed at night, only their damage may be noticed, not the insects. Female beetles lay their eggs in the soil or in flowers in late summer or fall. The emerging larvae crawl down into the soil to spend the winter, or they mature and pass the winter in plant debris. The larvae of some beetles feed on plant roots before maturing in the fall or spring.

Solution: Spray infested plants with an insecticide containing *diazinon*, *acephate* (Orthene®), or *malathion*. Make sure that your plant is listed on the product label.

Japanese beetle

Damaged zinnias. Inset: Japanese beetle (life size).

Problem: Leaf tissue has been eaten between the veins, making the leaves appear lacy. Flowers are eaten. Metallic green-and-bronze winged beetles, ½ inch long, feed in clusters on the flowers and foliage.

Analysis: As its name suggests, the Japanese beetle (*Popillia japonica*) is native to Japan. It was first seen in New Jersey in 1916 and has since become a major pest in the eastern United States. It feeds on hundreds of species of plants. The adult beetles are present from June to October. They feed only in the daytime and are most active on warm, sunny days. The female beetles live for 30 to 45 days. Just before they die, they lay their eggs just under the soil surface in lawns. The grayish white grubs that hatch from these eggs feed on grass roots. As the weather turns cold in the late fall, the grubs burrow 8 to 10 inches into the soil, where they hibernate. When the soil warms up in the spring, the grubs migrate back to the surface and resume feeding. They pupate and, in late May or June, reemerge as adult beetles.

Solution: Spray infested plants with an insecticide containing *acephate* (Orthene®), *carbaryl*, or *malathion*. Make sure that your plant is listed on the product label.

Aster leafhopper

Aster leafhopper (4× life size).

Problem: Spotted, pale green insects up to ⅛ inch long hop or fly away quickly when the plant is touched; nymphs crawl away sideways like crabs. The leaves are stippled and may be yellowing.

Analysis: Aster leafhopper (*Macrosteles fascifrons*), also known as the *six-spotted leafhopper*, feeds on many vegetable and ornamental plants. It generally feeds on the undersides of leaves, sucking the sap, which causes stippling. This leafhopper transmits aster yellows, a plant disease that can be damaging. Leafhoppers at all stages of maturity are active during the growing season. They hatch in the spring from eggs laid on perennial weeds and ornamental plants. Even areas where winters are so cold that the eggs cannot survive are not free from infestation because leafhoppers migrate from warmer regions on wind currents in the spring.

Solution: Spray plants with an insecticide containing *acephate* (Orthene®), *diazinon*, or *malathion*. Check to make sure that your plant is listed on the product label. Avoid spraying when flowers and bees are present. Eradicate weeds—especially thistles, plantains, and dandelions—that may harbor leafhopper eggs and aster yellows.

INSECTS (continued)

LEAVES, FLOWERS CHEWED

Greenhouse whitefly

Whiteflies on fuchsia (2× life size).

Problem: Tiny, white, winged insects 1/16 inch long feed mainly on the undersides of leaves.  Nonflying, scalelike larvae covered with white waxy powder may also be on the undersides of leaves. When the plant is touched, insects flutter rapidly around it. Leaves may be mottled and yellow. In warm-winter areas, black mold may cover leaves. The plant may grow poorly.

Analysis: Greenhouse whitefly (*Trialeurodes vaporariorum*) is a common pest of many garden and greenhouse plants. The four-winged adult lays eggs on the undersides of leaves. The larvae are the size of a pinhead, flat, oval, and semitransparent, with white waxy filaments radiating from the body. They feed for about a month before changing to the adult form. Both larval and adult forms suck sap from the leaves. The larvae are more damaging because they feed more heavily. Adults and larvae cannot fully digest all the sugar in the plant sap, and they excrete the excess in a fluid called *honeydew*, which often drops onto the leaves below. A sooty mold fungus may develop on the honeydew. In warm-winter areas, these insects can be active year-round, with eggs, larvae, and adults present at the same time. Whiteflies are unable to live through freezing winters. Spring reinfestations in freezing-winter areas come from migrating whiteflies and from infested greenhouse-grown plants placed in the garden.

Solution: Spray infested plants with an insecticide containing *acephate* (Orthene®), *diazinon*, or *malathion*. Make sure that your plant is listed on the product label. When spraying, be sure to cover both the upper and lower surfaces of the leaves.

Nocturnal insects

Cineraria damaged by nocturnal insects.

Problem: Holes appear in the leaves and flowers. Some of the leaves, stems, and flowers may be sheared off. No insects are seen feeding on the plants during the day. When affected plants are inspected at night with a flashlight, insects may be seen feeding on the foliage and flowers.

Analysis: Several kinds of insects feed on plants only at night, including some beetles, weevils, and caterpillars and all earwigs and cutworms. Beetles are hard-bodied insects with tough, leathery wing covers. The wing covers meet in the middle of the back, forming a straight line. Weevils are a type of beetle with a long snout. Earwigs are reddish brown, flat, elongated insects up to 1 inch long with pincers projecting from the rear of the body. Caterpillars and cutworms are smooth or hairy, soft-bodied worms. These insects usually hide in the soil, in debris, or in other protected locations during the day.

Solution: To control beetles and weevils, several different insecticides are used, including some containing *malathion* or *diazinon* for beetles, and some containing *malathion*, *chlorpyrifos*, and *carbaryl* for weevils; look up the entry for your specific plant. To control caterpillars, use ORTHO Orthene Systemic Insect Control or ORTHO Isotox Insect Killer. To control earwigs, spray with ORTHO Diazinon Ultra Insect Spray or dust with ORTHO Ant-Stop Ant Killer Dust around plants, flower pots, house and porch foundations, and woodpiles. To control cutworms, use ORTHO Diazinon Ultra Insect Spray. Always make sure that your plant is listed on the product label.

Snails and slugs

Snails feeding on hosta (1/4 life size).

Problem: Holes are chewed in the leaves, or entire leaves may be sheared from the stems.  Flowers are partially eaten. Silvery trails wind around on the plants and soil nearby. Snails or slugs may be seen moving around or feeding on the plants, especially at night. Check for them by inspecting the garden at night by flashlight.

Analysis: Snails and slugs are mollusks and are related to clams, oysters, and other shellfish. They feed on a wide variety of garden plants. Like other mollusks, snails and slugs need to be moist all the time. For this reason, they avoid direct sun and dry places and hide during the day in damp places, such as under flowerpots or in thick ground covers. They emerge at night or on cloudy days to feed. Snails and slugs are similar in appearance, except that the snail has a hard shell into which it withdraws when disturbed. Slugs lay white eggs encased in a slimy mass in protected places. Snails bury their eggs in the soil, also in a slimy mass. The young look like miniature versions of their parents.

Solution: Scatter ORTHO Bug-Geta Snail & Slug Killer in bands around the areas you wish to protect. Also scatter the bait in areas where snails or slugs might be hiding, such as in dense ground covers, weedy areas, compost piles, or pot storage areas. Before spreading the bait, wet down the treated areas to encourage snail and slug activity that night. Repeat the application every 2 weeks for as long as snails and slugs are active.

SEEDLINGS DIE

Wilted seedlings

Damping-off of marigold seedlings.

Problem: Seedlings die soon after emerging from the soil and are found lying on the ground.

Analysis: Seedlings may wilt and die from lack of water or from disease.

1. Dehydration: Seedlings are succulent and have shallow roots. If the soil dries out even an inch or so below the surface, seedlings may die.

2. Damping-off: Young seedlings are very susceptible to damping-off, a plant disease caused by fungi. Damping-off commonly occurs in wet soil with a high nitrogen level. Damping-off can be a problem when the weather remains either cold or cloudy and wet while seeds are germinating or if seedlings are too heavily shaded.

Solution: The numbered solutions below correspond to the numbered items in the analysis.

1. Do not allow the soil to dry out completely. Water when the soil surface starts to dry slightly. During warm or windy weather, you may need to water several times per day.

2. Allow the soil surface to dry slightly between waterings. Start seeds in Scotts Starter Fertilizer. Add Miracle-Gro Plant Food after the seedlings have produced their first true leaves. Protect seeds during germination by coating them with a fungicide containing *captan* or *thiram*. Add a pinch of fungicide to a packet of seeds (or ½ teaspoon per pound), and shake well to coat the seeds with the fungicide.

Germination problems

Damping-off of petunia.

Damping-off of snapdragon.

Problem: Seedlings fail to emerge.

Analysis: Seeds may fail to emerge for several reasons.

Solution: The numbered solutions below refer to the numbered items in the analysis.

1. Dehydration: Once the seeds have started to grow, even before they emerge from the soil they will die easily if allowed to dry out.

1. Do not allow the soil to dry out completely. Check seedbed and seed flats at least once a day. Water when soil surface starts to dry.

2. Damping-off: Germinating seedlings are very susceptible to damping-off, a plant disease caused by fungi. These fungi inhabit most soils, decaying the young seedlings as they emerge from the seed. Damping-off is favored by wet, cool soil.

2. Allow the soil surface to dry slightly between waterings. Start seeds in Scotts Starter Fertilizer. Add Miracle-Gro Plant Food after the seedlings have produced their first true leaves. Delay planting in spring until the soil warms. Use seed treated with a registered fungicide.

3. Slow germination: Seeds of different plants vary considerably in the amount of time they require to germinate.

3. Check the list of flower germination periods on page 351 to see if your seeds germinate slowly.

4. Poor seed viability: Seeds that are old, diseased, or of inferior quality may fail to germinate.

4. Purchase seeds from a reputable nursery or seed company. Plant seeds packed for the current year.

5. Wrong planting depth: Seeds of different flowers vary in their planting depth requirements. If planted too deeply or shallowly, the seeds may fail to germinate.

5. Plant seeds at the proper depth. Follow the instructions on the seed packet or consult with a reputable nursery.

6. Seeds washed away: If a seedbed is flooded or watered with a forceful spray, the seeds may wash away. Heavy rains can also wash seeds away.

6. Water seedbeds gently. Do not allow the water to puddle and run off. Use a watering can or hose nozzle that delivers a gentle spray.

7. Cold weather: Cold weather may delay seed germination considerably or prevent germination entirely.

7. Even though germination may be delayed, many seeds will probably sprout when the weather warms. The next year, plant seeds later in the season, after soil has warmed up.

SEEDLINGS DIE (continued)

Snails and slugs on seedlings	Cutworms on seedlings	Animal pests on seedlings

Seedlings sheared off by snails.

Cutworm damage to petunia seedlings.

Celery seedlings eaten by mice.

Problem: Seedlings are sheared off and eaten, with only the stems emerging from the ground. Silvery trails wind around on the plants and soil nearby. Snails or slugs may be seen moving around or feeding on the plants, especially at night. Check for them by inspecting the garden after dark with a flashlight.

Analysis: Snails and slugs are mollusks and are related to clams, oysters, and other shellfish. They feed on a wide variety of garden plants. Like other mollusks, snails and slugs need to be moist all the time. For this reason, they avoid direct sun and dry places and hide during the day in damp places, such as under flowerpots or in thick ground covers. They emerge at night or on cloudy days to feed. Snails and slugs are similar in appearance, except that the snail has a hard shell into which it withdraws when disturbed. Slugs lay white eggs encased in a slimy mass in protected places. Snails bury their eggs in the soil, also in a slimy mass. The young look like miniature versions of their parents.

Solution: Apply ORTHO Bug-Geta Snail & Slug Killer in bands around the areas you wish to protect. Also use the bait in areas where snails or slugs might be hiding, such as in dense ground covers, weedy areas, compost piles, or pot storage areas. Before applying the bait, wet down the treated areas to encourage snail and slug activity that night. Repeat the application every 2 weeks for as long as snails and slugs are active.

Problem: Seedlings are chewed or cut off near the ground. Gray, brown, or black worms, 1½ to 2 inches long, may be found about 2 inches deep in the soil near the base of the damaged plants. The worms coil when disturbed.

Analysis: Several species of cutworms attack plants in the vegetable garden. The most likely pests of seedlings planted early in the season are the surface-feeding cutworms. A single surface-feeding cutworm can sever the stems of many young plants in one night. Cutworms hide in the soil during the day and feed only at night. In the South, cutworms may also attack fall-planted seedlings.

Solution: Apply ORTHO Diazinon Ultra Insect Spray; ORTHO Bug-Geta Plus Snail, Slug & Insect Killer; or ORTHO Dursban® Lawn & Garden Insect Control around the base of undamaged plants when stem cutting is observed. Because cutworms are difficult to control, applications may need to be repeated at weekly intervals. Before transplanting in the same area, apply a preventive treatment of one of the above products and work it into the soil. Cultivate the soil thoroughly in late summer and fall to expose and destroy eggs, larvae, and pupae. Further reduce damage with cutworm collars around the stem of each plant. Make collars out of stiff paper or aluminum foil bent into a cylinder or out of tin cans or paper cups with the bottoms removed. These collars should be at least 2 inches high and surround the plant stem fairly closely when pressed into the soil.

Problem: Plants and seedlings may be partially or entirely eaten. Mounds of soil, ridges, or tunnels may be clustered in the yard. There may be tiny holes in the soil and small, dry, rectangular brown pellets on the ground near the damaged plants. Various birds and animals may be seen feeding in the garden, or their tracks may be around the damaged plants.

Analysis: Several different animals feed on flowers. Pocket gophers (found primarily in the West), field mice, rabbits, and deer cause major damage by eating seedlings or mature plants. Certain species of birds feed on seedlings. Moles, squirrels, woodchucks, and raccoons are generally less damaging but may also feed on flower roots, bulbs, and seeds. Even if these animals are not directly observed, their presence is usually evidenced by their tunnels, burrows, droppings, and tracks.

Solution: Fences, cages or screens, traps, repellents, or baits will protect most garden plants from animal damage or at least greatly reduce the damage. You can protect seedlings from squirrels, rabbits, and birds by placing cages made out of 1-inch chicken wire over small seedling beds.

AGERATUM (Flossflower)

Gray mold

Gray mold, a widespread fungal disease.

Problem: Brown spots and blotches appear on the leaves and possibly on the stems. As the disease progresses, a fuzzy brown or grayish mold forms on the infected tissue. Gray mold and spots may appear on the flowers, especially during periods of cool, wet weather. The leaves and stems may be soft and rotted.

Analysis: Gray mold, a widespread plant disease, is caused by a fungus (*Botrytis* species). The fungus initially attacks foliage and flowers that are weak or dead, causing spotting and mold. The fuzzy mold that develops is composed of millions of microscopic spores. Once gray mold has become established on plant debris and weak or dying leaves and flowers, it can invade healthy plant tissue. The fungus is spread by splashing water and by bits of infected plant debris that land on the leaves. Crowded plantings, rain, and overhead watering enhance the spread of the disease. Cool temperatures, moisture, and high humidity favor gray mold growth.

Solution: Spray infected plants with ORTHO Multi-Purpose Fungicide Daconil 2787® Plant Disease Control, repeating according to label directions as long as mold is visible or the weather remains favorable for the disease. The key to avoiding problems is good sanitation. Remove old flowers and dying or infected leaves and stems. Clean up and destroy plant debris. Avoid wetting the foliage. Water plants at soil level (instead of from overhead) and space them far enough apart so that air can circulate between them.

Tobacco budworm & corn earworm

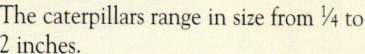
Corn earworms (⅛ life size).

Problem: Striped green, brown, or yellow caterpillars are chewing holes in leaves and buds. The caterpillars range in size from ¼ to 2 inches.

Analysis: Tobacco budworm and corn earworm (*Heliothis* species), closely related caterpillars, are the larval stages of night-flying moths. In addition to feeding on many ornamental plants, the caterpillars of both moths are major agricultural pests. The corn earworm especially is considered one of the most destructive pests of corn in the country. The moths survive the winter as pupae in the soil, emerging in the spring to lay their light yellow eggs singly on the undersides of leaves. The caterpillars hatch in 2 to 8 days and feed for several weeks on leaves and buds, then crawl into the soil and pupate. The new adults emerge 1 to 3 weeks later to begin the cycle again, laying eggs in the evenings and on warm, overcast days. In cooler areas of the country, the caterpillars are present from early spring to the first frost. In warmer areas, the feeding caterpillars are present year-round.

Solution: Control corn earworms and tobacco budworms with ORTHO Orthene Systemic Insect Control, applied when the caterpillars first appear. Repeat every 10 to 14 days as needed. Deep cultivating in the fall and winter will help destroy some of the overwintering pupae.

ALCEA (Hollyhock)

Rust

Rust, a common disease of hollyhock

Problem: Yellow or orange spots appear on the upper surfaces of leaves in the early spring. Grayish brown pustules develop on the undersides of leaves and possibly on the stems. These pustules may turn dark brown to black as the growing season progresses. Severely infected leaves shrivel, turn gray or tan, and hang down.

Analysis: Hollyhock rust (*Puccinia malvacearum*) is a fungus that causes the most serious and widespread disease of hollyhock. The fungus spends the winter as spores on living plant tissue and plant debris. Infection starts in the early spring as soon as conditions are favorable for plant growth. Splashing water and air currents spread the spores to healthy plants. Several weeds known as cheeseweeds or mallows (*Malva* species) are frequently infected with rust and are a source of spores. Rust is also favored by wet conditions.

Solution: Spray with ORTHO Multi-Purpose Fungicide Daconil 2787® Plant Disease Control or with a fungicide containing *triadimefon* in the spring as soon as first signs of infection are noticed. Spray foliage thoroughly, being sure to cover both the upper and lower surfaces of the leaves. The fungicide protects the new, healthy foliage but will not eradicate the fungus on diseased leaves. Spray once every 7 to 10 days or as often as necessary to protect new growth until the end of the growing season. Remove and destroy all infected foliage and any nearby cheeseweeds in the fall when the plant has stopped growing and again in early spring. Pick off and destroy infected plant parts during the growing season. Water early in the day so that foliage will dry thoroughly.

ANTIRRHINUM (Snapdragon)

Rust

Rust on snapdragon foliage and stem.

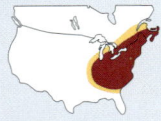

Problem: Pale yellow spots appear on the upper surfaces of leaves. Reddish brown pustules of spores develop on the undersides of leaves. Often these pustules form concentric circles. Spore masses may be on the stems. Severely infected leaves dry up. The plant is stunted and may die prematurely.

Analysis: Rust, a common disease of snapdragons, is caused by a fungus (*Puccinia antirrhini*). Wind and splashing water spread the fungal spores. Rust can survive only on living plant tissue and as spores on seed. It does not persist on dead plant parts. Plants must be wet for 6 to 8 hours before the fungus can infect the leaf surface. The disease thrives in moist conditions, cool nights (50° to 55°F), and warm days (70° to 75°F). Temperatures in the 90°F range and higher kill the spores.

Solution: Spray infected plants with ORTHO Multi-Purpose Fungicide Daconil 2787® Plant Disease Control or with a fungicide containing *mancozeb* at intervals of 5 to 8 days. Avoid wetting the foliage. To give foliage a chance to dry out, water in the morning rather than in the late afternoon or evening. Pick off and destroy infected plant parts during the growing season. Remove all snapdragon plants at the end of the growing season to prevent infected plants from reinfecting new plantings. Space plants far enough apart to allow good air circulation. Plant resistant varieties.

Root and stem rot

Root and stem rot, a fungal disease.

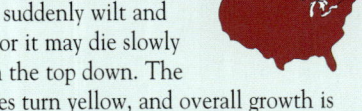

Problem: The plant may suddenly wilt and die, or it may die slowly from the top down. The leaves turn yellow, and overall growth is stunted. There may be lesions on the stems. The roots are decayed.

Analysis: Root and stem rot is caused by several fungi that thrive in waterlogged, heavy soils. They can attack the plant stems and roots directly or enter them through wounds. Infection causes the stems and roots to decay, resulting in wilting, yellowing leaves, and plant death. These fungi are spread by infested soil and transplants, contaminated equipment, and moving water. Many of these organisms also cause *damping-off* of seedlings (see page 91).

Solution: Let the soil dry out between irrigations. Minimize drought stress while drying out soil by moving containers into shade or erecting shade cloth over beds, by spraying with antitranspirant, and by constructing a tent of clear plastic over plants in shade. Resume watering when signs of drought stress appear. Improve soil drainage. Before planting the following year, apply a fungicide containing *PCNB* to the soil and work it in to a depth of 6 inches.

ASTILBE

Root nematodes

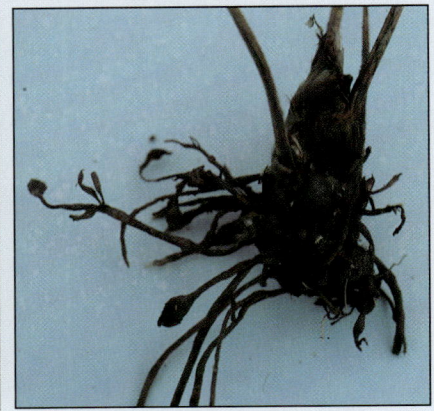

Root nematode damage on Astilbe.

Problem: The plant is stunted and growing poorly. Leaves are curled and may be yellowed. The plant may be pulled out of the ground easily. Roots are sparse, short, and dark. Small knots (1/16 inch) may be visible on the roots.

Analysis: Root nematodes are microscopic worms that live in the soil. They feed on plant roots, damaging and stunting them or causing them to become enlarged. The damaged roots can't supply sufficient water and nutrients to the plant, and the plant becomes stunted or slowly dies. Nematodes are found throughout North America, especially in areas with moist, sandy loam soil. They can move only a few inches per year on their own, but they may be carried long distances by soil, water, tools, or infested plants. Laboratory testing of roots and soil is the only positive method to confirm the presence of nematodes. Contact your local county extension office for sampling instructions and addresses of testing laboratories. Problems such as poor soil structure, drought stress, overwatering, nutrient deficiency, and root rot can produce symptoms of decline similar to those caused by nematodes. Root weevils, such as the black vine weevil, may cause similar symptoms. Eliminate these problems as causes before sending soil and root samples for testing.

Solution: No chemicals available to homeowners kill nematodes in planted soil. Nemetodes can be controlled before planting, however, by soil fumigation or solarization. To solarize, cultivate and wet an area thoroughly before covering it with clear polyethylene film. Leave in place 4 to 6 weeks.

GROWING GUIDE

Adaptation: Throughout North America.

Flowering Time:
Tuberous: Early to late summer.
Fibrous: Spring to late fall.

Light:
Tuberous: Filtered light to full shade.
Fibrous: Full shade to half-day sun. Dark-colored plants can tolerate full-day sun.

Planting Time:
Tuberous: Start indoors in the spring in zones 1 through 9. To determine your zone, see the map on page 336. You may start tuberous begonias outdoors in zone 10.
Fibrous: Start all year round in zones 9 and 10. Start in spring through summer in zones 2 through 8.

Soil:
Well drained, rich in organic matter. pH 6.0 to 7.0.

Fertilizer:
Tuberous: Fertilize with Miracid Plant Food every 2 weeks at half the recommended rate. Stop fertilizing about 6 weeks before the first fall frost.
Fibrous: Fertilize with a balanced plant food according to label directions.

Water:
How much: Apply enough water to wet the soil 6 to 8 inches deep.
How often: Water when the soil 1 inch below the surface is moist but not wet.

Gray mold

Gray mold, a common fungal disease.

Problem: Brown spots and blotches appear on the leaves and possibly on the stems. As the disease progresses, a fuzzy brown or grayish mold forms on the infected tissue. Gray mold and spots often appear on the flowers, especially during periods of cool, wet weather. The leaves and stems may be soft and rotted.

Analysis: Gray mold, a widespread plant disease, is caused by a fungus (*Botrytis cinerea*) that is found on most dead plant tissue. The fungus initially attacks foliage and flowers that are weak or dead, causing spotting and mold. The fuzzy mold that develops is composed of millions of microscopic dark spores. Once gray mold has become established on plant debris and weak or dying leaves and flowers, it can invade healthy plant tissue. The fungus is spread by splashing water and by bits of infected plant debris that land on the leaves. Cool temperatures, moisture, and high humidity favor gray mold growth. Crowded plantings, rain, and overhead watering also enhance the spread of the disease. Infection is a greater problem in spring and fall, when temperatures are lower. In warm-winter areas where freezing is rare, gray mold can be a year-round problem.

Solution: Control gray mold with a fungicide containing *chlorothalonil* (Daconil 2787®) or *mancozeb*. Spray every 1 to 2 weeks for as long as the mold is visible. Clean up plant debris, and remove dying or infected leaves, stems, and flowers. Provide enough space between plants to allow good air circulation. Try to avoid wetting the foliage when watering.

Leaf spots

Leaf spot can be caused by several fungi.

Problem: Spots and blotches appear on the leaves. The spots may be yellow, red, tan, gray, or brown. They range in size from barely visible to ¼ inch in diameter. Several spots may join together to form blotches. Leaves may be yellowing and dying. Leaf spotting is most severe in warm, humid weather.

Analysis: Begonias are susceptible to several fungi that cause leaf spots. Some of these fungi may eventually kill the plant or weaken it so that it becomes susceptible to attack by other organisms. Others merely cause spotting that is unsightly but not harmful. These fungi are spread by splashing water, wind, insects, tools, and infected transplants and seed. They survive the winter in diseased plant debris. Some leaf spot organisms affect a large number of plants. Most of these fungi do their greatest damage during mild weather (50° to 85°F). Infection is favored by moist conditions.

Solution: Spray with ORTHO Multi-Purpose Fungicide Daconil 2787® Plant Disease Control or a fungicide containing *mancozeb* every 7 to 10 days. Because leaf spots are favored by warm, humid conditions, it is important to spray frequently during these periods. The fungicides protect the new, healthy foliage. They will not eradicate the fungus on leaves that are already infected, however. Clean up and destroy infected leaves and debris. Water early in the day to allow the leaves to dry by nightfall.

BEGONIA (continued)

Bacterial leaf spot

Bacterial leaf spot on begonia.

Problem: Small, blisterlike spots appear on the leaves. The spots are translucent and turn brown with yellow, translucent margins. The spots enlarge and run together, giving the leaf a blotched appearance. Sometimes a slimy substance oozes from the infected areas, turning light brown as it dries. Infected leaves often die prematurely. If the stems become infected, the entire plant may collapse.

Analysis: Bacterial leaf spot is caused by bacteria (*Xanthomonas campestris* pv. *begoniae*) that infect tuberous and fibrous begonias. The slimy substance that oozes from infected lesions is composed of bacterial cells that can live for 3 months or more. The bacteria are spread by splashing water, contaminated equipment, and infected transplants. Infection thrives in high humidity. Localized leaf infection causes early leaf drop. If the plant's water-conducting tissue is infected, the whole plant softens and collapses.

Solution: Spray with a copper-based compound at intervals of 7 to 10 days to prevent the spread of the disease. Infected tissue will not recover, but new growth and healthy leaves will be protected. Cut off and discard infected plant parts. Wash your hands thoroughly and disinfect pruning shears after working on infected plants by dipping the shears after each cut into a solution of 1 part chlorine bleach to 9 parts water. Remove and destroy severely infected plants and the soil surrounding them. Avoid wetting or splashing leaves. Space plants far enough apart to allow good air circulation.

Leaf nematode

Leaf nematode damage.

Problem: Angular brown leaf blotches develop first on the lower leaves, then on the upper. The blotches enlarge, and eventually the leaves curl up, wither, and drop off. The plant is stunted, and new leaf buds may not develop.

Analysis: This plant condition is caused by leaf nematodes (*Aphelenchoides olesistus*)—microscopic worms that live and feed inside the leaf tissue. Infestation occurs when the foliage is wet. Nematodes migrate in the thin film of water on the outside of the leaf to infect healthy tissue. They are spread from plant to plant by splashing water and are most severe in warm, humid areas. Leaf nematodes can survive for 3 or more years in plant debris and in the soil.

Solution: Remove and destroy severely infested plants. Pick off and destroy all infested leaves and the two leaves directly above them. As much as possible, avoid wetting the foliage. Inspect new plants carefully to be sure they are not diseased, and do not plant in infested soil. Spray weekly with an insecticide containing *malathion* or *dimethoate* using a spreader-sticker, until the symptoms stop spreading.

Mealybugs

Mealybugs (½ life size).

Problem: Oval, white insects up to ¼ inch long cluster in white, cottony masses on stems and leaves. Leaves may be deformed and withered. The infested leaves may be shiny and sticky or may be covered with a sooty mold.

Analysis: Several species of mealybugs feed on begonias. Mealybugs damage plants by sucking sap, causing leaf distortion and death. The adult female mealybug may produce live young or may deposit her eggs in white, fluffy masses of wax. The immature mealybugs, called *nymphs,* are very active and crawl all over the plant. Soon after the nymphs begin to feed, they exude filaments of white wax that cover their bodies, giving them a cottony appearance. As they mature, their mobility decreases. Mealybugs cannot fully digest all the sugar in the sap, and they excrete the excess in a fluid called *honeydew.* Mealybugs can be spread when they are brushed onto uninfested plants or when young, active nymphs crawl to nearby plants. They may also be spread by the wind, which can blow egg masses and nymphs from plant to plant. Mealybug eggs and some adults can survive the winter in warm climates. Spring reinfestations in colder areas come from infested new plants placed in the garden.

Solution: Spray infested plants with ORTHO Orthene Systemic Insect Control. Spray at intervals of 7 to 10 days until the mealybugs are gone. Gently hose down plants to wash off honeydew. Remove and destroy severely infested leaves and plants.

CALENDULA

Cabbage looper

Cabbage loopers feeding (⅓ life size).

Problem: Foliage and flower buds are chewed. Leaves have ragged edges and irregular or round holes. Green caterpillars up to 1½ inches long with white stripes feed on the leaves.

Analysis: Cabbage loopers (*Trichoplusia ni*) are destructive caterpillars that feed on many garden ornamentals and vegetables. As the weather warms in spring, the brownish adult moth lays its tiny, pale green eggs at night on the upper surfaces of leaves. The eggs hatch into active green larvae, which feed extensively on buds and foliage for 2 to 4 weeks. Looper damage can occur from early spring through late fall. The caterpillars spend the winter as pupae attached to plant leaves.

Solution: Spray with ORTHO Isotox Insect Killer or ORTHO Orthene Systemic Insect Control when the caterpillars first appear. Repeat 10 to 14 days later if reinfestation occurs. In the fall, remove plant debris and weeds that may harbor pupae. The bacterial insecticide *Bacillus thuringiensis* (Bt) is effective when sprayed on young loopers.

CALLISTEPHUS
(China Aster)

Aster yellows

Aster yellows, spread by leafhoppers.

Problem: Leaf veins pale and may lose all their color. Part or all of the foliage yellows. The flowers are distorted and may turn green. The plant may grow many thin stems bearing pale, spindly leaves. The plant is usually dwarfed.

Analysis: Aster yellows is caused by phytoplasmas, microscopic organisms similar to bacteria. The phytoplasmas are transmitted from plant to plant by leafhoppers. The symptoms of aster yellows are more severe and appear more quickly in warm weather. Although the disease may be present in the plant, aster yellows may not manifest its symptoms in temperatures of 55°F or less. Aster yellows infects many ornamental plants, vegetables, and weeds. For a list of plants susceptible to aster yellows, see page 352.

Solution: Aster yellows cannot be eliminated entirely, but it can be controlled. Remove and destroy infected China asters and other ornamentals showing aster yellows infection. To remove sources of infection, eradicate nearby weeds that may harbor aster yellows and leafhopper eggs. Spray leafhopper-infested plants with ORTHO Orthene Systemic Insect Control. Repeat the spray whenever leafhoppers are seen.

CANNA

Bud rot

Bud rot attacks young leaves and flowers.

Problem: The newly opened young leaves may be partially or entirely blackened, or they may be covered with tiny white spots. The older leaves may be distorted and are often covered with yellow or brown spots and streaks. In many cases, the flower buds turn black and die before they open. Entire stalks are often decayed. A sticky substance may coat infected leaf tissue.

Analysis: Bud rot is caused by bacteria (*Xanthomonas* species) that usually attack the young canna leaves and flowers while they are still curled in the buds. The bacteria can spread from the leaves and flowers into the stems, causing plant death. Some of the infected tissue may exude a sticky ooze filled with bacteria. The bacteria are spread by splashing water and rain and by direct contact with equipment, hands, and insects. Wet conditions enhance the spread of the disease. Bud rot survives through the winter in diseased rhizomes, contaminated soil, and infected plant debris to reinfect young plants the following spring.

Solution: There are no effective chemical controls for this disease. To control bud rot, it is important to reduce excess moisture around the plants. Water in the morning so the foliage will dry out during the day. Try to avoid wetting the foliage. Space plants far enough apart to allow good air circulation. Pick off infected leaves and flowers. Remove severely diseased plants and the soil immediately surrounding them. Clean up plant debris. Plant only healthy plants and rhizomes.

CENTAUREA (Bachelor's Button, Dusty Miller)

CHRYSANTHEMUM (Shasta Daisy, Marguerite, Mum)

Root and stem rot

Root and stem rot on dusty miller.

Problem: Leaves turn yellow, wilt, and eventually die. The roots and lower part of the stems may be soft and rotten. There may be white fungal strands on infected stems and around the base of the plant.

Analysis: Root and stem rot is caused by any of several fungi, also known as *water molds*, that persist indefinitely in the soil. These fungi thrive in waterlogged, heavy soil. Some of them attack the plant stems at the soil level, while others attack the roots. Infection causes the roots and stems to decay. This results in wilting, then yellowing leaves, and eventually the death of the plant. These fungi are generally spread by infested soil and transplants, contaminated equipment, and splashing or running water. Many of these organisms also cause damping-off of seedlings.

Solution: Allow the soil around the plant to dry out. Minimize drought stress by placing containers in shade or erecting shade cloth over beds; by spraying with antitranspirant; and by covering with a clear plastic tent if plants are in shade. Resume watering when signs of drought stress appear. Remove and discard severely infected plants. Avoid future root rot problems by planting in well-drained soil.

GROWING GUIDE

Adaptation:
Throughout North America.

Flowering Time:
Chrysanthemums: Late summer to late fall.
Shasta daisies: Late spring to late summer.
Marguerites: Spring to fall.

Light: Full sun.

Soil:
Any good garden soil. pH 6.0 to 7.5.

Fertilizer:
Fertilize with Scotts All-Purpose Plant Food or Miracle-Gro Plant Food according to label directions.

Water:
How much: Apply enough water to plants in the ground to wet the soil 8 to 12 inches deep.
How often: Water when the soil 1 inch below the surface is just barely moist.

Handling:
Chrysanthemums: Pinch plants frequently during the spring to encourage bushy growth.
Shasta daisies: Divide clumps every 2 to 3 years. Pinch off old flowers to encourage continued bloom.

Leggy growth

Weak, leggy growth on garden mum.

Problem: Mums are leggy, and many of the stems are thin and spindly. Some plants topple and may need to be tied.

Analysis: Legginess in mums can be caused by two things.
1. Natural growth pattern: Most mum varieties grow tall and leggy naturally.
2. Too much shade: Mums are sun-loving plants. Chrysanthemums planted in a shaded area produce thin, leggy growth even when pinched back. In shade, the plants may not flower well.

Solution: The numbered solutions below correspond to the numbered items in the analysis. To prevent leggy growth, follow these guidelines:
1. Plants that are leggy may be pinched back to encourage bushier growth as long as they have not yet formed flower buds. The following year, when new plants are 6 to 8 inches tall, carefully pinch or nip off the young growing tips just above a leaf. The tiny side bud located between this leaf and the stem will grow into a new branch. Every 2 weeks, pinch back all the new growing points that have formed as a result of the previous pinching. Stop pinching the plant by August to let flower buds develop. Purchase short-growing mum varieties.
2. Move or transplant chrysanthemums to a location that receives at least 4 or 5 hours of direct sun daily.

Rust

Rust on Chrysanthemum *foliage*

Problem: Pale spots appear on the upper surfaces of leaves. Chocolate brown pustules of spores form on the undersides of leaves and on the stems. Infected leaves may wither and fall prematurely. The plant is stunted and may die.

Analysis: Chrysanthemum rust, a common disease of chrysanthemums, is caused by a fungus (*Puccinia tanaceti*). Wind and splashing water spread the fungal spores. Rust can survive only on living plant tissue and as spores on seed; it does not persist on dead plant parts. Plants must remain wet for 6 to 8 hours before the fungus can infect the leaf surface. The disease thrives in moist conditions, cool nights, and warm days. Temperatures of 90°F and higher, if maintained for 24 hours, kill the spores. Rust can survive the winter on old infected plants.

Solution: Spray infected plants with ORTHO Multi-Purpose Fungicide Daconil 2787® Plant Disease Control or a fungicide containing sulfur or *triadimefon* at weekly intervals as soon as the disease is noticed; continue spraying throughout the growing season when weather conditions are favorable for disease. To allow foliage to dry out more quickly, water in the morning rather than in the late afternoon or evening. Pick off and destroy infected plant parts during the growing season. Space plants far enough apart to allow good air circulation. Remove and destroy all infected plants in the fall to keep them from reinfecting new plantings. The following year, grow only rust-resistant varieties.

Leaf nematode

Foliar nematode damage.

Problem: Fan-shaped or angular yellow brown to gray leaf blotches develop progressively upward from the lower leaves. The blotches join together, and the leaf turns brown or black. The leaf then withers, dies, and hangs down along the stem. The plant is stunted, and new leaf buds do not develop. In the spring, young, succulent, leafy growth becomes thickened, distorted, and brittle.

Analysis: The cause of this damage is a microscopic worm called a leaf nematode (*Aphelenchoides ritzema-bosi*) that lives and feeds inside the leaf tissue. The nematode is restricted in its movement by larger leaf veins. This confined feeding range creates the angular shape of the blotch. When the foliage is wet, the nematode migrates in the thin film of water on the outside of the leaf to infect healthy tissue. This pest is spread from plant to plant by splashing water. It penetrates the plant tissue by entering through small breathing pores on the underside of the leaf. Leaf nematodes are most damaging in the wet-summer, warmer regions of the country. They can survive for 3 years or more in plant debris and in the soil.

Solution: Remove and destroy severely infested plants. Pick off and destroy all infested leaves and the two leaves directly above them. Avoid wetting the foliage as much as possible. Check new plants carefully to be sure they are not diseased, and do not replant them in infested soil. Spray with an insecticide containing *malathion* using a spreader-sticker. Spray weekly until the symptoms stop spreading.

Thrips

Thrips damage.

Problem: Silvery white streaks and flecks appear on the leaves and flowers. Damaged leaves may become papery and distorted, dropping prematurely. Many shiny black specks are scattered on leaf surfaces. The leaves and flowers may be distorted and brown. Damage may appear in one location, then slowly spread over the plant. To ascertain if thrips are present, shake a blossom or tap a leaf over a sheet of paper. Tiny (⅟20-inch) insects resembling brown or straw-colored wood slivers can be seen. When disturbed, they hop or fly.

Analysis: Several species of thrips attack chrysanthemums, daisies, and many other garden plants. Thrips are generally found in protected locations such as the insides of the leaf and flower buds, where they feed by rasping the soft plant tissue, then sucking the released plant sap. The injured tissue dies and turns white, causing the characteristic streaking of the leaves and flowers. In cold climates, thrips feed and reproduce from spring until fall, then hibernate through the winter. In warm-winter climates, thrips feed and reproduce all year. These pests reach their population peak in late spring to midsummer. They are especially troublesome during prolonged dry spells.

Solution: Thrips can be kept under control but not eliminated entirely. Spray plants before they bloom with ORTHO Orthene Systemic Insect Control or ORTHO Malathion 50 Plus Insect Spray two or three times at weekly intervals. Repeat the spray if reinfestation occurs. Pick off and destroy old infested leaves and flowers. Keep beds and borders free of weeds.

CHRYSANTHEMUM (Shasta Daisy, Marguerite, Mum) (continued)

Verticillium wilt

Verticillium wilt caused by soil-borne fungus.

Problem: Leaves yellow, wilt, and die, starting with the lower leaves and progressing up the plant. Older plants may be stunted. Leaf wilting and death often affect only one side of the plant. Flowering is poor. Dark brown areas may be on the infected stems. When the stem is sliced open near the base of the plant, dark streaks and discoloration of the water-conducting inner stem tissue are revealed.

Analysis: Verticillium wilt affects many ornamental plants. It is caused by a soil-inhabiting fungus (*Verticillium* species) that persists indefinitely on plant debris or in the soil. The disease is spread by contaminated seeds, plants, soil, equipment, and groundwater. The fungus enters the plant through the roots and spreads up into the stems and leaves through the water-conducting vessels in the stems. These vessels become discolored and plugged. This plugging cuts off the flow of water to the leaves, causing leaf yellowing and wilting.

Solution: No chemical control is available; the best solution is to plant varieties that are resistant to *Verticillium*. When pruning, wash your hands and dip pruning shears after each cut into a solution of 1 part chlorine bleach to 9 parts water. It is best to destroy infected plants. *Verticillium* can be removed from the soil only by fumigation or solarization techniques. To solarize soil, cultivate and wet an area thoroughly before covering it with clear polyethylene film during the hottest time of the year. Leave in place 4 to 6 weeks.

Mosaic virus

Mosaic virus is spread by aphids.

Problem: Leaves are mottled, and the leaf veins may turn pale. Plants may be dwarfed and bushy. Flowers are small and may have brown streaks. Leaves, stems, and flowers may be deformed.

Analysis: Mosaic virus is caused by several different viruses. Mosaic is primarily transmitted from plant to plant by aphids. The symptoms of mosaic virus can vary considerably in their severity depending on the type of virus and variety of chrysanthemum. Viruses can be transmitted to chrysanthemums from many weeds and ornamental plants. Some plants may be infected with mosaic virus without showing the typical symptoms.

Solution: Infected plants cannot be cured. They should be immediately removed and destroyed. Spray aphid-infested plants in the area with ORTHO Diazinon Ultra Insect Spray, ORTHO Orthene Systemic Insect Control, or ORTHO Malathion 50 Plus Insect Spray. Repeat the spray at intervals of 7 days as often as necessary to keep the aphids under control. For spot-treatment of a few plants use ORTHO Rose & Flower Insect Killer according to label directions. To reduce the numbers of plants that may harbor viruses, keep your garden free of weeds.

Aster yellows

Aster yellows is spread by leafhoppers.

Problem: Leaf veins pale and may lose all their color. Part or all of the foliage turns yellow. Leaf edges may turn brown. The flowers are dwarfed and distorted and may turn green. The plant may grow many thin stems bearing pale, spindly leaves. The plant is generally stunted.

Analysis: Aster yellows is caused by phytoplasmas, microscopic organisms similar to bacteria. The phytoplasmas are transmitted from plant to plant primarily by leafhoppers. The symptoms of aster yellows are more severe and appear more quickly in warm weather. Although the disease may be present in the plant, aster yellows may not manifest its symptoms in temperatures of 55°F or less. The disease infects many ornamental plants, vegetables, and weeds. For a list of plants susceptible to aster yellows, see page 352.

Solution: Aster yellows cannot be eliminated entirely, but it can be controlled. Remove and destroy infected ornamental plants. To remove sources of infection, eradicate nearby weeds that may harbor aster yellows and leafhopper eggs. Spray leafhopper-infested plants with an insecticide containing *acephate* (Orthene®), *diazinon*, or *malathion*. Repeat the spray whenever leafhoppers are seen.

COLEUS

Mealybugs

Mealybugs (5× life size).

Problem: Oval, white insects up to ¼ inch long cluster in white, cottony masses on the stems and leaves. Leaves may be deformed and withered. The infested leaves are often shiny and sticky. Ants may be present.

Analysis: Several species of mealybugs feed on coleus. Mealybugs damage plants by sucking sap, causing leaf distortion and death. The adult female mealybug may produce live young or may deposit her eggs in white, fluffy masses of wax. The immature mealybugs, called *nymphs*, are very active and crawl all over the plant. Soon after the nymphs begin to feed, they exude filaments of white wax that cover their bodies, giving them a cottony appearance. As they mature, their mobility decreases. Mealybugs cannot fully digest all the sugar in the sap, and they excrete the excess in a fluid called *honeydew*, which coats the leaves. Ants may feed on the honeydew. Mealybugs are spread by the wind, which may blow egg masses and nymphs from plant to plant. Ants may also move them, or young, active nymphs can crawl to nearby plants. Mealybug eggs and some adults can survive the winter in warm climates. Spring reinfestations in colder areas come from infested new plants placed in the garden.

Solution: Spray infested plants with ORTHO Orthene Systemic Insect Control. Spray at intervals of 7 to 10 days until the mealybugs are gone. Gently hose down plants to knock off mealybugs and wash off honeydew. Remove and destroy severely infested leaves and plants.

CYMBIDIUM

Mosaic virus

Mosaic virus disfigures foliage.

Problem: Leaves are mottled or streaked. Pale rings may develop on the foliage. As the leaves grow older, black or brown stripes develop along the leaf veins, and irregular, sunken blotches may form. The flowers may be marred with dark green or light-colored rings or streaks.

Analysis: Mosaic virus is caused by several closely related viruses. The symptoms of mosaic infections vary in their severity depending on the strain of virus and the cymbidium variety. Viral infections generally do not kill the plant but may greatly reduce its overall vigor and beauty. Mosaic persists in the plant indefinitely. Cuttings or divisions made from the diseased plant will also be infected. If diseased plants are touched or pruned, the virus can be transferred to healthy plants on contaminated hands, knives, and other pruning equipment. Aphids and other insects may also transmit these viruses.

Solution: No chemicals control or eliminate viral diseases. Discard all infected plants. Wash your hands and dip pruning shears after each cut into a solution of 1 part chlorine bleach to 9 parts water. Purchase only healthy plants. Keep aphids under control.

DAHLIA

Tuber rot

Tuber rot often infects damaged roots.

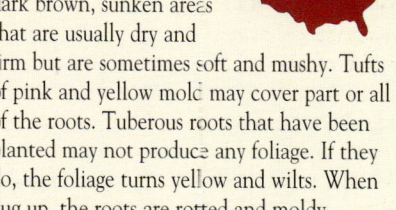

Problem: Tuberous roots in storage develop dark brown, sunken areas that are usually dry and firm but are sometimes soft and mushy. Tufts of pink and yellow mold may cover part or all of the roots. Tuberous roots that have been planted may not produce any foliage. If they do, the foliage turns yellow and wilts. When dug up, the roots are rotted and moldy.

Analysis: Tuber rot is caused primarily by two common soil-inhabiting fungi (*Fusarium* and *Botrytis* species). The fungi generally don't infect the tuberous roots unless the roots are wounded. If the roots are damaged when they are dug out of the ground, the fungi will penetrate the wounds and rot the tissue. The roots rot rapidly when they are stored in warm, humid conditions. If tuberous roots suffer frost damage while they are in storage, they will also be susceptible to fungal invasion. Sometimes tuberous roots in storage are contaminated, but the fungal decay has not progressed far enough to be noticed. When they are planted the following spring, they may not produce foliage. If they do produce foliage, the fungus causes wilting, yellowing, and eventually death of the plant.

Solution: Infected roots cannot be saved. To prevent tuber rot the following year, dig the roots carefully after they have matured fully. Discard any roots that show decay. Handle them carefully to prevent injuries. Store the roots in peat moss in a cool, dark place that is safe from frost.

DAHLIA (continued)

Wilt disease

Wilt disease on Dahlia.

Problem: The lower leaves turn yellow, wilt, and die; or all of the foliage may turn yellow and then wither. Older plants may be stunted. Yellowing and wilting often affect only one side of the plant. Flower heads droop. There may be dark brown areas on the infected stem. When the stem is sliced open near the base of the plant, dark streaks and discoloration are seen on the inner water-conducting stem tissue. The root system may be partially or entirely decayed.

Analysis: Wilt disease affects many ornamental plants. It is caused by either of two soil-inhabiting fungi (*Verticillium dahliae* or *Fusarium* species) that persist indefinitely on plant debris or in the soil. The disease is spread by contaminated seeds, plants, soil, and equipment. The fungus enters the plant through the roots and spreads into the stems through the water-conducting vessels. The vessels become discolored and plugged. This plugging cuts off the flow of water to the leaves, causing leaf yellowing and wilting.

Solution: No chemical control is available. It is best to destroy infected plants. *Verticillium* and *Fusarium* can be removed from the soil only by fumigation or solarization. To solarize soil, cultivate and wet an area thoroughly before covering it with clear polyethylene film during the hottest time of the year. Leave in place 4 to 6 weeks. The best solution is to use plants that are resistant to *Verticillium* and *Fusarium*.

DELPHINIUM

Cyclamen mite

Cyclamen mite damage.

Problem: Flower buds are deformed and blackened. They may not open, or the flowers may be distorted and shriveled. The leaves are curled, cuplike, wrinkled, thickened, and brittle and may have a purplish discoloration. The plant may be stunted to only a quarter of its normal size.

Analysis: Cyclamen mite (*Steneotarsonemus pallidus*), a microscopic plant pest, is a member of the spider family. Cyclamen mites are $\frac{1}{100}$ of an inch long and can be seen only with a magnifying glass. Although the mites are not visible to the unaided eye, their damage is very distinctive. Mites generally live and feed in leaf and flower buds and rarely venture out onto exposed plant surfaces. These pests spread by crawling from one overlapping leaf to another. They are also spread on contaminated tools, clothing, and hands. Cyclamen mites are seldom active during the hot summer months and are most injurious from the early spring until June and again in late summer, with the greatest damage occurring in periods of high humidity.

Solution: Spray infested plants with ORTHO Isotox Insect Killer or an insecticide containing *dicofol*; respray two more times at intervals of 7 to 10 days. Spray the foliage thoroughly, covering both the upper and lower surfaces of the leaves. Space plants far enough apart so that their foliage doesn't overlap. This prevents the mites from spreading. Wash your hands and tools after working on an infested plant to prevent spreading mites to healthy plants.

DIANTHUS (Carnation, Pink)

GROWING GUIDE

Adaptation: Throughout North America. Carnations may be grown as annuals in zones 3 to 8, but may need winter protection to be grown as perennials in these areas. They do not need protection in zones 9 and 10. To determine your climate zone, see the map on page 336.

Flowering Time: Late spring and summer.

Light: Full sun.

Planting Time: Spring, after all danger of frost is past.

Soil: Well drained. pH 6.5 to 7.5.

Fertilizer: Fertilize with Scotts All-Purpose Plant Food or Miracle-Gro Plant Food according to label directions.

Water:
How much: Apply enough water to wet the soil to the bottom of the root zone. For pinks and sweet william, wet the soil 4 to 6 inches deep. For carnations, wet the soil 8 to 10 inches deep.
Containers: Apply enough water to container plants so that about 10 percent of the water drains from the bottom of the container.
How often: Water when the soil is just barely moist.

Bacterial wilt

Bacterial wilt in carnation.

Problem: Stems, or sometimes entire plants, wilt. Leaves dry, turn yellow, and die. Roots are often rotted. Cracks may appear around the base of the stem, with yellow streaks extending up the length of the stem. When the stems are sliced open, yellowish to brownish discolorations of the stem tissue are revealed. The infected interior portions of the stem are sticky.

Analysis: Bacterial wilt, a disease of carnations and pinks, is caused by a bacterium (*Pseudomonas caryophylli*) that lives in the soil. The bacteria penetrate the plant stems through wounds or cuts in the roots or the base of the stem. Once inside, the wilt organisms multiply and clog the water-conducting stem tissue, causing the plant to wilt and die. The bacteria can also move down into the root system, causing decay. A sticky fluid, which coats infected stems and roots, contains millions of bacteria. The bacteria are spread to other plants by water, handling, contaminated soil, plant debris, and contaminated equipment. Bacterial wilt damage increases as the temperature grows warmer.

Solution: Once a plant is infected, it cannot be cured. It is best to remove and destroy all infected plants. Clean up plant debris. If you've been handling infected plants, wash your hands thoroughly with soap and hot water and disinfect any con-taminated tools by soaking in a solution of 1 part chlorine bleach and 9 parts water. Do not replant healthy carnations or pinks in contaminated soil. Avoid damage to plants when cultivating.

Stem rot

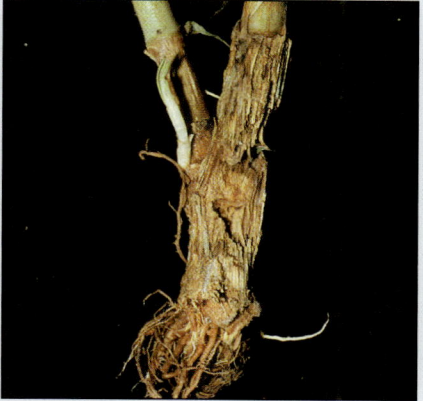

Rhizoctonia *stem rot, a fungal disease.*

Problem: The leaves turn pale and wilt, sometimes very suddenly. The lower leaves are rotted. The stem is slimy and decayed, and minute black pellets may be just barely visible around the base of the plant.

Analysis: Stem rot is caused by a fungus (*Rhizoctonia solani*) found in most soils. It penetrates the plant at or just below the soil level, rotting through the outer stem bark into the inner stem tissue. Unlike the soft, outer stem rot, the inner stem tissue becomes dry and corky when infected. As the rot progresses up the stem, the lower leaves rot, the foliage pales and withers, and the plant may die. *Rhizoctonia* thrives in warm, moist conditions.

Solution: If all the foliage is wilted, it is best to replace the plant. Plants not so severely affected can sometimes be saved but will often worsen and die. An effective method to help control the disease is to let the soil dry out between waterings. Minimize water stress while drying soil by placing containers in shade or erecting shade cloth over beds; spraying with antitranspirant; and covering with a tent of clear plastic if plants are in shade. Resume watering when signs of drought stress appear. Before planting the following year, spray or dust the soil with a fungicide containing *PCNB*.

Virus

Virus in carnation.

Problem: The leaves are mottled or have yellow to reddish spots, rings, or streaks parallel to the leaf veins. The lower leaves may turn yellow. Sometimes blotches are on the leaves. The flowers may be streaked or blotched with light or dark colors, or coloring may be uneven.

Analysis: Several viruses infect carnations. Viral infections are not generally very harmful to the plant. In fact, symptoms of infection may not show up unless several viruses are present in the plant at the same time. In severe infections, however, viruses can cause the lower leaves to turn yellow by suppressing the development of chlorophyll in the leaf tissue. As a result, the leaves produce less food, causing the plant to be weakened. Certain viruses are spread by aphids, which feed on diseased plants and transfer the virus to healthy plants. Other viruses can be spread when the virus is transferred to healthy plants on contaminated hands or equipment.

Solution: Once a plant is infected, no chemical will control the virus. Remove weak and stunted plants. Wash your hands thoroughly and disinfect pruning shears after working on infected plants by soaking the shears in a solution of 1 part chlorine bleach and 9 parts water. Keep the aphid population under control by spraying infested plants with an insecticide containing *malathion* or *acephate* (Orthene®). Purchase only healthy plants.

DIANTHUS *(continued)*

Alternaria leaf spot

Alternaria leaf spot on Dianthus.

Problem: Dark purple spots surrounded by yellow-green margins appear on the leaves and stems. Sunken, grayish brown dead areas develop in the center of the spots. Individual spots enlarge and merge to form blotches. Infected leaves turn yellow, blacken, and then die. Lesions develop on stems, especially at the bases. Flowers may be spotted. The leaves at the tips of infected stems may become mottled, turn yellow, and wilt. The entire plant may eventually wilt and die.

Analysis: Alternaria leaf spot is caused by a fungus (*Alternaria saponariae*) that infects carnations, pinks, and sweet William. Infection is most severe in wet, humid conditions. Spores are spread by wind and splashing water. Infection occurs when spores germinate on wet leaves, stems, or petals. The fungus survives as spores on plant debris.

Solution: Pick off and destroy infected plant parts and clean up and destroy plant debris. Spray with ORTHO Multi-Purpose Fungicide Daconil 2787® Plant Disease Control or a fungicide containing *captan* according to label directions. Water in the morning so the foliage will have a chance to dry out. Try to avoid wetting the foliage when watering. Thin out dense plantings to allow air circulation. Fertilize with Scotts All-Purpose Plant Food or Miracle-Gro Plant Food according to label directions.

DIGITALIS (Foxglove)

Foxglove anthracnose

Foxglove anthracnose, a fungal disease.

Problem: Light or purplish brown spots up to ⅛ inch in diameter appear on the leaves. The spots are circular or angular and have purplish margins. Often black, rough areas develop in the centers of the leaf spots. Sunken lesions may occur on the leaf veins and stems, and severely infected leaves turn yellow, wither, and drop off. Often plants are stunted and die, especially during periods of warm, moist weather. Seedlings may wilt and die.

Analysis: Foxglove anthracnose is caused by a fungus (*Colletotrichum fuscum*) that infects only foxgloves. The fungal spores are spread from plant to plant by splashing water or rain, insects, animals, and contaminated tools. If the leaves are wet, the spores germinate and infect the leaf tissue, creating spots and lesions. Anthracnose is favored by warm temperatures and moist conditions. The fungus survives the winter in diseased plant debris and infected seed. This fungus also causes damping-off of foxglove seedlings.

Solution: No fungicides are registered for this disease on foxglove. If practical, infected leaves may be picked off. Water in the morning so the foliage will have a chance to dry out. Try to avoid wetting the foliage when watering. Remove and destroy plant debris at the end of the growing season. The following year, plant foxglove in a different bed, spaced far enough apart to allow good air circulation.

DIMORPHOTHECA (African daisy)

Wilt disease

Wilt disease caused by soil-inhabiting fungi.

Problem: The lower leaves turn yellow, then wilt and die; or all of the foliage may turn yellow and then wither. Flower heads may droop. Often, the plant is affected only on one side. When the stem is sliced open near the base of the plant, dark streaks and discoloration of the inner water-conducting stem tissue are revealed. The root system may be partially or entirely rotted.

Analysis: Wilt disease affects many ornamental plants. It is caused by either of two soil-inhabiting fungi (*Verticillium albo-atrum* or *Fusarium* species) that persist indefinitely on plant debris or in the soil. The disease is spread by contaminated seeds, plants, soil, and equipment. The fungus enters the plant through the roots and spreads up into the stems and leaves through the water-conducting vessels in the stems. The vessels become discolored and plugged. This plugging cuts off the flow of water to the leaves, causing leaf yellowing and wilting.

Solution: No chemical control is available. It is best to destroy infected plants. *Verticillium* and *Fusarium* can be removed from the soil only by fumigation or solarization. To solarize soil, cultivate and wet an area thoroughly before covering it with clear polyethylene film during the hottest time of the year. Leave in place 4 to 6 weeks.

GLADIOLUS

GROWING GUIDE

Adaptation: Throughout North America.

Flowering Time: Summer to fall.

Light: Full sun.

Planting Time: From late winter or spring, when all danger of frost is past, to midsummer.

Soil: Any good, well-drained garden soil. pH 5.5 to 7.5.

Fertilizer: Fertilize with Scotts Bulb Food according to label directions.

Water: How much: Apply enough water to wet the soil 6 to 8 inches deep.
How often: Water when the soil 1 inch below the surface is just barely moist.

Handling:
For continuous summer bloom, make successive plantings at 2-week intervals from mid-April to mid-July. For cut flowers, harvest when the bottom flower is open. When cutting flowers, allow at least four leaves to remain on the plant.

Gladiolus thrips

Gladiolus thrips damage.

Problem: Silvery white streaks appear on flowers and foliage. The leaves turn brown and die.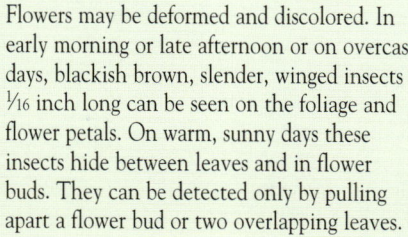
Flowers may be deformed and discolored. In early morning or late afternoon or on overcast days, blackish brown, slender, winged insects ⅟16 inch long can be seen on the foliage and flower petals. On warm, sunny days these insects hide between leaves and in flower buds. They can be detected only by pulling apart a flower bud or two overlapping leaves.

Analysis: Gladiolus thrips (*Taeniothrips simplex*), one of the most common pests of gladiolus plants, also feeds on many other garden ornamentals. Both the immature and adult thrips feed on plant sap by rasping the plant tissue. The injured tissue turns white, causing the characteristic streaking and silvering of the leaves and flowers. The adult female thrips inserts her eggs into growing plant tissue; the emerging young mature within 2 to 4 weeks. Thrips actively feed and reproduce from spring until the first frost of fall. They cannot survive freezing temperatures. In warm-winter climates, adult thrips hibernate in the soil until spring. In cold-winter climates, they overwinter by hibernating on gladiolus corms (the "bulbs" of the gladiolus plant) in storage. Corms infested by thrips turn brown and corky and may fail to grow, or they may produce only stunted, poor-quality flowers and foliage.

Solution: Spray infested plants with ORTHO Orthene Systemic Insect Control. Repeat the spray at intervals of no less than 7 days if reinfestation occurs. Dust corms with an insecticide containing *malathion* before storing. Discard brown, corky corms.

Viruses

Virus streaking in Gladiolus.

Problem: Leaves and flowers are streaked, spotted, or mottled. The leaves may also be yellowing, stiff, or thickened. Often the plant blooms prematurely, and the flowers only partially open and then fade rapidly. The entire plant may be dwarfed, although sometimes only the flower spike is stunted.

Analysis: Several plant viruses infect gladiolus. Depending on weather conditions and plant variety, the symptoms of infection can vary from barely noticeable to quite severe. Viral infections rarely cause a plant to die but can weaken it seriously. The virus increases in the corms (the "bulbs" of the gladiolus plant) year after year. Successive plantings from diseased corms provide flowers of poor quality. Viruses are spread by aphids and leafhoppers. These insects feed on diseased plants and transfer the virus to healthy plants at subsequent feedings.

Solution: Once a plant is infected, no chemical will control the virus. To prevent spread of the disease to healthy plants, remove and destroy infected plants. Infected corms cannot be reused even if the stem growth is removed. Keep the aphid population under control by spraying plants with ORTHO Diazinon Ultra Insect Spray or ORTHO Orthene Systemic Insect Control. Because two of the viruses that infect gladiolus are very common on vegetables in the bean and cucumber families, where practical, avoid planting gladiolus near beans, clover, cucumbers, squash, melons, or tomatoes.

GLADIOLUS (continued)

Fusarium yellows

Fusarium yellows on Gladiolus.

Problem: Foliage and flower spikes are stunted, and flowers may be small and faded. Yellowing starts on the leaf tips and spreads through the entire plant, which finally dies. When the dying plant is pulled out of the ground, the roots are found to be rotted and the corm (the "bulb" of the gladiolus plant) spotted with firm, circular, brown or black lesions. In some cases, the corm appears normal. When it is sliced open, however, brown, discolored inner tissue is revealed.

Analysis: Fusarium yellows, a common and widespread disease of gladiolus plants and corms, is caused by a soil-inhabiting fungus (*Fusarium oxysporum* f. *gladioli*). The fungus may penetrate and rot the corms in storage or in the ground. Wet soils and warm temperatures (70°F and higher) favor the rapid development of this disease. Corms in storage are sometimes contaminated, but the fungal decay may not have progressed far enough to be noticed. When these corms are planted the following spring, they may not produce foliage if severely infected. If they do produce feeble growth, it soon turns yellow and dies. The fungus survives in diseased corms and soil for many years. Corms that have been removed from the soil prematurely are especially susceptible to infection.

Solution: Destroy all plants and corms that show signs of infection. Dig corms only when they have fully matured. Do not replant healthy corms in soil in which diseased plants have grown. Store them in a dry, cool (40° to 50°F) place.

Scab

Scab, a bacterial disease that attacks corms.

Problem: Sunken brown to black lesions on the corm (the "bulb" of the gladiolus plant) are covered with a shiny, varnishlike material and are encircled by raised, brittle rims. Later in the season, after the corms have been planted, many tiny, raised, reddish brown specks develop on the bases of the emerging leaves. These specks become soft, elongated dead spots, which may in wet weather be covered with and surrounded by a shiny, oozing material. The leaves usually fall over.

Analysis: Scab, caused by bacteria (*Pseudomonas marginata*), earns its name from the scablike lesions it produces on the gladiolus corms. The bacteria penetrate the corm tissue, usually where the corm has been injured by soil insects or bulb mites, and then move up into the stem base, producing a soft, watery rot. This decay causes the leaves to fall over. The shiny, varnishlike spots that form on the leaves and corms contain millions of bacteria. Wet, heavy soil and warm temperatures favor the rapid development of this disease. The bacteria can live for several years in infected corms and plant debris. The bacteria are usually spread by insects but may also be spread by splashing water and by contaminated corms, soil, and tools. Severely infected plants may die.

Solution: No chemical controls this disease. Destroy infected corms. Remove and destroy infected plants. Plant healthy corms in well-drained soil where diseased gladioli have not grown. Control soil insects.

Neck rot

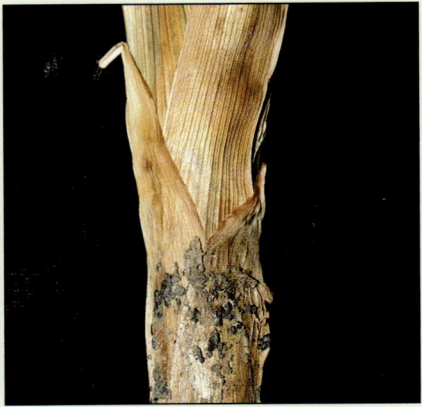

Neck rot, a fungal disease that attacks corms.

Problem: Foliage turns yellow and dies prematurely. Leaf bases are rotted and may be shredded. Black fungal pellets the size of pepper grains may cover the decayed leaf bases and husks of the corms (the "bulbs" of the gladiolus plant). Dark brown to black sunken lesions occur on the corms. These dry, corky lesions may enlarge and join together, destroying the entire corm.

Analysis: Neck rot is caused by a fungus (*Stromatinia gladioli*) that attacks corms either in storage or in the soil after planting. After the initial infection, the decay spreads up into the leaf bases, killing the leaves prematurely. Corms that are planted in cold, wet soil or stored in moist conditions are most susceptible to neck rot. The fungus is spread by contaminated soil and corms. The tiny black fungal pellets that form on infected tissue can survive for 10 years or more in the soil.

Solution: Discard infected corms and plants. Plant in well-drained soil. Dig corms before the onset of cold, wet weather and store them in a cool, dry place. If you wish to replant in areas where infected corms have been growing, fumigate or solarize the soil. To solarize soil, cultivate and thoroughly wet an area before covering it with clear polyethylene film during the hottest time of the year. Allow the plastic film to remain in place for 4 to 6 weeks.

Penicillium corm rot

Penicillium corm rot, a fungal disease.

Problem: Corky, reddish brown, sunken lesions ½ inch or larger in diameter appear on the corm (the "bulb" of the gladiolus plant). In cool, moist conditions, the rotted areas of the corm become covered with a blue green mold. Infected corms that have been planted may not produce any foliage. Foliage that is produced turns yellow and wilts. The corms are rotted and moldy.

Analysis: Penicillium corm rot is caused by a common fungus (*Penicillium gladioli*) that is often found on corms in storage. The fungus infects the corm through wounds or abrasions that occur usually when the corm is dug. The problem rarely occurs when corms are properly cured after harvest. After the initial infection, the rot spreads throughout the corm and into the stem tissue. The corms rot rapidly when they are stored in warm, humid conditions. The fungus forms masses of blue-green spores and tiny brown fungal pellets that can survive dry conditions and extremes of temperatures to invade healthy corms. If mildly infected corms are planted, they may or may not produce foliage, depending upon the severity of the infection.

Solution: Dig corms carefully only when the gladiolus leaves have turned entirely yellow in the fall. Destroy all corms showing decay. Handle healthy-looking corms carefully to prevent injuries. Dry and cure corms for a week at 85° to 90°F immediately after digging. Dip them in a fungicide solution containing *captan* before storing and again before planting. Store cured corms in a dry, cool (40° to 45°F) location.

Short stems

Short stems from inadequate chilling.

Problem: Flower stalks are very short, and flowers may be smaller than normal. Sometimes only the tip of the flower stalk emerges and blooms at ground level. No signs of insects or disease are present, and the foliage appears healthy.

Analysis: Short stems are the result of inadequate chilling. Hyacinth bulbs contain embryonic flowers and stems. A minimum of 6 weeks' exposure to cool temperatures (45° to 50°F) during the winter stimulates the stem cells to elongate, causing the bud to emerge from the ground. During cool spring weather (50° to 55°F), the stems continue to elongate to their full length, at which point the flowers mature and open. In warm-winter areas, hyacinth stems often fail to elongate properly because of inadequate chilling. Also, during unseasonable spring hot spells (when air temperature reaches 70°F or higher), hyacinth flowers are stimulated by heat to mature and open before the stems have entirely emerged from the ground.

Solution: You cannot do anything to increase the length of the hyacinth stem once the flower has matured. If warm spring temperatures are common in your area, plant hyacinths in locations where they will receive either direct sun only in the morning or late afternoon or filtered light. The lower air and soil temperatures in such areas will help to increase stem lengths.

Bacterial soft rot

Bacterial soft rot; diseased bulb on left.

Problem: Bulbs that have been planted may not produce any foliage, or, if foliage is produced, the flower stalk may not form. Sometimes the flower stalk develops but the flowers open irregularly and rot off. The entire stalk may rot at the base and fall over. If pulled gently, the leaves and flower stalk may lift entirely off the bulb, which is soft, rotted, and filled with a white, thick, foul-smelling ooze.

Analysis: Bacterial soft rot is caused by bacteria (*Erwinia carotovora*) that infect hyacinth bulbs both in storage and when planted in the ground. The bacteria initially penetrate and decay the upper portion of the bulb. The disease then progresses upward into the leaves and flower stalks and down through the bulb and roots. The thick ooze that accompanies the decay is filled with millions of bacteria. Bulbs that are infected before they are planted produce little, if any, growth. Even well-established, healthy plants may decay rapidly after they are infected, sometimes within 3 to 5 days. The bacteria survive in infected plant debris and bulbs and are spread by contaminated insects and tools and by diseased bulbs and plants. Soft rot thrives in moist conditions. If bulbs freeze while they are in storage, they are especially susceptible to infection.

Solution: No cure exists for this disease. Remove and destroy all bulbs and plants showing signs of decay. Store bulbs in a dry, cool (40° to 45°F) location. Plant only healthy bulbs in well-drained soil. Do not overwater.

IMPATIENS (Balsam)

Leaf spots

Leaf spot is most severe in wet weather.

Problem: The leaves are spotted or blotched with brown spots that may range in size from barely visible to ¼ inch in diameter. Several spots may join to form blotches. The leaves may turn yellow and die. Leaf spotting is most severe in wet weather.

Analysis: Several fungi cause leaf spots. Some of these will eventually kill the plant, while others merely cause spotting that is unsightly but not harmful. These fungi are spread by splashing water, wind, insects, and contaminated tools. Fungal strands or spores survive the winter in plant debris. Most of the leaf spot organisms do their greatest damage in moist conditions and temperatures of 50° to 85°F.

Solution: Spray infected plants with a fungicide containing *captan* at intervals of 7 to 10 days. Make sure that your plant is listed on the product label. Because leaf spots thrive in warm, wet conditions, it is important to spray conscientiously during these periods. Pick off spotted, diseased leaves, and clean up and destroy plant debris.

IRIS

GROWING GUIDE

Adaptation:
Rhizomatous: Zones 5 through 10.
Bulbous: Zones 7 through 10.
Irises may be grown in colder areas if heavily mulched. To determine your zone, see the map on page 336.

Flowering Time:
Rhizomatous: Early summer.
Bulbous: Late spring through early summer.

Light: Full sun.

Planting Time:
Rhizomatous: Summer.
Bulbous: Fall.

Soil: Well drained. pH 6.0 to 7.5.

Fertilizer: Fertilize with Scotts All Purpose Plant Food or Miracle-Gro Plant Food according to label directions.

Water: How much: Apply enough water to wet the soil to 1½ feet deep.
How often: Water when the soil 1 inch below the surface is just barely moist.

Handling: Mulch irises after the first hard fall frost. Divide rhizomatous iris clumps every 3 to 5 years.

Leaf spot

Leaf spot on leaves, flower stalks and buds.

Problem: Tiny brown spots from ⅛ to ¼ inch in diameter appear on the leaves. The spots have distinct reddish borders and may be surrounded by water-soaked margins that later turn yellow. After the plant has flowered, the spots enlarge rapidly and may join together to form blotches. Spotting is most severe in wet weather. The leaves die prematurely.

Analysis: Leaf spot is caused by a fungus (*Mycosphaerella macrospora*) that infects only irises and a few other closely related plants. This fungus attacks the leaves and occasionally the flower stalks and buds. It will not affect iris roots, bulbs, or rhizomes (elongated underground stems). Infection occurs early in the season, but spots do not appear until flowering. Although the fungus does not directly kill the plant, several years of repeated infection result in premature leaf death each summer, greatly reducing rhizome and bulb vigor. Some varieties of iris suffer leaf dieback even when they are only lightly spotted, while others can be covered with spots before they start to die. When the leaves are wet, or during periods of high humidity, the fungal spots produce spores that spread to other plants by wind or splashing water. The fungus survives winters in old infected leaves and debris.

Solution: Spray plants with ORTHO Multi-Purpose Fungicide Daconil 2787® Plant Disease Control or with a fungicide containing *mancozeb* or *triadimefon*. Repeat the spray every 7 to 10 days until the foliage starts to die back. Clean up and destroy plant debris and clip off diseased foliage in the fall. Spray in the spring as soon as new growth appears, and repeat four to six more times at intervals of 7 to 10 days.

Rust

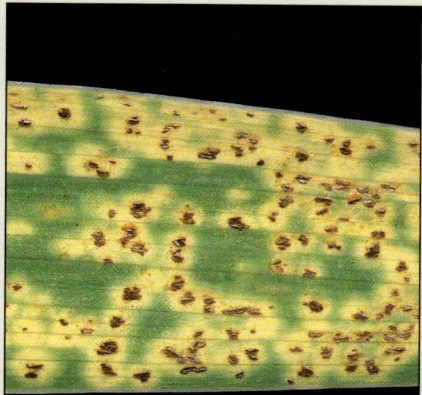

Rust, a fungus spread by water and wind.

Problem: Rust-colored, powdery pustules appear on both sides of the leaves. Later in the season, these pustules turn dark brown. Severely infected leaves may die prematurely, but the plant is seldom killed.

Analysis: Several closely related fungi (*Puccinia* species) cause rust. The rust-colored pustules are composed of millions of microscopic spores. Some of the spores spend the winter on iris leaves that have not died back entirely, while others overwinter on other kinds of plants. Infection usually starts in the spring as soon as conditions are favorable for plant growth. Splashing water and wind spread the spores to healthy plants. Because iris varieties vary greatly in their susceptibility to rust, some leaves may be killed prematurely, while others may not be affected. Rust thrives in wet weather.

Solution: Spray infected plants with a fungicide containing *mancozeb* or *triadimefon*. Repeat the spray two or more times at intervals of 7 to 10 days. Remove and destroy old and dying iris leaves in the fall. Water in the morning to allow the foliage a chance to dry out before nightfall. Plant rust-resistant varieties, if available.

Crown rot

Crown rot is caused by a widespread fungus.

Problem: Leaves of bearded and other irises grown from rhizomes (elongated underground stems) die, starting with the leaf tips and progressing downward. Leaf bases and possibly the rhizomes are dry, brown, and rotted. Leaves of Dutch irises and others grown from bulbs are stunted, turn yellow, and die prematurely. The leaves and stems at the soil level are rotted, and the bulbs are soft and crumbly. White matted fungal strands cover the crown (where the stem meets the roots) and soil, surrounding both rhizomatous and bulbous irises. Small tan to reddish brown pellets the size of mustard seeds form on the infected plant tissue and soil.

Analysis: Crown rot is caused by a widespread fungus (*Sclerotium rolfsii*). It decays and kills the leaf and stem bases, the bulbs, and often part or all of the rhizomes. Crown rot is spread by moving water, diseased transplants, infested soil, and contaminated tools. The fungal pellets can survive for many years in dry soil and extremes of temperature to reinfect healthy plants when conditions are suitable. Crown rot is most severe in overcrowded plantings with warm temperatures (70°F and up) and moist conditions.

Solution: Remove and destroy infected plants, bulbs, and rhizomes and the soil immediately surrounding them to 6 inches beyond the diseased area. Drench the area with a fungicide containing *PCNB*. The following year, repeat the soil drench at planting time and again when new growth is showing. Plant in well-drained soil with roots covered and rhizomes showing through the top of the soil. Plant only healthy bulbs and rhizomes. Thin out overcrowded plantings.

Iris borer

Iris borer damage. Inset: Iris borer (½ life size).

Problem: Dark streaks, water-soaked spots, and possibly slits develop in new leaves in the spring to early summer. Leaf edges may be chewed and ragged. By midsummer, the foliage is wilting and discolored. Leaf bases are loose and rotted. Rhizomes (elongated underground stems) are often filled with holes and may be soft and rotted. Pink caterpillars, from 1 to 2 inches long, are feeding inside the rhizomes.

Analysis: Iris borer, the larva of a night-flying moth (*Macronoctua onusta*), is the most destructive insect pest of iris. In the fall, the adult moth lays 150 to 200 eggs in old leaf and flower stalks. The eggs hatch in late April or early May. Emerging larvae initially feed on the leaf surfaces, producing ragged leaf edges and watery feeding scars. They then bore into the inner leaf tissue and gradually mine their way down into the rhizome, on which they feed throughout the summer. The damaged rhizome is susceptible to bacterial soft rot. The larvae leave the rhizome, pupate in the soil, and emerge as adult moths in the fall.

Solution: Apply ORTHO Borer & Leaf Miner Spray weekly from the time first growth starts until the beginning of June. To kill the borers in lightly infested rhizomes, poke a wire into borer holes. In May and June, squeeze the leaves in the vicinity of feeding damage to kill feeding borers inside. Destroy heavily infested plants and rhizomes. Clean up and destroy plant debris by April to eliminate overwintering eggs.

IRIS *(continued)*

Bacterial soft rot

Bacterial soft rot in bearded iris.

Problem: Leaves turn yellow, wilt, and eventually die. Dieback often starts at the leaf tips and progresses downward. The entire leaf cluster may be found lying on the ground. If pulled gently, the leaf cluster sometimes lifts off the rhizome. Leaf bases and rhizomes are often rotted and foul-smelling.

Analysis: Bacterial soft rot is caused by a bacterium (*Erwinia carotovora*). It is a serious and common disease of bearded and other rhizomatous irises. The bacteria enter the plant through wounds in the leaves and rhizomes, which are frequently made by iris borers. As infection develops, the plant tissue decays into a soft, foul-smelling mass. Finally, the plant dies and the inner rhizome tissue disintegrates. Infection and rapid decay thrive in moist, dark conditions. These bacteria live in the soil and in plant debris. They are spread by contaminated plants and rhizomes, soil, insects, and tools.

Solution: Remove and destroy all diseased plants; they cannot be cured. Discard diseased rhizomes before planting. If only a small portion of the rhizome is infected, you may possibly save it by cutting off the diseased portion. Avoid wounding the rhizomes when digging them. After dividing rhizomes, let the wounds heal for a few days before replanting. Plant irises in a sunny, well-drained location. Plant in well-drained soil with roots covered and rhizomes showing through the top of the soil. Clean up plant debris in the fall. Control iris borers.

LATHYRUS (Sweet Pea)

Powdery mildew

Powdery mildew. Inset: Close-up.

Problem: Grayish white powdery spots and patches develop on the stems and on both surfaces of the leaves. Leaves eventually turn yellow and wither.

Analysis: Powdery mildew is caused by a fungus (*Erysiphe polygoni*) that can be severe on sweet peas. Powdery mildew thrives in both humid and dry weather. The powdery patches consist of fungal strands and spores. The spores are spread by the wind to healthy plants. The fungus saps plant nutrients, causing yellowing and sometimes death of the leaves, especially older leaves. A severe infection may kill the plant. Since this powdery mildew attacks many kinds of plants, the fungus from a diseased plant may infect other types of plants in the garden. For a list of plants susceptible to powdery mildew, see page 343. Under conditions favorable to it, powdery mildew can spread rapidly through a closely spaced planting.

Solution: Spray infected plants with a fungicide containing *thiophanate-methyl* or *triadimefon*, or dust them with sulfur. Remove severely infected plants. Spray or dust at regular intervals of 7 to 10 days or as often as necessary to protect new growth. Clean up plant debris. Plant in a site with good air circulation and exposure to early morning sun.

LILIUM (Lily)

Root and bulb rots

Root and bulb rot. Diseased plant on right.

Problem: Plants are stunted and wilting, and the lower leaves turn yellow. The tips of the lower leaves may be dying and brown, and dead patches may appear along leaf edges. Flower buds may wither and fail to open. The bulbs and roots are rotted.

Analysis: Root and bulb rot problems are common with lilies. These plant diseases are caused by various fungi (*Rhizoctonia, Phytophthora, Pythium, Fusarium,* and *Cylindrocarpon* species). These fungi attack and decay the bulbs and roots, causing stunting, wilting, and eventually death of the foliage and flowers. These bulb and root rot organisms live in the soil and stored bulbs and thrive in wet soil. Sometimes bulbs in storage are lightly infected, but the fungal decay hasn't progressed far enough to be easily noticed. When planted, these bulbs may rot so quickly that they do not produce any foliage.

Solution: Remove and destroy infected plants. Check all bulbs carefully and discard any that are moldy, rotted, or dry and crumbly. Plant in well-drained soil. Store bulbs in a cool (35° to 40°F), dry location.

Viruses

Virus on Oriental lily, spread by aphids.

Problem: The leaves are mottled and streaked light and dark green, and the plant may be stunted and dying. The leaves may be spotted with tiny yellow, brown, or gray flecks. These flecks are elongated and run parallel to the leaf veins. Plants with these flecks are often stunted and have small, streaked flowers that do not open fully. Their leaves may be twisted or curled and frequently die prematurely, starting from the bottom of the plant.

Analysis: Several virus diseases of lilies cause mottling or flecking of the foliage. *Mosaic viruses* produce leaf mottling and discoloration. Depending on the species or variety, the symptoms of infection can be mild or severe. *Fleck* is produced if a plant is simultaneously infected by the *symptomless lily virus* and the *cucumber mosaic virus* (which may or may not produce mottling by itself). Leaf flecking is usually accompanied by stunting and poor-quality flowers and foliage. The plant is generally disfigured. Viruses remain in infected bulbs year after year, so successive plantings of diseased bulbs will produce only poor-quality flowers and foliage. All of these viruses are spread by aphids, which pick up the virus while feeding on diseased plants and then transmit it to healthy plants at later feedings.

Solution: No chemical controls viruses. Remove and destroy infected plants. Control aphids by spraying infested plants with an insecticide containing *malathion* or *acephate* (Orthene®). Respray if reinfestation occurs. Plant mosaic-resistant or immune lilies.

Leaf scorch

Leaf scorch is most prevalent in acid soils.

Problem: Brown semicircular or crescent-shaped areas develop along leaf margins. Leaf tips may be brown. Usually the lower leaves are affected first. When the pH is tested, the soil is found to be acidic.

Analysis: Leaf scorch is a condition that may develop in lilies when they are growing in acid soil with a pH lower than 6.5. Toxic amounts of aluminum and manganese salts become available and are absorbed by plant roots in acid soils. Leaf scorch is most likely to occur when the plant is not receiving adequate or balanced supplies of nutrients, such as nitrogen and calcium. During the season of rapid growth, significant temperature changes can also cause leaf scorch.

Solution: To decrease soil acidity, add ground dolomitic limestone according to the chart on page 337. Do not overlime acid soil, because availability of some nutrients may be decreased when the soil pH is too high. Fertilize with Scotts Bulb Food according to label directions.

Root and stem rot

Root and stem rot in wet, heavy soil.

Problem: Leaves and stems turn yellow, wilt, and die. The lower leaves and stems may be soft and rotted. White fungal strands may grow around the base of the plants.

Analysis: Root and stem rot is caused by any of several different fungi that persist indefinitely in the soil. These fungi thrive in waterlogged, heavy soil. Some of them attack the plant stems at the soil level, while others attack the roots. Infection causes the roots and stems to decay, resulting in wilting, then yellowing leaves and eventually the death of the plant. These fungi are spread by infested soil and transplants, contaminated equipment, and splashing or running water. Some of these organisms also cause damping-off of seedlings.

Solution: Allow the soil around the plants to dry out. Minimize drought stress by placing containers in shade or erecting shade cloth over beds; spraying with antitranspirant; and by covering with a tent of clear plastic if plants are in shade. Resume watering when signs of drought stress appear. Remove and discard severely infected plants. Avoid future problems by planting in well-drained soil.

NARCISSUS (Daffodil, Jonquil)

GROWING GUIDE

Adaptation: Throughout North America.

Flowering Time: Spring.

Light: Full sun is best. Daffodils will tolerate light shade, but may stop blooming after several years.

Planting Time: Fall.

Soil: Any good well-drained garden soil. pH 6.0 to 8.0.

Fertilizer: When planting, mix 1 teaspoon of bone meal or superphosphate 0–20–0 into the soil at the bottom of the planting hole. During the growing season, fertilize with Scotts Bulb Food according to label directions.

Water: How much: Apply enough water to wet the soil 1½ feet deep.
How often: Water when the soil 1 inch below the surface is just barely moist. Stop watering the bulbs after the foliage has turned yellow.

Handling: Continue to water and fertilize plants regularly until the foliage turns yellow, then remove the dead leaves.

Failure to bloom

Narcissus *failing to bloom.*

No flowers due to weak growth.

Problem: Foliage is healthy but may be sparse. Few or no flowers are produced. Flowers that are produced may be smaller than normal.

Analysis: Daffodils fail to flower for several cultural reasons.

Solution: The numbered solutions below refer to the numbered items in the analysis.

1. Overcrowding: Bulbs multiply each year, producing larger clumps the following spring. If the bulbs are not divided and transplanted every few years, they become overcrowded.

1. Divide bulbs when flower production drops off. As a general rule, divide bulbs every 3 to 4 years.

2. Too much shade: Daffodils planted in a shaded spot usually bloom well the first year. They require a sunny location for continued flowering over a long period of time, however. The leaves use light to manufacture food, which is stored in the bulbs for the next year's growth and flowering.

2. Grow daffodils in a location where they will receive 4 or more hours of full sun. Transplant daffodils from the shade to a sunny location any time after flowering. Try to keep the soil and roots immediately surrounding the bulb intact when transplanting.

3. Overheating: If bulbs are stored at high temperatures (80°F and up), the flower embryo inside the bulb will be killed. Leaves grow in the spring, but flowers are not produced.

3. Store bulbs at a cool temperature (55° to 60°F) and in a well-ventilated location.

4. Undersized bulbs: If bulbs are much smaller than normal, they may produce only foliage in the first or second year. Undersized bulbs don't store enough food to produce both leaves and blossoms, but the bulbs will grow larger until they finally do produce flowers.

4. Purchase only large, healthy bulbs. Fertilize with Scotts Bulb Food in fall and when shoots emerge in spring.

5. Foliage removed too soon: After a daffodil flowers, the remaining foliage continues to use the sun's rays to manufacture food for new bulbs and the next year's flowers. If the foliage is removed before it has a chance to die back naturally, the new bulbs may not have enough food stored to produce a flower.

5. Let the foliage turn yellow before removing it. To reduce the unsightly effect of yellowing foliage, trim off the yellow portions of the leaf tips, making several cuts until the entire leaf is removed. Or plant the bulbs in a bed of dense-foliaged plants, such as daylilies, that will hide the bulb leaves as they yellow and die.

Fusarium bulb rot

Fusarium bulb rot; decayed with few roots.

Problem: Leaves turn yellow, and the plant is stunted and dies prematurely. If the bulb is unearthed, few or no roots may be seen. Bulbs in storage develop a chocolate or purple brown spongy decay that is especially noticeable when the outer fleshy bulb scales are pulled away. White to pink fungal strands may grow on the bulbs.

Analysis: Fusarium bulb rot is caused by a fungus (*Fusarium oxysporum* var. *narcissi*) that attacks both growing plants and bulbs in storage. Growing plants are infected through their roots, while stored bulbs may be infected through wounds or abrasions in the bulb tissue. Infected bulbs that are planted continue to decay in the ground and produce few or no roots and stunted, yellowing foliage. The fungus can persist in the soil for many years and is spread by contaminated bulbs, soil, and tools. Generally, bulb rot is most destructive when soil temperatures reach 60° to 75°F. The disease is most common in warm climates where temperatures rarely drop below freezing and in daffodils that are forced for indoor winter use.

Solution: Discard diseased plants and bulbs and the soil surrounding the bulbs for 6 inches. Dig bulbs carefully to prevent wounds. Store them in a cool (55° to 60°F), well-ventilated place. Do not replant healthy bulbs in an area where diseased plants have previously grown. In warmer climates, use fusarium-resistant cultivars. Harvest before soil warms in the spring. Avoid planting in warm soil.

Narcissus bulb fly

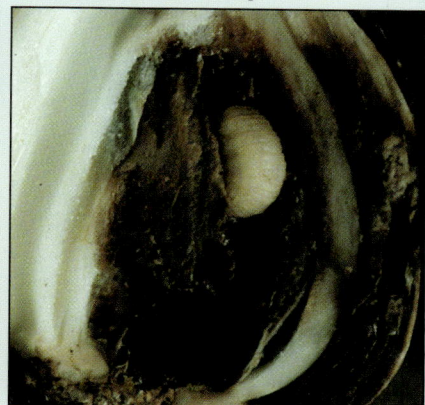

Narcissus bulb fly larva (life size).

Problem: Narcissus and daffodil bulbs feel soft and spongy and produce little or no growth after they are planted. Foliage that does emerge is yellow and stunted and looks grassy. No flowers are produced. In the spring, when daffodils start to bloom, flying insects that resemble small bumblebees (½ to ¾ inch long) hover around the plants. These black, hairy insects have bands of yellow, buff, or orange around their bodies.

Analysis: Narcissus bulb fly (*Merodon equestris*), a member of the fly family, occasionally attacks other flower bulbs. In the spring, the adult fly lays its eggs on the leaf bases and soil immediately surrounding the plant. The emerging larvae tunnel through the soil to the bulb and feed on the bulb tissue throughout the summer, making it soft and pulpy. The larvae spend the winter in the bulb as wrinkled, plump, grayish white to yellow maggots ½ to ¾ inch long. In the spring, they either remain in the bulb or move out into the surrounding soil to pupate. After 1 to 2½ months, the adult bulb fly emerges and starts the egg-laying cycle again.

Solution: Check all bulbs carefully before planting. If they are soft or spongy, discard them. In May, drench the foliage and surrounding soil with an insecticide containing *trichlorfon* to kill the adults and emerging larvae. Make sure that daffodils are listed on the insecticide label.

PAEONIA (Peony)

GROWING GUIDE

Adaptation:
Herbaceous peonies: Zones 3 through 7.
Tree peonies: Zones 3 through 9. To determine your zone, see the map on page 336.

Flowering Time: Spring.

Light: Full sun, or half-day sun in hot, dry areas.

Planting Time: Plant peonies in the fall, placing the eyes of the tubers 2 inches below the soil surface.

Soil: Well drained, high in organic matter. pH 6.0 to 7.5.

Fertilizer: When planting, mix a half-cup of superphosphate 0-20-0 into the soil for each plant. During the growing season, fertilize with Scotts Bulb Food according to label directions.

Water: How much: Apply enough water to wet the soil 1½ to 2 feet deep.
How often: Water when the soil 1 inch below the surface is just barely moist.

Handling: Divide herbaceous peony clumps every 6 to 10 years. When cutting flowers, leave 3 or 4 leaves on the stems. To ensure an ample bloom the next year, pick only a third of the blooms for cut flowers.

PAEONIA (Peony) *(continued)*

Gray mold

Gray mold attacking peony flower.

Problem: New shoots wilt and die. The bases of the wilted stems are brownish black and rotted. Young flower buds turn black and wither. Older buds and open flowers turn soft and brown and develop a gray or brown fuzzy covering in wet weather. This fuzzy growth, which may develop on all infected plant parts, is distinctive. Its presence helps distinguish this disease from phytophthora (see at right), with which it is often confused. Irregular brown lesions or patches form on the leaves. In severe cases, the plant base and roots may decay.

Analysis: Gray mold, a common disease of peonies, is caused by a fungus (*Botrytis paeoniae* or *B. cinerea*). Gray mold is most serious in the wet, cool conditions of early spring. Fungal growth on the stems, leaves, and flowers causes spotting, blackening, and decay. The fuzzy growth that forms on infected tissue is composed of millions of tiny spores. This fungus is spread by wind, by splashing rain or water, or by contaminated plants, soil, or tools. The fungus forms small black pellets that survive in plant debris and in the soil for many years.

Solution: Remove and destroy all decayed or wilting plant parts. Clean up plant debris during the growing season and again in the fall. In the spring, spray emerging shoots with a fungicide containing *mancozeb*. Repeat the spray two more times at intervals of 5 to 10 days. Respray if weather favorable for the disease continues.

Failure to bloom

Peony failing to bloom.

Problem: Peonies fail to produce flower buds, or they produce buds that fail to develop into flowers.

Analysis: Peonies may fail to bloom for several reasons.
1. Crown (where the stems meet the roots) buried at wrong depth: Peonies planted too deep or too shallow often fail to bloom.
2. Immature transplants: Peony roots that have been divided and transplanted usually do not flower for at least 2 years. If the divisions were extremely small, the plants may not flower for as long as 5 years.
3. Crowded plantings: Old, established peony clumps eventually become overcrowded and stop producing flowers.
4. Too much shade: Peonies stop blooming when they are heavily shaded by trees, tall shrubs, or buildings.
5. Lack of fertilization: Peonies that are not fed enough will not bloom.

Solution: The numbered solutions below correspond to the numbered items in the analysis.
1. Carefully dig up and reset the crown so that the eyes are 1½ to 2 inches below the soil surface.
2. With time, the young plants will mature and start flowering.
3. Peonies need to be divided after 6 to 10 years or anytime when flower production starts to drop off. Dig up and divide old clumps into divisions containing three to five eyes. These divisions may be replanted.
4. Transplant to a sunny location.
5. Apply Scotts All-Purpose Plant Food in early spring and work it in lightly around each plant.

Phytophthora blight

Peonies infected with phytophthora blight.

Problem: New shoots wilt and turn black. Flowers, buds, leaves, and stems shrivel and turn dark brown and leathery. Black lesions several inches long often appear on lower sections of the stem. The plant pulls up easily. Roots are black and rotted. The fuzzy growth characteristic of gray mold (see at left) does not occur in this disease.

Analysis: Phytophthora blight, a disease of peonies and many other plants, is caused by a fungus (*Phytophthora cactorum*) that is common in most soils. Like gray mold, this fungus thrives in the cool, wet conditions of early spring. Its spores can survive in the soil and in plant debris for many years. Initially the fungus attacks either the roots or the developing shoots at the soil level, causing shoot wilting and a dark decay of the stem tissue. Wherever the fungus is splashed onto the plant, it may cause lesions, spots, and a typical brown, leathery decay. This disease is spread by splashing rain or water and by contaminated plants, soil, and tools. It is most serious in heavy, poorly drained soils.

Solution: Remove and destroy plants with decayed roots. Pick off and destroy infected plant parts. Clean up plant debris. Spray the foliage and drench the base of the plant with a fungicide containing *mancozeb*. Spray three times at intervals of 5 to 10 days. Reapply the spray if infection recurs. Thin out overcrowded plants. Plant peonies in well-drained soil.

PELARGONIUM (Geranium)

GROWING GUIDE

Adaptation: Throughout North America.

Flowering Time: Spring and summer. In zones 9 and 10, some pelargoniums bloom throughout the year. To determine your zone, see the map on page 336.

Light: Full sun or light shade.

Planting Time: Spring, when all danger of frost is past; any time of year in zones 9 and 10.

Soil: Well drained. pH 6.0 to 8.0.

Fertilizer: Fertilize with Scotts All Purpose Plant Food or Miracle-Gro Plant Food according to label directions.

Water: How much: Apply enough water to plants in the ground to wet the soil 8 to 10 inches deep.
Containers: Apply enough water so that 10 percent of the water drains from the bottom of the container.

Handling: Remove old flower clusters to encourage continuing bloom.

Edema

Edema on ivy geranium.

Problem: Water-soaked spots appear on the leaves. Eventually, these spots turn brown and corky. Affected leaves may turn yellow and drop off. Corky ridges may form on the stems and leafstalks. In most cases, the soil is moist and the air is cool and humid.

Analysis: Edema is not caused by a pest but is the result of an accumulation of water in the plant. When the soil is moist or wet and the atmosphere is humid and cool, water is absorbed rapidly from the soil and lost slowly from the leaves, resulting in an excess amount of water in the plant. This excess water causes cells to burst. The ruptured cells eventually form spots and ridges. Edema occurs most frequently in greenhouses and in late winter and early spring during cloudy weather.

Solution: Plant geraniums in soil that drains well, and avoid overwatering them.

Alternaria leaf spot

Alternaria leaf spot showing spots with rings.

Problem: Dark brown, irregularly shaped spots appear on the leaves. The spots range in size from barely visible to ⅓ inch in diameter. Larger spots may contain several dark concentric rings. Spots may be surrounded by a diffuse yellow halo. Spotting occurs mostly on the older leaves, although new growth may also be affected. Severely infected leaves shrivel, turn black, and fall off. Only the leaves are affected.

Analysis: Alternaria leaf spot, a disease of geraniums, is caused by a fungus (*Alternaria alternata*) that thrives in prolonged cool, moist conditions and soils low in fertility. Infection rarely kills the plant but can weaken and disfigure it. The fungus forms spores on the diseased leaves; these spores are readily blown or splashed to healthy leaves. If the leaf surface is wet, the spores will germinate and initiate new infections. The fungus survives on plant debris and is spread by the wind and splashing water.

Solution: Pick off and destroy infected leaves and clean up debris. Fertilize with Scotts All-Purpose Plant Food or Miracle-Gro Plant Food according to label directions. If infection is severe and persistent, spray the plants with ORTHO Multi-Purpose Fungicide Daconil 2787® Plant Disease Control or a fungicide containing *mancozeb, ferbam,* or basic copper sulfate. Spray every 7 to 10 days until spotting has diminished.

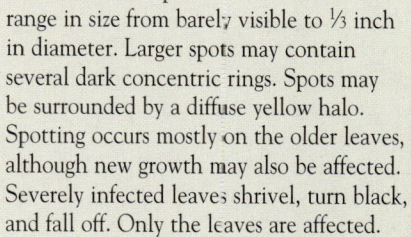

PELARGONIUM (Geranium) *(continued)*

Bacterial blight

Bacterial blight showing sunken leaf spots.

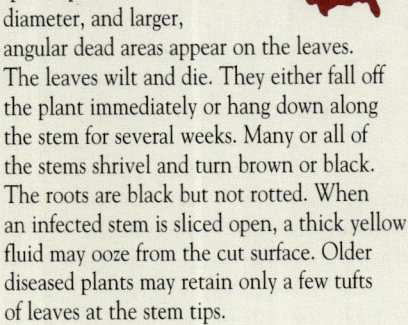

Problem: Tan to brown circular, sunken spots, up to ¼ inch in diameter, and larger, angular dead areas appear on the leaves. The leaves wilt and die. They either fall off the plant immediately or hang down along the stem for several weeks. Many or all of the stems shrivel and turn brown or black. The roots are black but not rotted. When an infected stem is sliced open, a thick yellow fluid may ooze from the cut surface. Older diseased plants may retain only a few tufts of leaves at the stem tips.

Analysis: Bacterial blight, a common and widespread disease of geraniums, is caused by bacteria (*Xanthomonas campestris* pv. *pelargonii* or *Pseudomonas* species). The disease develops most rapidly when the plants are growing vigorously and during periods of warm, moist weather. The bacteria decay the leaf tissue, causing small spots and lesions, then may penetrate throughout the entire plant, causing wilting and rotting. The bacteria can live in plant debris and in the soil for 3 months or more. Plants are not always killed by this disease. They often remain weak, stunted, and disfigured. The bacteria can be present in plants that show no symptoms. Cuttings taken from such plants carry the disease. Plants may suddenly collapse when environmental conditions favor the disease.

Solution: Remove and destroy infected plants. Clean up plant debris. Avoid overhead watering. Disinfect contaminated tools. Wash your hands thoroughly after handling infected plants. Purchase only healthy plants. Don't make cuttings from plants that have shown symptoms.

Black stem rot

Black stem rot, a common fungal disease.

Problem: Dark lesions form at the base of the stems. These lesions enlarge and turn black and shiny. The blackening progresses up the stem. The leaves wilt and drop, and the plant may eventually die.

Analysis: Black stem rot, a common disease of geraniums, is caused by a fungus (*Pythium* species) that lives in the soil. Black stem rot thrives in wet, poorly drained soil. The fungus attacks the stems at the soil level, then spreads upward. The stems decay and the foliage wilts, shrivels, and eventually dies. Black stem rot is spread by contaminated soil, transplants, and tools.

Solution: Remove and destroy infected plants. If they have been growing in containers, throw out the soil in which they grew. Wash and disinfect contaminated tools and pots using a solution of 1 part chlorine bleach and 9 parts water. Plant healthy geraniums in well-drained soil and let plants dry out between waterings.

Caterpillars

Geranium budworm (2× life size).

Problem: Irregular or round holes appear in the leaves and buds. Leaves, buds, and flowers may be entirely chewed off. Worms or caterpillars are feeding on the plants.

Analysis: Many species of moth or butterfly larvae feed on geraniums and other garden plants. Some common caterpillars include budworms, hornworms, and loopers. Usually, the adult moths or butterflies begin to lay their eggs on garden plants with the onset of warm spring weather. The larvae that emerge from these eggs feed on the leaves, flowers, and buds for 2 to 6 weeks, depending on weather conditions and species. Mature caterpillars pupate in cocoons attached to leaves or buildings or buried in the soil. One generation or several overlapping generations may occur during the growing season. The last generation of caterpillars in the fall survives the winter as pupae. Adult moths and butterflies emerge the following spring.

Solution: Spray infested plants with ORTHO Isotox Insect Killer or ORTHO Orthene Systemic Insect Control. Repeat the spray if reinfestation occurs, allowing at least 7 to 10 days between applications. The bacterial insecticide *Bacillus thuringiensis* (Bt) may also be used.

PETUNIA

GROWING GUIDE

Adaptation: Throughout North America

Flowering Time: Summer and fall.

Light: Full sun.

Planting Time: Spring, when all danger of frost is past.

Soil: Well-drained.

Fertilizer: Fertilize with Scotts All-Purpose Plant Food or Miracle-Gro Plant Food according to label directions.

Water: How much: Apply enough water to plants in the ground to wet the soil 6 to 8 inches deep.
Containers: Apply enough water so that about 10 percent of the water drains from the bottom of the container.
How often: Water when the soil 1 inch below the surface is just barely moist.

Handling: If plants become rangy, pinch back about half of the growth to force additional bushy growth.

Gray mold

Gray mold on petunia blossom.

Problem: Gray or brown spots appear on the flowers, especially during periods of wet weather. Brown spots and blotches may appear on the leaves and stems. As the disease progresses, a fuzzy brown or grayish mold may form on the infected tissue.

Analysis: Gray mold, a widespread plant disease, is caused by a fungus (*Botrytis* species) that is found on most dead plant tissue. The fungus initially attacks foliage and flowers that are weak or dead, causing spotting and sometimes mold. The fuzzy mold that may develop is composed of millions of microscopic spores. Once gray mold has become established on plant debris and weak or dying leaves and flowers, it can invade healthy plant tissue. The fungus is spread by the wind, splashing water, or infected pieces of plant tissue contacting healthy tissue. Cool temperatures and high humidity favor gray mold growth. Crowded plantings, rain, and overhead watering also enhance the spread of the disease. Infection is more of a problem in the spring and fall, when temperatures are lower. In warm-winter areas where freezing is rare, gray mold can be a year-round problem.

Solution: Spray infected plants with ORTHO Multi-Purpose Fungicide Daconil 2787® Plant Disease Control at regular intervals of 10 to 14 days for as long as the weather is favorable for the disease. Remove infected flowers and leaves, and clean up plant debris. Try to avoid wetting the flowers when watering.

Caterpillars

Caterpillars feeding on petunia.

Problem: Irregular or round holes appear in the leaves and buds. Leaves, buds, and flowers may be entirely chewed off. Smooth or hairy caterpillars, up to 4 inches in length, are feeding on the plants.

Analysis: Numerous species of moth or butterfly larvae feed on petunias, including cutworms, budworms, armyworms, hornworms, and loopers. As a rule, the adult moths or butterflies start to lay eggs on garden plants with the onset of warm weather in spring. The larvae that emerge feed on leaves, flowers, and buds for 2 to 6 weeks, depending on weather conditions and species. Mature caterpillars pupate in cocoons attached to leaves and structures or in the soil. One generation or several overlapping generations may occur during the growing season. The last generation of caterpillars in the fall survives the winter as pupae. Adult moths and butterflies emerge the following spring.

Solution: Spray infested plants with ORTHO Isotox Insect Killer, ORTHO Orthene Systemic Insect Control, or the bacterial insecticide *Bacillus thuringiensis* (Bt). Repeat the spray if reinfestation occurs, allowing at least 7 to 10 days between applications. Bt is most effective when caterpillars are young.

PETUNIA *(continued)*

RANUNCULUS

Cutworms

Cutworm damage on petunia.

Problem: Young plants are chewed or cut off near the ground. Many leaves may be sheared from the stems. Gray, brown, or black worms, 1½ to 2 inches long, may be found about 2 inches deep in the soil near the base of the damaged plants. The worms coil when disturbed.

Analysis: Several species of cutworms attack petunias and many other flowers and vegetable plants. The most likely pests of young petunia plants set out early in the season are surface-feeding cutworms and climbing cutworms. Climbing cutworms shear the leaves off older plants. Cutworms hide in the soil during the day and feed only at night. Adult cutworms are dark, night-flying moths with bands or stripes on their forewings.

Solution: Apply ORTHO Bug-Geta Plus Snail, Slug & Insect Killer or ORTHO Diazinon Ultra Insect Spray around the base of undamaged plants when stem cutting is observed. Because cutworms are difficult to control, treatments may need to be repeated at weekly intervals. Before transplanting into a new area, apply a preventive treatment of ORTHO Diazinon Ultra Insect Spray, ORTHO Dursban® Lawn & Garden Insect Control, or the bacterial insecticide *Bacillus thuringiensis* (Bt), and work it into the soil. Cultivate the soil thoroughly in late summer and fall to expose and destroy eggs, larvae, and pupae. Further reduce damage with cutworm collars, which should be at least 2 inches high and surround the plant stem fairly closely when pressed into the soil. Make them out of stiff paper, aluminum foil, tin cans, or paper cups.

Bird damage

Bird damage to Ranunculus *seedling.*

Problem: Tender young leaves are torn. Seedlings may be entirely eaten. Birds may be feeding in the garden, or their tracks may be found around the damaged plants.

Analysis: Birds are fond of ranunculus and frequently eat the tender parts of the plants. Individual birds may develop the habit of feeding on ranunculus and visit the plants every day.

Solution: Protect emerging shoots and young transplants with cages or coverings made of 1-inch-mesh chicken wire. Cages about 10"×10"×24" are self-supporting. Larger cages may need to be reinforced with heavy wire. Cheesecloth cages supported with stakes, wire, or string may also be used.

Ranunculus mosaic virus

Mosaic virus on Ranunculus *foliage.*

Problem: Leaves are mottled yellow green. Plants may be stunted, and flowers may be smaller than normal. In some cases, the petals are streaked.

Analysis: Ranunculus mosaic virus infects ranunculus plants and tubers. The severity of infection varies from plant to plant. Mosaic does not kill ranunculus but does greatly reduce the plants' overall vigor and beauty. The virus is spread by aphids that feed on diseased plants, then transmit the virus to healthy plants when they feed again. Mosaic persists in the plant indefinitely. Tubers obtained from diseased plants are also infected.

Solution: No chemicals control virus diseases. Discard infected plants. Prevent the spread of the virus by keeping the aphid population under control.

SALVIA

Verticillium wilt

Verticillium wilt. Diseased plant on right.

Problem: Leaves yellow, wilt, and die, starting with the lower leaves and progressing up the plant. Older plants may be stunted. Yellowing and wilting often affect only one side of the plant. Flowering is poor. Dark brown areas may be seen on the infected stems. When the stem is sliced open near the base of the plant, dark streaks and discoloration of the inner water-conducting stem tissue are seen.

Analysis: Verticillium wilt is caused by a soil-inhabiting fungus (*Verticillium* species) that persists indefinitely on plant debris or in the soil. The disease is spread by contaminated seeds, plants, soil, and equipment. The fungus enters the plant through the roots and spreads up into the stems and leaves through the water-conducting vessels in the stems. The vessels become discolored and plugged. This plugging cuts off the flow of water to the leaves, causing leaf yellowing and wilting.

Solution: No chemical control is available. It is best to destroy infected plants. *Verticillium* can be removed from the soil only by fumigation or solarization. To solarize soil, cultivate and thoroughly wet an area before covering it securely with clear polyethylene film during the hottest time of the year. Allow the plastic film to remain in place for 4 to 6 weeks. The best solution is to use plants that are resistant to verticillium.

TAGETES (Marigold)

Wilt and stem rot

Wilt and stem rot on marigold.

Problem: Leaves wilt and die. The lower stems have a dark, water-soaked appearance. They eventually shrivel and turn brown near the soil line. The plant pulls up easily to reveal rotted roots. The plant usually dies within 1 to 3 weeks.

Analysis: Wilt and stem rot is caused by a widespread fungus (*Phytophthora cryptogea*) that persists indefinitely in the soil. The fungus attacks the roots, then spreads up into the stems. As the roots and stems decay, the leaves wilt and turn yellow and the plant dies. The fungus thrives in cool, waterlogged soils. This disease is spread by contaminated soil, transplants, equipment, and moving water. African marigolds (*Tagetes erecta*) are quite susceptible, but French marigolds (*T. patula*) and other dwarf varieties are resistant to this fungus.

Solution: Remove and discard infected plants and the soil immediately surrounding them. Let the soil dry out between waterings. To minimize drought stress, place containers in shade or erect shade cloth over beds; spray with antitranspirant; and cover with clear plastic tents if plants are in shade. Resume watering when signs of drought stress appear. Plant marigolds in well-drained soil. Drench the flower bed in which diseased plants have been growing with a fungicide containing *captan* to help reduce the severity of the disease. Plant resistant French and dwarf marigolds.

TROPAEOLUM (Nasturtium)

Leaf spots

Leaf spot on nasturtium.

Problem: Spots and blotches appear on the leaves. The spots may be yellow, red, tan, gray, or black. They range in size from barely visible to ¼ inch in diameter. Several spots may join together to form blotches. Some of the leaves may be yellow and dying. Leaf spotting is most severe in warm, humid weather.

Analysis: Nasturtiums are susceptible to several fungi that cause leaf spots. Some of these fungi may eventually kill the plant or weaken it so that it becomes susceptible to attack by other organisms. Others merely cause spotting that is unsightly but not harmful. These fungi are spread by splashing water, wind, insects, tools, and infected transplants and seed. They survive the winter in diseased plant debris. Most of these fungi do their greatest damage during moist, mild weather (50° to 85°F).

Solution: Picking off the diseased leaves generally gives adequate control. If infection is severe, spray during wet periods with a copper-containing fungicide, such as one with basic copper sulfate or *mancozeb*. Clean up debris.

TULIPA (Tulip)

GROWING GUIDE

Adaptation: Throughout North America. In zones 9 and 10, bulbs must be prechilled (see Handling below for details). To determine your zone, see the map on page 336.

Flowering Time: Spring.

Light: Full sun to filtered light.

Planting Time: Fall.

Soil: Well-drained. pH 6.0 to 7.5.

Fertilizer: When planting, add 1 teaspoon of bone meal or superphosphate 0–20–0 to the bottom of the planting hole and mix it into the soil. During the growing season fertilize with a balanced plant food according to label directions.

Water:
How much: Apply enough water to wet the soil 1 to 1½ feet deep.
How often: Water when the soil an inch below the surface is just barely moist.

Handling: In zones 9 and 10, precool bulbs for 15 weeks before planting. Store bulbs in paper bags in the refrigerator crisper. Tulips usually do not flower as well the second or third year after planting. Either replace them or dig, separate, precool if necessary, and replant.

Failure to bloom

Old planting of tulips with poor blooming.

Undersized bulbs can cause failure to bloom.

Problem: Tulip bulbs produce healthy foliage but fail to bloom.

Analysis: Healthy tulips may fail to bloom for several reasons.

Solution: The numbered solutions below refer to the numbered items in the analysis.

1. Lack of cooling: To flower properly, tulip bulbs require a minimum exposure of 15 weeks to cool temperatures (40° to 50°F) in fall and winter. Cooling stimulates the embryonic flower stem within the bulb to elongate and emerge from the ground.

1. In warm-winter areas (zones 9 and 10; see page 336 for zone map), precool bulbs before planting, or buy precooled bulbs. Postpone planting until mid-December.

2. Foliage removed too soon: After a tulip flowers, the remaining foliage continues to manufacture food for new bulbs and flowers for the next year.

2. Let foliage turn yellow before removing it.

3. Lack of nutrients: When tulips are grown in infertile soil for more than 1 season, they form small, poor-quality bulbs. Such bulbs produce sparse foliage and few, if any, flowers.

3. Add Scotts Bulb Food when planting, when the new leaves appear on the plants in the spring, and again after flowers bloom.

4. Undersized bulbs: Bulbs smaller than 2½ inches in circumference may not contain an embryonic flower. They will produce only foliage for 1 to 2 years, until they are large enough to produce a flower.

4. Purchase only large, healthy bulbs from a reputable nursery or mail-order company.

5. Old plantings: Tulip flowers are largest and most prolific the first spring after newly purchased bulbs have been planted. After flowering, the bulb usually disintegrates and several small "daughter bulbs" form. Often these daughter bulbs are too small to provide many flowers. Depending on the variety, a planting of tulips generally continues to flower for only 2 to 4 years. Tulips are especially short-lived in warm-winter areas (zones 9 and 10; see page 336 for zone map).

5. Replace old tulips with fresh bulbs. You may also dig up, separate, and replant old bulbs. Often they do not flower again for at least 1 year. Unless your soil and climate conditions are ideal for growing tulips, the bulbs and flowers will never be as large and prolific as they were the first year. You can prolong the flowering life of a tulip bed by planting the bulbs deeper than usual. Place them 12 inches deep in the soil rather than the usual 6 inches. The soil must be well-drained to prevent rot.

Failure to grow

Poor growth can have several causes.

Diseased plants.

Root and bulb rot

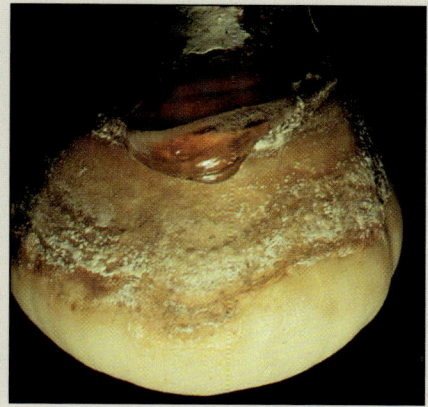

Bulb rot in tulip caused by soil-borne fungus.

Problem: Tulip bulbs do not produce any growth in the spring.

Analysis: Improper cultural techniques, diseases, and animal pests can all contribute to lack of growth in tulips.

Solution: The numbered solutions below correspond to the numbered items in the analysis.

1. Lack of cooling: Tulip bulbs require a period of cooling to develop properly. They need to spend at least 15 weeks below 50°F in order to perform ideally. If newly purchased bulbs are not precooled, or if soil temperatures remain at 55°F or above during the winter, root formation, flower emergence, and the production of new daughter bulbs for future flowering are inhibited.

1. In warm-winter areas (zones 9 and 10; see page 336 for zone map), refrigerate newly purchased bulbs for 15 weeks. For details, see "Handling" on page 120.

2. Foliage removed too soon: After a tulip flowers, the remaining foliage continues to manufacture food for the developing new bulbs and the following year's flowers. If the foliage is removed before it has a chance to turn yellow naturally, the new bulbs will either be very small or will not form at all.

2. Allow foliage to turn yellow before removing it.

3. Lack of fertilization: After the first year, tulips do not continue to perform well when the soil they are planted in is infertile. After several years, they stop producing growth.

3. Fertilize emerging tulips with Scotts Bulb Food in the spring and once a month after the plants have flowered until the foliage dies back.

4. Root rot: Infected tulip bulbs planted in heavy, poorly drained soil frequently decay.

4. Before planting, discard discolored, spongy, or moldy bulbs. Plant in well-drained soil. See "Root and bulb rot" at right.

5. Rodents: Mice, pocket gophers, and other rodents may feed on tulip bulbs. Dig in the area where the bulbs were planted and check for underground tunnels and half-eaten bulbs, both of which indicate rodent damage.

5. The most effective method of protecting tulips is to plant them in baskets made of ¼-inch wire mesh. Traps or baits may also be used.

Problem: Foliage is sparse and stunted. Often the leaves turn red, wilt, and die. When the plant is dug up, rotted bulbs are revealed. The bulbs may be either mushy or firm and chalky. Usually they are covered with a white, pink, or gray mold. Reddish brown to black pinhead-sized pellets may be on the bulb husks and leaf bases and in the soil immediately surrounding the plant.

Analysis: Root and bulb rot, a disease of tulips, is caused by several different soil-inhabiting fungi that attack and decay the bulbs and roots. Some of these fungi form tiny pellets on the bulbs and in the soil. These pellets survive through dry conditions and extremes of temperature. Bulb and root rots thrive in wet, poorly drained soils. Bulbs injured during digging or storing are especially susceptible to infection. Sometimes bulbs in storage are lightly infected, but the fungal decay hasn't progressed far enough to be easily noticed. When planted, these bulbs may rot so quickly that they don't produce any foliage.

Solution: Remove and destroy infected plants and the soil immediately surrounding them. Check and discard infected bulbs before planting. Avoid wounding bulbs when cultivating around them or handling them. Plant in a well-drained location. Do not replant tulips in infested soil for at least 3 years.

TULIPA (Tulip) *(continued)*

VERBENA

Botrytis blight

Botrytis blight. Inset: Infected bulbs.

Problem: Light to dark-colored spots appear on the leaves and flowers. The spots enlarge to form extensive gray blotches, which may cover the entire leaf and flower. During periods of cool, moist weather, a fuzzy brown or grayish mold forms on the infected tissue. Many of the leaves and stems are distorted, and they often rot off at the base. Dark, circular, sunken lesions appear on infected bulbs. Dark brown pinhead-sized pellets form on the bulb husks.

Analysis: Botrytis blight, a common disease of tulips, is caused by a fungus (*Botrytis tulipae*). The fungus persists through the winter and hot, dry periods as tiny fungal pellets in the soil, plant debris, and bulbs. In the spring, these pellets produce spores that attack foliage and flowers, causing spotting, decay, and mold. Wounded, weak, and dying plant tissues are especially susceptible to infection, as are tulip bulbs that are injured when dug up to be stored. The fungus is spread by splashing water. Botrytis blight is most serious during periods of cool, moist weather.

Solution: Remove and destroy diseased plants, leaves, flowers, and debris. Before planting tulip bulbs, check them for signs of infection, and discard diseased bulbs. Start spraying emerging plants when they are 4 inches tall with a fungicide containing *mancozeb*. Use a spreader-sticker when spraying. Spray plants every 5 to 7 days until the flowers bloom. Remove tulip flowers just as they start to fade, and cut off the foliage at ground level when it turns yellow. Rotate plants to a new location the following year.

Viruses

Tulip showing viral color breaking.

Problem: Flowers are streaked, spotted, or mottled in an irregular pattern. The leaves may also be streaked or mottled with light green or white. The plant may be stunted and low in vigor.

Analysis: Several plant viruses commonly infect tulips, causing a characteristic streaking or mottling of the flowers and foliage. Stunted growth may accompany the infection. Viral infections rarely cause a plant to die but can weaken it seriously. The viruses increase in the bulbs year after year. Successive plantings from diseased bulbs yield infected flowers and foliage of poor quality. Some viruses are spread by aphids. These insects feed on diseased plants and transfer viruses to healthy plants at subsequent feedings.

Solution: Once a plant is infected, no chemical will control the virus. To prevent the spread of the virus to healthy tulips, remove and destroy infected plants and all the bulbs associated with that plant. Keep the aphid population under control by spraying with ORTHO Orthene Systemic Insect Control. Because these viruses may also infect lilies, avoid planting tulips near lilies. Parrot tulips may exhibit showy streaked patterns, but the streaking is genetic in origin and cannot be transferred to other tulips and lilies.

Powdery mildew

Powdery mildew, a common fungal disease.

Problem: Grayish white powdery spots and patches cover the leaves and stems, primarily the upper surfaces of leaves. Infected leaves eventually turn yellow and wither.

Analysis: Powdery mildew is caused by a fungus (*Erysiphe cichoracearum*) that thrives in both humid and dry weather. The powdery patches consist of fungal strands and spores. The spores are spread by the wind to healthy plants. The fungus saps plant nutrients, causing yellowing and sometimes death of the leaves. A severe infection may kill whole plants. Since this powdery mildew attacks many kinds of plants, the fungus from a diseased plant may infect other types of plants in the garden.

Solution: Spray infected plants with a fungicide containing *thiophanate-methyl*. Spray at regular intervals of 10 to 12 days or as often as necessary to protect new growth. This fungicide protects the new, healthy foliage but will not eradicate the fungus on leaves that are already infected. Remove and destroy severely infected plants. Where practical, pick off diseased leaves. Clean up and destroy plant debris. Plant in a site with good air circulation and exposure to early morning sun.

VIOLA (Pansy, Violet)

Spindly growth and poor flowering

Spindly growth and poor flowering in pansy.

Problem: Leaves are small and thin, and stems are long and spindly. Flowering is poor, and flowers are small.

Analysis: Several cultural problems may contribute to spindly growth.
1. Failure to remove old flowers: If fading flowers remain on the plant, only a few small new flowers are produced. When the old flowers are allowed to remain, the plant uses its energy to develop seed instead of to produce new flowers.
2. Inadequate light: Pansies and violas grow lanky and flower poorly when planted in deep shade. They require at least strong filtered light to grow compactly, and they flower most profusely in full sun during mild weather.
3. Old age: Pansies and violas are perennials and theoretically last from year to year. In cold-winter climates, freezing temperatures kill them. In warm-winter areas, they often last for a year or more but usually start to produce lanky, unattractive growth after the first growing season.

Solution: The numbered solutions below correspond to the numbered items in the analysis.
1. Remove faded flowers.
2. Grow plants in full sun, part-day sun, or strong filtered light.
3. Treat pansies and violas as annuals. Plant them in the spring (or fall in zones 9 and 10; see page 336 for zone map) and replace them when they start to decline in the summer. Pinch back one-third of the spindly stems to one-third of their height to rejuvenate rangy plants.

Root and stem rot

Root and stem rot caused by soil-borne fungi.

Problem: Leaves turn yellow, wilt, and die. The roots and lower stems are soft and rotten. There may be white fungal strands on infected stems and around the base of the plant.

Analysis: Root and stem rot is caused by several fungi that persist indefinitely in the soil. These fungi thrive in waterlogged, heavy soils. Infection causes the stem and roots to decay, resulting in wilting, yellowing leaves, and the death of the plant. These fungi are generally spread by infested soil and transplants, contaminated equipment, and splashing or running water.

Solution: Remove dead and dying plants. It is important to allow the soil to dry between irrigations; root and stem rots are encouraged by waterlogged conditions. Improve soil drainage.

ZINNIA

Alternaria blight

Alternaria blight in Zinnia foliage.

Problem: Reddish brown circular or irregular spots up to ½ inch in diameter

appear on the leaves. The centers of the spots may turn grayish white. The blossoms are also often spotted. Severely infected leaves, stems, and flowers turn brown and die. Dark, sunken lesions may be at the base of the stems.

Analysis: Alternaria blight, a common and widespread disease of zinnias, is a leaf spot caused by a fungus (*Alternaria zinniae*). The disease thrives in moist conditions. The fungal spores are spread from plant to plant by wind and splashing water. The fungus survives on infected debris in the soil and on contaminated seed. This fungus also causes damping-off of seedlings.

Solution: Remove dying plants. Spray infected plants with ORTHO Multi-Purpose Fungicide Daconil 2787® Plant Disease Control or a fungicide containing *mancozeb*. Pick off infected leaves and flowers. Clean up and destroy plant debris. Avoid overhead watering.

Trees, Shrubs, and Vines

Sharing the planting of a tree ties generations together like few other acts.

Unless trees or shrubs start browning drastically or suddenly fall over, many gardeners tend to take them for granted. The shade, color, and serenity trees provide may not be as conspicuous as brilliantly flowering anemones or a startlingly large squash.

Given optimal surroundings, trees live hundreds or even thousands of years. But sometimes something goes wrong: Insect pests, virus and fungus infections, grading changes, too much or too little water, and pollution take their toll. Unlike flower problems, which can be seen merely by looking down, tree problems may begin at the towering top or be hidden amongst the branches. By the time you notice the damage, it may be severe. But, with knowledge and persistence, most trees and shrubs can be saved.

STAYING ALERT FOR SYMPTOMS

Part of effective shrub and tree management is maintaining an awareness of their health. Make a visual inspection each week as you go about your normal gardening chores. Distorted leaf growth is often the first eye-level symptom. Leaves may be crinkled, rolled, or otherwise different than normal. Leaves may or may not change color. Some develop brown spots, others develop red highlights out of season. Since leaf distortion is a clear sign that something is bothering the tree, it is time to play gardener-detective before the problem progresses.

REMEDYING GROUND POLLUTION

If the tree is receiving water but looks as if it is suffering from drought, ground pollution may be the problem. Road salt is one possible cause. Improper herbicide application is another. The wind may carry herbicide from weeds, the intended targets, to garden plants. Damage may also result when spray equipment used to apply an herbicide is not cleaned thoroughly before being used to apply an insect control. Whatever the cause, this sort of damage usually appears several days after herbicide spraying.

Symptoms include puckered leaves or twisted needles. Leaves may seem off-color, yet still be green. If garden plants receive a significant dose of herbicide, distorted new growth may continue throughout the growing season.

To remedy ground-pollution effects, give soil around the tree several ample waterings to flush out chemicals. Prune off twigs or branches with distorted leaves. Apply fertilizer to prevent tree stress and aid recovery.

To prevent future problems, apply herbicide on a windless day and aim the sprayer carefully; follow label directions. Use one sprayer for herbicides and another for pesticides.

MITIGATING AIR POLLUTION

Air pollution is a worsening problem over which the average gardener has little control. Because it begins so gradually, its initial effects are subtle. Yet pollution paves the way for a host of insect annoyances—including the seemingly ever-present bark beetles, which home in on weak or injured trees.

Most air pollution falls into one of two categories: ozone and smog.

Ozone: This type of air pollution does considerable damage throughout the United States and is the prime culprit for pollution damage on the East Coast. Though the ozone layer in the outer atmosphere is essential to our well-being, ozone at lower elevations is a threat to plants and animals. Ozone forms when the gases produced by an industrialized society combine once they get into the air.

Ozone enters trees and shrubs through leaf pores. Once in foliage, it destroys cell membranes, causing them to collapse. Cell membrane death shows up as white to tan leaf markings. In pines, new needles are flecked with yellow. Older needles may be deep yellow and smaller than normal. Needles may drop. If just some leaves or needles are affected, tree and shrub growth slows. Blossoms may fall. If enough foliage is affected, as has happened in heavily polluted areas, entire trees die.

Smog: On the West Coast, smog is the type of air pollution that causes the most severe damage. The damage-causing component of smog is peroxyacetyl nitrate (PAN), which affects shrubs and smaller plants to a greater degree than it affects trees. Young, rapidly growing spring foliage is especially sensitive to smog.

Smog enters leaves through pores. The undersides of leaves turn silver as a result of internal damage. In severe cases, damaged leaves turn light beige and die.

Pollution solutions: No magic can eliminate air pollution, but you can alleviate the damage it causes to your home landscape. If possible, plant European white birch, gray dogwood, American arborvitae, winged euonymus, or Norway maple—these species are more smog- and ozone-tolerant than others. Apply fertilizer regularly, and make certain that watering is sufficient to reach all tree roots. Avoid overwatering, because sitting in soggy soil makes some species prone to pollution damage, particularly in warm climates. Make certain that all or some of the damage is not attributable to insects, disease, or nutrient deficiencies, which are problems you can counter directly.

PROVIDING FERTILIZER

Nutrient deficiencies tend to affect a rapidly growing young tree more than a mature one. For this reason, aid young growth by applying fertilizer in the spring. Providing nutrients early will make the most of growing time so, when winter comes, the new growth has sufficient strength to withstand the cold.

Treating a sapling and a mature specimen to the same fertilizer program is not always

Balled and burlapped trees, their trunks wrapped for protection, are a common sight at nurseries everywhere.

wise. A mature ornamental tree with healthy leaf color and a growth rate appropriate for the variety needs little fertilizer. Overfertilizing can increase leaf density to the point where interior leaves do not get enough sun to survive. Any plants under such a thick leafy umbrella may also fail. In addition, the ground under the canopy remains moist, opening up the path for a variety of fungal infections.

ASSESSING FOLIAGE DROP

Leaf drop causes many gardeners to apply fertilizer with all speed. Falling foliage may be a symptom of tree illness or damage that fertilizer will alleviate; on the other hand, falling leaves may be a natural part of the life cycle. In autumn, for example, the level of green pigment (chlorophyll) in the leaves of deciduous trees changes in preparation for an annual foliage drop. Exquisite reds, golds, and oranges take the place of greens, then the leaves fall. Leaf drop in deciduous trees allows them to enter a resting, or dormant, state in which they can survive the winter.

Evergreens also shed leaves and needles as part of their life cycle. Usually the older leaves are the ones that fall, leaving room for new growth. Pines shed needles the second year after they appear; junipers hold needles 10 years or more before they drop. Evergreen foliage may fall throughout the year or in batches during particular seasons. Many evergreen trees in dry climates routinely lose some older foliage at the beginning of summer. Holly leaves drop in late winter.

If you are worried about what appears to be abnormally heavy leaf shedding, inspect the tree carefully. The most common cause of excessive leaf drop is insufficient sunlight. Perhaps the tree once received enough sun, but changes in surrounding growth have altered the light pattern. Sometimes the shape of the tree blocks sunlight from reaching lower leaves. Pruning may restore the tree's vigor. Always look for signs of insect damage, another cause of abnormal foliage drop, and take control measures if needed. Some climatic conditions, such as drought or heavy rainfall, may cause changes in the amount of leaf fall. Changes in grading or runoff direction can also cause shedding. Reduce tree stress by watering or correcting drainage, and the amount of foliage drop may return to normal.

SOLVING ROOT PROBLEMS

If the full branch spread of a tree is 25 feet wide, the roots reach out at least that far. The wide spread of trees' root zones often leads to problems.

Far-reaching roots: If the roots of your tree extend into your neighbor's yard and your neighbor does any construction, your tree may suffer damage over which you have little control. The practical way to avoid this problem is to research mature tree size before purchase, then pace off the eventual root zone to determine whether there will be encroachment on nearby lots. When in doubt, plant the smaller tree. Many exquisite shrubs, growing no more than 8 to 10 feet high, can be clipped or trained to look like trees.

Of course the opposite can happen: Roots from a neighbor's tree may invade your property and compete with your plants for nutrients. Some trees, such as willow and birch, can be aggressive in seeking water, and they can cause dry ground and mineral deprivation around them by removing the water and nutrients they need for survival. You may not be able to grow anything in these areas.

Two antidotes for far-reaching roots are root pruning and root barriers.

Root pruning: If the amount of needed pruning is extensive, hire a professional arborist to do the work. If the job requires cutting only a few small roots, do the pruning yourself. After the cutting, tree growth may slow for a bit but the tree will be unaffected otherwise.

Root barriers: If root pruning is not feasible, consider installing a root barrier. Dig a trench at least 3 feet deep just outside the area affected by the invading tree roots. The deeper the trench, the more effective the barrier will be. Cut roots that cross the trench. Insert a thin wall of concrete, sheet metal, or rolled roofing material into the trench, which will inhibit tree roots from reaching into the planted area. Backfill the trench. Even with the best barrier, you may have to remove persistent tree roots every 5 years or so.

If root competition is severe and uncontrollable, your only choice may be to install drought-tolerant plants. Compensate for nutrient loss caused by the invasive roots by applying fertilizer.

Roots in plumbing lines: Roots frequently clog plumbing lines. The problem tree may be quite a distance away, on your property or your neighbor's. Some trees are known for their invasive properties. Others become problems because the amount of water they get is inadequate for their needs, so they seek elsewhere for survival. The best way to prevent root-clogged lines is to thoroughly research any tree before putting it into curb strips, near house foundations, next to patios, in lawns, and near any plumbing line. If you are in the position of installing a new line, ask your supplier about pipe that is relatively rootproof.

If a drain is plugged with tree roots, a plumber needs specific equipment to clear it. Some firms specialize in sewer and drain line clearing. To prevent further plugging, pour 1 pound of copper sulfate crystals into the lowest entry point to the sewer line (such as a toilet or basement drain) at a time when drains and sewer lines are not being used (such as just before you go to bed). Copper sulfate is highly poisonous; use it with the utmost care. Flush the toilet or wash the crystals into the pipe with a bucket of water. The copper sulfate collects in the root mass and kills the roots, which rot in a few weeks. It is not circulated through the tree, so the tree itself is not harmed. You may have to repeat the treatment from time to time, as long as the invasive tree remains.

PRUNING AWAY EXCESS SHADE

Too much shade may be a neighbor's complaint, your own, or both. Researching ultimate tree growth before planting helps prevents this problem. Sometimes, however, a huge, problematic shade tree is part of newly purchased property.

Jobs for professionals: Limbs and large branches—with their thousands of food, water, and sunlight-gathering leaves— are a vital part of a tree's survival system. Cutting limbs off without planning can result in severe shock that may injure or even kill a mature tree. Hire a licensed tree surgeon or arborist to prune limbs. Your garden center or nursery may be able to recommend a local professional. Before the surgeon starts pruning, discuss the specifics of the job to ensure that the final result suits your landscape plan and budget.

Prune-it-yourself jobs: Home gardeners with the proper equipment and technique can undertake limited branch pruning. The rewards will be immediate and long-lasting: In addition to removing undesirable shade, the pruning is likely to slow new growth so you won't have to prune again soon. For growth limitation, do not prune in early spring because the nutrients that would normally feed the removed area will be diverted into other branches, forcing them into a growth spurt. Prune later in the season, when growth has slowed.

In general, you can remove any branch that crosses another, dead branches, branches growing inward toward the trunk, and branches growing downward. Cut small sections off a large branch before you go after the whole. If a branch is really heavy with growth, its falling weight may tear bark all the way to the trunk.

Do not get so carried away with pruning that you eliminate the tree. More trees are ruined by incorrect pruning than by complete neglect. After you have cut the most obvious of the branches mentioned previously, stand back; take a rest; and perhaps finish the job another day, when you have a fresh supply of objectivity.

Overzealous pruners can do more than harm the tree; they can hurt themselves. Take the time for safety precautions. Pay attention to the cutting angle so the limb falls away from you. Do only pruning that allows you to keep both feet on the ground. Pruning shears and saws with extension handles let you reach up quite a distance.

In general, leave pruning cuts open to the air—sealing them may trap fungi inside the tree. If a wound does not appear to be healing well, however, apply a pruning sealer. Available as liquids or foams, sealers protect the tree from dehydration and prevent excess sap flow, which could be an invitation to insects. Apply sealer evenly over the entire cut.

TREATING BARK WOUNDS

Bark protects a tree by covering the xylem and phloem, the system that conveys water, minerals, sugar, and protein. If bark wounds do not heal quickly and properly, the xylem and phloem are exposed to sun and wind and will quickly dry out and die. This deprives the tree of vital nutrition.

Bark wounds can be caused by insects, animals, pruning, or mechanical injury. Monitor bark wounds carefully. If the wound does not heal quickly and properly, apply a

pruning sealer as described in the pruning section.

TREATING CANKERS

A tree canker is a lesion caused by bacteria or fungi. The infected wood is discolored, sunken, and oozing. If cankers are widespread, consult an arborist for treatment. If cankers are few and isolated, try treating them yourself.

Remove cankered branches by pruning off the branch at the trunk or at least 6 inches below the canker. If the canker is on the main trunk or a main branch, use a chisel and a sharp knife to remove the lesion. Remove all discolored and oozing wood and bark. Sterilize equipment after each cut by dipping it in a solution of 1 part household bleach and 9 parts water, to avoid infecting other tree segments. After the area dries, apply pruning sealer. If cankers are extensive and treatment is ineffective, the tree may have to be removed.

REMOVING TREES

A tree may grow too large for its surroundings, be too diseased to save, or may have to make way for new construction. Removing a large tree is a job for a licensed professional, who has the correct equipment for cutting and climbing.

Such a job takes a professional's consideration as well as equipment. The tree's weight must be lessened by judicious pruning before it is felled. Limbs must be cut

so that they fall without injuring people, pets, plants, equipment, roofs, and power lines.

After a tree has been removed, stump treatment is necessary.

TREATING AND REMOVING STUMPS

The aggressiveness that may have caused you to remove the tree may be evident in its regrowth. Within 30 minutes of tree removal, treat freshly cut stumps with ORTHO Brush-B-Gon Poison Ivy, Poison Oak & Brush Killer, which acts systemically. Before you begin, remove suckers from the sides of the stump. Then paint or daub the brush killer over the entire stump surface.

If some time has passed before stump treatment, use a hatchet to make multiple notches around the stump, angling downward into the bark. Cut through the bark without removing it. Pour brush killer into the notches, following label directions. If the herbicide tends to leak from the stump and desirable plants are nearby, cover the stump with a plastic bag secured at the stump base.

You may want to remove a stump altogether. This is best done when the stump, roots and all, is dead. A stump is dead if it fails to resprout the season after the tree has been cut. Digging out the stump is quite a job. If you undertake it, avoid back injury by lifting and prying safely. A landscape contractor with proper equipment can remove a stump in a short time.

When you obtain an estimate for tree removal, ask if the price includes grinding the stump. Burying stumps can cause fungal problems later.

Securely wrapping a tree for the trip home from the nursery helps reduce the amount of dehydrationh the wind causes as well as physically protects the leaves from shredding or tattering.

PLANTING AND TRANSPLANTING TREES

Many trees fail due to root injury and lack of preparation when planting or transplanting. The tree begins to die back from the outer portions inward and from the top down. With care, however, the home gardener can move small trees successfully. A tree with a trunk up to 1 foot in diameter is in the plant-it-yourself category. The more mature the tree, the more difficult it is to move, though professional tree movers can transplant even large trees successfully.

Deciduous trees are best moved in spring, before leaves begin to appear, or in fall, after leaves drop. Evergreen trees are best planted in September.

Hole preparation: Prepare the receiving hole before moving the tree. The root system of the tree to be moved probably spreads as far as the branches; the hole needs to be twice that width. Dig the hole deep enough to allow the tree to remain at its current level. If soil is sandy or heavy clay, improve it with organic material.

If soil turns out to be hardpan, you have a problem. Sometimes referred to as shallow soil or caliche, hardpan is formed by extensive compaction from construction or it may be the natural soil formation in the region. Hardpan may feel like bedrock, but it is not. Bedrock is actually solid rock. If

you find it, plant elsewhere or do container gardening.

When you hit hardpan, your shovel feels like it has hit brick. Instead of moving easily through soil, you scrape dirt away a bit at a time. Eventually, if you scrape down far enough, you may get through the hardpan. But it can extend for quite a distance.

You must provide good soil to allow root penetration to the full depth and width of eventual root growth. This will be approximately equal to the branch spread of the full-grown tree.

Even if you are able to complete the hole, the remaining hardpan can present a problem for the tree planted in it. Rain or irrigation water tends to remain in hardpan for hours or even days. Standing water blocks nutrient absorption and invites fungi and bacteria. To compensate, you should make the hole even larger than normal, difficult as that may be.

If you decide to finish the digging yourself, drill through soil with an auger or posthole digger. Break up the hardpan to a depth of 2 feet for trees, 1 foot for shrubs. A less strenuous option is to hire a contractor to plow the hardpan and complete the digging. Do not put hardpan back into the planting hole. Backfill with topsoil mixed with organic material.

Do not plant a tree in uncorrected hardpan. Not only will it not prosper, its roots will remain shallow, and a heavy windstorm could topple it.

If the work or expense of planting in hardpan does not coincide with your definition of recreational gardening, remember the option of planting the tree in a container. Drought-tolerant species that perform adequately in restricted growing places include silk trees, redbuds, dwarf mungo pines, and strawberry trees.

Tree preparation for transplanting: Cut back about one third of the tree before digging it out. This lessens nutrient need until roots take hold and reduces the area exposed to drying sun and wind. Do not just cut from the top down; prune broken and crossing branches and branches that are too close to each other.

To excavate a tree 10 feet high, dig a trench at least 4 feet in diameter and about 18 inches deep. Generally, roots spread as far as branches. If thick roots obstruct the space, move farther away from the trunk and, with a digging fork, carefully remove anchoring soil from around the exposed

roots. Leave as much soil around the roots as possible. Sway the plant gently to loosen the tree's hold.

The move: With a ball of earth surrounding the roots, immediately move the tree into the receiving hole. Do not let the roots dry out. If the move will take some time, use a trunk spray to slow down water loss or wrap the trunk of a thin-barked tree, such as birch, with burlap. Leave the burlap on the tree for the first year after transplanting.

Continuing care: With careful transplanting, you have given the tree a boost toward a long healthy life in the new site. The transplant still needs your special attention, however.

Place a 4-inch-deep layer of mulch around the transplanted tree to further conserve moisture. To prevent fungal diseases encouraged by moisture buildup, keep mulch away from the trunk. Don't overwater or overfertilize—that may stimulate top growth that the traumatized roots cannot handle.

■ **Sucker pruning:** Suckers may appear around the base of a transplanted tree. If this happens and no top growth is evident, the tree has sustained damage that may hinder its survival. In such a situation, suckers indicate root injury and the failure to prune the treetop sufficiently before transplanting. Keeping suckers diligently pruned may help the tree to recover.

■ **Trunk support:** In heavy wind areas, a transplanted trunk may need support. There are two widely used methods of wind staking. The first is to hammer a strong stake near the center of the hole, then set the tree close to the stake. Since this method presents the possibility of the trunk rubbing the stake, some gardeners prefer a two-stake method. This involves placing a stake at either side of the tree but not touching it. These stakes should be aligned at a right angle to the usual wind direction. Either rot-resistant wood stakes or metal ones serve the purpose.

In the past, experts thought using tall stakes was necessary to prevent any tree movement. Current studies indicate that some trunk movement encourages eventual sturdiness. Using short stakes might be an effective compromise. Allow some room for tree sway when installing the stakes, and orient the tree so that the side with the most branches faces into the wind. Place the largest root in the direction of the wind for strongest support. Loosely connect each stake to the tree with a nonabrasive tape in a figure-eight pattern. Do not use rope or wire, which will cut into the tree.

Another option, often used for larger trees, is to hammer 3 or 4 pegs into the ground several feet from the tree base. The tree is then secured to the pegs with shielded wires. A piece of old garden hose makes an excellent shield, or purchase shielded holding wires designed for trunk support. After the first year, if the tree is flourishing and seems well-established, remove the stakes of guide wires. Remove all restraints by the second year.

Trunk injury may occur when wires or other restraints are left on a growing tree. Sometimes the restraints become so embedded in the tree that they are almost invisible. Since the nutrient system for the tree is directly under the bark, such a restraint begins to choke off circulation. New leaves may be small and discolored, twigs and then larger branches may die, and tree growth may slow. The entire tree above the restraint can die. Removing an embedded restraint can be difficult. However, the deeper it gets, the more circulation it impedes. Use wire cutters at exposed sites. If you cannot free the embedded material, consult an arborist.

PREVENTING ROOT PROBLEMS

Girdling is a common root malady that can take two forms: container girdling and girdling roots.

■ **Container girdling:** If the roots of a tree you are considering for purchase meander out of the container or wrap around the outside of the container, choose another plant. This root condition, called container girdling, is a visible sign of problems ahead. So are circling roots at the soil surface. To check for roots circling within the container, stick your finger in the top 2 to 3 inches near the trunk. Brush away a bit of topsoil. Roots that look or feel damaged, broken, or tightly compacted might be circling within the limited space of the container.

If you do buy a containerized tree with a tight or circling rootball, the tree may still be salvageable. Remove any ties around the rootball. Carefully move the roots apart so they spread out normally in the planting

hole. To accomplish this, you may have to cut and remove some of the circling roots. Removing one quarter of the roots in the outer inch of the rootball should not damage the tree. Add and firm backfill gradually to get the best root contact.

Girdling roots: A girdling root is any root that wraps itself completely around the trunk, either above soil level or just below it. Girdling roots may occur if trees are grown in hardpan. They are also common when containerized trees are placed in too small a hole or if the tree suffered from container girdling before it was planted.

As the trunk enlarges, the rootball tightens, effectively cutting off circulation. Tree leaves are small and discolored. They may drop out of season. Twigs, then larger branches, may die. If the girdling is below the soil surface, the problem may become severe before the cause is discovered. One clue to below-ground girdling is a trunk that narrows where it enters the soil—a normal trunk exhibits a slight flare at that point. A severely constricted trunk may break in a heavy windstorm.

The girdling root must be completely or partially removed to open up tree circulation. Since you are removing part of the root system, which provides the tree with nutrients, compensate for the loss by pruning back weak, crowded, or excess branches. If constriction at the base is severe, stake the tree for a few years to prevent wind breakage. In serious cases that require professional attention, the tree surgeon can sometimes save a girdled tree by using a procedure called bridge grafting.

TREATING GALLS

Sometimes a tree or shrub develops an odd-looking bump or bumps. They can be small or very large. They may appear to be part of the tree or completely separate, almost like a piece of fruit. These strange plant growths are called galls (see page 139). They occur because a foreign substance has been injected into the tree. As a protective measure, the tree surrounds the foreign substance with firm tissue.

Gardeners tend to worry about gall formation. But aside from changing the appearance of a plant, most galls do no serious damage. Small plants may be stunted if the gall blocks the flow of nutrients, but if the plant is ill, something else is usually

causing the problem. The exception is crown gall, which can grow so rapidly that nutrient flow is markedly decreased. A tree with crown gall will usually survive for years but will decline slowly and eventually die.

Each gall is specific to the tree type. In addition, each gall is specific to the insect, bacterium, or fungus causing it. Oaks can develop 805 different galls; rose bushes can develop 133; and maple trees can develop 48. Nematodes cause nodule-like cysts on plant roots. Mite infestation may cause bladder gall and spindle gall. Bacteria cause euonymus crown gall and wartlike galls on oleander. Fungi cause camellia leaf gall; the white thick azalea leaf gall; and gall rust, the large, rough, round orange swellings on pine trunks and branches.

Insects cause most gall formations. Psyllids cause the cylindrical nipple-like growths of hackberry leaf gall. Aphids cause the green pineapple galls seen on spruce and the white cottony galls found on fir. Other gall-forming insects include gall midges, gall wasps, caterpillars, beetles, and thrips.

Galls caused by insects begin when an adult female places an egg in a plant bud. Plant cells begin to close around the egg. Some insects even inject a toxin that stimulates rapid and abnormal cell growth, hastening development of the gall. The egg hatches, and the larva begins development. The plant cells, responding to the larva within, begin to enlarge, forming the gall.

When purchasing container-grown trees or shrubs, ask the nursery or garden center sales person if you can lift the root ball from the pot to inspect the roots. Healthy roots have good color and are not so dense as to be pot bound. Avoid plants with girdling roots near the trunk.

When the larva completes its growth, it chews its way out of the gall.

A tiny hole remains; so does the gall. Some galls are home to just one larva, others to many.

Oak galls, or oak apples, are most commonly seen as round balls growing on twigs. However, oak galls can be round, spiny, star-shaped, flat, or long, and may be found on leaves and branches as well as twigs. Occasionally the leaf galls slow nutrient use by leaves, resulting in discoloration and premature drop. Some twig galls can cause twig dieback.

Nothing can cure a tree of galls. If they are causing a problem or if you simply do not like their appearance, prune away the affected growth. To help prevent recurrence, cut off the galls and destroy them before the adult insects emerge in spring. If galls are especially numerous or unsightly, controlling the insects or diseases causing them may be necessary. Identification of the insect or disease responsible can be difficult. Take samples to a garden center, or consult an arborist.

CONTROLLING INSECT PESTS

The battle with insect pests seems as old as time. Modern research has provided a new strategy for the fight, however, and understanding the enemy will help protect your garden.

Integrated pest management:
Today, integrated pest management (IPM) is the buzzword for gardeners seeking healthy trees and shrubs, with an economy of labor and a concern for the environment. IPM calls for planting resistant species, using natural controls, supplying fertilizer and water for maximum plant strength, and using pesticides only where necessary and always according to label directions.

Beneficial insects such as ladybugs, lace wings, syrphids (also called flowerflies or hover flies), mantises, and parasitic wasps are part of any IPM program. Encourage their presence by providing thick shrubs. These provide not only shade, but protection from predators such as birds. If you have several large shrubs on your property and want to nurture beneficial bugs, leave the bases of the shrubs untrimmed. That slightly untidy area is a favorite trysting spot for helpful insects, and under its protection they climb up the shrub

or tree, often to deposit eggs. Leafy branches then cover both eggs and larvae and keep them high enough off the ground to avoid many predators. (Be aware, however, that this also provides cover for unwanted pests such as rodents. If you are having problems with mice or rats, you may be forced to trim shrubs further.)

Many garden centers offer a variety of beneficial insects for you to release in your garden. Once released, they are free to go where they please, but if you have provided appropriate living conditions, it is likely that they will remain to feed on pests.

Hopping insects: Trees are afflicted with their share of these common pests. Some, such as leafhoppers, are extremely small. Others are weirdly shaped, such as treehoppers. And some are large and noisy, such as cicadas.

Cicadas infest both shade and fruit trees. Most of the damage they do results from egg-laying, which damages twigs. The twigs then turn brown and drop. Nymphs chew tree roots, and their feeding may eliminate both flowers and fruit. Adults suck sap from limbs and twigs. Broods of the periodical cicada may be large and destructive or small and merely an annoyance. Cicadas usually appear where they have appeared before, because trees harbor eggs from previous generations. As many as 40,000 cicadas have been known to infest a single tree.

Each brood may spend 13 or 17 years underground, feeding on tree roots, before appearing in masses aboveground. The species with the 13-year life cycle is found primarily in eastern states; the 17-year species is found primarily in southeastern states. Cicadas are wedge-shaped and black-bodied. They have red-orange eyes and red wing edges. An adult's full size is about 1½ inches long. Males make the annoying high-pitched droning that seems to go on forever. The females puncture twigs with knifelike egg-laying organs. Each female deposits up to 600 eggs within the twig.

Cicada young, or nymphs, resemble brown ants when young. They drop to the ground and enter soil. For the 13 or 17 years they take to complete development, nymphs feed on tree roots. In May or June of a breeding year, they crawl up from the ground on tree trunks or almost any other available object. They change to adults and begin breeding.

Cicada control consists of monitoring

cicada outbreaks. Where heavy populations are known to occur, avoid planting new trees while cicadas are visibly present. Prune damaged twigs where the female has laid eggs. To protect small trees, cover them with netting or cheesecloth during cicada outbreaks. *Carbaryl* sprays are effective against cicadas. Spray when the males first begin to drone, and respray after one week.

Plant bugs: Small shield-shaped insects, plant bugs often live with leafhoppers. These insects are extremely mobile—they both run and fly when disturbed. They feed by sucking sap, and several species inject toxins as they feed. The tree parts attacked wither and die.

The plant-bug category includes many species. The four-lined plant bug is greenish yellow with black stripes and 1/16 inch long. It infests many ornamental trees and shrubs as well as crop plants. Leaves develop tan to reddish brown spots. The spots may join together to totally discolor the leaf. The sycamore plant bug, 1/8 inch long, is common on sycamore as well as ash, mulberry, and hickory trees. Feeding on leaves causes yellowish or reddish spots. Sometimes holes appear where dead leaf tissue has dropped out. The yucca-plant bug appears only on yucca, causing stippled leaves covered with black waste matter. The adult of the species is blue-black with reddish head and throat. Its nymphs are bright scarlet and may be plentiful on leaves.

Control plant bugs with a *malathion* or *carbaryl* spray when you first notice damage in spring. Make certain to cover both the top and undersides of leaves. Repeat as necessary.

Leaf-destroying insects: Leaves plagued by insects may seem as if they are punctured with shotgun pellets, cut with scissors, the victims of a hole punch, or suddenly transformed into fine lace. Sometimes leaves disappear almost overnight. Leaf-destroying insects include leafcutter bees, rose slugs, cankerworms, Japanese beetles, alder flea beetles, leaf beetles (see page 196), two-banded taxus weevils, bagworms, snails, slugs, mimosa webworm, holly leafminer, tent caterpillars (see page 175), stain moth caterpillars, woolly-bear caterpillars, and gypsy moth caterpillars. See the indicated pages for identification and control.

Insect pests of pines: The increasing use of pines for reforestation, conservation, city areas, yard ornamentation, and holiday trees has caused an increase in pine-infesting insects. Many times the problem is a direct result of planting pines in clusters rather than mixing species in a natural way. Pine moths, pine sawflies, and other pests move quickly from one pine to another. Insect populations quickly build up to numbers natural predators cannot control. In general, avoid planting new pines near areas that contain pest-infested trees. Do not use bluegrass in pine-planted areas. Bluegrass encourages field mice, which may girdle pine trees.

Regular inspection of pines is crucial. Trouble signs include off-color foliage, unusual leaf drop, distorted growth, and insect presence. If caught in time, problems can be eased or eliminated. If you remove damaged pine wood, destroy it—prunings are an invitation to bark beetle invasion.

LEAF-DESTROYING INSECTS

Leafcutter bees (*Megachile* species)**:** Stout-bodied, hairy. Can be black, green, purple, or metallic blue. Females cut precise circles or ovals from leaf edges. Valuable pollinators. Control not recommended.

Rose slugs (*Endelomyia aethiops*)**:** About ½ inch long, greenish white to dark green, velvety or covered with bristles. Skeletonize topsides of leaves; leaves turn brown. Damage can be mistaken for rust. Control with *carbaryl* on foliage at first sign of feeding. Repeat as necessary.

Cankerworms (*Alsophila pometaria* and *Paleacrita vernata*)**:** Inch-long caterpillars, wingless moth adult females. Chew leaves and can defoliate entire trees. Tree banding with sticky substance such as tanglefoot helps prevent caterpillars from crawling up trunk. Spray with *carbaryl* (Sevin®), *acephate* (Orthene®), or *diazinon* in late April to early May. Respray as leaves expand.

Woolly-bear caterpillars (*Diacrisia virginica*)**:** Black or yellow fuzzy 2-inch-long caterpillars. Black type feeds mainly on weeds, but yellow type is destructive to desirable plants. May skeletonize shrubs in fall. If numbers are small try hand-picking; otherwise, try a *Bacillus thuringiensis* (Bt) or *carbaryl* spray.

Taxus weevils (*Otiorhynchus sulcatus*)**:** White larvae underground; ⅜-inch black adults with long snouts. Notch leaf edges as if trimmed with a ticket punch. May also chew bark. Spray foliage and soil with *acephate* (Orthene®). Respray as necessary.

Holly leafminers (*Phytomyza ilicis*)**:** Minute black flies. Multiple feeding punctures resemble pinpricks and distort leaves. Begin spraying with *diazinon* as soon as you notice flies.

Alder flea beetle (*Altica ambiens*)**:** Hopping, greenish blue or shiny dark blue adults, ⅕ inch long. Produce numerous small holes or pitted areas on leaves. Remove weeds and debris where beetles overwinter. Spray with *carbaryl* when you first notice damage in spring. Respray as necessary.

PINE-DESTROYING INSECTS

Pine engraver beetles (*Ips* species)**:** Brown or black beetles, ⅛ inch long. Create small circular holes in branches or trunk. White pine and spruce particularly susceptible. Serious problem in California. Needles turn yellow, then red, then die. Spray foliage and trunks with *lindane* to form a protective coating. If more than half of foliage has yellowed, remove the tree.

Pitch-mass borers (*Synanthedon pini*)**:** Small moth with black metallic wings and orange-banded black body. Large globular masses of pitch emerge from tree, particularly in whorl areas. Remove pitch masses, where borers pupate. Spray foliage and trunk with *lindane*.

Zimmerman pine moths (*Dioryctria zimmermani*)**:** Reddish gray adults, 1 inch long; white to reddish yellow larvae, ¾ inch long. Browning and wilting of new growth. Resin masses near whorls. Prune infested branches before August; destroy them. Spray foliage with *lindane* in mid-April and mid-August.

Pine webworms (*Tetralopha robustella*)**:** Adults are 1 inch long and have purple-black forewings and smoky black hind wings; larvae are yellow-brown, ¾ inch long, and have stripes along their sides. Brown, globular nests of silk, old needles, and brown sawdustlike waste matter. Remove and

Liquidambar styraciflua, *sometimes called Sweet Gum, is native to riversides in the South. It is a beloved street tree throughout much of the United States, notable for its excellent fall color.*

destroy nests. Spray with *Bacillus thuringiensis* (Bt) or *carbaryl* before protective webbing appears.

Insects attacking pine or spruce include the white pine weevil (see page 182), pales weevil, Zimmerman pine moth, pitch-mass borer, European pine shoot moth (see page 181), Eastern pine shoot borer, pine webworm, Nantucket pine tip moth (see page 181), pine root collar weevil, European pine sawfly, redheaded pine sawfly, white pine sawfly (for sawflies see page 181), irregular pine scales, pine tortoise scales, southern pine beetles, pine engraver beetles, pine tube moth, white pine aphid (see page 180), pine needle miner, and Saratoga spittlebug (see page 181). Spider mites (see pages 177 and 192) and nematodes (see pages 155 and 179) also attack pine.

In general, supply optimum water and fertilizer to make trees less appealing to these pests. Frequent inspection will help in early detection, which will allow you to solve the problem by pruning infested branches or spraying with appropriate chemicals before you lose the entire tree.

NO FLOWERS

Few or no flowers

Flowerless dogwood in dark location.

Flower buds pruned off crape myrtle.

Buds die or drop

Drought stress.

Problem: Plants fail to bloom, or they bloom only sparsely and sporadically.

Problem: Many or all of the buds or flowers die or drop off.

Analysis: Plants produce few or no buds or flowers for any of several reasons.

Solution: The numbered solutions below refer to the numbered items in the analysis.

Analysis: Buds may die or drop for any of several reasons.

1. Juvenility: Plants, like people, must reach a certain age or size before they are able to reproduce. They will not develop flowers or fruit until this time.

1. Plants will eventually begin to flower if they are otherwise healthy and adapted to the area. The juvenile stage in some trees and vines may last 15 years.

1. Transplant shock: Whenever a tree or shrub is transplanted, it goes through a period of shock. Dormant plants usually recover more quickly and are injured less than growing plants. Even when transplanted properly, however, dormant plants may still lose some of their buds. Plants that have begun growth or are in bloom often drop many of their flower buds or flowers shortly after transplanting. Some buds may remain on the plant but not open.

2. Inadequate winter cooling: In order to produce flowers, many plants must undergo a period of cooling during winter. The plant must be exposed for a certain number of hours to temperatures between 30° and 45°F. The number of hours needed varies from species to species. If the cooling requirement is not satisfied, flowering will be delayed and reduced, and flower buds may drop off. This is a common problem when plants adapted to cold climates are grown in the South.

2. Plant trees and shrubs adapted to your area. Consult your local garden center or your county extension office.

2. Cold or frost injury: Flower buds or flowers may be killed by cold or freezing temperatures. Many or all of them either fail to open or drop off. Cold injury occurs during the winter when temperatures drop below the lowest point tolerated by buds of that particular plant species. Frost injury is caused by an unseasonal cold snap, in either fall or spring, which damages buds, developing flowers, and tender shoots of growing plants.

3. Improper pruning: If a plant is pruned improperly or too severely, flower and fruit production can be reduced or, in some cases, prevented. Drastic pruning, especially on young plants, stimulates a flush of green growth that inhibits flowering. Flowering is also reduced if flower buds are pruned off.

3. Prune lightly, at a time when no flower buds are present.

3. Drought: Flowers or flower buds dry and drop off when there is a temporary lack of moisture in the plant. This may be caused by dry soil, minor root injuries, or anything else that disrupts water movement to the top of the plant.

4. Nutrient imbalance: Plants overfertilized with nitrogen tend to produce a flush of green growth. Some plants do not make flowers while they are growing vigorously.

4. Do not overfertilize plants or make a heavy application of nitrogen before flowering.

4. Insects: Certain insects, such as thrips and mites, feed on flower buds. When infestations are heavy, their feeding kills flower buds, causing them to dry and drop off. Some infested buds may open but be distorted.

5. Shade: Flowering plants require a certain amount of light to produce flowers. If these plants are grown in inadequate light, they produce few or no flowers.

5. Thin out shading trees, or move plants to a sunnier area. For a list of shade-loving trees and shrubs, see page 355.

Bud drop caused by cold injury.

Solution: The numbered solutions below refer to the numbered items in the analysis.

1. Whenever possible, transplant trees and shrubs during the dormant season. Avoid wounding the roots when planting, and do not let the plant dry out. Apply an antidesiccant spray to plants a few days before transplanting.

2. Plant trees and shrubs adapted to your area. Consult your local garden center or your county extension office. Protect shrubs and small trees from early or late cold snaps by covering them with burlap or a plastic tent. Placing an electric light bulb underneath the covering offers heat for additional protection.

3. Water trees and shrubs regularly. Most plants recover from minor root injuries. Frequent shallow waterings and light fertilization may speed recovery. Avoid wounding plants.

4. Control insects with chemicals. For more information on thrips and mites and their controls, look under your specific plant in the index.

Excess flowers and fruits

Messy fruit drop.

Problem: Trees and shrubs drop flowers, seedpods, or fruit, creating unwanted messy litter.

Analysis: Most trees and shrubs produce flowers or flower structures that develop into seedpods or fruit. Some plants, such as juniper and boxwood, produce inconspicuous flowers and fruit. Others, such as ornamental crabapple, olive, sweet gum, horse chestnut, and glossy privet, produce many conspicuous flowers, seedpods, or fruits. The dropping flowers and fruits of such plants create litter that may detract from the beauty of the landscape and increase time spent in garden upkeep.

Solution: If available, select and plant male-flowering varieties, which produce no fruit. Prevent flower and fruit production by spraying with a compound containing the growth regulator *NAA* (*napthalene acetic acid*) when flower buds are forming. Contact your local County Extension office to determine this period for your particular plant. Make sure that your plant is listed on the product label, and follow directions carefully. If plants are small enough to be moved, transplant them to a location where their flower and fruit drop will not be a nuisance. If spraying is impractical, replace messy trees and shrubs with plants that do not produce litter.

Beetles

Long-horned beetle (2× life size).

Problem: Shiny or dull, hard-bodied insects with tough, leathery wing covers appear on the plant. The wing covers meet in the middle of the back, forming a straight line down the insect's body. Leaf tissue is chewed, notched, or eaten between the veins, making the leaves appear lacy. The bark may be chewed, or flowers may be eaten. Holes may be found in branches or in the trunk.

Analysis: Many types of beetles feed on ornamental trees and shrubs. In most cases, both larvae (grubs) and adults feed on the plants, so damage is often severe. The insects spend the winter as grubs inside the plant or in the soil or as adults in bark crevices or in hiding places on the ground. Adult beetles lay eggs on the plant or on the soil during the growing season. Depending on the species, the grubs may feed on foliage, mine inside the leaves, bore into stems or branches, or feed on roots. Beetle damage to leaves rarely kills the plant. Grubs feeding inside the wood or underground are much more damaging and often kill branches or the whole plant.

Solution: Grubs feeding in the soil or inside the plant are difficult to detect and control. Control measures are often aimed at the adults. Several insecticides may be used to control these pests. Make sure that your plant is listed on the label. Look under your specific plant in the alphabetical section beginning on page 157.

INSECTS (continued)

Gypsy moth

Gypsy moth larvae

Gypsy moth and egg masses (2× life size).

Problem: Leaves are chewed; the entire tree is often defoliated by late spring or early summer. Large (up to 2½ inches long), hairy, blackish caterpillars with rows of red and blue spots on their backs are feeding on the leaves, hiding under leaves or bark, or crawling on buildings, cars, or other objects outdoors. Some people are allergic to the hairs. Insect droppings accumulate underneath the infested tree. Trees defoliated for several consecutive years, especially weak ones, may be killed.

Analysis: The gypsy moth (*Lymantria dispar*) is a general feeder, devouring more than 450 species of plants. Gypsy moth populations fluctuate from year to year. When moths are low in number, they prefer oaks as their host. When their numbers increase, the moths defoliate entire forests and spread to other trees and shrubs. Repeated, severe defoliation weakens trees and reduces plant growth. Defoliated deciduous trees are rarely killed unless already in a weakened condition. They are more susceptible to attack by other insects and by plant diseases that may kill them, however. Gypsy moths are also an extreme nuisance in urban areas and in parks and campgrounds. The overwintering masses of eggs, covered with beige or yellow hairs, are attached to almost any object outdoors. The eggs hatch from mid- to late spring. The tiny larvae crawl to trees, where they feed, or drop on silken threads to be carried by the wind to other plants. As the caterpillars mature, they feed at night and rest during the day. The larvae may completely cover sides of houses or other objects during these resting periods. When population levels are high, the insects feed continually on the tree, and large amounts of excrement accumulate beneath it. The larval hairs may cause allergies. The larvae pupate in sheltered places, and dark brown male or white female moths emerge in midsummer.

Solution: If the insects are bothersome or if trees are weak or unhealthy either from the previous year's gypsy moth feeding or from drought, mechanical damage, other insects, or plant diseases, treatment with an insecticide is required. Apply the insecticide from the beginning of hatching until the larvae are 1 inch long, about the blooming period of *Spiraea vanhouttei*. Cover the tree thoroughly. Contact a professional arborist for large trees. Spray smaller trees with the bacterial insecticide *Bacillus thuringiensis* (Bt) or with ORTHO Orthene Systemic Insect Control when tiny larvae are first noticed. Repeat the spray at weekly intervals if damage continues. Homeowners can reduce infestations by destroying egg masses during winter months. During the spring, when larvae are feeding, place burlap bands on trees, leaving the bottom edges free. Larvae will crawl under these flaps to hide during the day. Collect and destroy them daily. Keep trees healthy. Fertilize regularly, and water during periods of drought. When planting trees, choose species that are less favored by the gypsy moth. (For a list of these plants and of plants favored by the gypsy moth, see page 357.) Trees less favored by the insects are damaged only slightly by larval feeding. In addition, when interplanted with more favored hosts, they may reduce damage by preventing a large buildup of insects in the area. It is a federal offence to transport items that have eggs or larvae attached to them.

Japanese beetle

Japanese beetle (2× life size).

Problem: Leaf tissue is chewed between the veins, giving the leaves a lacy appearance. The entire plant may be defoliated. Metallic green-and-bronze winged beetles, ½ inch long, feed in clusters on the plant.

Analysis: As its name suggests, the Japanese beetle (*Popillia japonica*) is native to Japan. It was first seen in New Jersey in 1916 and has since become a major pest in the eastern United States. It feeds on hundreds of different species of plants. The adult beetles are present from June to October. They feed only in the daytime and are most active on warm, sunny days. The female beetles live for 30 to 40 days. Just before they die, they lay their eggs just under the soil surface in lawns. Grayish white grubs soon hatch and feed on grass roots. As the weather turns cold in the late fall, the grubs burrow 8 to 10 inches down into the soil, where they remain dormant for the winter. When the soil warms up in the spring, the grubs move back up near the soil surface and resume feeding on roots. They soon pupate, and in late May or June, they reemerge as adult Japanese beetles.

Solution: Control the adults with ORTHO Orthene Systemic Insect Control or ORTHO Malathion 50 Plus Insect Spray in late May or June. Make sure that your plant is listed on the product label. Repeat the spray 10 days later if damage continues. In the fall, apply Milky Spore Disease to your lawn to control the following year's grubs. The following year, begin spraying as the adults emerge, about the same time as Queen Anne's lace is blooming.

Leaf-feeding caterpillars

Looper (2× life size).

Problem: Caterpillars are clustered or feeding singly on the leaves. The surface of the leaf is eaten, giving the remaining tissue a lacy appearance, or the whole leaf is chewed. Sometimes the leaves are webbed. The tree may be completely defoliated. Damage appears anytime between spring and fall.

Analysis: Many species of caterpillars feed on the leaves of trees and shrubs. Depending on the species, the moths lay their eggs from early spring to midsummer. The larvae that hatch from these eggs feed singly or in groups on buds, on one leaf surface (these are called *skeletonizers*), or on the entire leaf. Certain caterpillars web leaves together as they feed. In some years, damage is minimal because of unfavorable environmental conditions or control by predators and parasites. When conditions are favorable, however, entire plants may be defoliated. Defoliation weakens the plants because no leaves are left to produce food. When heavy infestations occur several years in a row, branches or entire plants may be killed.

Solution: Spray with ORTHO Diazinon Ultra Insect Spray, ORTHO Orthene Systemic Insect Control, or the bacterial insecticide *Bacillus thuringiensis* (Bt) when damage is first noticed. Spray the leaves thoroughly. Repeat the spray if the plant becomes reinfested. Make sure that your plant is listed on the product label.

Tent caterpillars and fall webworm

Western tent caterpillars (¼ life size).

Problem: In the spring or summer, silk nests appear in the branch crotches or on the ends of branches. Leaves are chewed; branches or the entire tree may be defoliated. Groups of caterpillars are feeding in or around the nests.

Analysis: Tent caterpillars (*Malacosoma* species) and fall webworm (*Hyphantria cunea*) feed on many ornamental trees. In the summer, tent caterpillars lay masses of eggs in a cementing substance around twigs. They hatch early the next spring as the leaves unfold, and the young caterpillars construct their nests. On warm, sunny days, they emerge from the nests to devour the surrounding foliage. In mid- to late summer, brownish or reddish moths appear.

The fall webworm lays many eggs on the undersides of leaves in the spring. In early summer, the young caterpillars make nests over the ends of branches, inside which they feed. As the leaves are devoured, the caterpillars extend the nests over more foliage. Eventually the entire branch may be enclosed with this unsightly webbing. The caterpillars drop to the soil to pupate. Up to four generations occur between June and September. Damage is most severe in the late summer.

Solution: Spray with ORTHO Isotox Insect Killer, ORTHO Orthene Systemic Insect Control, or the bacterial insecticide *Bacillus thuringiensis* (Bt). Make sure that your plant is listed on the product label. The bacterial insecticide is most effective against young caterpillars. Remove egg masses found in winter.

Bagworm

Bagworm case on honeylocust (life size).

Problem: Leaves are chewed; branches or the entire tree may be defoliated. Hanging from the branches are carrot-shaped cases, or "bags," from 1 to 3 inches long. The bags are constructed from interwoven bits of dead foliage, twigs, and silk. When a bag is cut open, a tan or blackish caterpillar or a yellowish grublike insect may be found inside. A heavy attack by bagworms may stunt trees.

Analysis: Bagworms (*Thyridopteryx ephemeraeformis*) eat the leaves of many trees and shrubs. The larvae hatch in late May or early June and immediately begin feeding. Each larva constructs a bag that covers its entire body; as the larva develops, it adds to the bag. The worm partially emerges from its bag to feed. When all the leaves are eaten off the branch, the bagworm moves to the next branch, dragging its bag along. By late August, the larva spins silken bands around a twig, attaches a bag permanently, and pupates. In the fall, the winged male moth emerges from his case, flies to a bag containing a female, mates, and dies. The female bagworm spends her entire life inside her bag. After mating, she lays 500 to 1,000 eggs and dies. The eggs spend the winter in the mother's bag.

Solution: Spray with ORTHO Orthene Systemic Insect Control, the bacterial insecticide *Bacillus thuringiensis* (Bt), or an insecticide containing *diazinon* between late May and mid-July to kill the young worms. Older bagworms are more difficult to control. Repeat the spray after 10 days if leaf damage is still occurring. Handpick and destroy bags in winter to reduce the number of eggs.

INSECTS (continued)

SCALES OR POWDER ON PLANT

Aphids

Aphids on hawthorn (life size).

Problem: Tiny (⅛-inch) green, yellow, black, brownish, or gray soft-bodied insects cluster on the bark, leaves, or buds. Some species are covered with white, fluffy wax. The insects may have wings. Leaves are discolored and may be curled and distorted. They sometimes drop off. A shiny or sticky substance may coat the leaves. A black, sooty mold often grows on the sticky substance. Plants may lack vigor, and branches sometimes die. Ants may be present.

Analysis: Many types of aphids infest ornamental trees and shrubs. They do little damage in small numbers. They are extremely prolific, however, during a cool growing season. Damage occurs when the aphid sucks the juices from the plant. Sap removal often results in scorched, discolored, or curled leaves and reduced plant growth. A severe infestation of bark aphids may cause branches to die. Aphids are unable to digest fully all the sugar in the plant sap, and they excrete the excess in a fluid called *honeydew,* which often drops to cover anything beneath the tree or shrub in a sticky film. A sooty mold fungus may develop on the honeydew, causing the leaves to appear black and dirty. Ants feed on this sticky substance and are often present where there is an aphid infestation.

Solution: Spray with ORTHO Isotox Insect Killer, ORTHO Orthene Systemic Insect Control, ORTHO Malathion 50 Plus Insect Spray, or an insecticide containing *diazinon* when damage is first noticed. Make sure that your plant is listed on the product label. Repeat the spray if the plant becomes reinfested.

Scales

Lecanium scale on redbud (life size).

Problem: Crusty or waxy bumps or clusters of somewhat flattened scaly bumps cover the leaves, stems, branches, or trunk. The bumps can be scraped or picked off; the undersides are usually soft. Leaves turn yellow and may drop. In some cases, a shiny or sticky substance coats the leaves. A black, sooty mold often grows on the sticky substance.

Analysis: Many types of scales infest trees and shrubs. They lay their eggs on leaves or bark, and in spring to midsummer the young scales, called *crawlers,* settle on the leaves, branches, or trunk. The small (1/10-inch), soft-bodied young feed by sucking sap from the plant. The legs usually atrophy, and a hard crusty or waxy shell develops over the body. Female scales lay their eggs underneath their shell. Some species of scales are unable to digest fully all the sugar in the plant sap, and they excrete the excess in a fluid called *honeydew.* A sooty mold fungus may develop on the honeydew, causing the leaves to appear black and dirty. An uncontrolled infestation of scales may kill a plant after 2 or 3 seasons.

Solution: Spray with ORTHO Isotox Insect Killer, ORTHO Orthene Systemic Insect Control, or an insecticide containing *diazinon* when the young are active. Early the following spring, before new growth begins, spray the trunk and branches with ORTHO Volck Oil Spray to control overwintering insects. Use Volck Oil Spray only when temperatures will remain above 40°F for 24 hours following the treatment.

Cottony cushion scales / mealybugs / woolly adelgids

Pine bark adelgid on white pine (¼ life size).

Problem: The undersides of leaves and the stems, branch crotches, or trunk are covered with white, cottony masses. Leaves may be curled, distorted, and yellowing. Knotlike galls may form on the stems or trunk. Sometimes a shiny or sticky substance coats the leaves. A black, sooty mold may grow on the sticky substance. Twigs and branches may die.

Analysis: Cottony cushion scales, mealybugs, pine bark adelgids, and woolly adelgids all produce white, waxy secretions that cover their bodies. This visual similarity makes separate identification difficult for the home gardener. Young insects are usually inconspicuous on the host plant. Their bodies range in color from yellowish green to brown, blending in with the leaves or bark. As the insects mature, they exude filaments of white wax, giving them a cottony appearance. Mealybugs and scales generally deposit their eggs beneath the waxy covering or among the white, fluffy masses. Damage is caused by the withdrawal of plant sap from the leaves, branches, or trunk. Because the insects are unable to digest fully all the sugar in the plant sap, they excrete the excess in a fluid called *honeydew,* which often drops onto the leaves or plants below. A sooty mold fungus may develop on the honeydew, causing the leaves and twigs to appear black and dirty.

Solution: Spray with ORTHO Isotox Insect Killer or ORTHO Orthene Systemic Insect Control. The folllowing spring, before growth starts, spray with ORTHO Volck Oil Spray. Use Volck Oil Spray only when temperatures will remain above 40°F for 24 hours following the treatment.

Powdery mildew

Powdery mildew on London plane tree.

Problem: Leaves, flowers, and young stems are covered with a thin layer or irregular patches of a grayish white powdery material. Infected leaves may turn yellowish or reddish and drop. Some leaves or branches may be distorted. In late fall, tiny black dots (spore-producing bodies) are scattered over the white patches like grains of pepper.

Analysis: Powdery mildew is caused by any of several fungi that thrive in both humid and dry weather. Some fungi attack only older leaves and plant parts; others attack only young tissue. Plants growing in shady areas are often severely infected. The powdery patches consist of fungal strands and spores. The spores are spread by the wind to healthy plants. The fungus saps plant nutrients, causing discoloration and sometimes the death of the leaf. Certain powdery mildews also cause leaf or branch distortion. Since these powdery mildews often attack many different kinds of plants, the fungus from a diseased plant may infect other plants in the garden.

Solution: Several fungicides—including ORTHO Multi-Purpose Fungicide Daconil 2787® Plant Disease Control and those containing *triforine* (Funginex®) and *cycloheximide*—are used to control powdery mildew. For control suggestions, look under your specific plant in the alphabetical section beginning on page 157.

Leaf and stem rusts

Rust on Oregon grapeholly.

Problem: Yellow, orange, red, or black powdery pustules appear on the upper or lower surfaces of leaves or occasionally on the bark. The powdery material can be scraped or rubbed off. Leaves are discolored or mottled yellow to brown. Leaves may become twisted and distorted and may dry and drop off. Infected stems may be swollen or blistered or may develop oblong or hornlike galls up to 2 inches long.

Analysis: Many different species of rust fungi infect trees and shrubs. Some rusts produce spore pustules on leaves or stems, and others produce galls or hornlike structures on various parts of the plant. Most rusts attack only one species or a few related species of plants, although some rusts require two plant species to complete their life cycles. Part of the life cycle is spent on the tree or shrub and part is spent on other plants. In most cases, the symptoms produced on the two hosts are different. Wind and splashing water spread rust spores to healthy plants. When conditions are favorable (moisture and moderate temperatures, 55° to 75°F), the spores germinate and infect the tissue.

Solution: Several fungicides, including ORTHO Multi-Purpose Fungicide Daconil 2787® Plant Disease Control and those containing *triforine* (Funginex®), *ferbam*, *mancozeb*, and *cycloheximide*, may be used to control rust. Look under the entry for your plant in the alphabetical section beginning on page 157 to determine which fungicide is appropriate. Some rust fungi are harmless to the plant and do not require control measures. Rake up and destroy leaves in the fall.

Sooty mold

Sooty mold on camellia.

Problem: A black, sooty mold grows on the leaves and twigs. It can be completely wiped off the surfaces. Cool, moist weather hastens the growth of the mold.

Analysis: These common black molds are found on a wide variety of plants in the garden. They are caused by any of several fungi that grow on the sugary material left on plants by aphids, mealybugs, scales, whiteflies, and other insects that suck sap from the plant. The insects are unable to digest all the sugar in the sap, and they excrete the excess in a fluid called *honeydew*, which drops onto the leaves below. The honeydew may also drop out of infested trees and shrubs onto plants growing beneath them. The sooty mold fungi develop on the honeydew, causing the leaves to appear black and dirty. Sooty molds are unsightly but are fairly harmless because they do not attack the leaf directly. Extremely heavy infestations prevent light from reaching the leaf, so the leaf produces fewer nutrients and may turn yellow. The presence of sooty molds indicates that the plant or a nearby plant is infested with insects.

Solution: Wipe sooty molds from the leaves with a wet rag, or rain will eventually wash them off. Prevent more sooty mold from growing by controlling the insect that is producing the honeydew. Inspect the leaves and twigs above the sooty mold to find out which insects are present. For control of aphids, scales, and mealybugs, see page 136.

SPOTS ON LEAVES

Spots on Leaves

Entomosporium leaf spot on Photinia.

Leaf spots on Liquidambar.

Problem: Spots and blotches appear on the leaves and flowers.

Analysis: Several diseases, insects, and environmental factors cause spots and blotches on leaves and flowers.

Solution: The numbered solutions below correspond to the numbered items in the analysis.

1. Fungal leaf spot: Spots caused by fungi are often small and circular and may be found on all of the leaves. Sometimes only the older or younger leaves are affected. The spots range in size from barely visible to ¾ inch in diameter. They may be yellow, red, tan, gray, brown, or black and often have a definite margin. Spots sometimes join together to form blotches. Often the leaves turn yellow and die. Infection is usually most severe during moist, mild weather (50° to 85°F).

1. Spray plants with ORTHO Multi-Purpose Fungicide Daconil 2787® Plant Disease Control or a fungicide containing *maneb, mancozeb,* or *zineb* when new growth begins. Repeat at intervals of 2 weeks for as long as the weather remains favorable for infection. Make sure that your plant is listed on the product label. Raking and destroying leaves in the fall may help control the fungus.

2. Insects: Several types of insects, including lace bugs, leafhoppers, mites, plant bugs, and thrips, cause spotting of leaves. Leaves may be spotted brownish, yellow, or white, or they may be completely discolored. Sometimes the insects are visible, feeding on the lower or upper surfaces of leaves.

2. These insects can be controlled with various insecticides. For effective chemicals, look under the entry for your plant in the alphabetical section beginning on page 157.

3. Sunburn: Shade-loving plants placed in a sunny location develop spots and blotches on the leaves most directly exposed to the sun. Sun-loving plants also develop sunburn symptoms if they are allowed to dry out. Initially, sunburned leaves develop a whitish or yellowish bleached appearance between the veins. Large, dark blotches form on the damaged tissue. Leaves not directly exposed to the sun remain green and uninjured.

3. Where practical, pick off the injured leaves and plant parts. Check to see whether your plant is adapted to sun or shade by looking it up in the alphabetical section beginning on page 157. Provide shade or transplant shade-loving plants. (For a list of common shade plants, see page 355.) Water plants regularly, especially on hot, sunny, or windy days.

GALLS OR GROWTHS

Mushrooms and conks

Conk on the trunk of a bigleaf maple.

Problem: Mushrooms appear around the base of the tree. Or white, yellow, gray, or brownish growths, usually hard and woody, protrude from the trunk. The plant may appear unhealthy.

Analysis: Mushrooms and conks (hard, woody growths that protrude from tree trunks) are the reproductive bodies of fungi. Most mushroom fungi live on decaying matter. When conditions are favorable, the fungi produce mushrooms and conks containing spores that are spread by the wind. Several different mushroom fungi decay the heartwood of living trees. Most of these grow only in older wood, which they enter through wounds. (For information on heart rot, see page 150.) Mushrooms or conks usually appear annually in the dead portions of the trees. Some conks may remain attached to the wood for years. *Armillaria mellea,* a fungus that causes a disease called *armillaria root rot, mushroom root rot, oak root fungus,* or *shoestring root rot,* invades healthy roots. In the fall or winter, mushrooms appear around the base of the plant, growing on the infected roots.

Solution: By the time conks or mushrooms appear on the trunk, it is too late to do anything about the wood rot. Inspect the tree to determine the extent of decay (contact a professional arborist if necessary). Trees or branches with extensive decay are dangerous and should be removed. Keep plants vigorous by fertilizing and watering regularly. The spread of armillaria root rot may be inhibited if the rot is in only part of the roots. For more information on this root rot, see page 187.

Galls or growths on leaves branches or trunks

Galls on Virginia witch hazel.

Leaf galls on willow.

Problem: Swellings, thickenings, and growths develop on the leaves, shoots, branches, or trunk. Plants with numerous galls on branches or the trunk may be weak, and leaves may be yellowing. Branches may die.

Analysis: These growths can be caused by three factors.

1. Fungal leaf or stem gall: Several fungi, including rust fungi, cause enlargement and thickening of leaves and shoots. Affected plant parts are usually many times larger than normal and are often discolored and succulent. Some leaf or stem galls turn brown and hard with age. The galls are unsightly but rarely harmful to the plant. Fungal galls are most severe when spring weather is wet.

2. Bacterial crown gall: This plant disease is caused by a soil-inhabiting bacterium (*Agrobacterium tumefaciens*) that infects many ornamentals, fruits, and nuts in the garden. The bacteria enter the plant through wounds in the roots or the base of the trunk (the crown). The galls disrupt the flow of water and nutrients up the roots, stems, and trunk, weakening and stunting the top growth. Galls do not usually kill the plant.

3. Insect galls: Many different types of insects cause galls by feeding on plant tissue or by injecting a toxin into the tissue during feeding. As a result of this irritation, blisters or growths of various shapes form on leaves, swellings develop on roots or stems, and buds and flowers grow abnormally. Most gall-forming insects cause minor damage to the plant, but the galls may be unsightly.

Solution: The numbered solutions below refer to the numbered items in the analysis.

1. Pick off and destroy affected parts as soon as they appear. If galls are a problem this year, spray next spring with a fungicide containing *ferbam, mancozeb,* or *maneb* just before the buds open. Add a spreader-sticker to the spray. Repeat the spray 2 weeks later.

2. Infected plants cannot be cured. They often survive for many years, however. To improve the appearance of shrubs with stem galls, prune out and destroy affected stems below the galled area. Disinfect pruning shears after each cut by dipping in a solution of 1 part chlorine bleach and 9 parts water. Destroy severely infected shrubs. Consult a professional horticulturist to remove galls from valued trees. The bacteria will remain in the soil for at least 2 to 3 years. For a list of plants resistant to crown gall, see page 354.

3. Many gall-forming insects require no controls. If you feel the galls are unsightly, however, or if the galls cause dieback, control measures may be necessary. For recommended control measures, look under the entry for your plant in the alphabetical section beginning on page 157.

Leafy mistletoe

Leafy mistletoe.

Problem: Leafy olive green plants up to 4 feet across are attached to the branches. The tufts are most noticeable during the winter on trees without their leaves. Affected branches are often swollen. Some may break from the weight of the plants. Branches beyond the growth occasionally die.

Analysis: Leafy mistletoes (*Phoradendron* species) are semiparasitic plants that manufacture their own food but depend on their host plant for water and minerals. The plant produces sticky seeds that are spread by birds from one tree to another, or by falling from higher to lower branches. The seeds germinate almost anywhere but penetrate only young, thin bark. The rootlike attachment organs of the plant penetrate the tree's water-conducting vessels, which they tap for nutrients and water. At the point of attachment, the branch or trunk swells, sometimes to two or three times its normal size. Growth of mistletoe is slow at first, but after 6 or 8 years, plants may be 3 feet across. Trees heavily infested with mistletoe may be weakened and sometimes die.

Solution: Leafy mistletoe's rootlike attachment organs may spread through host tissue up to 1½ feet from the swollen area. They must be removed or the mistletoe will resprout. Prune limbs 18 inches below the point of mistletoe attachment. To prevent the spread of seeds, remove tufts before seeds form in the spring. If it is impractical to prune the tree limbs, remove the mistletoe and wrap the infected areas with black plastic to kill the resprouting mistletoe.

GALLS OR GROWTHS (continued)

Algae, lichens, and mosses

Lichen.

Problem: Brown, gray, green, or yellow crusty, soft, or leaflike growths develop on trees in moist forested areas. The growths are usually found on the lower or shaded parts of trunks and branches.

Analysis: Algae, lichens, and mosses are sometimes mistaken for plant diseases, especially if the tree they are attached to appears unhealthy. They do not harm the plant, however. Most algae grow where moisture is abundant, on the lower, shaded side of the trunk. They appear only as a green color that is not very noticeable on the bark. Lichens are a combination of green algae and fungi. They range in color from brown to green and appear crusty or leaflike. They are sensitive to air pollution and are found only in areas where the air is clean. True mosses are small green plants growing in a mat and having tiny leaves and stems. They are abundant in moist areas and are much more apparent than algae. Spanish moss is a flowering plant (in the pineapple family) that is very noticeable hanging from branches of Southern trees.

Solution: Algae, lichens, and mosses do not harm the plant, but they may be unsightly. Control them by pruning away surrounding vegetation to increase the amount of light and air flow, which will reduce the moisture in the soil and air around the plant and discourage their growth.

Dwarf mistletoe

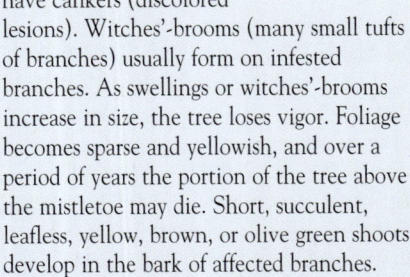

Dwarf mistletoe. Inset: Seeds.

Problem: Twigs and small branches of conifers are swollen and have cankers (discolored lesions). Witches'-brooms (many small tufts of branches) usually form on infested branches. As swellings or witches'-brooms increase in size, the tree loses vigor. Foliage becomes sparse and yellowish, and over a period of years the portion of the tree above the mistletoe may die. Short, succulent, leafless, yellow, brown, or olive green shoots develop in the bark of affected branches.

Analysis: Dwarf mistletoe (*Arceuthobium* species), a parasitic plant, infests many conifers. Dwarf mistletoe lacks a normal root system and true leaves. It relies on its host plant to supply most of the nutrients it requires. In this respect it is different from leafy mistletoe (see page 139), which depends on its host plant only for water and minerals. In midsummer, dwarf mistletoe spreads by sticky seeds that are explosively discharged for distances up to 50 feet. Seeds land on needles and then slide down to the bark when the needles are moistened. When seeds germinate, usually the following spring, rootlike structures penetrate the bark. The structures form a network in the branches, causing swellings and the formation of cankers. Dwarf mistletoe weakens the tree as it saps nutrients and water and distorts the growth of branches. Within 1 to 3 years after infestation, aerial shoots are produced that are from ½ inch to 4 inches long.

Solution: Remove heavily infested trees. Prune off branches of less severely infested trees, making the cuts at the trunk. Remove new infestations as they appear.

Witches'-broom

Witches'-broom on hackberry.

Problem: A dense tuft of small, weak twigs develops on a branch. Leaves on the tuft may be smaller than normal and off-color. The branches are weak and unhealthy.

Analysis: A witches'-broom is a dense proliferation of twig growth, usually caused by an insect, a plant disease, or mistletoe (see page 139). The witches'-broom looks messy but does not harm the plant. The insect or disease that caused it, however, may be harmful.

Solution: If the witches'-broom affects the appearance of the plant, prune it off. It may be difficult for the home gardener to identify the cause of brooming. If witches'-brooms continue to develop, contact a professional arborist or your local county extension office.

HOLES OR TRAILS IN LEAVES

Leafrollers

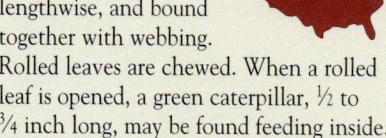

Leafroller on linden.

Problem: Leaves are rolled, usually lengthwise, and bound together with webbing. Rolled leaves are chewed. When a rolled leaf is opened, a green caterpillar, ½ to ¾ inch long, may be found feeding inside. Flower buds may also be chewed.

Analysis: Several leafrollers feed on the leaves and buds of woody ornamentals. Some species feed on only one plant. Others feed on many plants in the garden. Leafrollers are the larvae of small (up to ¾-inch) brownish moths. The insects spend the winter as eggs or larvae on the plant. In the spring the larvae feed on the young foliage, sometimes tunneling into and mining the leaf first. They roll one or more leaves around themselves, binding the leaves together with a silken webbing, then feed within the rolled leaves. This provides protection from weather, parasites, and chemical sprays. Some leafrollers mature in summer and have several generations each year. Other leafrollers have only one generation. In the fall, either the larvae mature into moths and lay overwintering eggs, or they spend the winter inside the rolled leaf.

Solution: Spray with ORTHO Isotox Insect Killer, ORTHO Orthene Systemic Insect Control, ORTHO Diazinon Ultra Insect Spray, or the bacterial insecticide *Bacillus thuringiensis* (Bt) in the spring when leaf damage is first noticed. For insecticides to be most effective, they should be applied before larvae are protected inside the rolled leaves. Check the plant periodically in the spring for the first signs of infestation.

Leafminers

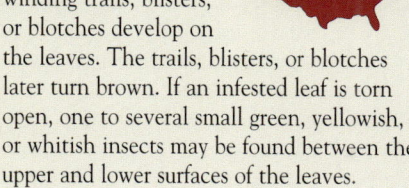

Leafminer damage to holly.

Problem: Green or whitish translucent winding trails, blisters, or blotches develop on the leaves. The trails, blisters, or blotches later turn brown. If an infested leaf is torn open, one to several small green, yellowish, or whitish insects may be found between the upper and lower surfaces of the leaves.

Analysis: Leafminers are the larvae of flies, moths, beetles, or sawflies. Adult female moths lay their eggs on or inside the leaves, usually in early to late spring. The emerging larvae feed between the upper and lower leaf surfaces, producing blisters, blotches, or trails. The infested tissue stands out prominently against the normal green foliage as it turns whitish or light green to brown. The insects pupate inside the leaves or in the soil and emerge as adults. Some adults also feed on the leaves, chewing holes or notches in them.

Solution: Control of leafminers is difficult because they spend most of their lives protected inside the leaves. Insecticides are usually aimed at the adults. Once leafminers are noticed in the leaves, inspect the foliage periodically, or check with your local county extension office to determine when the adults emerge. Then spray with ORTHO Orthene Systemic Insect Control, ORTHO Isotox Insect Killer, or ORTHO Diazinon Ultra Insect Spray. Make sure that your plant is listed on the product label.

Nocturnal insects

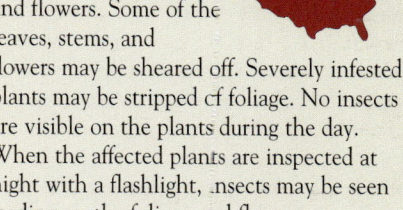

Weevil damage to rhododendron.

Problem: Holes or notches appear in leaves and flowers. Some of the leaves, stems, and flowers may be sheared off. Severely infested plants may be stripped of foliage. No insects are visible on the plants during the day. When the affected plants are inspected at night with a flashlight, insects may be seen feeding on the foliage and flowers.

Analysis: Several types of insects feed on plants only at night, including weevils, other beetles, and caterpillars. Weevils are a type of beetle with a long snout. Beetles are hard-bodied insects with tough, leathery wing covers. The wing covers meet in the middle of the back, forming a straight line. Caterpillars are smooth or hairy soft-bodied worms. Nocturnal insects usually hide in the soil, debris, or other protected places during the day.

Solution: Control these insects with insecticides. For more information on nocturnal insects and the specific insecticides that control them, look up beetles on page 133 and caterpillars on page 135.

HOLES OR TRAILS IN LEAVES (continued)

WILTING

Snails and slugs

Slug damage on hosta.

Problem: Irregular holes with smooth edges are chewed in the leaves. Leaves may be sheared off entirely. Silvery trails wind around the plants and soil nearby. Snails and slugs may be seen moving around or feeding on the leaves, especially at night. Check for them by inspecting the garden at night by flashlight.

Analysis: Snails and slugs are mollusks and are related to clams, oysters, and other shellfish. They feed on a wide variety of plants, including ornamentals and vegetables. Like other mollusks, snails and slugs need to be moist all the time. For this reason, they avoid direct sun and dry places and hide during the day in damp places, such as under flowerpots or in thick ground covers. They emerge at night or on rainy days to feed. Snails and slugs are similar, except that the snail has a hard shell into which it withdraws when disturbed. Slugs lay white eggs encased in a slimy mass in protected places. Snails bury their eggs in the soil, also in a slimy mass. The young look like miniature versions of their parents.

Solution: Apply ORTHO Bug-Geta Snail & Slug Killer around the trees and shrubs you wish to protect. Also apply the bait in areas where snails or slugs might be hiding, such as in dense ground covers, weedy areas, compost piles, or pot-storage areas. Before spreading the bait, wet down the treated areas to encourage snail and slug activity that night. Repeat the application every 2 weeks for as long as snails and slugs are active.

Lack of water

Wilting Philadelphus.

Problem: The plant wilts often, and the soil is frequently or always dry. The leaves or leaf edges may turn brown and shrivel.

Analysis: Water in the soil is taken up by plant roots. It moves up into the stems and leaves and evaporates into the air through microscopic breathing pores in the surfaces of the leaves. Water pressure within plant cells keeps the cell walls rigid and prevents the leaves and stems from collapsing. When the soil is dry, the roots are unable to furnish the leaves and stems with water, the water pressure in the cells drops, and the plant wilts. Most plants will recover if they have not wilted severely. Frequent or severe wilting, however, will curb a plant's growth and may eventually kill it.

Solution: Water the plant immediately. To prevent future wilting, follow the cultural instructions for your plant in the alphabetical section beginning on page 157.

Extreme heat or wind

Wilting dogwood.

Problem: The plant is wilting, but the foliage usually looks healthy. No signs of insects or disease are present, and the soil is moist. Wilting is most common on shrubs or plants with limited root systems.

Analysis: During hot, windy periods, small or young plants may wilt, even though the soil is wet. Wind and heat cause water to evaporate quickly from the leaves. If the roots can't absorb and convey water fast enough to replenish this loss, the leaves wilt.

Solution: Keep the plant well-watered during hot spells, and sprinkle it with water to cool off the foliage. The plant will usually recover when the temperature drops or the wind dies down. Provide shade during hot weather, and use temporary windbreaks to protect from wind. Plant shrubs adapted to your area.

Damaged trunk or stem

Leaf drop on dogwood caused by borer infestation.

Wilting rhododendron.

Planted too shallowly

Arborvitae planted too shallowly.

Problem: All or part of the plant is wilting, and the leaves may turn yellow, then brown, and die. Wounds or sunken lesions (cankers) are seen on the plant, or holes, surrounded by sap or sawdust, are found in the branches or trunk.

Analysis: Damage to the wood or bark disrupts water and nutrient movement through the plant, causing wilting.

1. Wounds: Any kind of mechanical injury that breaks roots, stems, or bark may cause the plant to wilt. A plant may be accidentally wounded by a motor vehicle, animal, or foot traffic; its roots may be damaged by cultivation, construction, or other soil disturbance. In severe cases, the plant dies.

2. Cankers: Cankers are sunken, dark-colored lesions that develop as a result of infection by fungi or bacteria. Cankers on small or young plants often cause the portion of the plant above the canker to wilt. Branches or the entire plant may eventually die.

3. Borers: Most borers are the larvae of beetles or moths. Many kinds of borers infest stems, branches, or trunks. The larvae feed by tunneling through the bark, sapwood, and heartwood, stopping the flow of nutrients and water in that area. Large trees and shrubs usually turn yellow and brown rather than wilt.

Solution: The numbered solutions below correspond to the numbered items in the analysis.

1. Thin out some of the branches and keep the plant well-watered. During hot weather, provide shade to reduce evaporation from the leaves. Prevent mechanical injuries to plants by being careful when working around the roots and stems. If necessary, place barriers around plants to prevent damage from vehicles, animals, and foot traffic.

2. Prune dying branches below the canker. Avoid wounding plants.

3. Prune out stems with borers. Keep the plant well-watered. Feed regularly with Scotts Evergreen, Shrub and Tree Food. For more information on borers and their controls, see page 147.

Problem: A recently planted tree or shrub wilts frequently. Roots or the rootball may be exposed.

Analysis: Newly planted trees and shrubs may wilt frequently if they are planted too shallowly. Plants that have been set in the ground at a higher level than they were originally growing wilt because the exposed part of the soil ball dries out quickly. This may kill the surface roots, especially if the soil washes away, exposing them.

Solution: Remove the plant, along with its soil ball, and replant it more deeply. The plant should be set at the same level as when it was growing in the pot or the ground before transplanting. Be careful not to plant too deeply since this can create an equally serious problem that may not show up for a few years. Water the plant thoroughly so the entire soil ball is moistened. Keep the plant well-watered until it becomes established. In poor drainage areas, plants may be set slightly high with a small amount of soil covering the rootball, as long as the soil is covered by a generous layer of mulch.

143

WILTING (continued)

FLUID ON BARK OR LEAVES

Dry rootball

Dry rootball of recently planted shrub.

Problem: The entire plant is wilting. The soil surrounding the plant is moist, but the rootball is dry. The tree or shrub has been planted recently.

Analysis: Plants that are sold in the nursery balled and burlapped are grown in fields. When the plants reach a size suitable for selling, they are dug up with a ball of soil around their roots. Sometimes the soil in which they are grown is extremely heavy. When the heavy soil is balled and burlapped, it sometimes shrinks as it dries or is compacted, and the ball becomes impermeable to water. After the plant is replanted, water runs off the outside of the ball rather than penetrating the soil, and the roots dry out. Rootballs of balled-and-burlapped plants or of container plants may also dry out if the soil in which the plant is set is much lighter or heavier than the soil in the rootball. The water runs into the lighter soil instead of moistening the soil around the roots; or the surrounding heavy clay soil draws water away from the light soil in the rootball, so that the rootball dries out.

Solution: To wet the rootball after planting, build a basin around the plant the diameter of the rootball. Keep water in the basin for 3 hours. Add a wetting agent (which can be purchased at your local nursery) to the water initially. Continue watering in the basin for 6 weeks. Water whenever the rootball (not the surrounding soil) is moist but not wet 1 inch below the surface. Before planting a balled and burlapped plant, water or soak the rootball before planting. If the soil texture in the rootball is very different from that of the surrounding soil, provide a transition zone, using a mix of rootball and native soil. Build a basin and apply water as instructed above for 6 weeks.

Transplant shock

Transplant shock.

Problem: The plant wilts, but the foliage usually looks healthy. No signs of insects or disease are present, and the soil is moist. The plant has been recently transplanted.

Analysis: Plants frequently wilt or stop growing for a while after being transplanted. Transplant "shock" is not related to shock in humans but is the result of roots being cut or injured during transplanting. Wilting occurs when the injured roots are unable to supply the plant with enough water, even when the soil is wet.

Solution: To reduce the water requirement of the plant, prune off one-fourth to one-third of the branches. Water the plant well until it becomes established. If necessary, provide shade during hot weather. In the future, transplant when the tree or shrub is dormant, if possible, and when the weather is cool, in early morning, in the late afternoon, or on a cloudy day.

Slime flux

Slime flux on poplar.

Problem: Sour-smelling sap oozes from wounds, cracks, and branch crotches, mainly during the growing season. The sap drips down the bark and dries, causing unsightly gray streaks. Some wilting and leaf scorch may occur on affected branches. Insects are attracted to the sour-smelling ooze. Elms and poplars are commonly infected.

Analysis: Slime flux, also called *wetwood*, is caused by several bacteria. The bacteria infect the heartwood and ferment, producing abnormally high sap pressure and forcing the fermented sap, or *flux*, out of wounds, cracks, or crotches in the tree. Flux is especially copious when the tree is growing rapidly. Large areas of the bark may be coated with the smelly, bacteria-laden sap, which dries to a grayish white color. Also, wounds do not heal, and the bark is unsightly. A tree with this problem is often under water stress, which may cause drought damage (wilting, scorched leaves, and dieback) to the branches. The problem may persist for years.

Solution: There are no controls. To avoid the unsightly stained bark, bore a slightly upward-slanting drainage hole into the water-soaked wood below each oozing wound. Insert a ½-inch-diameter plastic tube just until it stays firmly in place. (If the tube penetrates the water-soaked wood inside the tree, it will interfere with drainage.) The tube will carry the dripping sap away from the trunk, but it will not cure the disease. Disinfect tools after pruning infected trees by dipping tools after each cut in a solution of 1 part chlorine bleach and 9 parts water.

Oozing sap

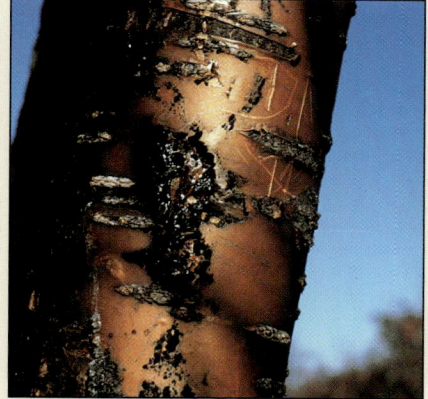
Oozing sap on cherry.

Oozing sap on Coulter pine.

Honeydew

Honeydew on maple.

Problem: Beads of amber-colored or whitish sticky sap appear on healthy bark. Or sap oozes from patches of bark, cankers, wounds, or pruning cuts.

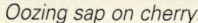

Analysis: Oozing sap, also called *gummosis*, occurs in all trees and shrubs to some degree. One or a combination of the following factors cause it.

1. Natural tendency: Certain species of plants have a tendency to ooze sap. Frequently, small beads of sap form on healthy bark of these plants.

2. Environmental stress: Plants that are stressed because they are growing in overly wet soil may produce large quantities of sap, even though they are not diseased. Also, many plants respond to changes in weather conditions or soil moisture by oozing profusely.

3. Mechanical injury: Most plants ooze sap when the bark is wounded. This is especially noticeable on maple and birch. If these trees are injured during the fall, they will ooze a large amount of sap the following spring.

4. Disease: Plants respond to certain fungal and bacterial infections by forming cankers—dark, sunken areas that gum profusely. Gummosis, or oozing sap, is one of the initial signs of infection.

5. Borer damage: Many types of insects bore holes into bark. Sap oozes from these holes. The tunnels these insects bore in the wood often become infected by decay organisms.

Solution: The numbered solutions below correspond to the numbered items in the analysis.

1. As long as the bark appears healthy, there is nothing to worry about.

2. If your plant is growing in wet, poorly drained soil, allow the soil to dry out between waterings. Provide for drainage away from trunks and roots. If oozing sap occurs as a result of rapid changes in weather and soil moisture, reduce the effects of stress on the plant by keeping it healthy. Maintain health and vigor by fertilizing and watering regularly.

3. Avoid mechanical injuries to the plant. Stake, tie, and prune properly.

4. Remove badly infected branches and cut out cankers. Keep the plant vigorous by fertilizing and watering regularly.

5. Borers are difficult to control once they have burrowed into the wood. For more information on borers and their controls, see page 147.

Problem: A shiny or sticky substance coats the leaves and sometimes the twigs. Insects may be found on the leaves directly above, and ants, flies, or bees may be present. A black, sooty mold often grows on the sticky substance.

Analysis: Honeydew is a sweet, sticky substance secreted by aphids, mealybugs, psyllids, whiteflies, and certain scales. These sucking insects cannot fully digest all the sugar in the plant sap, and they excrete the excess in a fluid called *honeydew*, which drops onto the leaves directly below and adheres to anything beneath the tree or shrub. Lawn furniture or cars under infested plants may be stained. Ants and certain flies and bees feed on honeydew and may be found around the plant. A sooty mold fungus often develops on the sticky substance, causing the leaves and twigs to appear black and dirty. The fungus does not infect the leaf but grows superficially on the honeydew. Extremely heavy fungus infestations may prevent light from reaching the leaf, reducing food production.

Solution: Wipe honeydew from the leaves with a wet rag, or hose it off. Eventually the rain will wash it off. Prevent more honeydew by controlling the insect that is producing it. Inspect the leaves and twigs above the honeydew to find out what type of insect is present. For control instructions, see the following pages: for aphids, mealybugs, and scales, page 136; for psyllids, page 161.

HOLES OR CRACKS IN BARK

Bark-feeding animals

Rodent damage to crabapple.

Porcupine damage to elm.

Sapsuckers

Sapsucker holes.

Problem: Bark has been chewed or gnawed from the trunk and lower branches. In some cases, the trunk is entirely girdled. Deer, rabbits, mice, or squirrels may have been seen in the yard, or their tracks may be evident on the ground or snow. Damage is usually most severe during the winter, when other food sources are scarce.

Analysis: Several animals chew on tree bark.
1. Deer: These animals feed on leaves, shoots, buds, and bark. They feed by pulling and twisting the bark or twig tissue, leaving ragged or twisted twig ends or patches of bark. Generally they feed on the lower branches and upper trunk. The males may also damage plants by rubbing their antlers on the trunk and branches.
2. Rabbits: These animals chew on the bark at the base of the trunk. They chew bark and twigs off cleanly, leaving a sharp break. The damaged trunk is often scarred with paired gouges left by the rabbit's front teeth. Rabbits generally feed no more than 2 feet above the ground or snow level. They damage small or young plants more severely.
3. Field mice, or voles: These animals damage plants by chewing off the bark at the base of the trunk just at or slightly above or below ground or snow level. They may girdle the trunk, often killing the plant. Mice leave tiny scratches in the exposed wood. Some species of mice feed on plant roots, causing the slow decline and death of the plant.
4. Squirrels: These animals damage trees and shrubs by wounding the bark. Red squirrels feed on maple sap in the spring. The resulting bark wounds are V-shaped. Canker disease fungi sometimes invade the wounds, weakening or killing the tree. Some squirrels feed on bark when food is scarce in the winter. Other squirrel species use bark and twigs for building nests.
5. Porcupines: During the winter, porcupines eat tree bark, sometimes stripping large patches of bark high above the ground. If the trunk is girdled, the top of the tree may die.

Solution: Various methods may be used to exclude or control deer, rabbits, mice, and squirrels in the garden. These methods usually involve protecting the plants with fencing and tree guards or controlling the animals with traps. Trapping or killing these animals is illegal in some states. Contact your state Department of Fish and Game to determine local regulations.

Problem: Rows of parallel holes, ¼ inch in diameter, appear on the trunk. Sap often oozes from the holes, and portions of the surrounding bark may fall off. When damage is severe, part or all of the tree is killed. Yellow-bellied or red-breasted birds may be seen pecking on the tree.

Analysis: Two species of sapsuckers, members of the woodpecker family, feed on tree bark and sap. The red-breasted sapsucker is found in the Pacific Northwest. The yellow-bellied sapsucker is common throughout much of the United States. Sapsuckers peck into many trees before finding a suitable one that has sap with a high sugar content. Once the birds find a favorite tree, they visit it many times per day and feed on it year after year. Portions of the bark often fall off after sapsuckers have pecked many holes. If the trunk is girdled, the tree above the damaged area dies. Sometimes disease organisms enter the holes and damage or kill the tree.

Solution: It is difficult to prevent sapsucker damage to trees. Wrapping the damaged trunk with burlap or smearing a sticky material (such as the latex used for ant control) above and below the holes may inhibit new pecking damage. For a list of trees most commonly attacked by sapsuckers, see page 353.

Borers

Borer emergence holes.

Problem: Foliage on a branch or at the top of the tree may be sparse; eventually the twigs and branches die. Holes or tunnels are apparent in the trunk or branches. Sap or sawdust surrounds or drips from the holes. The bark may die over the tunnels and slough off, or knotlike swellings may be found on the trunk and limbs. Weakened branches break during wind- or snowstorms. Weak, young, or newly transplanted trees may be killed.

Analysis: Borers are the larvae of beetles or moths. Many kinds of borers attack trees and shrubs. Females lay their eggs in bark crevices throughout the summer. The larvae feed by tunneling through the bark, sapwood, and heartwood, stopping the flow of nutrients and water in that area by cutting the conducting vessels; branch and twig dieback result. Sap flow acts as a defense against borers if the plant is healthy. When the insect burrows into the wood, tree sap fills the hole and drowns the insect. Factors that weaken the tree—such as mechanical injuries, transplanting, damage by leaf-feeding insects, and poor growing conditions—make it more attractive to egg-laying females.

Solution: Cut out and destroy all dead and dying branches, and remove severely infested young plants. Spray or paint the trunk and branches with an insecticide containing *chlorpyrifos.* Timing of spraying and pruning is critical. Contact your local county extension office for the best time to spray and prune trees in your area. Maintain plant health and vigor by watering and fertilizing regularly. To limit borer breeding sites, remove dead and dying trees promptly.

Cankers

Canker on Ceanothus.

Problem: Sunken, oval, or elongated dark-colored lesions (cankers) develop on the trunk or branches. The bark at the edge of the canker may thicken and roll inward. In some cases, sticky, amber-colored sap oozes from the canker. Foliage on infected plants may be stunted and yellowing. Some of the leaves may turn brown and drop off. Twigs and branches may gradually die, and the plant may eventually be killed.

Analysis: Many species of fungi and bacteria cause cankers and dieback. Infection usually occurs through injured or wounded tissue. Bark that has been damaged by sunscald, cold, pruning wounds, or mechanical injury is especially susceptible. Some decay organisms infect the leaves first, then spread down into healthy twigs. Cankers form as the decay progresses. Some plants produce a sticky sap that oozes from the cankers. The portion of the branch or stem above the canker may die from the clogging of the water and nutrient-conducting vessels in the branch. Cankers that form on the trunk are the most serious and may kill the tree. The plant may halt the development of a canker by producing callus tissue, a growth of barklike cells, to wall off the decay.

Solution: Remove badly infected branches and cut out cankers. Avoid wounding the plant. Keep the plant vigorous by fertilizing and watering regularly.

Bark shedding

Bark shedding on madrone.

Problem: Bark is cracking or peeling, usually on the older branches and trunk.

Analysis: The shedding or cracking of bark is often noticeable and may be of concern to people not familiar with this natural process. The outer bark changes over the lifetime of a tree. The bark of young trees is live tissue, usually smooth and relatively soft. As the trees mature, the bark dies and hardens, sometimes becoming rough. Trunks and branches increase in diameter with age. The increase in girth causes the outer bark of many plants to crack in a variety of patterns. With some tree species, such as white birch, cracking develops to such an extent that the bark peels and falls off. Newly exposed bark is often smooth and lighter in color than the bark that was shed. Some trees, such as sycamore and shagbark hickory, characteristically have loose outer bark. The bark is constantly in the process of peeling and shedding.

Solution: This process is normal. No controls are necessary.

HOLES OR CRACKS IN BARK *(continued)*

Sunscald

Sunscald.

Sunscald on dogwood.

Problem: Patches of bark die, crack, and later develop into cankers. The dead bark eventually sloughs off, exposing undamaged wood. The affected bark area is always on the southwest side of the tree. Trees with dark bark are likely to be more severely affected. The cracks and cankers develop in either summer or winter.

Analysis: When a tree growing in a deeply shaded location is suddenly exposed to intense sunlight, or when a tree is heavily pruned, the southwest side of newly exposed bark is injured by the rapid change in temperature. This may develop when a forested area is excessively thinned or when a tree is moved from a shaded nursery to an open area such as a lawn.

1. Summer sunscald: With intense summer heat, exposed bark is killed and a canker develops, usually revealing the undamaged wood beneath the bark. Within several seasons, the tree may break at the cankered area and topple. Summer sunscald is most severe when the soil is dry.

2. Winter sunscald: Bark injury develops with rapid changes in bark temperature from cold nights to sunny winter days. Exposed bark becomes much warmer than the air during the day but cools rapidly after sunset. This rapid temperature change often results in bark cracking and, later, cankering. Trees with thin, dark bark are most severely affected.

Solution: Once the bark is injured, you cannot do anything about sunscald. Wrap the trunks of recently exposed or newly transplanted trees with tree-wrapping paper, available in nurseries. White interior latex paint or whitewash is also effective. The wrap or paint should be left on for at least 2 winters. Remove the wrap for the spring and summer months to prevent it from harboring plant diseases and insects. Reapply the paint the second season if it has washed off. Trees will eventually adapt to increased exposure by producing thicker bark. Give trees, especially recently transplanted trees, adequate water in the summer and, if necessary, in the fall. Water transplants when the top 2 inches of the rootball are dry. Remove badly infected branches and cut out cankers. Avoid wounding the plant. Keep the plant vigorous by fertilizing and watering regularly.

Lightning damage

Lightning-damaged tree.

Problem: Part or all of the tree suddenly turns brown and dies. No external signs of damage may appear, or a strip of bark may be burned or missing from the entire length of the trunk. In less severe cases, trees survive for several years or recover completely. Sometimes tops of trees or branches explode, leaving a jagged stub. A lightning storm has occurred recently.

Analysis: Tall trees, trees growing in open locations, and trees growing in moist soil or along riverbanks are susceptible to damage by lightning. Lightning damage is variable. Some trees die suddenly from internal damage or burned roots without any external sign that lightning has struck. Other trees burst into flames or explode when struck. Sometimes only a strip of bark is burned or missing from the trunk and the tree recovers. Some species of trees are more resistant to lightning bolts than others. (For lists of trees resistant or susceptible to lightning injury, see page 353.) Some scientists believe that trees high in starch, deep-rooted species, and decaying trees are more susceptible to damage than trees high in oils, shallow-rooted species, or healthy trees.

Solution: Remove all loose and injured bark. To reduce or prevent damage, water trees during dry spells. Remove severely damaged trees. Valuable old trees can be protected with lightning conductors. Contact a professional arborist.

TWIGS OR BRANCHES BREAK

Limb breakage

Limb breakage caused by weak fork.

Limb breakage caused by snow load.

Frost cracks

Frost cracks.

Problem: Healthy branches break and fall, usually during storms or high winds. Some may drop in the middle of the day during a hot spell.

Analysis: Several different environmental factors cause limb breakage.

1. Weak fork: The angle between a branch and the trunk, called a fork, is normally greater than 45 degrees in most species. If the angle is much less than this, bark is sometimes trapped between the branch and the trunk, preventing the wood from growing together at that point. This weakens the branch. As the branch and trunk increase in length, the additional weight causes the fork to split at the weak junction. A large portion of the tree may fall. Some trees that develop weak forks break more readily than others because of their growth habits and brittle wood.

2. Wind: Branches may fall during high winds, especially in areas where there are tornadoes, hurricanes, and other forms of extreme winds. Moderate winds often hasten the dropping of limbs weakened by injury, insects, or plant disease.

3. Sudden limb drop: Large limbs sometimes drop during the middle of the day for no apparent reason. This usually occurs on hot, calm days. The cause is not known.

4. Snow and ice: Plants heavily coated with snow or ice may lose large limbs because of the additional weight. Evergreen trees, or deciduous trees with leaves still attached, are most susceptible because of the greater surface to which the snow or ice can adhere.

Solution: If the break is a split, or if at least one-third of the bark at the break is intact, the limb can be bolted back in place. If less than one-third of the bark is intact, or if the branch has fallen off the tree, prune off the remaining branch stub. To prevent further damage, brace or cable trees, and knock off snow and ice constantly to prevent buildup. In areas with high winds, prune back some of the branches to reduce the wind load. In the future, do not plant trees with brittle wood that breaks easily. (For a list of trees with weak forks and brittle wood, see page 353.)

Problem: Longitudinal cracks develop on the trunk, usually on the south and southeast and sometimes on the west sides. The cracks generally close during the growing season.

Analysis: Frost cracks develop from the expansion and shrinkage of bark and wood during periods of wide temperature fluctuations. This causes internal mechanical stress, which causes already weakened or decayed areas of the bark and outer wood to split open. The sudden break is often accompanied by a loud noise like a gunshot. Cracks usually heal during the growing season, but they may remain partially open after the weather warms or reopen during the following winter.

Solution: If a large crack fails to heal, a rod or bolt may be installed to hold the tree together. Consult a professional arborist. Plant trees adapted to your climate. Protect young trees with tree-wrapping paper in late fall, or by whitewashing the trunk.

TWIGS OR BRANCHES BREAK *(continued)*

DISCOLORED LEAVES

Twig pruners and twig girdlers

Twig girdler (2× life size).

Problem: Small, cleanly cut twigs, ¼ to 2 inches in diameter, lie under the tree in the fall. The tree is often abnormally bushy. Small (up to 1-inch), whitish larvae may be found inside the fallen twigs.

Analysis: Several species of wood-boring beetles cause unsightly damage to trees by altering their natural form. In midsummer to fall, wood-boring beetles lay their eggs in the wood of small twigs. The *twig pruner* larvae tunnel toward the base of the twigs, eating all but the outer bark. In the fall, they back into the hollowed-out twigs. High winds cause the nearly severed twigs containing the larvae to break and drop to the ground. Adult *twig girdlers* lay eggs in twigs, then chew a circle around the outside of the twigs. The girdled twigs die and break off. The eggs in the fallen twig are able to develop without being hindered by the flow of sap through the twig. Both twig pruner and twig girdler larvae mature in the twigs on the ground. The damage to the tree is the result of excessive pruning on the branch tips. Several new side shoots develop where the twigs break off, causing abnormal bushiness and an unnatural shape.

Solution: Chemical control is not usually necessary or practical. Gather and destroy all severed twigs in the late fall, when the insects are inside them. Also gather severed twigs from nearby trees. This practice usually controls the pest. If necessary to protect small trees, spray with an insecticide containing *chlorpyrifos* in late summer to kill the adults before they lay eggs. Repeat two times at 30-day intervals.

Heart rot

Heart rot.

Problem: Branches break and fall, usually during storms. The wood in the area of breakage is discolored and often spongy. Mushroomlike growths may be found on the wood.

Analysis: Heart rot is caused mainly by fungi. Decay organisms rot deadwood (such as fallen trees) as part of nature's recycling process. Several of them may also invade live trees through wounds. Healthy, vigorous trees may stop the spread of decay by producing cells that wall off the invaded area. Old trees with many wounds are rarely very resistant to microorganisms, and decay spreads through the wood. The decay does not usually kill the living tissue of the tree, so branches and leaves are kept alive. But internal decay reduces the strength of affected limbs. During a storm, weakened branches fall. Some of the decay organisms develop yellowish to brown, mushroomlike growths, called *conks,* on the outside wood in areas where decay is present. (For more information on mushrooms and conks, see page 138.)

Solution: Cut off the remaining branch stub flush against the larger branch or tree trunk. Inspect the rest of the tree to determine the extent of decay (you may need to contact a professional arborist). Branches or the entire plant should be removed if decay is extensive. As much as possible, avoid wounding plants. Keep them vigorous by fertilizing and watering regularly.

Salt burn

Salt burn on mock orange.

Problem: The tips and edges of older leaves turn dark brown or black and die. The rest of the leaf may be lighter green than normal. The browning or blackening can develop in both dry and wet soils, but it is more severe in dry soil. In the worst cases, leaves drop from the plant.

Analysis: Salt burn is common along the seashore and in areas of low rainfall. It also occurs in soils with poor drainage, in areas where salt has been used to melt snow and ice, and in areas where too much fertilizer has been applied. Excess salts dissolved in the soil water accumulate in the leaf tips and edges, where they kill the tissue. These salts also interfere with water uptake by the plant. This problem is rare in areas of high rainfall, where the soluble salts are leached from most soils. Poorly drained soils also accumulate salts because they do not leach well; much of the applied water runs off instead of washing through the soil. Fertilizers, most of which are soluble salts, also cause salt burn if too much is applied or if they are not diluted with a thorough watering after application.

Solution: In areas with low rainfall, leach accumulated salts from the soil with an occasional heavy watering (about once a month). If possible, improve drainage around the plants. Follow package directions when using fertilizers; several light applications are better than one heavy application. Water thoroughly afterward. Avoid the use of bagged steer manure, which may contain large amounts of salts.

Leaf rusts

Cedar-apple rust on hawthorn.

Problem: Leaves are discolored or mottled yellow to brown. Yellow, orange, red, or blackish powdery pustules appear on the leaves. The powdery material can be scraped off. Leaves may become twisted and distorted and may dry and drop off. Twigs may also be infected. Plants are often stunted.

Analysis: Many species of leaf rust fungi infect trees and shrubs. Some rusts require two plant species to complete their life cycles. Part of the life cycle is spent on the tree or shrub and part on various weeds, flowers, or other woody trees or shrubs. Rust fungi survive the winter as spores on or in living plant tissue or in plant debris. Wind and splashing water spread the spores to healthy plants. When conditions are favorable (with moisture on the leaf in the form of rain, dew, or fog and with moderate temperatures, 54° to 74°F), the spores germinate and infect the tissue. Leaf discoloration and mottling develop as the fungus saps plant nutrients. Some rust fungi produce spores in spots or patches, while others develop into hornlike structures.

Solution: Several fungicides, including those containing *triforine* (Funginex®), *chlorothalonil* (Daconil 2787®), *ferbam*, *maneb*, *zineb*, or *cycloheximide*, may be used to control rust. Look under the entry for your plant in the alphabetical section beginning on page 157 to determine which fungicide is appropriate. Some rust fungi are fairly harmless to the plant and do not require control measures. Where practical, remove and destroy infected leaves as they appear. Rake up and destroy leaves in the fall.

Leaf scorch

Leaf scorch on maple.

Leaf scorch on birch.

Problem: Leaf edges and the tissues between the veins turn tan or brown. Leaves are scorched and have a dry appearance. Brown areas often increase in size until little green is left except around the center vein. Dead leaves may remain attached to the plant, or they may drop. Leaf scorch is most severe in the upper branches and when the soil is dry.

Analysis: Scorch is caused by a lack of water in the leaves. This lack can result from any of several causes.

Solution: The numbered solutions below correspond to the numbered items in the analysis.

1. Extreme heat and wind: Excessive evaporation of moisture from the leaves can cause leaf scorch. In hot or windy weather, water evaporates rapidly from the leaves. If the roots cannot absorb and convey water fast enough to replenish this loss, the leaves turn brown and wither. This usually occurs in dry soil, but leaves can also scorch when the soil is moist and temperatures are near 100°F for extended periods. Young plants with limited root systems are most susceptible.

1. To prevent further scorch, deep-water plants during periods of hot weather to wet down the entire root space. Because of their limited root systems, recently transplanted trees and shrubs should be watered more often than established plants. Water them when the rootball is dry 2 inches below the surface. If the leaves scorched when the soil was moist, provide shade during hot weather and screens for protection from wind—or transplant to a protected area.

2. Winter burn: Winter burn is similar to scorch from intense heat or wind except that it occurs during warm, windy days in late winter. In colder climates, water cannot be replaced by the roots because the soil is frozen, resulting in leaf desiccation. Conifers are most susceptible, especially those planted in exposed areas. Symptoms may not appear until spring.

2. Provide windbreaks and shelter for plants growing in cold, windy regions. Covering smaller plants with burlap helps prevent leaf drying. If necessary, water in late fall or winter to ensure adequate soil moisture. Mulch plants after they are dormant to reduce the depth of frost penetration into the soil.

3. Damaged roots: Trees and shrubs may develop scorched and yellow leaves, early fall color, and dieback after the roots have been injured.

3. For more information on damaged roots, see pages 155 and 156.

4. Underwatering: Many plants survive even when regularly underwatered. They do not function normally, however, and the leaves frequently burn and wilt over the entire plant.

4. Do not let the soil dry out to the point where leaves scorch. Check the moisture needs of your plant under its cultural information in the alphabetical section beginning on page 157.

DISCOLORED LEAVES (continued)

Spider mites

Spider mite damage and webbing.

Problem: Leaves are stippled yellow, white, or bronze and are dirty. A silken webbing is sometimes found on the leaves or stems. New growth may be distorted, and the plant may be weak and stunted. To determine if a plant is infested with spider mites, examine the bottoms of the leaves with a hand lens. Or hold a sheet of white paper underneath an affected leaf or branch and tap it sharply. Green, red, or yellow specks the size of pepper grains will drop to the paper and begin to crawl around.

Analysis: Spider mites, related to spiders, are major pests of many plants. They cause damage by sucking sap from leaves and buds. As a result of their feeding, the plant's green leaf pigment disappears, producing the stippled appearance. While they feed, many mites produce a fine webbing over the foliage that collects dust and dirt. Some mites are active throughout the growing season but they thrive especially in dry weather with temperatures of 70°F and above. Other mites, especially those infesting conifers, are most prolific in cooler weather. They are most active in the spring and sometimes fall and during warm periods in winter in mild climates. By the onset of hot weather, these mites have usually caused their maximum damage.

Solution: Spray with ORTHO Isotox Insect Killer, ORTHO RosePride Orthenex Insect & Disease Control, or horticultural oil when damage is first noticed. Repeat the spray two more times at intervals of 7 to 10 days. Make sure that your plant is listed on the product label.

Lace bugs

Lace bug and droppings (4× life size).

Problem: The upper surfaces of leaves are mottled or speckled yellow, gray, or white and green. The mottling is distinguished from other insect damage—such as that caused by mites or leafhoppers—by the shiny, hard, black droplets found on the undersides of damaged leaves. Small (⅛-inch), light or dark, spiny, wingless insects or brownish insects with clear lacy wings may be visible around the droplets. Plant growth is usually stunted. Damage occurs in spring and summer.

Analysis: Many species of lace bugs feed on trees and shrubs. Each species usually infests only a few related types of plants. Depending on species, lace bugs spend the winter as adults in protected areas on the plant or as eggs in leaf veins or cemented to the lower surface of the leaf in a crusty brown material. Both the spiny, wingless, immature insects and the lace-winged adults suck sap from the undersides of leaves. The green leaf pigment disappears, resulting in the characteristic speckling or mottling. As lace bugs feed, droplets of black excrement accumulate around them. Damage is unsightly, and food production by the leaf is reduced, resulting in loss of plant vigor.

Solution: Spray with ORTHO Diazinon Ultra Insect Spray or ORTHO Orthene Systemic Insect Control when damage first appears in the spring. Cover the undersides of leaves thoroughly. Repeat the spray 7 to 10 days later. A third application may be necessary if the plant becomes reinfested in midsummer.

Leafhoppers

Leafhopper damage to dogwood.

Problem: Leaves are stippled white, yellow, or light green, and leaves and stems may be distorted. Sometimes the plant has a burned appearance. Infested leaves may drop prematurely. When infestations are severe, twigs and small branches sometimes die. Whitish or green wedge-shaped insects, up to ½ inch long, hop and fly away quickly when the plant is touched.

Analysis: Many species of leafhoppers infest ornamental trees and shrubs. Some leafhoppers cause only minor damage to the leaves, while other species severely retard plant growth. Leafhoppers usually spend the winter as eggs in bark slits made by the females, although some may overwinter in the South and migrate north in the spring. Injury to the bark from egg laying may kill twigs. When the weather warms in the spring, young leafhoppers emerge and settle on the undersides of leaves, where they suck out the plant sap, causing the stippling and distortion. Severely infested leaves often drop in midsummer. Some leafhoppers cause a condition known as *hopperburn*. Insect feeding causes distortion and gives the leaves a burned appearance. Several generations of leafhoppers may occur each year.

Solution: Spray with ORTHO Isotox Insect Killer, ORTHO Orthene Systemic Insect Control, ORTHO Diazinon Ultra Insect Spray or with an insecticide containing *malathion* when damage is first noticed. Cover the lower surfaces of the leaves thoroughly. Make sure that your plant is listed on the product label.

Thrips

Greenhouse thrips damage to coffee plant.

Problem: Young leaves are severely curled and distorted. Parts of the leaf may die and turn black, or the entire leaf may drop from the plant. Or leaves are flecked and appear bleached or silvery, often becoming papery and wilted. Shiny black spots may cover the surfaces. Flowers and buds may have white streaks or be brown and distorted. Minute (1/25-inch) white or yellow spindle-shaped insects and black or brown winged insects are barely visible either inside the distorted leaves and flowers or on the undersides of leaves. Heavily infested plants may be stunted.

Analysis: Thrips are a common pest of many garden and greenhouse plants. Some species cause leaf or flower distortion; others cause a flecking of the leaves, producing a bleached appearance. Thrips feed by rasping the soft plant tissue, then sucking the released plant sap. Some leaf thrips leave unsightly black, varnishlike spots of excrement around the areas where they feed. The black or brown adults have wings. They can spread rapidly by flying to new plants, or the wind may blow them long distances. They lay their eggs either on the plant or in surrounding weeds. The young are yellow or white and spindle-shaped. Some thrips can transmit diseases.

Solution: Spray the leaves or buds and flowers with ORTHO Isotox Insect Killer, ORTHO Orthene Systemic Insect Control, or ORTHO Malathion 50 Plus Insect Spray. Make sure that your plant is listed on the product label. Remove and destroy infested buds and flowers.

Overwatering or poor drainage

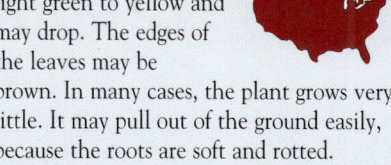

Overwatering damage to yew hedge.

Problem: Leaves turn light green to yellow and may drop. The edges of the leaves may be brown. In many cases, the plant grows very little. It may pull out of the ground easily, because the roots are soft and rotted. The soil is frequently or constantly wet.

Analysis: Overwatering and poor drainage are serious, common problems that often kill plants. Roots require air to function normally. Air is contained in tiny pores in the soil. When the soil is watered, air is forced out of the soil pores and replaced with water. If water cannot drain out of the soil, or if it is constantly reapplied, the soil pores remain filled with water. The roots cannot absorb the oxygen they need in such saturated conditions, and they die. As the roots rot, the root system is less able to supply the plant with nutrients and water, resulting in starvation and eventually in the death of the plant.

Solution: Do not apply water so frequently that the soil is constantly wet. Depending on the particular requirements of your tree or shrub (see cultural information for your specific plant in the alphabetical section beginning on page 157), allow the soil to dry partially or completely between waterings. If your soil drains poorly, improve drainage.

Lack of nitrogen

Nitrogen-deficient fuchsia.

Problem: Leaves turn yellow and may drop, beginning with the older leaves. New leaves are small, and growth is slow.

Analysis: Nitrogen, one of the most important nutrients for plant growth, is deficient in most soils. Nitrogen is essential in the formation of green leaf pigment and many other compounds necessary for plant growth. When short on the nutrient, plants take nitrogen from their older leaves for new growth. Poorly drained, overwatered, compacted, and cold soils are often infertile. Plants growing in these soils often show symptoms of nitrogen deficiency. Various soil problems and other nutrient deficiencies may also cause leaf discoloration.

Solution: For a quick response, spray the leaves and the soil beneath the plant with liquid or water-soluble fertilizer. Feed plants regularly with Scotts Evergreen, Shrub and Tree Food. Add organic amendments to compacted soils and those low in organic matter, and improve drainage in poorly drained soils. Do not keep the soil constantly wet.

DISCOLORED LEAVES
(continued)

Iron deficiency

Iron-deficient azalea.

Problem: New leaves are pale green or yellow. The veins may remain green, forming a Christmas-tree pattern on the leaf. Old leaves remain green. In extreme cases, new leaves are all-yellow and stunted.

Analysis: Plants frequently suffer from deficiencies of iron and other minor nutrients, such as manganese and zinc, elements essential to normal plant growth and development. Deficiencies can occur when one or more of these elements is depleted in the soil. Often these minor nutrients are present in the soil, but alkaline (pH above 7.0) or wet soil conditions cause them to form compounds that cannot be used by the plant. Alkalinity can result from overliming or from lime leached from cement or mortar. Regions where soil is derived from limestone or where rainfall is low usually have alkaline soils.

Solution: To correct the iron deficiency, spray the foliage with a chelated iron fertilizer, and apply the fertilizer to the soil around the plant. Apply soil sulfur or ferrous sulfate to lower the pH. Maintain an acid pH by fertilizing with Scotts Azalea, Camellia, Rhododendron Food. When planting in an area with alkaline soil, add a handful of soil sulfur, or add enough peat moss to make up 50 percent of the amended soil, and mix well.

INVADING ROOTS

Surface roots

Roots cracking pavement.

Surface roots in a lawn.

Problem: Roots are exposed on the surface of the soil or make bumps by growing just beneath it. Or roots crack and raise the pavement. Exposed roots may be lumpy and galled.

Analysis: Several factors can cause roots near the surface to expand.

Solution: The numbered solutions below correspond to the numbered items in the analysis.

1. Surface roots in lawns: If plants receive only light irrigation on the soil surface, roots in this upper zone will expand, pushing above the surface. Plants growing in lightly irrigated lawns often have shallow roots.

1. In addition to watering the lawn, deep-water trees in lawns every 2 weeks, to a depth of 3 to 4 feet. Cover the roots by slowly raising the level of the lawn (1 inch per year) with soil topdressing.

2. Waterlogged soil: Roots need oxygen to grow and develop. Waterlogged soil has little oxygen available for root growth because the soil pores are filled with water. The only available oxygen is near the soil surface, so the surface roots develop most.

2. If the soil is waterlogged from overwatering, cut back on watering. If necessary, improve drainage around the plant. Cover the exposed roots with 2 to 4 inches of soil. Before planting new trees and shrubs, make sure drainage is adequate.

3. Natural tendency: Some plant species are more likely than others to develop surface roots.

3. For a list of plants likely to develop surface roots, see page 354.

4. Compacted soil: Trees and shrubs growing in compacted soil develop surface roots.

4. Loosen compacted soil with a crowbar. Before planting in compacted soil, loosen the soil.

5. Confined roots: Plants growing in areas with limited root space, such as in containers, often have roots growing on the soil surface.

5. Plant shrubs or small trees adapted to growing in confined root areas (see page 354).

6. Planting strips: Trees growing in planting strips adjacent to lawns frequently crack the sidewalks that separate them from the lawns. All of the available water and food is beyond the walk, in the lawns. The roots that extend under the walk into the lawn expand rapidly, cracking the walk.

6. Sever small roots that are pushing up pavement. If possible, avoid cutting large roots. In the future, plant large trees on the lawn side of the sidewalk and use shrubs in planting strips.

BRANCHES DIE

Fire blight

Fire blight on crabapple.

Problem: Blossoms and leaves of some twigs suddenly wilt and turn black as if scorched by fire. Leaves curl and hang downward. Tips of infected branches may hang down in a "shepherd's crook." The bark at the base of the blighted twig becomes water-soaked, then dark, sunken, and dry; cracks may develop at the edge of the sunken area. In warm, moist spring weather, drops of brown ooze appear on the sunken bark.

Analysis: Fire blight is caused by a bacterium (*Erwinia amylovora*) that is very destructive to many trees and shrubs. The bacteria spend the winter in the sunken areas (cankers) on the branches. In the spring, the bacteria ooze out of the cankers onto the branches and trunk and are carried by insects to the plant blossoms. The bacteria spread rapidly through the plant tissue in warm (65°F or higher), humid weather. Bees visiting these infected blossoms spread the disease. Rain and tools may also spread the bacteria.

Solution: During spring and summer, prune out infected branches about 12 inches beyond any visible discoloration and destroy them. Disinfect pruning tools by dipping after each cut in a solution of 1 part chlorine bleach and 9 parts water. A protective spray of a bactericide containing basic copper sulfate or streptomycin applied before bud-break in the spring helps prevent infection. Repeat at intervals of 5 to 7 days until the end of bloom. Don't add too much nitrogen fertilizer in the spring and early summer. Nitrogen forces succulent growth, which is more susceptible to fire blight. In summer or fall, after the disease stops spreading, prune out any remaining infected branches.

WEAK OR DYING PLANT

Lawn mower and string trimmer blight

Lawn mower blight on oak.

Problem: Leaves are small and discolored and often drop prematurely. Twigs may die. The plant is stunted and in a general state of decline. Bark at the base of the trunk is wounded. The tree is planted in a lawn.

Analysis: Trees growing in lawns may be severely injured by slight but repeated injuries to the bark from lawn mowers or string trimmers. Lawn mower blades and trimmer strings may also slice through the bark into the wood. Nutrients and water cannot pass through the damaged part of the trunk to reach the top of the tree. In severe cases, young trees are killed. These wounds are often entry points for disease-producing organisms that may also kill the tree.

Solution: Prune off dying twigs and branches, and fertilize with Scotts Evergreen, Shrub and Tree Food. To prevent additional damage to the trunk, kill all grass around the base of the tree with an herbicide. Or install a tree guard to eliminate the need for edging and trimming grass around the trunk.

Root nematodes

Root nematode damage.

Problem: Leaves are small and discolored and often drop prematurely. Eventually twigs and then larger branches die. The plant is stunted and in a general state of decline.

Analysis: Root nematodes are microscopic worms that live in the soil. They feed on plant roots, damaging and stunting them or causing them to become enlarged. The damaged roots can't supply sufficient water and nutrients to the aboveground plant parts, and the plant is stunted or slowly dies. Nematodes are found throughout the United States, especially in southern areas with moist, sandy loam soil. They can move only a few inches per year on their own, but they may be carried long distances by soil, water, tools, or infested plants. Laboratory testing of roots and soil is the only positive method for confirming the presence of nematodes. Contact your local county extension office for sampling instructions and addresses of testing laboratories. Soil and root problems— such as poor soil structure drought stress, overwatering, nutrient deficiency, and root rots—can also produce similar symptoms of decline. Eliminate these problems as causes before sending soil and root samples for testing.

Solution: No chemicals available to homeowners kill nematodes in planted soil, but they can be controlled before planting by soil fumigation or solarization. To solarize soil, cultivate and thoroughly wet an area before covering it with clear polyethylene film during the hottest time of the year. Allow the plastic film to remain in place for 4 to 6 weeks. Mulch, water, and fertilize with Scotts Evergreen, Shrub and Tree Food to minimize stress.

WEAK OR DYING PLANT (continued)

Grade change

Tree death caused by grade change.

Tree decline due to construction and grade change.

Root weevil larvae

Root weevil damage to azalea.

Problem: Leaves turn yellow and may drop. Branches die. The tree declines, usually over a period of several years, and may die. The soil level under the tree was recently changed, either raised or lowered.

Analysis: Raising or lowering the level of the soil (the grade) around trees can be very damaging. If the grade has been raised, the tree emerges from the soil in a straight line (no flare at the base of the trunk). If the grade has been lowered, roots are exposed.

1. Grade raised: A large quantity of soil dumped around a tree can suffocate the roots by cutting off their supply of air and water. The extent of damage depends on the kind of tree, its age and condition, the type and depth of fill, and how much of the root system is covered. Young, healthy trees are much more tolerant than old trees, and soil containing gravel or sand causes less severe injury than heavy clay soil. If only a portion of the root system is covered, or if the fill is relatively shallow (less than 3 inches of porous soil or less than 1 to 2 inches of clay soil), the tree will be weakened but usually will not die. Severe decline symptoms (progressive dieback from the top down) often do not occur for several years. Insects or diseases may kill the weakened plant sooner than it would have been killed otherwise.

2. Grade lowered: Many roots may be severed when soil is removed from around a tree, and exposed roots will dry out. The number and size of the roots severed determines the extent of damage. Cutting large roots close to the trunk is more likely to kill the tree than cutting the ends of roots. The plant is often unstable and may blow down in a strong wind.

Solution: Once decline symptoms have developed, significant damage has already occurred. The numbered solutions below correspond to the numbered items in the analysis.

1. By the time symptoms caused by raising the grade are noticed, it is usually too late to do anything. If fill has been around the tree for less than one growing season and no symptoms have yet appeared, remove the fill if possible. If the tree is in a severe state of decline, remove it. If decline symptoms are not severe and fill is less than 12 inches deep, therapeutic treatments may save the tree. Remove all dead and dying branches and remove the soil from around the base of the trunk. Dig holes to the original soil level every few feet over the entire root area (under the branches) and place 6-inch bell tiles in the holes. In the future, protect valuable trees to be filled over by installing tile pipes in a thick bed of gravel covered with a minimal amount of fill. For a list of trees that are severely, moderately, or rarely damaged by fill, see page 353.

2. If a tree is in a severe state of decline caused by lowering the grade, remove the tree. Cable unstable trees to a stable object, or remove them. For trees with symptoms, prune off damaged roots and torn bark. Cut back the top growth so it is in balance with the remaining roots (if 20 percent of the roots are damaged, cut off 20 percent of the branches). Water trees during dry periods.

Problem: Leaves are small and discolored and often drop prematurely. Eventually twigs and then larger branches die. The plant is stunted and in a general state of decline. If soil is removed from around the base of the plant, exposing some roots, bark on the roots or small rootlets are seen to be chewed. White grubs may be found in the soil around the roots.

Analysis: Root weevil larvae, called *grubs*, infest the roots of many ornamental plants. The damage caused by the white, legless grubs is often so gradual that the insects are well established before injury is apparent. If the grubs remain undetected, the plant may die abruptly with the onset of hot, dry weather. Female weevils lay eggs at the soil line near plants during the summer. The emerging grubs burrow into the soil. They feed on the roots in the fall and then spend the winter in the soil. Most root weevils cause their major damage in the spring. Their feeding girdles the roots and stems, disrupting the flow of nutrients and water through the plant and causing the roots and the top of the plant to die.

Solution: Discard dying plants. Because the grubs are in the soil, they are difficult to kill. To prevent the next generation of weevils from causing damage, eliminate the adults. Check periodically for notched leaves, or search for feeding adults in the leaf litter around the plant. Disturbed adults often "play dead" for a few minutes before moving. You may also search for them on the plants with a flashlight after dark. Spray the foliage and the ground under the plant with ORTHO Orthene Systemic Insect Control or ORTHO Isotox Insect Killer.

ABIES (Fir)

Spruce budworms	Balsam twig aphid	Douglas-fir tussock moth

Spruce budworm (2× life size).

Balsam twig aphid damage.

Tussock moth damage. Inset: Larva (life size).

Problem: Needles on the ends of branches are chewed and webbed together. In mid-July, 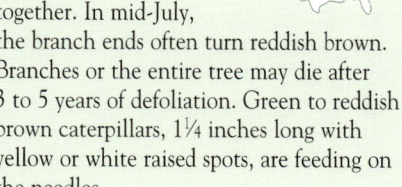 the branch ends often turn reddish brown. Branches or the entire tree may die after 3 to 5 years of defoliation. Green to reddish brown caterpillars, 1¼ inches long with yellow or white raised spots, are feeding on the needles.

Analysis: Spruce budworms (*Choristoneura* species) are very destructive to ornamental spruce, fir, and Douglas fir and may infest pine, larch, and hemlock. The budworm is cyclical. It comes and goes in epidemics about 30 years apart. The moths are small (½ inch long) and grayish, with bands and spots of brown. The females lay pale green eggs in clusters on the needles in late July and August. The larvae that hatch from these eggs crawl to hiding places in the bark or in lichen mats; or they are blown by the wind to other trees, where they hide. The tiny larvae spin a silken case and hibernate until spring. In May, when the weather warms, the caterpillars tunnel into needles. As they grow, they feed on opening buds; later they chew off needles and web them together. The larvae feed for about 5 weeks, pupate on twigs, and emerge as adults.

Solution: When buds have fully expanded in late May or early June, spray with the bacterial insecticide *Bacillus thuringiensis* (Bt) or ORTHO Orthene Systemic Insect Control.

Problem: The youngest needles are curled and twisted, with their lighter underside turned upward. These needles drop from the plant prematurely. Some needles are killed. Young twigs are twisted, and bark is roughened. Shoots may become saturated with a shiny, sticky secretion so that needles adhere to one another. A black, sooty mold may grow on this sticky substance. Clustered on the shoots are tiny (⅛-inch), waxy, bluish gray adult or pale green, immature soft-bodied insects with woolly posterior secretions.

Analysis: The balsam twig aphid (*Mindarus abietinus*) does little damage in small numbers. Aphids are extremely prolific, however, and populations can rapidly build up to damaging numbers. Damage occurs when the aphid injects its saliva into the plant as it sucks the juices from the fir shoots. The aphid is unable to digest fully all the sugar in the sap, and it excretes the excess in a fluid called *honeydew*. The honeydew often drops onto the shoots or other plants below. A sooty mold fungus may develop on the honeydew, causing the fir leaves to appear black and dirty. Late in the summer, females lay several small (⅒-inch) eggs, covered with tiny rods of white wax, in bark crevices. The eggs are conspicuous and are a useful index of the amount of injury that may occur during the following year.

Solution: Kill aphids by spraying with ORTHO Orthene Systemic Insect Control or ORTHO Malathion 50 Plus Insect Spray at bud-break in late April or early May. Repeat the spray in 2 weeks if the tree becomes reinfested.

Problem: Much of the foliage is eaten, starting at the top of the tree and progressing downward. 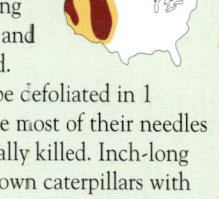 The entire tree may be defoliated in 1 season. Trees that lose most of their needles a second year are usually killed. Inch-long hairy, gray or light brown caterpillars with black heads and tufts of orange hairs on their backs may be feeding on the needles.

Analysis: Douglas-fir tussock moth (*Orgyia pseudotsugata*) populations are cyclical. Every 7 to 10 years, forest populations build up to epidemic proportions. When this occurs, true firs, Douglas firs, spruce, pine, and larch may be completely defoliated. In cities, damaging numbers may be found every year. In mid- to late summer the hairy, wingless female moths lay their eggs in a frothy substance covered with a layer of hairlike scales. When the eggs hatch the following spring, the caterpillars begin feeding on the new needles at the top of the tree. As the younger foliage is devoured the caterpillars move downward, feeding on older needles. Large numbers of tan excrement pellets accumulate around the base of the tree. Since conifers do not replace their old needles, defoliated trees are often killed after 2 seasons. Less severely damaged trees may be killed later by bark beetles. In August, the caterpillars pupate to emerge as adults.

Solution: When you first notice damage or caterpillars in May or early June, spray with an insecticide containing *acephate* (Orthene®) or with the bacterial insecticide *Bacillus thuringiensis* (Bt). Repeat the spray 2 weeks later if damage continues.

ACER (Maple, box elder)

GROWING GUIDE

Adaptation: Zones 2 through 10. To determine your zone, see the map on page 336. Protect from drying winds.

Light: Full sun to part shade.

Soil: Any good, deep, well-drained garden soil. When choosing a planting site, pick an area that will handle the ultimate spread and height of the tree.

Fertilizer: Fertilize with Scotts Evergreen, Shrub and Tree Food according to label directions.

Water:
How much: Apply enough water to wet the soil 3 to 4 feet deep.
How often: Maples prefer moist soil. Water when the soil is moist but no longer wet 4 inches below the surface.

Pear thrips

Pear thrips damage on sugar maple.

Problem: Leaves are small, mottled yellow and brown, and distorted. Blisterlike scars may be found on the veins. In moderate infestations, the leaves in the crown are yellow and sparse, a condition resembling the damage caused by late frost. When damage is severe, the tree is defoliated in spring but produces new leaves in June or July.

Analysis: Pear thrips (*Taeniothrips inconsequens*) infest a variety of forest trees but have recently become a serious pest of sugar maples. The insects damage the foliage by piercing the leaves and sucking the plant juices. Pear thrips have slender, brownish bodies less than 1/16 inch long. They emerge from the soil in April or May and migrate to the expanding buds of the trees, where they lay eggs in the buds and feed on the young foliage. Larvae hatch from the eggs within 2 weeks, feed until early June, and then drop to the soil to pupate. Pear thrips populations are cyclical, building up over several years to damaging numbers, then dropping again.

Solution: No control for this pest is presently available. Researchers are studying the problem. Contact your local county extension office for advice.

Bladder gall mite and spindle gall

Galls caused by bladder gall mites.

Problem: In early spring, maple leaf tissue develops irregular, spherical, or bladderlike growths, known as galls, on the upper surfaces of the leaves. Leaves next to the trunk and on large branches are most affected. The galls are yellowish green at first but later turn pinkish to red and finally black. If the galls are numerous, leaves become deformed, and some turn yellow and drop prematurely.

Analysis: Bladder galls, which grow on silver maples, and spindle galls, which grow on sugar maples, are caused by tiny bladder gall mites (*Vasates quadripedes*) and spindle gall mites (*V. aceriscrumena*), which are too small to be seen with the naked eye. Each gall contains many mites. The mites congregate on buds just before they open in the spring. As the buds open, each mite punctures and enters a leaf on the underside, injecting a growth-promoting substance that causes abnormal tissue formation. A gall encloses the mite, with an opening remaining on the underside. The mites feed and females lay eggs inside their galls. The eggs hatch, and as they mature, the young mites crawl out through the opening and infest new leaves. In July, mite activity stops and the mites migrate to the bark to spend the winter.

Solution: Once galls are formed, you cannot do anything about them. The galls cause no serious injury, however. If you wish to prevent unsightly leaves the following year, spray buds, branches, and trunk with a dormant oil spray, ORTHO Isotox Insect Killer, or ORTHO RosePride Orthenex Insect & Disease Control before the buds start to open in the spring.

Cottony maple scales and mealybugs

Cottony cushion scales on maple (¼ life size).

Problem: The undersides of leaves, stems, or branch crotches are covered with white, cottony, cushionlike masses. Leaves turn yellow and may drop prematurely. Sometimes a shiny or sticky substance coats the leaves. A black, sooty mold often grows on the sticky substance. Numerous side shoots sometimes grow out of an infested crotch area. Twigs and branches may die back.

Analysis: Cottony maple scales and mealybugs are common on maples throughout the United States. The similarities in appearance of these insects make identification difficult. The young insects are yellowish brown to green. They feed throughout the summer on the stems and undersides of the leaves. Damage is caused by the withdrawal of plant sap from leaves and branches. The insects are unable to digest fully all the sugar in the sap, and they excrete the excess in a fluid called *honeydew*. The honeydew often drops onto leaves or onto plants, cars, and lawn furniture below. A sooty mold fungus may develop in the honeydew, causing the maple leaves and other plants to appear black and dirty. If the insects are not controlled, heavily infested branches may die after several seasons.

Solution: Apply ORTHO Diazinon Ultra Insect Spray, ORTHO Orthene Systemic Insect Control, or ORTHO Volck Oil Spray in midsummer when the young are active. The following spring, when trees are dormant, apply a dormant spray with lime sulfur mixed with spray oil.

Maple anthracnose

Anthracnose on Norway maple.

Problem: Irregular, light brown spots of dead tissue appear on the leaf from late May to August. They develop during or just following cool, wet, humid weather. Many spots occur along the veins. They may enlarge and run together, causing the death of the entire leaf. Partially killed leaves appear sunscorched, but this disease is distinguished from sunscorch by the presence of dark dots (spore-producing structures) barely visible on the underside of the leaves. The spore-producing structures develop while the leaves are still on the tree. Sunken reddish oval areas often develop on the infected twigs.

Analysis: Maple anthracnose is caused by two related fungi (*Discula* spp. and *Kabatiella apocrypta*) that spend the winter on fallen leaves or in sunken cankers on twigs in the tree. During cool, rainy weather, spores are blown and splashed onto newly expanding and young leaves. Dead spots develop on the leaf where the fungus enters the tissue. The spots expand, and the fungus can kill the leaf in rainy seasons, causing premature defoliation. The tree will grow new leaves if defoliation takes place in spring or early summer. When the tree is severely affected for successive years, the fungus will enter and kill branches.

Solution: Trees affected by this disease for a single year do not require a chemical control. Rake and burn old leaves and prune out dead twigs below the canker on the bark. This reduces the amount of disease the following year. If the following spring is wet and humid, spray valuable specimens with a fungicide containing *maneb* when the leaves uncurl. Repeat the treatment two more times at intervals of 2 weeks.

Verticillium wilt

Verticillium wilt. Inset: Infected stem.

Problem: The leaves on a branch turn yellow at the margins, then brown and dry. During hot weather, the leaves may wilt. New leaves may be stunted and yellowish. The infected tree may die slowly, branch by branch, over several seasons—or the whole tree may wilt and die within a few months. Some trees may recover. The tissue under the bark on the dying side shows dark streaks, which may be very apparent or barely visible when exposed. To examine for streaks, peel back the bark at the bottom of the dying branch.

Analysis: Verticillium wilt affects many ornamental trees and shrubs. It is caused by a soil-inhabiting fungus (*Verticillium* species) that persists indefinitely on plant debris or in the soil. The disease is spread by contaminated seeds, plants, soil, equipment, and groundwater. The fungus enters the tree through the roots and spreads up into the branches through the water-conducting vessels in the trunk. The vessels become discolored and plugged. This plugging cuts off the flow of water and nutrients to the branches, causing leaf discoloration and wilting.

Solution: No chemical control is available. Fertilize with Scotts Evergreen, Shrub and Tree Food to stimulate vigorous growth. Remove all deadwood. Do not remove branches on which leaves have recently wilted. These branches may produce new leaves in 3 to 4 weeks or the following spring. Remove dead trees. If replanting in the same area, plant trees and shrubs that are resistant to verticillium.

159

AESCULUS (Horse-chestnut, buckeye)

Summer leaf scorch

Summer leaf scorch on horsechestnut.

Problem: During hot weather, usually in July or August, leaves turn brown around the edges and between the veins. Sometimes the whole leaf dies. Many leaves may drop during late summer. This problem is most severe on the youngest branches. Trees generally do not die.

Analysis: In hot weather, water evaporates rapidly from the leaves. If the roots can't absorb and convey water fast enough to replenish this loss, the leaves turn brown and wither. This usually occurs in dry soil, but leaves can also scorch when the soil is moist. Horsechestnut trees vary in their susceptibility; one tree may be very susceptible while the tree next to it may show no sign of scorch. Drying winds, severed roots, limited soil area, or low temperatures can also cause scorch.

Solution: To prevent further scorch, water trees deeply during periods of hot weather to wet down the entire root space. Add mulch over the root system. Water newly transplanted trees whenever the rootball is dry 1 inch below the surface. Scorch on trees in moist soil cannot be controlled. Plant trees adapted to your climate.

BETULA (Birch)

Bronze birch borer

Dieback. Inset: Adult borer (½ life size).

Problem: Leaves are yellowing, and foliage is sparse at the top of the tree. Side growth on the lower branches is increased. Twigs and branches may die. The leaves on these branches turn brown but don't drop. D-shaped holes and ridges are found on the trunk and branches. Swollen ridges are packed with sawdust. Weak, young, or newly transplanted trees are usually killed.

Analysis: The bronze birch borer (*Agrilus anxius*) is the larva of an iridescent olive brown beetle about ½ inch long. For about 6 weeks in summer, adult beetles lay eggs in bark crevices, usually around a wound. The larvae that hatch from these eggs are white and have flat heads. They bore into the wood just beneath the bark. The feeding and tunneling of the larvae stop the flow of water by cutting the conducting vessels; branch and twig dieback result. If the tree is healthy, sap flow helps defend against borers; when the insect burrows into the wood, tree sap fills the hole and drowns the insect. Factors that weaken the tree—such as poor growing conditions, transplanting, and mechanical injuries—make it more attractive to female beetles.

Solution: Cut out and destroy all dead and dying branches. Severely infested young trees should be removed. In spring, spray or paint the trunk and branches with ORTHO Borer & Leaf Miner Spray to kill young larvae before they burrow into the wood. Repeat two times at intervals of 2 weeks. Maintain tree vigor by watering and fertilizing regularly. European white birch and cutleaf ornamental varieties are most susceptible. Plant resistant varieties if possible.

BUXUS (Boxwood)

GROWING GUIDE

Adaptation: Zones 5 through 10. To determine your zone, see the map on page 336. Protect from drying winds.

Light: Full sun or partial shade, especially in hot climates.

Soil: Well drained, rich in organic matter.

Fertilizer: Fertilize with Scotts Evergreen, Shrub and Tree Food according to label directions.

Water: How much: Apply enough water to wet the soil 1 to 2 feet deep.
How often: Boxwood will not tolerate drought. Water when the soil is moist but not wet 2 inches below the surface.

Boxwood leafminer

Boxwood leafminer larvae (5× life size).

Problem: The leaves are puckered or blistered. The lower surface of the leaf is spotted yellow; the upper surface is green  at first and then flecked brown and yellow. Leaves may drop prematurely. Growth is poor and the plant is rangy. Twigs may die back if the plant is infested more than 1 year. When the leaf is torn open, two or more small (⅛-inch), yellowish maggots or brownish pupae may be found between the upper and lower surfaces of the leaf.

Analysis: The boxwood leafminer (*Monarthropalpus buxi*) is one of the most serious pests of boxwood. The larvae spend the winter in the leaf. When the weather warms in spring, they feed on the tissue between the leaf surfaces. In late April or May a tiny (¹⁄₁₀-inch), gnatlike orange fly emerges from the pupal case inside the leaf. The emerging flies swarm around the plant in early morning, mating and laying eggs in the leaves. New blisters develop in midsummer from feeding by this next generation of larvae. When the weather turns cold, the larvae become inactive until the following spring.

Solution: Leafminer control is most effective when insecticides are applied just before eggs are laid in late spring. Spray with ORTHO Orthene Systemic Insect Control or with an insecticide containing *chlorpyrifos* in late April or early May. If eggs are laid before a control is applied, an insecticide containing *dimethoate* applied in late June may kill young miners. Plant resistant varieties if possible.

Boxwood psyllid

Boxwood psyllid damage.

Problem: Terminal leaves are cupped and yellowing. Buds inside the cupped leaves are  often dead. No new growth occurs on branch tips with damaged leaves. When the cupped leaves are peeled open in early May, a tiny (¹⁄₁₆-inch), grayish green, immature insect is found inside. It is usually covered with a white, waxy material. Damage begins in early spring when buds first open. Small (⅛-inch) flies with transparent wings are sometimes seen jumping on leaves or flying around the plant from late May until the end of summer. Leaves may be covered with a shiny, sticky substance or with a dark powder.

Analysis: The boxwood psyllid (*Psylla buxi*) is prevalent in temperate regions of the country where boxwood is grown. Japanese boxwood is more severely attacked than English boxwood. The immature psyllid feeds by sucking the juices from growing leaves, resulting in the yellowing and cupping. As it feeds, it secretes a white, waxy material that protects it from parasites and chemical sprays. The insect is unable to digest all the sugar in the juices, and it excretes the excess as *honeydew*, a sticky substance that covers the leaves. A black, sooty mold often grows on the honeydew. The insect matures in early summer, and the female fly lays her eggs in the base of buds in the fall, where they remain until the following spring.

Solution: Control with ORTHO Orthene Systemic Insect Control or an insecticide containing *malathion* when damage is first noticed in early spring; repeat 2 weeks later. Spray the plant thoroughly to penetrate the waxy secretions and leaf buds.

Volutella canker and blight

Volutella blight.

Problem: In the spring before new growth appears, leaves on the tips of affected 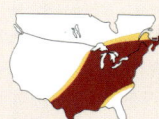 branches turn pale green, then bronze, and finally tan or straw colored. The bark may loosen and peel off at the base of infected stems and branches, revealing areas of darkened, discolored wood; the entire twig or stem eventually dies. Creamy pink pustules appear on the undersides of infected leaves that have survived the winter. Later in the season, new growth may turn yellow or tan and develop pustules, especially if the weather is wet.

Analysis: Volutella canker and blight is caused by a fungus (*Volutella buxi*) that attacks both Japanese and English boxwood. Plants are more susceptible to the disease if they have been weakened by winter injury, poor growing conditions, or insect infestation. The fungus survives the winter on infected stems, leaves, and plant debris. Wind and splashing water spread the spores to healthy leaves and twigs. In the early spring, cankers form in twigs and branches, resulting in dieback. The fungus can continue to blight new growth throughout the growing season as long as conditions remain moist.

Solution: Fungicides are not effective in controlling this disease. Remove shrubs that are dying. Prune out and destroy infected twigs and branches. Clean up accumulated plant debris. Maintain plants in good health.

161

CAMELLIA

Camellia yellow mottle leaf virus

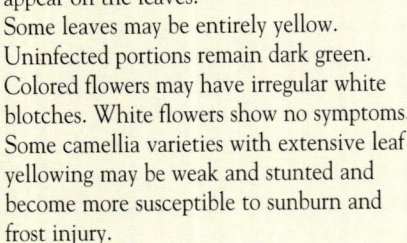

Camellia virus.

Problem: Irregular yellow splotches of various sizes and shapes appear on the leaves. Some leaves may be entirely yellow. Uninfected portions remain dark green. Colored flowers may have irregular white blotches. White flowers show no symptoms. Some camellia varieties with extensive leaf yellowing may be weak and stunted and become more susceptible to sunburn and frost injury.

Analysis: Camellia yellow mottle leaf virus is transmitted by propagating from an infected plant or by grafting from an infected plant to a healthy one. This generally occurs in the nursery where the plant is grown. Sometimes the virus disease is intentionally transmitted to create variegated flowers. The disease is usually fairly harmless unless there is extensive leaf yellowing. Yellowing results from the suppression of chlorophyll development by the virus. The leaves produce less food, causing the plant to be weakened.

Solution: Once the plant is infected, no chemical will control the virus. There is no danger of the virus spreading to other camellias unless you graft with infected plants. Remove excessively weak camellias. Buy only healthy plants.

Sooty mold

Sooty mold on Camellia japonica *foliage.*

Problem: A black, sooty mold grows on leaves and twigs. It can be completely wiped off the surfaces.

Analysis: This common black mold is caused by several species of fungi that grow on the sugary material left on plants by aphids, mealybugs, scales, whiteflies, and other insects that suck sap from the plant. The insects are unable to digest all the sugar in the sap, and they excrete the excess in a fluid called *honeydew*, which drops onto the leaves below. The honeydew may also drop out of infested trees and shrubs onto camellias growing beneath them. Sooty mold is unsightly but is fairly harmless because it does not attack the leaf directly. Extremely heavy infestations prevent light from reaching the leaf, so the leaf produces fewer nutrients and may turn yellow. The presence of sooty mold indicates that the camellia or another plant near it is infested with insects.

Solution: Hose off sooty mold or wipe it from the leaves with a wet rag, or rain will eventually wash it off. Prevent more sooty mold from growing by controlling the insect that is producing the honeydew. Inspect the leaves and twigs above the sooty mold to find out what type of insect is present. For control of aphids, mealybugs, and scales, see page 136.

Catalpa sphinx

Catalpa sphinx caterpillars (½ life size).

Problem: Small holes are chewed in the upper surfaces of leaves. Large yellow-and-black-striped caterpillars with a sharp "horn" at the tail end feed in groups on young leaves. As the caterpillars develop (to a length of 1 to 3 inches), they spread throughout the tree and feed singly on leaf edges. The tree may be completely defoliated. Damage occurs from May to August.

Analysis: The catalpa sphinx caterpillar (*Ceratomia catalpae*) is the larval stage of a large night-flying moth that is seldom seen. The moth passes the winter as a pupa in the ground. In spring, the moth emerges and the female lays her eggs on young catalpa leaves. The eggs hatch and the larvae feed for several weeks. If the larvae are left uncontrolled, they develop into moths, producing a second generation of caterpillars that may completely defoliate the tree by mid- to late summer.

Solution: Control with ORTHO Orthene Systemic Insect Control, ORTHO Isotox Insect Killer, or the bacterial insecticide *Bacillus thuringiensis* (Bt) when damage is first noticed in the spring. Repeat in midsummer if the tree becomes reinfested. In fall, clean up debris beneath the tree to reduce the number of overwintering pupae.

CELASTRUS
(Bittersweet)

Euonymus scale

Euonymus scale (4× life size).

Problem: Yellow or whitish spots appear on the upper surfaces of leaves. Leaves may drop, and the plant may become bare by midsummer. In severe cases, stems die back. The stems and undersides of leaves are covered with dark brown, oyster-shell-shaped, crusty bumps (females) or with soft, white, ridged, elongated scales (males). The bumps and scales can be scraped off.

Analysis: Euonymus scale (*Unaspis euonymi*), a serious pest of bittersweet and other ornamental shrubs and vines, is found throughout the country. The scales spend the winter on the twigs and branches of bittersweet, laying their eggs in spring. In late spring to early summer the young scales, called *crawlers*, settle on leaves or stems. The small (1/10-inch), soft-bodied young feed by inserting their mouthparts and sucking sap. The legs atrophy, and a crusty or waxy shell develops over the body. The males are white and very noticeable on the leaves and stems. Mature female scales are brown and shaped like an oyster shell. They lay their eggs underneath the shell. The cycle may be repeated up to three times during the growing season. An uncontrolled infestation may kill the plant after 2 or 3 years. Other species of scales also infest bittersweet.

Solution: Spray with ORTHO Isotox Insect Killer or ORTHO Orthene Systemic Insect Control in early summer (late spring in the South), when the young are active. The following spring, before new growth begins, spray the trunk and branches with a dormant oil spray to control overwintering insects.

CORNUS (Dogwood)

Summer leaf scorch

Summer leaf scorch on dogwood.

Problem: During hot weather, usually in July or August, leaves turn brown at the edges and  between the veins. Sometimes the whole leaf dies. Many leaves may drop during late summer. This problem is most severe on the youngest branches. Trees do not generally die. Browning and withering can develop whether the soil around the roots is moist or dry.

Analysis: Leaf scorch is caused by excessive evaporation of moisture from the leaves. In hot weather, water evaporates rapidly from the leaves. If the roots can't absorb and convey water fast enough to replenish this loss, the leaves turn brown and wither. For optimum growth, dogwoods require moist soil. Leaf scorch is most severe when water is unavailable because the soil is dry. Scorch may also develop when the soil is moist, however, if the weather is extremely hot. Drying winds, severed roots, and limited soil area can also cause scorch.

Solution: To prevent further scorch, water trees deeply during periods of hot weather to wet down the entire root space. If practical, apply a 3- to 4-inch-deep mulch over the root system. Water newly transplanted trees whenever the rootball is dry 2 inches below the surface. Scorch on trees in moist soil cannot be controlled. Plant trees adapted to your climate. In hot summer areas, plant dogwoods in partial shade.

Flatheaded apple tree borer

Borer damage on flowering dogwood.

Problem: Branches or entire trees die, usually during hot weather. Large patches of bark may  become sunken and discolored and may ooze droplets of moisture. If the tree is not killed, these bark patches may die and flake off, leaving bare wood.

Analysis: The flatheaded apple tree borer (*Chrysobothris femorata*) attacks weakened and newly planted trees. Besides dogwood, it also attacks most hardwoods. Adults emerge in late spring and early summer to lay eggs under loose bark, usually on the sunny side of the tree. These eggs hatch into white, legless larvae with flattened heads that eventually reach 1 inch in length. The larvae bore into the inner bark and sapwood, disrupting the flow of water to the top of the tree. The bark over their meandering tunnels dies and eventually sloughs off. When enough of the circumference of the trunk is tunneled, the tree dies. Newly planted dogwoods are particularly susceptible, often being killed within 2 or 3 years.

Solution: If you discover damaged bark while the larva is still present, try to kill the larva with a knife or by probing with a piece of wire. Be careful not to cause more damage than the larva is causing, though. If possible, prune out infested branches. The following spring, kill newly hatched borers with ORTHO Borer & Leaf Miner Spray. Begin spraying about the time mockorange (*Philadelphus coronarius*) blooms. Repeat two or three times at 3-week intervals. Keep trees healthy and growing vigorously. Water during drought. Protect newly planted dogwoods with a trunk wrap for their first couple of years.

CORNUS (Dogwood)
(continued)

CRATAEGUS (Hawthorn)

Dogwood borer

Dogwood borer damage. Inset: Adult (life size).

Problem: In midsummer, the leaves turn red and drop prematurely; eventually twigs or branches die back. Bark sloughs off around holes in a swollen area on the trunk or at the base of branches. Late in the summer, a fine sawdust may drop from the holes. Young trees are usually killed.

Analysis: The dogwood borer (*Synanthedon scitula*), also known as the *pecan borer*, is the larva of a brownish, ½-inch-long, clearwing moth. The borer infects flowering dogwood, pecan, and many other ornamental and fruit trees. The moths are active from May until September. The moth lays its eggs on the bark, usually near a wound or old borer injury. After the eggs hatch, the ½-inch-long white larvae with brown heads find an opening in the bark. They feed in the wood just under the bark, girdling the branches and causing the dieback. Several other borers also infest dogwood. For more information on borers, see page 147.

Solution: Spray or paint the trunk and branches with ORTHO Borer & Leaf Miner Spray, beginning in the spring when saucer magnolia and bridalwreath spirea are in bloom. Repeat four more times at monthly intervals. For more precise spray timing, use pheromone traps. Spray the first time 1 week after moths are caught. Feed the tree regularly with Scotts Evergreen, Shrub and Tree Food to maintain vigor. To prevent borer entrance, avoid pruning during the summer months when the moths are present, and avoid wounding the trunks and branches. You can also reduce damage by inserting a fine wire up the entry hole to kill the larva.

Fire blight

Fire blight, a destructive bacterial disease.

Problem: Blossoms and leaves of infected twigs suddenly wilt and turn black as if scorched by fire. Brown or blackened leaves cling to the branches. The bark at the base of the blighted twig becomes water-soaked, then dark, sunken, and dry; cracks may develop at the edge of the sunken area. In warm, moist spring weather, drops of brown ooze appear on the sunken bark. Young trees may die.

Analysis: Fire blight is caused by a bacterium (*Erwinia amylovora*) that is destructive to hawthorn and many other related plants. (For a list of susceptible plants, see page 341.) The bacteria spend the winter in the sunken cankers on the branches. In spring, the bacteria ooze out of the cankers and attract bees and other insects. The bacteria spread rapidly through the plant tissue in warm (65°F or higher), humid weather. Insects visiting these infected blossoms later carry bacteria-laden nectar to healthy blossoms. Rain, wind, and tools may also spread the bacteria.

Solution: During spring and summer, prune out infected branches 12 to 15 inches beyond any visible discoloration and destroy them. Disinfect pruning shears after each cut by dipping in a solution of 1 part chlorine bleach and 9 parts water. A protective spray of a bactericide containing basic copper sulfate or *streptomycin* applied before bud-break in the spring will help prevent infection. Repeat at intervals of 5 to 7 days until the end of bloom. In the fall, prune out any remaining infected branches. When planting new trees, use resistant varieties.

Leaf Spots

Entomosporium *leaf spot.*

Problem: Spots and blotches appear on the leaves. The spots may be red, purple, yellow, brown, or black. They range in size from barely visible to ¼ inch in diameter. Several spots may join to form blotches. Infected leaves may die and drop, and if spotting is severe, the tree may defoliate prematurely. Leaf spotting is most severe in moist, humid weather.

Analysis: Several fungi, including *Fabraea thuemenii*, cause leaf spots on hawthorn. These spots are unsightly but rarely harmful to the plant. Severe, recurrent infection can cause repeated defoliation that may weaken the tree and reduce its flowering potential, however. The fungi are spread by wind and splashing water. Spots develop where the fungi enter the leaf tissue. The fungi survive the winter on twigs and in fallen leaves and plant debris. Most leaf-spotting fungi do their greatest damage in moist, mild to warm weather (between 50° and 85°F).

Solution: There is no way to get rid of spots once leaves are infected. To help prevent spotting the following year, clean up and destroy fallen leaves and other plant debris. In the spring when the leaves emerge, spray trees with ORTHO Multi-Purpose Fungicide Daconil 2787® Plant Disease Control or with a fungicide containing *captan*. Respray when the leaves are half grown and again when they are fully grown. Continue spraying at intervals of 10 to 14 days for as long as wet weather continues.

CYPRESS FAMILY

GROWING GUIDE

Adaptation:
Chamaecyparis species (falsecypress):
Zones 4 through 9. Plants should be protected from hot, dry winds.
Cupressocyparis and Cupressus species (cypress): Zones 5 through 10.
Thuja species (arborvitae, cedar):
Zones 2 through 9.
To determine your climate zone, see the map on page 336.

Light: Full sun.

Soil:
Chamaecyparis species: Needs good drainage.
Cupressocyparis and Cupressus species: Tolerate a wide variety of soils.
Thuja species: Any good garden soil.

Fertilizer: Fertilize with Scotts Evergreen, Shrub and Tree Food according to label directions.

Water: How much: Apply enough water to wet the soil 6 to 12 inches deep.
How often:
Chamaecyparis species: Grows best in moist soil, but will tolerate some drought. Water when the soil is moist, but not wet, 4 inches below the surface.
Cupressocyparis species: Water when the soil is barely moist 4 inches below the surface.
Cupressus species: Grows best if kept on the dry side. Water when the soil is dry 4 inches below the surface.
Thuja species: Will tolerate wet soil, but does best in moist soil. Water when the soil is moist, but not wet, 4 inches below the surface.

Twig and needle blight

Twig and needle blight damage on Western red cedar.

Problem: Needles, twigs, and branches turn brown. In some cases, the upper branches die  from the tips back; in other cases, the lower two-thirds of the plant dies. The needles often drop in late summer, starting at the tips and leaving the infected branches bare. Sometimes minute black dots appear on the dead needles and stems. This disease is most serious in wet weather or in shady locations. Plants may be killed.

Analysis: Several fungi cause twig and needle blight on plants in the cypress family. During wet weather, spores germinate on twigs and spread into the needles and twigs above and below the point of entrance, killing them. Reinfection may continue until the whole plant dies, or the plant may persist for many years in an unsightly condition. With some fungi, black spore-producing bodies develop and spend the winter on dead needles.

Solution: Prune out and destroy infected branches below the line between diseased and healthy tissue, making the cut into live tissue. Valuable specimens can be sprayed with a fungicide containing basic copper sulfate at weekly intervals throughout the growing season until new growth stops. Avoid overhead watering. Plant trees in areas with good air circulation and full sun.

Leaf browning and shedding

Leaf browning on cedar.

Problem: The older leaves, on the inside of the tree nearest the trunk, turn brown and drop. This condition may develop in a few days or over several weeks, in either spring or fall.

Analysis: Leaf browning and shedding is a natural process of coniferous evergreens similar to the dropping of leaves of deciduous trees. It is usually more pronounced on arborvitae (*Thuja*) than on other plants in the cypress family. Sometimes it takes place every year; in other cases, it occurs every second or third year. When growing conditions have been favorable the previous season, leaf shedding occurs over several weeks and is less noticeable. If the plant has been exposed to unfavorable conditions during the growing season, such as drying or a spider mite or insect infestation, however, leaf drop develops within a few days. Leaf drop is also caused by new growth shading older interior growth.

Solution: No chemical controls are necessary. Water regularly and feed with Scotts Evergreen, Shrub and Tree Food according to label directions. Provide full sun. Check plants for insects and mites during the growing season.

CYPRESS FAMILY *(continued)*

Leafminers and tip moths

Leafminer damage to arborvitae.

Problem: Leaf tips turn yellow, then brown and dry, contrasting sharply with the healthy green foliage. Damage is most severe for plants growing in shady areas. When a yellow leaf is torn open, a small (⅓-inch-long), greenish caterpillar with a dark head may be found inside. Gray or brownish moths with a ⅓-inch wingspread may be seen flying around the plant in April, May, or June.

Analysis: Several species of insects known as leafminers (*Argyresthia* species) in the eastern United States and tip moths on the West Coast infest arborvitae, cypress, and juniper. Damage is unsightly, but plants may lose more than half of their foliage and still survive. The larvae spend the winter inside the leaf tips. When the weather warms in late spring, adult moths emerge and lay eggs on the leaves. The eggs hatch, and the larvae tunnel into the leaf tips, devouring the green tissue. The tips above the point of entry yellow and die. The larvae feed until late fall or through the winter until early spring.

Solution: Spray with ORTHO Orthene Systemic Insect Control, ORTHO Isotox Insect Killer, or an insecticide containing *diazinon* when eggs are hatching in early summer, about the time black locust is in bloom. To kill adults, spray a second time a month later, when mountain laurel and Washington hawthorn are in bloom. Trim and destroy infested leaves in fall and spring.

Winter injury

Winter injury to arborvitae.

Problem: Leaves turn yellow at first, then rusty brown and dry. Twigs and branches may die back. The tree is growing in a climate where cold, dry, windy days are common or where plants may be exposed to late fall or early spring freezes. The soil may be frozen.

Analysis: Arborvitae are damaged by cold, drying winter winds, especially if temperatures are below freezing. These trees are commonly planted as windbreaks and in exposed areas where growing conditions may be unfavorable. Moisture is lost from the leaves more rapidly than it can be replaced by the root system. Cells in the leaves dry out and die. This condition is most pronounced when water is unavailable because the soil is dry or frozen. Leaves, along with twigs and branches, also die during early fall or late spring freezes when the plant is growing. Young succulent growth cannot withstand freezing temperatures.

Solution: Prune out dead twigs and branches. Provide shelter for plants growing in extremely cold areas. To avoid succulent growth in the fall, do not fertilize late in the season. During a dry fall, irrigate plants thoroughly to reduce winter injury. One or more applications of an antidesiccant spray beginning in late fall may help reduce damage.

EUONYMUS

Powdery mildew

Powdery mildew on Euonymus fortunei.

Problem: The surfaces of leaves are covered with a thin layer or irregular patches of a grayish white powdery mildew. Infected leaves are yellow and may drop prematurely. In late summer, tiny black dots (spore-producing bodies) are scattered over the white patches like ground pepper.

Analysis: Powdery mildew is caused by two species of fungi (*Oidium euonymi-japonici* and *Microsphaera alni*) that thrive in both humid and dry weather. Fungal strands and spores make up powdery patches or a thin powdery layer. The spores are spread by wind to healthy plants. The fungus saps plant nutrients, causing leaf yellowing and sometimes death of the leaf. In late summer and fall, the fungus forms small, black, spore-producing bodies that are dormant during the winter but produce spores to reinfect new plants the following spring. The fungus is especially devastating in low-light situations and is generally most severe in late summer and fall. Since *Microsphaera alni* attacks many kinds of plants, the fungus from a diseased plant may infect other types of plants in the garden.

Solution: Spray with ORTHO RosePride Funginex Rose & Shrub Disease Control or ORTHO RosePride Orthenex Insect & Disease Control when mildew is first noticed. Clean up plant debris in late summer.

Euonymus scale

Euonymus scale (2× life size).

Problem: Yellow or whitish spots appear on the leaves. Most of the leaves may drop by midsummer. Branches often die, and heavy infestations may kill the plant. The stems and lower surfaces of leaves are covered with somewhat flattened white, scaly bumps and dark brown oyster-shell-shaped bumps. In severe cases, the whole plant appears white.

Analysis: Many species of scales infest euonymus, but the most common and destructive is euonymus scale (*Unaspis euonymi*). It is especially damaging to evergreen euonymus. The narrow white scales are males, and the larger brown scales are females. The females spend the winter on the plant and lay their eggs in the spring. In late spring to early summer the young scales, called *crawlers*, settle on the leaves and twigs or are blown by the wind to other susceptible plants. The small (⅒-inch), soft-bodied young suck sap from the plant. The legs atrophy, and a shell develops, brown over the female or white over the male. Females lay their eggs underneath their shells.

Solution: Euonymus scales may be hard to detect until after they have caused serious damage. Check the plant periodically for yellow spotting and scales. During the winter, check the base of the plant for hidden overwintering females. Spray with ORTHO Isotox Insect Killer, ORTHO Orthene Systemic Insect Control, or ORTHO Volck Oil Spray in early June and mid-July. In the South, a third application may be necessary in early September. To control scales during the dormant season, spray with ORTHO Volck Oil Spray. For very heavy infestations, cut plants to the ground and spray new growth in June.

Crown gall

Crown gall on Euonumus fortunei.

Problem: Large, corky galls up to several inches in diameter appear at the base of the plant and on the stems and roots. The galls are rounded, with a rough, irregular surface. Plants with numerous galls are weak; growth is slowed and leaves turn yellow. Branches may die back.

Analysis: Crown gall is caused by soil-inhabiting bacteria (*Agrobacterium tumefaciens*) that infect many ornamentals, fruits, and vegetables in the garden. The bacteria are often brought to a garden initially on the roots of an infected plant and are spread with the soil and by contaminated pruning tools. The bacteria enter the shrub through wounds in the roots or at the base of the stem (the crown). They produce a substance that stimulates rapid cell growth in the plant, causing gall formation on the roots, crown, and sometimes branches. The galls disrupt the flow of water and nutrients up the roots and stems, weakening and stunting the top of the plant. Galls do not usually cause the shrub to die.

Solution: Crown gall cannot be eliminated from the shrub. Infected plants may survive many years, however. To improve the appearance of the plant, prune out and destroy affected stems below the galled area. Disinfect pruning shears after each cut by dipping in a solution of 1 part chlorine bleach and 9 parts water. Destroy severely infected shrubs. The bacteria will remain in the soil for 2 to 3 years. If you wish to replace the shrub soon, plant only resistant species. For a list of plants resistant to crown gall, see page 354.

Scales

Brown scale (4× life size).

Problem: Brown, crusty bumps; thick, white, waxy bumps; or clusters of somewhat flattened yellow or whitish scaly bumps cover the fronds. The bumps can be scraped or picked off. Fronds may turn yellow, and leaflets may drop. In some cases, a shiny, sticky substance coats the fronds. Scales are sometimes mistaken for reproductive spores produced by the fern. The round, flat, sometimes hairy spores are found only on the undersides of fronds, spaced at regular intervals. They are difficult to pick or scrape off.

Analysis: Many types of scales infest ferns. They lay their eggs on the fronds, and in spring to midsummer the young scales, called *crawlers*, settle down to feed. These small (⅒-inch), soft-bodied young feed by sucking sap from the plant. The legs usually atrophy, and a hard crusty or waxy shell develops over the body. Mature female scales lay their eggs underneath their shells. Some species of scales are unable to digest fully all the sugar in the sap, and they excrete the excess in a fluid called *honeydew*, which coats the fronds.

Solution: Spray with an insecticide containing *malathion* or with an insecticidal soap when the young are active. Some ferns are sensitive to *malathion*. Test the spray on a frond before spraying the entire plant. Wait a few days to see if the area turns brown. Discard severely infested plants.

FERNS (continued)

GLEDITSIA (Honeylocust)

Cottony cushion scales and mealybugs

Mealybugs (⅓ life size).

Problem: Leaflets or stems are covered with white, cottony, cushionlike masses. Leaflets turn yellow and may drop; fronds may die.

Analysis: Cottony cushion scales and mealybugs infest many plants in the garden. Mealybugs are usually found outdoors only in warmer climates, but they may be found on indoor plants anywhere. The visual similarities between these insects make identification difficult. They are conspicuous in late spring and summer because the females are covered with a white, cottony egg sac, containing up to 2,500 eggs. Females lay their egg masses on leaves and stems. The young insects that hatch from these eggs are yellowish brown to green. They feed throughout the summer, causing damage by withdrawing plant sap from the ferns. Some species of scales and mealybugs are unable to digest fully all the sugar in the plant sap, and they excrete the excess in a sticky fluid called *honeydew*.

Solution: Spray with an insecticide containing *carbaryl* or an insecticidal soap when insects are first noticed. Some ferns may be damaged. Test the spray on a small portion of the plant before using. Wait a few days to see if the area turns brown. Discard severely infested plants.

Scorch

Scorch on fern frond.

Problem: The tips and edges of fronds turn brown or black and die. Entire fronds may wilt, turn yellow or brown, and die. The problem usually occurs first on the youngest fronds. This condition usually develops when the soil is dry but may also occur when the soil is moist.

Analysis: Scorch is caused by excessive evaporation of moisture from the fronds. In hot weather, water evaporates rapidly from the fronds. Water loss is especially heavy when the weather is also dry and windy. If the roots can't absorb and convey water fast enough to replenish this loss, the fronds turn brown or black and wither. Scorch usually occurs when the soil is allowed to dry out, but ferns can also suffer from scorch when the soil is moist if the weather is exceptionally hot and dry. Winds, severed roots, limited soil area, and low temperatures can also cause scorch.

Solution: To prevent further scorch, water plants thoroughly. Ferns need to be kept moist. They must be watered frequently and deep enough so that the soil doesn't dry out. During periods of exceptionally hot, dry weather, keep ferns wet by gently hosing down or sprinkling the fronds several times a day to reduce scorch damage. Grow ferns in shady areas that are protected from strong winds. Keep beds mulched.

Mimosa webworm

Mimosa webworm damage.

Problem: Small clumps of leaves bound together with silk threads are scattered over the tree. The upper surfaces of the leaves are skeletonized. The leaves turn brown and die, causing infested trees to look as if they have been scorched by fire. Small (up to 1-inch), pale gray or brown caterpillars with five white stripes are feeding inside the silken nests or hanging from the trees on threads. Small trees or the 'Sunburst' variety of thornless honeylocust may be completely defoliated by late summer.

Analysis: The mimosa webworm (*Homadaula anisocentra*) feeds only on honeylocust and mimosa trees. The webworm passes the winter as a pupa in a white silken cocoon. The cocoon is found in sheltered places such as crevices in the bark of the infested tree, in soil and debris beneath the tree, or under the siding of nearby buildings. The moth emerges in the spring and lays her eggs. When the eggs hatch, the larvae feed on the leaflets for several weeks. A second generation of webworms hatches in August. The larvae of this second generation are more damaging because they are usually quite numerous. In warmer areas, a third generation hatches in September.

Solution: When webbing first appears, spray with ORTHO Isotox Insect Killer, ORTHO Bug-B-Gon Ready-Spray, ORTHO Orthene Systemic Insect Control, or the bacterial insecticide *Bacillus thuringiensis* (Bt). (Bt is only effective when webbing first appears.) Repeat in August, when Queen Anne's lace is blooming. Use high spray pressures to penetrate the webs and cover the tree thoroughly. In the fall, rake up and burn debris from under infested trees, or turn over the soil and bury the leaves.

Honeylocust pod gall midge

Pod gall midge damage on honeylocust.

Problem: Green, globular, podlike galls, ⅛ inch in diameter, develop on new leaflets in spring and early summer. The galls turn reddish, then brown, and many of the infested leaflets drop. Twigs or branches sometimes die back after several years of infestation. One to several whitish larvae, ¼ inch long, are feeding inside the gall.

Analysis: The larva of the honeylocust pod gall midge (*Dasineura gleditschiae*), a tiny orange to black fly, causes unsightly galls on honeylocust trees, especially the thornless varieties. The adult female midge begins laying eggs on new leaflets in the spring. When the eggs hatch, the larvae feed on the tissue, causing the leaflet to fold over them and form a pod gall. As the larvae develop inside, the galls turn brown. The flies emerge to lay more eggs. Honeylocust produces new leaves over a long period, so the cycle may repeat itself up to seven times annually. The galls do not usually damage the tree, but the ornamental value of the tree is reduced when the galled leaflets dry up and drop prematurely. Twigs sometimes die back after repeated attack, but new shoots normally form at the base of the dead twigs.

Solution: Once formed, the galls cannot be removed. Prune off dead twigs. Spray with an insecticide containing *diazinon* or *malathion*. Repeat treatments every 7 to 10 days until no more galls are formed. The following spring, begin spraying as soon as the honeylocust begins to leaf out.

Failure to bloom

Flowerless bigleaf hydrangea.

Problem: Hydrangeas fail to produce blooms in the spring.

Analysis: Hydrangeas may fail to bloom for several reasons.
1. Cold injury: Extreme winter temperatures or late spring cold snaps will kill hydrangea flower buds, which form during the late summer or fall.
2. Improper pruning: Because some hydrangea species produce flower buds in the late summer or fall, pruning in the winter or spring will remove these potential flowers.
3. Too much shade: Hydrangeas growing in deep shade may fail to form flower buds.

Solution: The numbered solutions below correspond to the numbered items in the analysis.
1. Plant hydrangeas in a protected spot in the garden. Protect them by placing a wire cylinder around each plant and then either filling the cylinder with loosely packed straw or covering it with burlap. Protect hydrangeas grown in containers by moving them to a cool basement during the winter.
2. Prune hydrangeas after they have finished blooming by cutting back the longer branches.
3. Expose the plants to brighter light by pruning away some of the surrounding vegetation, or transplant hydrangeas to a location that receives filtered or half-day sun or, in cool-winter areas, in full sun.

Sunburn

Sunburn on hydrangea.

Problem: During warm, sunny weather, leaves on the outside of the plant turn yellowish or brown in the center of the leaf tissue. Some leaves may drop.

Analysis: Hydrangeas are shade plants; their leaves are sensitive to the heat of the sun. Outside leaves facing the light turn yellow or brown when the shrub is planted in full sun. The injury is unsightly but does not damage the plant. Plants are more susceptible to sunburn when in dry soil. Hydrangea leaves may also scorch (die and turn brown around the edges) during hot weather. For more information on scorch, see page 151.

Solution: Provide some shade where the plant is growing, or move the plant to a shaded location. Keep plants adequately watered.

JUNIPERUS (Juniper)

GROWING GUIDE

Adaptation: Throughout North America

Light: Full sun (or partial shade in hot climates).

Soil: Junipers prefer sandy, well-drained soil, but they will thrive in almost any type of garden soil.

Fertilizer: Fertilize with Scotts Evergreen, Shrub and Tree Food according to label directions.

Water: How much: Apply enough water to wet the soil 1 to 3 feet deep, depending on the size of the juniper.
How often: Young junipers should be watered when the soil 4 inches below the surface is just barely moist. Established plants are drought resistant and require summer watering only in the hottest climates. Water when the soil 4 inches below the surface is dry. Do not plant junipers where they will receive water meant for a lawn.

Pruning: Some junipers grow tall (to 90 feet) or spread up to 15 feet. It is best to use a variety that will fit the area to be planted rather than depending on pruning. Older junipers will not sprout new growth from older wood, so never remove all the foliage from a branch. If a shrub is to be sheared, begin when the plant is young, and shear lightly and regularly.

Needle browning

Drought damage.

Salt burn damage.

Problem: Needles turn yellow and then brown and dry, either on the outer, lower branches; on the inside of the plant nearest the trunk; from the top of the plant down; or from the tips back. Sometimes only a branch or one side of the plant is affected. In other cases, the whole plant turns brown.

Analysis: Several different conditions may cause the needles to turn brown on junipers.

Solution: The numbered solutions below refer to the numbered items in the analysis.

1. Dog urine: When dogs urinate on the plant, the foliage on the outer, lower branches turns yellow, then brown, as if scorched. The salts in the urine burn the foliage.

1. If you suspect that a dog has recently urinated on a juniper, wash the foliage and thoroughly soak the ground around the plant to dilute the salts in the urine.

2. Natural leaf browning and shedding: The older needles, on the inside of the plant nearest the trunk, turn brown and drop off in the spring or fall. This is a natural process similar to the dropping of leaves of deciduous plants.

2. No controls are necessary for natural leaf browning and shedding.

3. Drought or winter injury: When the plant is damaged by drought or winter injury, needles gradually turn yellow, then brown or reddish from the top of the plant down and from the tips of the branches back. This may happen when the soil is dry and the plant is not getting enough water or when the soil is frozen in the winter.

3. Prune out dead twigs and branches. Provide adequate water during periods of extended drought, and shelter plants growing in windy locations. Water in late fall or early winter, if necessary, to ensure adequate soil moisture during the winter. Mulch plants after they are dormant to reduce the depth of frost penetration into the soil. Do not plant junipers in areas where they are not adapted.

4. Salt burn: Needles turn brown from the tips back. This condition may develop on one side of the plant only or on the whole plant. Salt burn is common in alkaline soils, when water has a high salt content, in soils with poor drainage, in overfertilized soils, or along roadsides where winter runoff contains road salts. The salts in the soil inhibit water and nutrient uptake, causing the needles to turn brown. Similar symptoms can occur in oceanfront plantings from wind-borne salt mist.

4. Prune off badly damaged areas. Avoid new injury by heavily irrigating plants once during the growing season. If you suspect that your water contains salts, have it analyzed through your county extension office or local water department. Do not overfertilize plants; follow package directions. Avoid planting in areas where road or sea salt may be a problem. Occasionally hose down oceanfront plantings to remove salt accumulations on the foliage.

Phomopsis twig blight

Phomopsis twig blight in juniper.

Problem: In the spring, needles, twigs, and smaller branches turn light brown to  reddish brown, then gray, gradually dying from the tips back. Plants less than 5 years old are often killed. Minute black dots may appear on the needles and stems when the needles have dried and turned grayish. The disease is most serious during wet weather or in shady, moist locations. This problem sometimes resembles drought damage. The difference is that the border between healthy tissue and dead tissue is sharp with this disease but gradual with drought. In addition, the disease is found only on isolated branches rather than uniformly throughout the plant, as with drought.

Analysis: Phomopsis twig blight is caused by a fungus (*Phomopsis juniperovora*). It is highly destructive to junipers, cryptomeria, chamaecyparis, and arborvitae throughout most of the United States. The spores are spread by splashing rain, overhead watering, insects, and tools. The fungus enters through wounds or healthy tissue, killing the stem and needles above and below the point of entrance. Black spore-producing structures develop and overwinter on dead needles.

Solution: Spray with a fungicide containing *mancozeb* when symptoms first appear in spring. Repeat at 7- to 10-day intervals as long as necessary. Prune out and destroy infected branches below the line between diseased and healthy tissue, making the cut into live tissue. Plant trees in areas with good air circulation and full sun. Plant resistant varieties (see page 356).

Root and crown rots

Juniper dieback.

Dieback caused by Phytophthora root rot.

Problem: Normal foliage color dulls, and the plant loses vigor. The foliage may wilt, or it may turn yellow or light brown. Major branches or the entire plant may die. The plant sometimes lives for many months in a weakened condition, or it may die quickly. The roots and lower stems are brownish and the roots are often decayed. There may be fine woolly brown strands on the roots and white powdery spores on the soil surface; or there may be fan-shaped plaques of white strands between the bark and wood of the roots and lower stems. Mushrooms may appear at the base of the plant in the fall.

Analysis: Root and crown rots on junipers are caused by several fungi that live in the soil and on living roots.

1. *Phytophthora* species: These fungi cause browning and decay on the roots, and browning of the lower stems. The plants usually die slowly, but young plants may wilt and die rapidly. The disease is most prevalent in heavy, waterlogged soils.

2. *Phymatotrichum omnivorum*: This fungus, also known as *cotton root rot* or *Texas root rot*, is a severe problem on many plants in the Southwest. The plant often wilts and dies suddenly. Older plants may die more slowly, showing general decline and dieback symptoms. Brown strands form on the roots and white powdery spores on the soil. The disease is most severe in heavy alkaline soils.

3. *Armillaria mellea*: This disease, also known as *shoestring root rot, mushroom root rot,* or *oak root fungus*, is identified by the presence of fan-shaped plaques of white fungal strands between the bark and the wood of the roots and lower stems. This fungus grows rapidly under wet conditions. For more information on *Armillaria*, see page 187.

Solution: The numbered solutions below correspond to the numbered items in the analysis.

1. Remove dead and dying plants. When replanting, use plants that are resistant to phytophthora. (For a list of resistant trees and shrubs, see page 342.) Improve soil drainage. Avoid overwatering junipers.

2. Remove dead and dying plants. When replanting, buy only resistant varieties. (For a list of resistant trees and shrubs, see page 345.) Before planting, increase the soil acidity by adding 1 pound of ammonium sulfate for every 10 square feet of soil. Make a circular ridge around the planting area and fill the basin with 4 inches of water. Repeat the treatment in 5 to 10 days. Improve drainage.

3. Remove dead plants. The life of a newly infected plant may be prolonged if the disease has not reached the lower stems. Expose the base of the plant to air for several months by removing several inches of soil. Prune off diseased roots. When replanting, use only resistant varieties. (For a list of plants resistant to *Armillaria*, see page 345.)

JUNIPERUS (Juniper) (continued)

Spruce spider mite and twospotted mite

Spider mite damage (on left).

Problem: The needles are dirty and stippled yellow. A fine silken webbing may be on the twigs. Needles may turn brown and fall off. To determine if a plant is infested with mites, hold a sheet of white paper underneath some stippled needles and tap the foliage sharply. Minute specks the size of pepper grains will drop to the paper and begin to crawl around. The pests are easily seen against the white background.

Analysis: Spider mites, related to spiders, are major pests of junipers and other evergreen trees and shrubs. They cause damage by sucking sap from the needles. As a result of their feeding, the plant's green leaf pigment disappears, producing the stippled appearance. Spruce spider mites (*Oligonychus ununguis*) are more prolific in cooler weather. They feed and reproduce primarily during spring and in some cases fall. By the onset of hot weather (70°F and up), these mites have caused their maximum damage. Twospotted mites (*Tetranychus urticae*) develop rapidly in hot, dry weather (70°F and up), so they can build up to tremendous numbers by midsummer.

Solution: Spray with ORTHO Bug-B-Gon Ready-Spray, ORTHO Isotox Insect Killer, or ORTHO Orthene Systemic Insect Control when damage is first noticed. Wet the foliage thoroughly, covering the tops and bottoms of branches and the plant interior. Repeat the application two more times at intervals of 7 days. Hose down plants frequently to knock off webs and mites.

Juniper scale

Juniper scale (4× life size).

Problem: The tree or shrub looks gray and off-color and has no new growth. Eventually the needles turn yellow. Branches and possibly the whole plant die back. The foliage is covered with clusters of tiny (⅛-inch), somewhat flattened, yellow-and-white scaly bumps. A shiny, sticky substance may coat the needles. A black, sooty mold often grows on the sticky substance.

Analysis: Juniper scale (*Carulaspis juniperi*) is found throughout the country on many types of juniper and also on cypress (*Cupressus* species only) and incense cedar. The female scales spend the winter on the plant. They lay their eggs in the spring, and in midsummer (late spring in the South) the new generation, called *crawlers*, settles on the needles. These small (¹⁄₁₀-inch), soft-bodied young feed by sucking sap from the plant. The legs atrophy, and a crusty shell develops over the body. Mature female scales lay their eggs underneath the shell. Juniper scales are unable to digest fully all the sugar in the plant sap and excrete the excess in a sticky fluid called *honeydew*. A sooty mold may develop on the honeydew. An uncontrolled infestation of scales may kill the plant in 2 or 3 seasons.

Solution: Prune out heavily infested branches. Spray with ORTHO Bug-B-Gon Ready-Spray, ORTHO Orthene Systemic Insect Control, or ORTHO Volck Oil Spray when the young are active in late spring or midsummer, about the time mountain laurel and Washington hawthorn are in bloom. Early the following spring, before new growth begins, spray the trunk and branches with ORTHO Volck Oil Spray.

Cedar-apple rust

Cedar-apple rust, spore-producing "horns".

Problem: In spring or early summer, brownish green swellings appear on the upper surfaces of needles. The galls enlarge until, by fall, they range in size from 1 to 2 inches in diameter. The galls turn chocolate brown and are covered with small circular depressions. The following spring, during warm, rainy weather, the small depressions swell and produce yellow or orange jellylike "horns" up to ¾ inch long. The galls eventually die, but they remain attached to the tree for a year or more. Infected twigs usually die.

Analysis: Cedar-apple rust is caused by one of several fungi (*Gymnosporangium* species) that infect both juniper and apple trees. It cannot spread from juniper to juniper, or from apple to apple, but alternates between the two. Wind-borne spores from apple leaves infect juniper needles in the summer. The fungus grows very little until the following spring, when the galls begin to form. The second spring, spores from the orange "horns" are carried by the wind to infect apple trees. Later, orange spots appear on the apples and upper surfaces of the leaves. (For more information on cedar-apple rust on apples, see page 174.) In spring or summer, spores are released and carried by the wind back to junipers. The entire cycle takes 18 to 20 months on juniper plus 4 to 6 months on apple trees.

Solution: Remove galls and destroy. When possible, do not plant junipers and apple trees within several hundred yards of one another. Spraying junipers with a fungicide containing *ferbam* or *mancozeb* may help prevent new infections from apple trees.

LIGUSTRUM (Privet)

Privet rust mite and privet mite

Privet rust mite damage.

Problem: The leaves turn bronze, brown, or yellow and cup downward. They may have a silvery stipple. Leaves drop prematurely, sometimes before discoloration occurs. The plant is weak and often stunted.

Analysis: Both privet rust mites (*Aculus ligustri*) and privet mites (*Brevipalpus obovatus*) are extremely small and generally cannot be seen without magnification. They cause damage by sucking sap from the leaf tissue. They feed on the undersides of leaves. As a result of their feeding, the leaf cups or curls under and often drops from the plant. The privet rust mite may feed only on the cells in the surface of the leaf, causing a bronze russeting or browning. The green leaf pigment is unaffected. Or the mite may feed deeper in the tissue, resulting in leaf yellowing caused by the disappearance of chlorophyll. Some types of privet drop their leaves before discoloration develops. The rust mite is most prolific during cool weather. It feeds and reproduces primarily during spring and fall. The privet mite is usually active throughout the growing season. By midsummer it may build up to tremendous numbers.

Solution: Spray with an insecticide containing *carbaryl* when damage is first noticed. Spray the foliage thoroughly, being sure to cover both the upper and lower surfaces of the leaves. Repeat the application two more times at intervals of 7 to 10 days.

Scales

Scales (2× life size).

Problem: Brownish crusty bumps or clusters of somewhat flattened yellowish, brown, reddish, gray, or white scaly bumps cover the trunk, the branches, or the undersides of leaves. The bumps can be scraped or picked off. Leaves turn yellow and may drop, and twigs or branches may die back. A shiny, sticky substance often coats the leaves. A black, sooty mold may grow on the sticky substance.

Analysis: Many types of scales infest privet. They lay their eggs on the leaves or bark, and in spring to midsummer the young scales, called *crawlers*, settle on leaves, twigs, or the trunk. These small (1/10-inch), soft-bodied young feed by sucking sap from the plant. The legs usually atrophy, and a hard crusty shell develops over the body. Mature female scales lay their eggs underneath their shell. Some species of scales that infest privet are unable to digest fully all the sugar in the plant sap, and they excrete the excess in a sticky fluid called *honeydew*. A sooty mold may develop on the honeydew, causing the privet leaves to appear black and dirty. An uncontrolled infestation of scales may kill twigs or branches after 2 or 3 seasons.

Solution: Spray with ORTHO Isotox Insect Killer when the young are active. Early the following spring, before new growth begins, spray the trunk and branches with ORTHO Volck Oil Spray to control overwintering insects.

MALUS (Crabapple)

GROWING GUIDE

Adaptation: Zones 2 through 9. To determine your zone, see the map on page 336. Most are not adapted to mild winter or desert areas.

Flowering Time: Spring.

Light: Full sun.

Soil: Adapted to a wide variety of soils, but grows best in rich, well-drained garden soil.

Fertilizer: Fertilize with Scotts Evergreen, Shrub and Tree Food according to label directions.

Water: How much: Apply enough water to wet the soil 3 to 4 feet deep.
How often: Water when the soil is barely moist 4 inches below the surface.

Insects and Diseases: Problems of crabapple are very similar to those of apple. For more information about plant problems of crabapple, see insects and diseases of apples in the section starting on page 214.

MALUS (Crabapple) *(continued)*

Fire blight

Fire blight.

Problem: The blossoms and leaves of infected twigs suddenly wilt and turn black as if scorched by fire. The leaves curl and hang downward. The bark at the base of the blighted twig becomes water-soaked, then dark, sunken, and dry; cracks may develop at the edge of the sunken area. In warm, moist spring weather, drops of brown ooze appear on the sunken bark. Young trees may die.

Analysis: Fire blight is caused by a bacterium (*Erwinia amylovora*) that is destructive to many trees and shrubs. (For a list of susceptible plants, see page 341.) The bacteria spend the winter in the sunken areas (cankers) on the branches. In the spring, the bacteria ooze out of the cankers. Insects that are attracted to this ooze become smeared with it, and when the insects visit a flower for nectar, they infect the plant with the bacteria. The bacteria spread rapidly through the plant tissue in warm (65°F or higher), humid weather. Insects visiting infected blossoms later carry bacteria-laden nectar to healthy blossoms. Rain, wind, and tools may also spread the bacteria. Tender or damaged leaves may be infected in midsummer.

Solution: Prune out infected branches 12 to 15 inches beyond any visible discoloration and destroy them. Disinfect pruning shears after each cut by dipping in a solution of 1 part chlorine bleach and 9 parts water. A protective spray of a bactericide containing fixed copper or *streptomycin* applied before bud-break in the spring will help prevent infection. Repeat at intervals of 5 to 7 days until the end of bloom. Avoid overfertilization, which causes lush growth more susceptible to fire blight.

Apple scab

Scab on crabapple.

Problem: Olive, velvety spots, ¼ inch or more in diameter, appear on the leaves. The tissue around the spots may be puckered. The leaves often turn yellow and drop. In a wet year, the tree may lose all of its leaves by midsummer. The fruit and twigs develop circular, rough-surfaced, olive green spots that eventually turn corky and black. The fruit is usually deformed.

Analysis: Apple scab is caused by a fungus (*Venturia inaequalis*). It is a serious problem on crabapples and apples in areas where spring weather is humid with temperatures ranging from 60° to 70°F. The fungus spends the winter in infected fallen leaves. In the spring, spore-producing structures in the dead leaves continuously discharge spores into the air. The spores are blown by the wind to new leaves and flower buds. If water is on the tissue surface, the fungus infects the tissue and a spot develops. More spores are produced from these spots and from twig infections from the previous year. The spores are splashed by the rain to infect new leaf and fruit surfaces. As temperatures increase during the summer, the fungus becomes less active.

Solution: To obtain adequate control of scab, apply protective sprays starting as soon as bud growth begins in the spring. Spray with ORTHO Multi-Purpose Fungicide Daconil 2787® Plant Disease Control. Repeat five to eight times at intervals of 7 to 10 days. Rake up and destroy infected leaves and fruit in the fall. When planting new trees, use resistant varieties.

Cedar-apple rust

Cedar-apple rust, a fungal disease.

Problem: Pale yellow spots appear on the leaves and fruit in mid- to late spring. These spots gradually enlarge, turn orange, and develop minute black dots. Small (¹⁄₁₆-inch) cups with fringed edges form on the lower surfaces of leaves. Infected leaves and fruit may drop prematurely; the fruit is often deformed.

Analysis: Cedar-apple rust is caused by a fungus called *Gymnosporangium juniperi-virginianae* that affects both crabapples and certain species of juniper and red cedar. This disease cannot spread from crabapple to crabapple, or from juniper to juniper, but must alternate between the two. In the spring, spores from brown and orange galls on juniper or cedar are blown up to 3 miles to crabapple trees. During mild, wet weather, the spores germinate and infect the leaves and fruit, causing spotting and premature leaf and fruit drop. During the summer, spores are produced in small cups on the undersides of leaves. These spores are blown back to junipers and cedars, causing new infections. For more information on cedar-apple rust on juniper, see page 172.

Solution: Cedar-apple rust cannot be controlled on the current season's foliage and fruit. The following spring, spray trees with ORTHO Multi-Purpose Fungicide Daconil 2787® Plant Disease Control when the flower buds turn pink, again when 75 percent of the petals have fallen from the blossoms, and once more 10 days later. If possible, do not plant crabapples within several hundred yards of junipers or red cedar. Infestation can take place from junipers as far away as 3 miles.

Tent caterpillars

Tent caterpillars feed on many kinds of trees.

Problem: In spring, silk webs appear in the crotches of trees or on the ends of branches.  The leaves are chewed and the tree may be completely defoliated. Groups of bluish or black hairy caterpillars with yellow or white stripes and blue or white spots are feeding in or around the webs.

Analysis: Tent caterpillars (*Malacosoma* species) feed on many ornamental and fruit trees in the garden. The insects are found in nearly all parts of the United States. In summer, tent caterpillars lay masses of 150 to 300 eggs in bands around twigs. The eggs hatch in early spring when leaves are beginning to unfold; the young caterpillars immediately begin to construct the webs. On warm, sunny days, they devour the surrounding foliage and may strip trees in just a few days. The caterpillars feed for 4 to 6 weeks and then pupate. In mid- to late summer, brownish or reddish moths emerge and lay the overwintering eggs.

Solution: Cut out and destroy large webs. Spray smaller webs with ORTHO Bug-B-Gon Ready-Spray, with an insecticide containing *malathion*, or with the bacterial insecticide *Bacillus thuringiensis* (Bt) when webs are first noticed. A bacterial insecticide is most effective against small caterpillars, so it is best to spray before nets are large. Use high-pressure spray equipment to penetrate the webbing. To prevent damage the following year, destroy the brown egg masses that encircle the twigs during the winter.

GROWING GUIDE

Adaptation: Zones 2 through 8. To determine your zone, see the map on page 336. Not adapted to desert areas.

Light: Full sun to partial shade.

Soil: Tolerates a wide variety of soils.

Fertilizer: Fertilize with Scotts Evergreen, Shrub and Tree Food according to label directions.

Water: How much: Apply enough water to wet the soil 3 to 4 feet deep.
How often: Water when the soil is barely moist 4 inches below the surface.

Pruning: Prune to maintain shape. If two leaders (tops) develop, remove one. For dense, bushy growth, remove a third of each year's new growth. When planting, allow enough room for spread and height—some spruce grow to 150 feet.

Spruce budworms

Spruce budworm larva (2× life size).

Problem: Needles on the ends of branches are chewed and webbed together. Brownish  caterpillars up to 1 inch long with yellow raised spots are feeding on the needles. Initial damage occurs on new foliage, often near the top of the tree. With severe infestations, older foliage may also be attacked. The whole tree may be defoliated. By midsummer, branch ends may turn reddish brown, becoming gray in the fall. Branches or the entire tree may die after 5 or more years of defoliation.

Analysis: Spruce budworms (*Choristoneura* species) are very destructive to ornamental spruce, fir, and Douglas fir and may infest pine, larch, and hemlock. The budworm is cyclical. It comes and goes in epidemics 10 or more years apart. The moths are small (½ inch long) and grayish, with bands and spots of brown. The females lay pale green eggs in clusters on the needles in late July and August. The larvae that hatch from these eggs crawl to hiding places in the bark or in lichen mats, or they are blown by the wind to other trees, where they hide. The tiny larvae spin a silken case and hibernate there until spring. In May, when the weather warms, the caterpillars tunnel into needles. As they grow, they feed on opening buds; later they chew off needles and web them together. The larvae feed for about 5 weeks, pupate on twigs, and emerge as adults.

Solution: The following spring, when new needles appear and caterpillars are about ½ inch long, spray with ORTHO Orthene Systemic Insect Control or the bacterial insecticide *Bacillus thuringiensis* (Bt).

PICEA (Spruce) (continued)

| Eastern spruce gall adelgid | Spruce needle miner | Pine needle scale |

Galls produced by adelgids.

Damaged needles. Inset: Larva (life size).

Pine needle scale (4× life size).

Eastern spruce gall adelgid

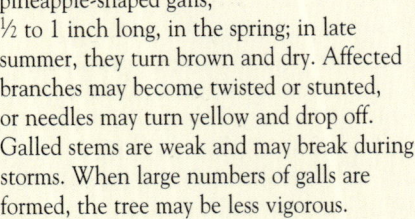

Problem: The ends of branches develop greenish purple, pineapple-shaped galls, ½ to 1 inch long, in the spring; in late summer, they turn brown and dry. Affected branches may become twisted or stunted, or needles may turn yellow and drop off. Galled stems are weak and may break during storms. When large numbers of galls are formed, the tree may be less vigorous.

Analysis: Eastern spruce gall adelgids (*Adelges abietis*) are closely related to aphids and are often referred to as aphids. This adelgid is most damaging to Norway spruce but may occasionally infest white, black, and red spruce. The insect spends the winter at the base of a terminal bud. When buds begin to grow in the spring, the adelgids lay clusters of several hundred eggs that are covered with white, waxy threads. The young that hatch from these eggs feed on developing needles. They suck the juices from the needles, inducing the formation of galls that enclose them. The adelgids live and feed in chambers inside the galls. In mid- to late summer, the galls turn brown and crack open. Adelgids that emerge lay eggs near the tip of the needles. The young that hatch from these eggs spend the winter at the base of the buds.

Solution: If possible, prune and destroy the greenish purple galls before midsummer. Spray with ORTHO Orthene Systemic Insect Control or an insecticide containing *chlorpyrifos* in the spring just before growth begins. It is usually impractical to spray large trees.

Spruce needle miner

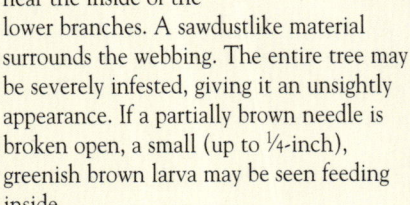

Problem: Groups of brown needles are webbed together, usually near the inside of the lower branches. A sawdustlike material surrounds the webbing. The entire tree may be severely infested, giving it an unsightly appearance. If a partially brown needle is broken open, a small (up to ¼-inch), greenish brown larva may be seen feeding inside.

Analysis: The spruce needle miner (*Endothenia albolineana*) is the larva of a small (½-inch), dark brown moth. The moth lays eggs in late spring to early summer on the undersides of old needles. The larvae that hatch from these eggs bore into the base of the needles, feeding on the interior. When the interior is consumed, the caterpillars cut off the needles at their base and web them together, forming a nest. The needle miners feed until the first frost and then enter a hollow needle, where they spend the winter. When the weather warms in spring, the larvae continue feeding until April or May. They pupate inside the webbed nest of needles and emerge as adults to lay more eggs. Several other types of needle miners may cause similar damage to spruce trees.

Solution: The following spring, before buds break, wash out infested needles with a strong stream of water from a garden hose. Gather the debris and destroy it. When moths and young larvae appear, usually in late May or early June, spray with ORTHO Orthene Systemic Insect Control. Repeat the treatment in 7 to 10 days.

Pine needle scale

Problem: Needles are covered with clusters of somewhat flattened, white, scaly bumps. When heavily infested, the foliage may appear completely white. The bumps can be scraped or picked off; the undersides are usually soft. Needles turn brown and eventually drop. Repeated severe infestations may kill young trees or weaken older trees.

Analysis: Pine needle scale insects (*Chionaspis pinifoliae*) may seriously damage spruce and pine trees and may infest fir, hemlock, and cedar. The scales survive the winter on the spruce needles as eggs beneath the dead mother scales. The eggs hatch in late spring, and the young scales, called *crawlers*, move to new green needles. The small (⅒-inch), soft-bodied young feed by inserting their mouthparts and sucking sap from the plant. The legs atrophy, and a crusty white shell develops over the body. Mature female scales lay their eggs underneath their shell in June or July. This next generation feeds throughout late summer and matures in fall. Females of this generation lay the overwintering eggs.

Solution: Spray young trees with ORTHO Isotox Insect Killer or ORTHO Bug-B-Gon Ready-Spray in late spring when the crawlers are active, at about the time lilacs begin to bloom. Early the following spring, before new growth begins and when the danger of frost is past, spray with a pesticide containing lime sulfur to kill the overwintering eggs. Inspect ornamental spruce twice a year for evidence of infestation. Older trees seldom require controls.

Spruce aphid and green spruce aphid

Spruce aphid damage to blue spruce.

Problem: Many of the older needles turn brown and drop. Only the newest needles remain green. The tree looks bare and sickly. Tiny (⅛-inch), green, soft-bodied insects may be seen feeding on the needles.

Analysis: Spruce aphids and green spruce aphids (*Elatobium abietinum* and *Cinara fornacula*) may be very destructive to spruce. The aphids appear in early spring, around February. They are extremely prolific, and populations can rapidly build up to damaging numbers during March and April. Damage occurs when the aphids suck the juices from the spruce needles. They usually remain on a single needle until it is almost ready to drop. By the time the needles turn brown and the damage is noticeable, the insect population has declined. A heavily damaged tree may require several years to recover and replace its lost foliage.

Solution: By the time the damage is noticed, it is usually too late to treat the tree during the current year. Spray with ORTHO Bug-B-Gon Ready-Spray, ORTHO Orthene Systemic Insect Control, or an insecticidal soap the following February or March. Repeat the spray 10 days later.

Spruce spider mite

Spruce spider mite damage.

Problem: Needles are dirty and stippled yellow. A silken webbing may be found on the twigs and needles. Needles usually turn brown and fall off. To determine if a tree is infested with mites, hold a sheet of white paper underneath some stippled needles and tap the foliage sharply. Minute dark green to black specks about the size of pepper grains will drop to the paper and begin to crawl around. The pests are easily seen against the white background.

Analysis: The spruce spider mite (*Oligonychus ununguis*) is one of the most damaging pests of spruces and many other conifers. These mites suck sap from the undersides of needles. As a result of their feeding, chlorophyll disappears, causing the stippled appearance. Spider mites first appear between April and June. In subtropical areas, mites may be active during warm periods in winter. Spruce spider mites are most active in the cool temperatures of spring and fall, becoming dormant in hot weather (over 90°F). Mites can build up to tremendous numbers during the growing season. Young spruce trees may die the first season. If left uncontrolled for several years, older trees may die, with symptoms progressing from the lower branches upward. Several other kinds of mites may infest spruce trees.

Solution: Spray with ORTHO Bug-B-Gon Ready-Spray or ORTHO Isotox Insect Killer, covering the foliage thoroughly. Repeat the spray two more times, 7 to 10 days apart. Additional sprays may be needed in early fall or spring if the tree becomes reinfested. To control dormant mites and eggs, spray with ORTHO Volck Oil Spray in late fall or early spring.

Canker and dieback

Dieback. Inset: White pitch on bark.

Problem: The needles on the branches nearest the ground turn brown and dry. Occasionally this condition develops first in the upper branches. The needles may drop immediately, or they may remain attached for a year. Eventually the entire branch dies back. Amber-colored pitch usually oozes from the infected area, becoming white as it dries. The infection may spread to the higher branches. To determine if the tree is infected, slice off the bark on a dead branch in the area where the diseased tissue and healthy tissue meet. Small black spore-producing bodies are found beneath the bark.

Analysis: Canker and dieback are caused by fungus (*Cytospora kunzei*) that is very destructive to Norway and Colorado blue spruce. The fungus enters the tree at a wound, killing the surrounding healthy tissue. A canker develops and expands through the wood in all directions. When the canker encircles a branch, the branch dies and the needles turn brown. Sap oozes from the dying branch. Eventually small black spore-producing bodies develop in the bark. Older (more than 15 years), weak, and injured trees are most susceptible to the disease.

Solution: Prune off and destroy dead or dying branches where the branch meets the trunk. Disinfect pruning shears by dipping after each cut in a solution of 1 part chlorine bleach and 9 parts water. Do not prune during wet weather. Avoid wounding trees with lawn mowers, tools, and other equipment. Keep trees vigorous by watering during dry spells and fertilizing every few years.

PINUS (Pine)

GROWING GUIDE

Adaptation: Throughout North America.

Light: Full sun.

Soil: Tolerates a wide variety of soils, but the soil should be well-drained.

Fertilizer: To avoid rank growth, do not fertilize pines heavily. When fertilizing, use Scotts Evergreen, Shrub and Tree Food according to label directions.

Water: Once established, pines require very little supplemental watering. Water young plants and potted plants when the soil is barely moist 2 inches below the surface. Established plants may require additional water during periods of extended drought, especially when growing in areas that normally receive water.

Pruning: Prune to maintain shape. Cut back candles (new growth before needles begin to emerge) at least half-way to slow growth or to increase bushiness. When planting, allow large species enough room for growth. Pines often drop many needles annually. This is a natural process of growth, not a cause for concern.

Needle cast

Needle cast. Inset: Fruiting structures.

Problem: The tips of the needles on the previous year's growth turn brown in winter. By spring, the infected needles are completely discolored, giving the tree a scorched appearance. Many needles may drop from the tree, leaving only the new green growth. Tiny black, elongated structures develop on the midrib of dead needles. The black structures may be swollen, with cracks down the middle. In severe cases, branch tips die back. Shaded parts of the tree are more frequently infected.

Analysis: Needle cast is caused by either of two fungi (*Hypoderma lethale* or *Lophodermium pinastri*). It is most severe on young pine trees, but older trees may be infected on the lower branches. In the summer, during wet weather, spores are released from the elongated black fruiting structures on infected needles. Splashing rain and wind may carry the spores several hundred feet. The fungus enters the tissue, but symptoms do not appear until early the following spring. Brown spots with yellow margins develop on the needles in March or April. The fungus grows through the tissue, and by late April or May the needles are completely brown. The needles drop, and the spores from these infections continue the cycle.

Solution: Remove and destroy fallen needles that collect in branch crotches and on the ground. If needle cast was serious in the spring, spray valuable specimens with ORTHO Multi-Purpose Fungicide Daconil 2787® Plant Disease Control starting in late July. Repeat the spray at intervals of 10 to 14 days through September. If trees are shaded, remove any shade-producing structures or plants where practical.

Bark beetles

Pitch tubes from turpentine beetle.

Problem: Needles in the crown of the tree turn yellow at first, then brownish orange to reddish brown. Small holes appear in the trunk. Tubelike masses of pitch, accompanied by boring dust, may also be present. The dust may be visible on spider webs at the base of the trunk. When the bark near the holes or pitch tubes is cut away, legless grubs with brown heads may be found in tunnels under the bark.

Analysis: Bark beetles (*Dendroctonus* species), about the size of rice grains, feed primarily on pine and occasionally on spruce and larch. Injured, weak, and dying trees are most susceptible to attack. Some species of bark beetles, known as turpentine beetles, create tubes of pitch on the lower bark of trees. Another species, the pine bark beetle, attacks the middle and upper trunk and does not create pitch tubes. Adult beetles of all species burrow under the bark, where they lay eggs. The larvae that hatch feed by tunneling through the bark. They form pupae in their tunnels and emerge as adults. A few beetles in a tree will not kill it; however, many pitch tubes indicate that enough beetles are present to kill or weaken the tree.

Solution: Severely infested trees should be removed and destroyed as soon as possible. Prune off infested limbs. If pitch tubes are present, smash them with a heavy rubber mallet to close the tunnel and squash the insects in the gallery beneath. Spray the trunk with ORTHO Borer & Leaf Miner Spray. Keep the tree in good health by watering it thoroughly every 4 to 6 weeks during dry months. Fertilize weakened trees with Scotts Evergreen, Shrub and Tree Food. Avoid injuring tree roots and trunks.

Root nematodes

Nematode damage.

Problem: The tree is yellowing and growing poorly. Branches die, or the entire tree turns brown and dies.

Analysis: Many types of root nematodes infect pines. Nematodes are microscopic worms (unrelated to earthworms) that live in the soil. Root nematodes feed on plant roots, damaging and stunting them. The damaged roots can't supply sufficient water and nutrients, and the plant is stunted or slowly dies. Root nematodes prefer moist, sandy loam soils. They can move only a few inches per year on their own, but they may be carried long distances by soil, water, tools, or infested plants. Testing roots and soil is the only positive method for confirming the presence of nematodes. Contact your local county extension office for sampling instructions and addresses of testing laboratories. Soil and root problems—such as poor soil structure, drought stress, nutrient deficiency, and root rots—can also produce symptoms of decline similar to those caused by nematodes. Eliminate these problems as causes before sending soil and root samples for testing. Another type of nematode that lives inside the conducting vessels causes similar aboveground symptoms. See "Pine wilt" on this page.

Solution: No chemicals available to homeowners kill nematodes in planted soil, abut they can be controlled before planting by soil fumigation or solarization. To solarize soil, cultivate and thoroughly wet an area before covering it with clear polyethylene film during the hottest time of the year. Allow the plastic film to remain in place for 4 to 6 weeks.

Pine wilt

Pine wilt nematode damage. Inset: Dying shoot.

Problem: Needles wilt and turn yellow, then brown, usually in late summer or fall. The dead needles remain on the branches. Some of the branches die, and in severe cases, the entire tree dies. Trees often die suddenly, sometimes within a few weeks. Reddish brown beetles mottled with white may be seen on the bark. These beetles are 1 inch long and have antennae longer than their bodies.

Analysis: Pine wilt is caused by microscopic nematodes (*Bursaphelenchus xylophilus*). These nematodes usually damage only certain species of pine, but they may also infest firs, spruces, and other conifers. Pine wilt is endemic to American forests, although it has been recognized as a problem only fairly recently. Pine wilt nematodes are spread by certain species of long-horned beetles, including the sawyer beetle (*Monochamus titillator*). The beetles usually attack weak and dying trees and transfer the nematodes from tree to tree while feeding. The nematodes damage the water-conducting vessels in the trunk and branches; this reduces or cuts off the flow of water through the tree. A diagnostic test is necessary to confirm the presence of pine wilt nematodes.

Solution: If you suspect pine wilt, contact your local county extension office. Remove and destroy all parts of infested trees down to ground level. Clean up tree branches and other debris. Maintain trees in good health to reduce the chances of beetle infestation. Protect valuable specimens by spraying in late spring with an insecticide containing *chlorpyrifos*. When planting, choose species resistant to pine wilt.

Spruce spider mite

Spruce spider mite damage.

Problem: Needles are dirty and stippled yellow. A silken webbing is sometimes found on the twigs and needles. Older needles at the base of the tree are usually attacked first. To determine if a tree is infested with mites, hold a sheet of white paper underneath some stippled needles and tap the foliage sharply. Tiny specks the size of pepper grains will drop to the paper and begin to crawl around. The pests are easily seen against the white background.

Analysis: The spruce spider mite (*Oligonychus ununguis*) is one of the most damaging pests of evergreen trees. These mites suck sap from the undersides of needles. As a result of their feeding, the tree's green leaf pigment disappears, producing the stippled appearance. This symptom may be mistaken for certain types of damage caused by air pollution. Spider mites first appear between April and June, hatching from eggs laid at the base of pine needles the previous fall. Mites can rapidly build up to tremendous numbers during the growing season. Young pine trees may die the first season. If left uncontrolled for several years, older trees sometimes die, with symptoms progressing from the lower branches upward. Several other species of mites also infest pines of the West Coast.

Solution: Wash webbing and mites from the tree with a strong spray of water. In spring or early fall, treat with ORTHO Isotox Insect Killer or ORTHO Bug-B-Gon Ready-Spray. Repeat the application two more times at intervals of 7 to 10 days. Additional sprays may be needed if the tree becomes reinfested.

PINUS (Pine) (continued)

Pine adelgid

Pine bark adelgids (life size).

Problem: The needles or trunk are covered with white, woolly masses. If the infestation is heavy, the tree appears to be covered with snow. Infested shoots may droop; the needles turn yellow and may die. Trees heavily infested for several years are usually stunted.

Analysis: Pine adelgids (*Pineus* species), small (⅛-inch), soft-bodied insects, are closely related to aphids and used to be called "woolly aphids." The adults are always covered with dense white filaments of wax. When this substance is removed, the insects appear purplish or green. Some species spend part of their lives on other types of evergreens, usually spruce, often producing galls on the branches. In early summer, the insects migrate to pines and suck sap from the needles. Other species spend their entire lives on pines, feeding and reproducing on the trunks. Those species that spend the winter on other types of plants produce a generation in the fall that flies to the winter host.

Solution: Control with ORTHO Orthene Systemic Insect Control, ORTHO Isotox Insect Killer, or ORTHO Malathion 50 Plus Insect Spray. Spray pines with infested trunks in April; spray pines with infested needles in late June. Cover the tree thoroughly. Repeat the spray if the plant becomes reinfested. In late winter, before spring growth begins, spray with dormant oil. Also spray other pines and spruces in the vicinity. Spray only when the temperature is expected to remain above 40°F for 24 hours after spraying.

Pine needle scale

Pine needle scale (life size).

Problem: Needles are covered with clusters of somewhat flattened, white, scaly bumps. When heavily infested, the foliage may appear completely white. The bumps can be scraped or picked off; the undersides are usually soft. Needles develop yellow mottling, turn brown, and eventually drop. Repeated severe infestations may kill young trees or weaken older trees.

Analysis: Pine needle scale insects (*Chionaspis pinifoliae*) may seriously damage pine and spruce trees and may infest fir, hemlock, and cedar. The scales survive the winter on pine needles as eggs beneath the dead mother scales. The eggs hatch in late spring, and the young scales, called *crawlers*, move to new green needles. The small (¹⁄₁₀-inch), soft-bodied young feed by inserting their mouthparts and sucking sap from the plant. The legs atrophy, and a crusty white shell develops over the body. Mature female scales lay their eggs underneath their shell in July. This next generation feeds throughout late summer and matures in fall.

Solution: In late spring when the crawlers are active, spray young trees with ORTHO Orthene Systemic Insect Control, ORTHO Bug-B-Gon Ready-Spray, or an insecticide containing *malathion*. Early the following spring, before new growth begins and when the danger of frost is past, spray with ORTHO Volck Oil Spray or a pesticide containing lime sulfur to kill the overwintering eggs. Inspect ornamental pines twice a year for evidence of infestation. Older trees seldom require controls.

Aphids

Aphids on white pine (life size).

Problem: Needles are discolored and may be deformed; many may drop from the tree. New growth is often slowed, and twigs may die. A shiny, sticky substance often coats the needles and branches. A black, sooty mold may grow on the sticky substance. Small (up to ¹⁄₁₆-inch) green, brown, or black soft-bodied insects cluster on the needles, twigs, or main stems of small trees. An uncontrolled infestation may kill young trees.

Analysis: Several different types of aphids (*Cinara* and *Eulachnus* species) infest the needles or bark of pines. Aphids do little damage in small numbers. They are extremely prolific, however, and populations can rapidly build up to damaging numbers during the growing season. Damage occurs when the aphid sucks the juices from the pine needles, growing tips, or bark. The aphid is unable to digest fully all the sugar in the sap, and it excretes the excess in a fluid called *honeydew*, which often drops onto the needles and bark below. A sooty mold may develop on the honeydew, causing the pine needles, bark, or other coated plants to appear black and dirty. Ants feed on this sticky substance and are often present where there is an aphid infestation.

Solution: Control with ORTHO Bug-B-Gon Ready-Spray, ORTHO Orthene Systemic Insect Control, ORTHO Malathion 50 Plus Insect Spray, or an insecticidal soap when aphids first appear. Repeat the spray if the tree becomes reinfested in mid- or late summer. Be sure to check the tree in the fall as well.

Pine spittlebug and Saratoga spittlebug

Pine spittlebug (3× life size).

Problem: A frothy mass of bubbles appears on the twigs at the base of the needles. A small (¼-inch), tan or green, wingless insect may be found inside the mass. Needles may turn yellow and drop off; black, sooty mold may grow on surrounding branches. Continuous heavy infestations of insects kill branches or cause the death of young or weak trees.

Analysis: The pine spittlebug (*Aphrophora parallela*) may cause serious injury to Scotch and white pines. The Saratoga spittlebug (*A. saratogensis*) kills branches of jack and red pines. Pine spittlebug adults are grayish brown, wedge-shaped insects, ½ inch long. The females lay their eggs at the base of buds in late summer. The eggs hatch the following May, and the young insects suck the sap from twigs and the main trunk. The bug excretes drops of undigested sap mixed with air, producing the frothy "spittle" that surrounds its body. Some of the excreted sap drops onto lower branches, which may be colonized by a black, sooty mold. The life cycle of the Saratoga spittlebug is similar, but the tan females lay their eggs on plants beneath the tree, especially on sweet fern. The adults migrate to trees in late June, feed until late fall, and then return to the low-growing plants to lay their eggs.

Solution: Spray with ORTHO Orthene Systemic Insect Control when insects are first noticed—in late May and again in July for the pine spittlebug, and in late June or early July for the Saratoga spittlebug. Use a high spray pressure.

Sawflies

Sawfly larvae (life size).

Problem: The needles are partially chewed, or the entire branch is defoliated. In some cases, only the younger needles are eaten. Usually, however, the older needles are preferred. Gray-green, tan, or black caterpillar-like larvae, up to an inch long, are found on the needles. Larvae may live as single individuals or in conspicuous colonies of more than 100.

Analysis: Many species of sawflies (*Neodiprion* species and *Diprion* species) infest pines. The dark, clear-winged adults are nonstinging wasps. The females insert rows of eggs in the needles with sawlike egg-laying organs. The larvae that hatch from these eggs feed singly or in groups on the needles. Eventually, entire needles are devoured. Small trees may be completely defoliated. The larvae then move to adjacent trees to feed. Some species of sawflies feed only in the spring or summer. Others are present throughout the growing season, producing five or six generations per year. When the larvae mature, they drop to the ground and spin cocoons. Most sawflies spend the winter in the soil, although several species overwinter as eggs on the needles.

Solution: Spray the needles with ORTHO Orthene Systemic Insect Control or ORTHO Isotox Insect Killer when damage or the insects are first noticed. Inspect the trees periodically during the growing season to detect infestations before severe defoliation occurs.

Pine tip and shoot moths

European pine shoot moth damage.

Problem: Branch tips turn yellow, then brown and dry. The dead branches contrast sharply with healthy green foliage. In the summer, pitch accumulates around the dead needles. Trees may appear bushier than normal, or they may be crooked and distorted. At the base of the needles or inside a brown, resin-coated tip, cream-colored to reddish brown worms, up to ¾ inch long, may be found feeding on the tissue. Young trees may die.

Analysis: Seven species of tip and shoot moths (*Rhyacionia* species) infest various pines in different parts of the country. The adult is a reddish brown and gray moth, up to 1 inch long. The moths fly at night but may be seen during the day if a branch is disturbed. They lay their eggs in mid- to late spring at the ends of branches. The larvae that hatch from these eggs bore into needles and buds, where they feed and mature. Depending on the species, pupation occurs in the mined-out area or in the soil around the base of the tree. Most species of tip moths produce one generation per year. The Nantucket pine tip moth, *Rhyacionia frustrana,* has as many as four or five generations yearly in warm climates.

Solution: Spray with ORTHO Orthene Systemic Insect Control or ORTHO Isotox Insect Killer in mid-April to early May. Repeat the spray in mid-May. If reinfestation occurs the same year, the Nantucket pine tip moth is probably involved. Repeated sprays every 4 weeks from early May to August may be necessary. If practical, prune out and destroy infested tips from October through January. Avoid pruning when moths are active.

PINUS (Pine) *(continued)*

| Pitch moths | Western and eastern gall rust | White pine weevil |

Pitch moth damage.

Western gall rust on Monterey pine.

White pine weevil damage.

Problem: One or more masses of sticky cream, yellow, or pinkish pitch appear on the trunk. These masses may be 2 or 3 inches wide and protrude 1 to 2 inches from the side of the trunk. The pitch masses are usually found in wounds or in branch crotches. When the pitch mass is scraped away, a larva up to 1 inch long may be found underneath.

Analysis: Pitch moths (*Vespamima* species) attack pine, spruce, and Douglas fir. The adults are clear-winged moths that resemble yellowjackets. They lay eggs during the spring and summer in the trunks and larger limbs, particularly at sites of recent trunk injury or where old pitch masses exist. The larva that hatches feeds on the inner bark for 1 to 2 years, pupates, and finally emerges as the adult moth during the summer. Usually there is one larva per pitch mass. Although pitch masses are unsightly, pitch moths do not usually threaten the life of a tree. Tree limbs may be weakened enough to break under the weight of snow, however.

Solution: Scrape away fresh pitch masses and kill the larvae. The larva can be found in the bark under a pitch mass or in the pitch mass itself. Avoid mechanical injury to trees. Confine pruning of larger limbs to fall and early winter months so injuries dry up before moths appear in the spring.

Problem: Rough, spherical swellings develop on branches or on the main trunk. In the spring, the swellings appear orange or yellow. Growth beyond the galls is often stunted, distorted, and off-color.

Analysis: Western and eastern gall rust are caused by two species of fungi (*Endocronartium harknessii* and *E. quercuum*). Western gall rust requires only one host to complete its life cycle; spores from one pine can infect another. Eastern gall rust requires both pine and oak to complete its life cycle. In early spring, orange or yellow spores are produced over the ruptured surfaces of the swellings (galls). The spores are blown and carried by wind and insects to susceptible trees. When moisture and temperatures are optimum, western gall rust spores infect pine tissue, causing an increase in the number and size of plant cells. Within 6 months to 1 year, swellings develop. The galls enlarge and produce spores after 1 to 2 years. Eastern gall rust spores infect only oak. Spores produced on the oak trees reinfect pines. The galls caused by both fungi interrupt the sap movement in the tree. They also stimulate witches'-brooms, dense stunted growth beyond the galls. If many of these develop, the tree becomes unsightly and weak, and limbs break during storms.

Solution: Where practical, prune off galled branches before the galls produce spores in early spring.

Problem: The main shoot at the top of a healthy tree stops growing and turns yellow in midsummer. The shoot tip usually droops, producing a "shepherd's crook." Several new shoots may develop from below the dying shoot so that the top of the tree is forked. In fall and winter the drooping shoot appears brown and dry. A white resin is seen on the bark, and small holes in the dead shoot. Trees are disfigured but not killed.

Analysis: The white pine weevil (*Pissodes strobi*) attacks the leaders of all pines, most spruces, and some firs. This small (⅕-inch), brown, snouted beetle with white patches spends the winter in dead plant material at the base of the tree. In the spring, just before new growth begins, it moves to the top of the tree to feed on the inner bark tissue. Eggs are then laid in small punctures in the bark. Resin droplets that ooze from the punctures later dry and turn white. The ¼-inch larvae that hatch from these eggs bore into the wood. The feeding cuts off the flow of water and nutrients through the stem, causing the shoot to droop and die. Several new shoots often develop from below the dead shoot, destroying the natural shape of the tree. In late summer, the larvae mature and return to the ground to spend the winter.

Solution: Prune out and destroy infested twigs in early summer before the beetles emerge. Train a side branch to replace the dead leader by pruning all but one of the new shoots to half the length of the newly selected leader. The following spring, spray the leader with ORTHO Borer & Leaf Miner Spray as soon as the buds begin to swell. Spray the top of the tree thoroughly.

PLATANUS (Sycamore)

Sycamore lace bugs

Sycamore lace bug damage.

Problem: The upper surfaces of leaves are mottled or speckled white and green. The mottling 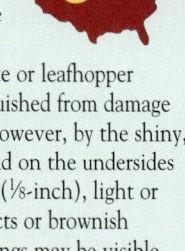 may be confused with mite or leafhopper damage. It can be distinguished from damage caused by other insects, however, by the shiny, hard, brown droplets found on the undersides of damaged leaves. Small (⅛-inch), light or dark, spiny, wingless insects or brownish insects with clear, lacy wings may be visible around the droplets. Foliage on severely infested trees may be completely white, then turn brown by mid-August.

Analysis: Two species of lace bugs (*Corythucha* species) infest sycamores and London plane trees. They survive the winter as adults in bark crevices or in other protected areas on the tree. When the buds begin to open in the spring, the adults attach their eggs to the undersides of leaves with a brown, sticky substance. The eggs hatch, and the spiny, wingless, immature insects—and later the brown, lacy-winged adults—suck sap from the undersides of leaves. The green leaf pigment disappears, resulting in the characteristic white-and-green mottling. As the lace bugs feed, droplets of brown excrement accumulate around them.

Solution: Spray young trees with ORTHO Bug-B-Gon Ready-Spray, ORTHO Isotox Insect Killer, or ORTHO Orthene Systemic Insect Control when damage first appears in spring. Cover the undersurfaces of the leaves thoroughly. Repeat 7 to 10 days later. It is important to spray early, preventing as much damage as possible. Inspect trees every 2 weeks during the growing season to catch infestations before severe damage is done.

Sycamore anthracnose

Anthracnose. Inset: Spore-producing bodies.

Problem: In the spring, buds or expanding shoots turn brown and die. Dead areas appear along the veins of young leaves. As the leaves mature, the spots may expand and cover them entirely. Most infected leaves drop from the tree. Later in the season, twigs and older leaves may be infected. Infected twigs hang on the tree or drop to the ground with the leaves. Larger limbs may die. Dark brown spore-producing bodies appear on the bark and dead leaves. The tree is often stunted and bushy.

Analysis: Sycamore anthracnose is caused by a fungus (*Apiognomonia veneta*) that is the most serious problem of sycamore and causes minor damage to the London plane tree. The fungus survives the winter on fallen leaves and twigs and in swollen cankers in the tree. During cool (below 55°F), wet weather, spores are blown and splashed onto buds, expanding shoots, and young leaves. The fungus enters the tissue and kills it, causing the buds and shoots to die back. The fungus moves down onto the twigs, and spores develop. The spores may infect mature leaves or any new growth on the tree, causing a sun-scorched appearance. Swollen, cracked cankers develop on infected twigs and branches. When the cankers encircle the wood, the limbs die.

Solution: Prune off and destroy infected twigs and dead branches. Gather and destroy fallen leaves and twigs. Feed and water regularly to keep the tree vigorous. In areas where spring is cool and moist, spray with ORTHO Multi-Purpose Fungicide Daconil 2787® Plant Disease Control when buds begin to grow in the spring. Repeat when leaves reach full size and again 2 weeks later.

POPULUS (Poplar, aspen)

GROWING GUIDE

Adaptation: Throughout North America.

Light: Full sun.

Soil: Tolerates a wide variety of soils. Roots are invasive.

Fertilizer: Fertilize with Scotts Evergreen, Shrub and Tree Food according to label directions.

Water: How much: Apply enough water to wet the soil 3 to 4 feet deep.
How often: Some poplars are drought-tolerant once established. However, most poplars prefer moist soils, and some even tolerate soils that are soggy or flooded.

Pruning: Prune off suckers and broken branches, and prune to maintain shape.

POPULUS (Poplar, aspen) *(continued)*

Leaf-feeding caterpillars

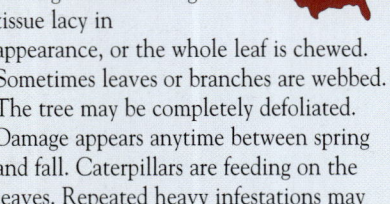

Satin moth caterpillars (life size).

Problem: The surface of the leaf is eaten, leaving the remaining tissue lacy in appearance, or the whole leaf is chewed. Sometimes leaves or branches are webbed. The tree may be completely defoliated. Damage appears anytime between spring and fall. Caterpillars are feeding on the leaves. Repeated heavy infestations may weaken or kill trees.

Analysis: Many species of caterpillars feed on poplar leaves wherever the trees are grown. Depending on the species, the moths lay their eggs from early spring to midsummer. The larvae that hatch from these eggs feed singly or in groups on buds, on one leaf surface (these are called *skeletonizers*), or on the entire leaf. Certain caterpillars web leaves together or web a branch as they feed. In some years, damage is minimal because of unfavorable environmental conditions or control by predators and parasites. When conditions are favorable, however, entire trees may be defoliated by late summer. Defoliation weakens trees because no leaves are left to produce food. When heavy infestations occur several years in a row, branches or entire trees may be killed.

Solution: Spray with ORTHO Isotox Insect Killer or with the bacterial insecticide *Bacillus thuringiensis* (Bt) when damage is first noticed. Cover the leaves thoroughly. Repeat the spray if the tree becomes reinfested.

Scales

Oystershell scales (¼ life size).

Problem: Brown, black, or red-orange crusty bumps or somewhat flattened brownish, white, or grayish scaly bumps cover trunks, stems, or the undersides of leaves. The bumps can be scraped or picked off; the undersides are usually soft. Leaves turn yellow and may drop. In some cases, a shiny, sticky substance coats the leaves. A black, sooty mold often grows on the sticky substance. Large portions of the tree may be killed if infestations are heavy.

Analysis: Many types of scales infest poplar. They lay their eggs on leaves or bark, and in spring to midsummer the young scales, called *crawlers*, settle on the leaves, twigs, and trunk. The small (¹⁄₁₀-inch), soft-bodied young feed by sucking sap from the plant. The legs usually atrophy, and a hard crusty shell develops over the body. Some species of scales that infest poplar are unable to digest fully all the sugar in the plant sap, and they excrete the excess in a sticky fluid called *honeydew*. A sooty mold may develop on the honeydew, causing the poplar leaves to appear black and dirty. An uncontrolled infestation of scales may kill the tree after 2 or 3 seasons.

Solution: Spray with ORTHO Isotox Insect Killer, ORTHO Bug-B-Gon Ready-Spray, or an insecticide containing *malathion* when the young are active. To control overwintering insects, spray with ORTHO Volck Oil Spray before growth begins in the spring.

Poplar borers

Damaged tree. Inset: Adult borer (½ life size).

Problem: Liquid oozes from holes in the bark. Swollen areas with holes in their centers develop on twigs, branches, or the trunk. Wood is honeycombed with irregular tunnels. Leaves may discolor and wilt, and branches die. A sawdustlike material and many broken twigs are usually found beneath the tree.

Analysis: At least five species of beetles (*Saperda* species) feed on poplars, causing galls to form. The inch-long, striped and spotted, brownish or gray beetles with long antennae appear in late spring or early summer. The females lay eggs in small holes gnawed in the bark of twigs and branches that are more than ½ inch in diameter. As the legless, whitish grubs hatch from these eggs, they tunnel into the wood. Excess tissue grows around the wound, resulting in a swollen area or gall. When infestations are severe, nearly all twigs and branches more than ½ inch in diameter have one or more galls. The galls weaken the twigs and branches, causing them to break and litter the ground during stormy weather. The grubs remain in the wood for 1 or 2 years. These borers select damaged or dying trees; healthy plants are seldom attacked.

Solution: Remove and destroy severely damaged trees. Do not store newly cut wood near other trees; borers may emerge and attack the trees. Spray the bark of less severely damaged trees with an insecticide containing *chlorpyrifos* in late May or early June. Repeat the treatment 2 weeks later.

Canker and dieback

Dieback. Inset: Canker.

Problem: Dark sunken areas appear on the twigs, branches, or trunk. Leaves on infected branches may be spotted, or they may be stunted and lighter green than normal. Twigs and branches are often killed. Young or weakened trees are most susceptible.

Analysis: Several fungi cause canker and dieback on poplars. Lombardy poplars are especially vulnerable. The fungi enter the tree through a wound or, in some cases, through the leaves, killing the surrounding healthy tissue. A dark sunken canker develops in the wood and expands through it in all directions. If the fungus infects the leaves first, it grows down through the leaf stems and forms cankers on the twigs. The canker cuts off the flow of nutrients and water to the twigs or branch, causing the leaves to turn yellow. Twig or branch dieback follow if the canker girdles the wood. The tree may wall off the spreading fungus by producing callus tissue, a rapid growth of barklike cells. If the expanding canker is stopped before it covers half the diameter of the trunk, the tree usually survives. The fungus may grow faster than the callus, however, or the tree may not produce a callus, resulting in the death of the branch or the whole tree.

Solution: Prune off dead twigs and small cankered branches, cutting well below the canker. Remove and destroy severely infected trees. To prevent the development of new cankers, avoid wounding trees. Keep trees vigorous by fertilizing and watering.

PYRACANTHA

Fire blight

Fire blight on Pyracantha.

Problem: Blossoms and leaves of infected twigs suddenly wilt and turn black as if scorched by fire. The leaves curl and hang downward. The bark at the base of the blighted twigs becomes water-soaked, then dark, sunken, and dry; cracks may develop at the edge of the sunken area. In warm, moist spring weather, drops of brown ooze appear on the sunken bark. Young plants may die.

Analysis: Fire blight is caused by a bacterium (*Erwinia amylovora*) that is very destructive to many trees and shrubs. (For a list of susceptible plants, see page 341.) The bacteria spend the winter in the sunken areas (cankers) on the branches. In the spring, the bacteria ooze out of the canker. Bees, flies, and other insects are attracted to the sweet, sticky ooze and become smeared with it. When the insects visit a flower for nectar, they infect it with the bacteria. The bacteria spread rapidly through the plant tissue in warm (65°F or higher), humid weather. Insects visiting these infected blossoms later carry bacteria-laden nectar to healthy blossoms. Tender or damaged leaves may be infected in midsummer.

Solution: During spring and mid- to late summer, prune out infected branches 12 to 15 inches beyond any visible discoloration and destroy them. Disinfect pruning shears after each cut by dipping in a solution of 1 part chlorine bleach and 9 parts water. A protective spray of a bactericide containing basic copper sulfate or *streptomycin* applied before bud-break in the spring will help prevent infection. Repeat at intervals of 5 to 7 days until the end of bloom. In the fall, prune out any remaining infected branches.

QUERCUS (Oak)

GROWING GUIDE

Adaptation: Throughout North America. There are oaks for all climates and conditions, including salt air, heat, wind, and moist areas.

Light: Full sun.

Soil: Any good, deep, well-drained garden soil. Some oaks do not tolerate alkaline soils (pH 7.0 and above).

Fertilizer: Some oaks benefit from periodic feedings under the outer branches. Fertilize with Scotts Evergreen, Shrub and Tree Food according to label directions.

Water: Some oaks grow best with ample water and will thrive in lawns. Others, such as the western native oaks, prefer drier conditions. They will decline and eventually die if overwatered. For a list of oaks and their water requirements, see page 356.
How much: Apply enough water to wet the soil 4 to 5 feet deep.
How often: Water young trees when the soil is barely moist 4 inches below the surface.
Oaks requiring ample water: Established trees should be watered when the soil is barely moist 4 inches below the surface.
Oaks preferring dry conditions: Never water around the trunk. Most of these oaks need no supplemental water once established.

Pruning: Prune to maintain shape and to remove dead wood. When planting, allow room for growth.

QUERCUS (Oak) *(continued)*

Oak leaf blister

Oak leaf blister on red oak.

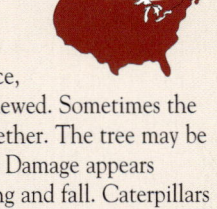

Problem: Puckered, circular areas, up to ½ inch in diameter, appear on the leaves in the spring. The blisterlike spots are yellowish green at first and later die and turn brown. The leaves usually remain attached to the tree.

Analysis: Leaf blister is caused by a fungus (*Taphrina caerulescens*) that is unsightly but rarely harms the tree. It is a problem on various species of oak, particularly red, black, scarlet, and live oaks. The fungus spends the winter in the bud scales on the tree. During cool, wet spring weather, it enters the developing leaves. Green blisters form where the fungus enters the tissue. The infected tissue eventually dies and turns brown. In the fall, the fungus produces overwintering spores. If the following spring is cool and wet, the cycle begins again.

Solution: If leaf blister was a problem the previous year and the weather the current spring is cool and wet, spray the tree with ORTHO Multi-Purpose Fungicide Daconil 2787® Plant Disease Control when the buds begin to swell, about 1 or 2 weeks before the leaves appear. Cover the entire tree thoroughly with the spray.

Borers

Borer larva and holes.

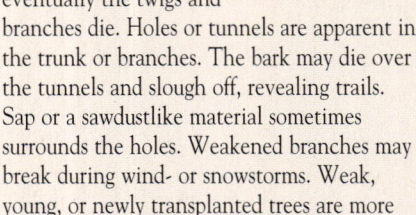

Problem: Foliage on a branch or at the top of the tree is sparse; eventually the twigs and branches die. Holes or tunnels are apparent in the trunk or branches. The bark may die over the tunnels and slough off, revealing trails. Sap or a sawdustlike material sometimes surrounds the holes. Weakened branches may break during wind- or snowstorms. Weak, young, or newly transplanted trees are more susceptible to injury and may be killed.

Analysis: Borers are the larvae of beetles or moths. Several kinds of borers attack oaks. Throughout the summer, females lay their eggs in bark crevices. The larvae feed on the bark, sapwood, and heartwood. This stops the flow of nutrients and water in that area by cutting the conducting vessels; branch and twig dieback results. Sap flow acts as a defense against borers if the tree is healthy; when the borer burrows into the wood, tree sap fills the hole and kills the insect. Factors that weaken the tree—such as mechanical injuries, transplanting, damage by leaf-feeding insects, and poor growing conditions— make it more attractive to egg-laying females.

Solution: Borers are difficult to control once they have burrowed into the wood. Cut out and destroy all dead and dying branches, and remove severely infected young trees. Spray the trunk and branches with ORTHO Borer & Leaf Miner Spray in May. Repeat the spray 2 weeks later and again in July and August. Maintain plant vigor by watering during periods of drought and fertilizing regularly.

Leaf-feeding caterpillars

Gypsy moth caterpillar (2× life size).

Problem: The surface of the leaf is eaten, giving the remaining tissue a lacy appearance, or the whole leaf is chewed. Sometimes the leaves are webbed together. The tree may be completely defoliated. Damage appears anytime between spring and fall. Caterpillars are feeding on the leaves. Repeated heavy infestations may weaken or kill trees.

Analysis: Many species of caterpillars feed on oak leaves wherever the trees are grown. Depending on the species, the moths lay their eggs from early spring to midsummer. The larvae that hatch from these eggs feed singly or in groups on buds, on one leaf surface (these are called *skeletonizers*), or on the entire leaf. Certain caterpillars web the leaves together as they feed. In some years, damage is minimal because of unfavorable environmental conditions or control by predators and parasites. When conditions are favorable, however, entire trees may be defoliated by late summer. Defoliation weakens trees because no leaves are left to produce food. When heavy infestations occur several years in a row, branches or entire trees may be killed.

Solution: Spray with ORTHO Isotox Insect Killer, ORTHO Orthene Systemic Insect Control, or the bacterial insecticide *Bacillus thuringiensis* (Bt). Repeat the spray if the tree becomes reinfested.

Pit scales

Oak pit scales (3× life size).

Problem: During the summer or early fall, leaves turn brown and twigs or branches die back.  Dead leaves usually remain attached to the branches throughout the winter. In the spring, new leaves may appear 3 weeks late on infested deciduous oaks. Repeated heavy infestations often kill young trees. Small (¹⁄₁₀-inch), somewhat flattened, green, golden, or brown scaly bumps cluster on the twigs and branches. The bark is pitted where these insects cluster.

Analysis: Several species of pit scales (*Asterolecanium* species) may seriously damage oaks. The scales lay their eggs in spring and summer. The young scales, called *crawlers,* that hatch from these eggs settle on new growth and the previous year's branches, not far from the parent. The small (¹⁄₁₀-inch), soft-bodied young feed by inserting their mouthparts and sucking sap from the plant. Pits develop where the scales feed. The legs atrophy, and a hard, crusty shell develops over the body. Mature female scales lay eggs underneath their shells. Several other types of scales also infest oak.

Solution: Spray with ORTHO Bug-B-Gon Ready-Spray or ORTHO Orthene Systemic Insect Control in mid-May to June when the young are active. To control overwintering insects, spray with ORTHO Volck Oil Spray in winter or spring. Cover the tree thoroughly.

Armillaria root rot

Cankers.

Fungal mats.

Problem: Leaves are stunted and yellow, and the foliage throughout the tree may be sparse. Branches eventually die. Weakened trees are most severely infected. Occasionally trees die suddenly without showing symptoms, but in most cases they die slowly over a period of several years. Honey-colored mushrooms, 2 to 5 inches in diameter, may grow singly or in clusters during the fall or winter on the lower trunk or on the ground near infected roots. If the soil is removed from around the base of the tree, black rootlike strands, about the diameter of pencil lead, are seen attached to the larger roots. A white fan-shaped growth occurs between the bark and wood of these larger roots and on the trunk just below the soil surface. The infected tissue has a mushroom odor.

Analysis: Armillaria root rot, also called *oak root fungus* or *shoestring root rot,* is caused by a fungus (*Armillaria mellea*) that rots the roots of many woody and nonwoody plants. Oaks are often lightly infected with this fungus for years with no damage. When the trees are under stress from drought, overwatering, physical injury, insects, or disease, however, they often succumb to *Armillaria.* The fungus is spread short distances (under a foot) through the soil by the rootlike fungal strands. When they contact susceptible plant roots, the strands penetrate the host if conditions are favorable. Once the fungus enters the bark tissue, it produces a white fan-shaped mat of fungal strands that invade and decay the tissue of the roots and lower trunk. The fungus spreads rapidly if the oak tree is in a weakened state. Water and nutrient uptake by the roots is inhibited, causing the foliage and branches to die. In the fall, mushrooms—the reproductive bodies of the fungus—often appear around infected trees. A closely related fungus (*Clitocybe tabescens*) found in the Southeast produces symptoms on oak similar to armillaria root rot.

Solution: Remove and destroy infected trees, including the stump and the root system. The fungus can live on the stump and roots for many years, infecting susceptible plants nearby. Healthy-appearing plants growing adjacent to diseased trees may already be infected. Check around the roots and lower stems for signs of the fungus. The life of a tree may be prolonged if it is not severely infected. Remove the soil from around the rotted parts of the roots and trunk. Cut out the diseased tissue down to healthy wood and allow the healthy wood to air-dry through the summer. Deep-watering is recommended if the tree needs water. Avoid surface watering, especially wetting of the crown and trunk root area. Cover the exposed parts before temperatures drop below freezing. When replacing trees that have been infected with *Armillaria,* use resistant plants. Avoid planting susceptible species in recently cleared forestlands where armillaria root rot is common. To inhibit disease development in an established oak tree, provide optimum growing conditions, avoid injuring the tree, and control pests and diseases.

187

RHODODENDRON (Azalea)

GROWING GUIDE

Adaptation: Zones 4 through 10 except in desert areas. To determine your zone, see the map on page 336. Azaleas are more tolerant of warm, dry climates than rhododendrons.

Flowering Time: Late winter to early summer. A few bloom in mid- to late summer.

Light: Partial shade (full sun in cool summer areas). Plants don't bloom well in deep shade.

Soil: Rich, well-drained, acid (pH 4.5 to 6.0) soil, high in organic matter and low in salts. When planting, add at least 50 percent peat moss to the soil, and keep 2 inches of mulch around the base of the plant.

Fertilizer: Fertilize with Scotts Azalea, Camellia, Rhododendron Food according to label directions.

Water:
How much: Apply enough water to wet the soil 1 to 2½ feet deep.
How often: Water when the soil is moist but not wet under the mulch.

Pruning: Prune just after flowering.
Azaleas: For a bushier plant, pinch off growing tips.
Rhododendrons: Remove spent flower clusters. Be careful not to break the new buds.

Salt burn

Salt burn.

Problem: Leaf edges are brown and dead. Browning usually occurs on older leaves first. This distinguishes the problem from wind burn, which develops on young, exposed leaves first. Leaves may be lighter green than normal. In severe cases, leaves drop off.

Analysis: Salt burn is most common in areas of low rainfall. It also occurs in soils with poor drainage and where too much fertilizer has been applied. Excess salts dissolved in the soil water accumulate in leaf edges, where they kill the tissue. These salts also interfere with water uptake by the plant. This problem is rare in areas of high rainfall, where the soluble salts are leached from most soils. Poorly drained soils do not leach well; much of the applied water runs off the surface instead of washing through the soil. Fertilizers, which are soluble salts, also cause salt burn if too much is applied or if they are not diluted with a thorough watering after application.

Solution: Salt burn damage does not disappear from the leaves, but injury can be avoided in the future. In areas of low rainfall, leach accumulated salts from the soil with an occasional heavy irrigation (about once a month). If possible, improve drainage around the plants by removing them and adding soil amendments. If plants are severely damaged, replace them with healthy plants. Follow package directions when using commercial fertilizers; water thoroughly after application. Avoid the use of bagged steer manure, which may contain large amounts of salt, on azaleas and rhododendrons.

Sunburn (sunscald)

Sunburn on azalea.

Problem: During warm, sunny weather, the center portion of the leaf bleaches to a tan or off-white color. Once the initial damage has occurred, the spot rarely increases in size. Injury is generally more severe on plants with light-colored flowers.

Analysis: Rhododendrons and azaleas are generally classified as shade plants. Their leaves are sensitive to the heat of direct sun, which kills the leaf tissue. Scalding occurs when the shrub is planted in full sun. The intense reflection from a light-colored, south-facing wall can also scald leaves. Damage may appear in only one hot summer day. The injury is unsightly but does not damage the plant. Weakened leaves are more susceptible to invasion by fungi and bacteria, however. Plants that do not receive enough water are more susceptible to sunscald.

Solution: Provide some shade where the plant is growing, or move the injured plant to a shaded location. Scalded leaves will not recover. Where practical, remove affected leaves. Don't let plants dry out during hot weather.

Iron deficiensy

Iron deficiency on rhododendron.

Problem: The leaves are pale green to yellow. The newest leaves may be completely yellow, with only the veins and the tissue right next to the veins remaining green. With progressively older leaves, only the leaf edges may be yellowing. The plant may be stunted.

Analysis: Iron deficiency is a common problem in acid-loving plants such as azaleas and rhododendrons. These plants prefer soil with a pH between 5.0 and 6.0. The soil is seldom deficient in iron, but iron is often found in an insoluble form that is not available to the plant, especially in soil with a pH above 7.0. A high soil pH can result from overliming or from lime leached from cement or mortar. Regions where soil is derived from limestone or where rainfall is low also have high-pH soils. Plants use iron in the formation of chlorophyll in the leaves. When iron is lacking, new leaves are yellow.

Solution: Spray the foliage with a chelated iron fertilizer, and apply the fertilizer to the soil around the plant to correct the iron deficiency. Lower the pH of the soil by treating it with aluminum sulfate and watering it in well. Maintain an acid pH by fertilizing with Scotts Azalea, Camellia, Rhododendron Food. When planting azaleas or rhododendrons, add enough peat moss to make up at least 50 percent of the amended soil. This is especially important if you live in an area where the soil is alkaline. Never lime the soil around azaleas or rhododendrons.

Wilt and root rot

Phytophthora root rot.

Problem: The young leaves are yellowish and wilting. Eventually the whole plant wilts and dies, even though the soil is moist. Dead leaves remain attached to the plant and are rolled along the midrib. The symptoms may develop over a few weeks or may take many months. Heavy, poorly drained soil favors disease development. The tissue under the bark close to ground level shows a dark discoloration when cut. To check for discoloration, peel back the bark at the bottom of the plant. A distinct margin separates white, healthy wood from dark, diseased wood.

Analysis: Wilt and root rot is caused by several soil-inhabiting fungi, also known as *water molds*. These fungi (*Phytophthora* and *Pythium* species) attack a wide variety of ornamental plants. The fungi destroy the roots and may work their way up the stem. If they girdle the stem, the plant wilts and dies. Very wet conditions favor the fungi, which are most common in heavy, poorly drained soils. Although azaleas and rhododendrons need constant moisture, they must also have good drainage.

Solution: No chemical control is available. Allow the soil to dry, while minimizing drought stress by providing shade, spraying with antitranspirant, and covering plants in shade with a clear plastic tent. Resume watering when signs of drought stress appear. Improve the drainage of the soil before replanting azaleas or rhododendrons in the same location. If drainage cannot be improved, plant in beds raised a foot or more above grade. Or plant shrubs that are resistant to wilt and root rot (see page 342 for a list).

Scales

Azalea bark scale (life size).

Problem: Clusters of somewhat flattened white, yellowish, brown, reddish, or gray scaly bumps cover the undersides of leaves, the young branches, or the branch crotches. The bumps can be scraped or picked off; the undersides are usually soft. Leaves may turn yellow and drop off, and branches may die back. The plant is killed when infestations are heavy.

Analysis: Many species of scales infest rhododendrons and azaleas throughout the country. Scales spend the winter on the trunk and twigs of the plant. They lay eggs in spring and in midsummer; the young scales, called *crawlers*, settle on various parts of the shrub. The small (1/10-inch), soft-bodied young feed by inserting their mouthparts and sucking sap from the plant. The legs usually atrophy, and a scaly or crusty shell develops over the body. Mature female scales lay their eggs underneath their shell. Leaf drop and twig dieback occur when scales completely cover the leaves and branches. An uncontrolled infestation may kill a plant after 2 or 3 seasons.

Solution: Spray with ORTHO Isotox Insect Killer, ORTHO Orthene Systemic Insect Control, or an insecticide containing *diazinon* or *malathion* in midsummer (late spring in the South) when the young are active. Early the following spring, before new growth begins, spray the branches and trunk with a dormant oil spray to control overwintering insects.

ROSA (Rose)

GROWING GUIDE

Adaptation: Throughout North America.

Flowering Time: Spring and summer (through fall in the South).

Light: Full sun. Plant in areas with good air circulation.

Soil: Any good, well-drained garden soil.

Fertilizer: Fertilize with Scotts Evergreen, Shrub and Tree Food according to label directions.

Water: How much: Apply enough water to wet the soil 1½ to 2 feet deep.
How often: Roses need plenty of water. Water when the soil is moist but not wet 4 inches deep.

Pruning: Remove dead or unhealthy wood. Remove branches that cross through the center of the plant. Prune off at least a third to half of last year's growth.

Few or no blooms

Lack of flowers caused by poor pruning.

Rose failing to bloom.

Problem: Plants fail to bloom or bloom only sparsely.

Analysis: Roses produce few or no buds or flowers for any of several reasons.

1. Too much shade: Roses grow and bloom best in full sun. They need at least 4 to 5 hours of direct sunlight for normal blooming.

2. Improper dormant pruning: Most rose varieties are grafted onto a rootstock. Tree roses and some climbing roses are grafted onto an intermediate trunkstock. If the hybrid canes are pruned off below the bud union, the rootstock or trunkstock will produce suckers that are flowerless or that produce flowers very different from the desired variety. Some climbing roses and many old-fashioned roses bloom from flower buds formed the previous season. Heavy pruning of such plants will remove all the buds.

3. Excessive or improper pruning during the growing season: If roses are excessively trimmed and pruned during the growing season, many or all of the developing flower buds may be inadvertently removed.

4. Old flowers left on plant: Roses do not produce as many new flowers when the old blooms are allowed to fade and form seeds.

5. Flushes of bloom: Many roses bloom in flushes. The first flush usually occurs in late spring, and the second flush occurs in late summer or early fall.

6. Diseased or infested plants: Roses that have been attacked by diseases or insects do not flower well.

Solution: The numbered solutions below refer to the numbered items in the analysis.

1. Thin out shading trees and shrubs, or transplant roses to a sunnier location. Replace them with shade-loving plants.

2. Do not prune roses below the bud union. Take special care when pruning climbing roses and standard tree roses since the bud union between the trunkstock and the grafted variety may be several feet from the ground. Prune old-fashioned roses lightly during the dormant season. If heavy pruning is needed, wait until after the plants have bloomed in spring.

3. During the growing season, prune roses only to shape them or to remove suckers and dead or dying growth. When cutting or removing flowers, leave at least two 5-leaflet leaves on the cane to ensure continued flower production.

4. Remove flowers as they begin to fade.

5. You cannot do anything to alter flushes; this is a natural plant cycle.

6. Look up the symptoms on pages 191, 192, 193, and 194 to determine the cause. Treat accordingly.

Flower thrips

Flower thrips damage.

Problem: Young leaves are distorted, and foliage may be flecked with yellow. Flower buds are deformed and usually fail to open. The petals of open blossoms, especially those of white or light-colored varieties, are often covered with brown streaks and red spots. If a deformed or streaked flower is pulled apart and shaken over white paper, tiny yellow or brown insects fall out and are easily seen against a white background.

Analysis: Flower thrips (*Frankliniella tritici*) are the most abundant and widely distributed thrips in the country. They live inside the buds and flowers of many garden plants. Both the immature and the adult thrips feed on plant sap by rasping the tissue. The injured petal tissue turns brown, and the young expanding leaves become deformed. Injured flower buds usually fail to open. Thrips initially breed on grasses and weeds. When these plants begin to dry up or are harvested, the insects migrate to succulent green ornamental plants. The adults lay their eggs by inserting them into the plant tissue. A complete life cycle may occur in 2 weeks, so populations can build up rapidly. Most damage to roses occurs in early summer.

Solution: Thrips are difficult to control because they continuously migrate to roses from other plants. Immediately remove and destroy infested buds and blooms. Spray with ORTHO Orthene Systemic Insect Control, ORTHO Isotox Insect Killer, or ORTHO Bug-B-Gon Ready-Spray three times at intervals of 7 to 10 days.

Black spot

Black spot.

Problem: Circular black spots with fringed margins appear on the upper surfaces of the leaves in the spring. The tissue around the spots or the entire leaf may turn yellow, and the infected leaves may drop prematurely. Severely infected plants may lose all of their leaves by midsummer. Flower production is often reduced, and quality is poor.

Analysis: Black spot is caused by a fungus (*Diplocarpon rosae*) that is a severe problem in areas where high humidity or rain is common in spring and summer. The fungus spends the winter on infected leaves and canes. The spores are spread from plant to plant by splashing water and rain. The fungus enters the tissue, forming spots the size of a pinhead. The black spots enlarge, up to ¾ inch in diameter, as the fungus spreads; spots may join to form blotches. Twigs may also be infected. Plants are often killed by repeated infection.

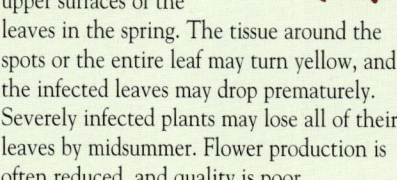

Solution: Spray with ORTHO RosePride Funginex Rose & Shrub Disease Control, ORTHO RosePride Orthenex Insect & Disease Control, or ORTHO Multi-Purpose Fungicide Daconil 2787® Plant Disease Control. Repeat the treatment at intervals of 7 to 10 days for as long as the weather remains wet. Spraying may be omitted during hot, dry spells in summer. Prune off infected canes. Avoid overhead watering. In the fall, rake up and destroy the fallen leaves. After pruning plants during the dormant season, spray with ORTHO Dormant Disease Control Lime-Sulfur Spray. The following spring, when new growth starts, begin the spray program again. Plant resistant varieties (for a list, see page 357).

Powdery mildew

Powdery mildew.

Problem: Young leaves, young twigs, and flower buds are covered with a thin layer of grayish white powdery material. Infected leaves may be distorted and curled, and many may turn yellow or purplish and drop off. New growth is often stunted, and young canes may be killed. Badly infected flower buds do not open properly. In late summer, tiny black dots (spore-producing bodies) may be scattered over the powdery covering like ground pepper.

Analysis: Powdery mildew is caused by a fungus (*Sphaerotheca pannosa* var. *rosae*). It is one of the most widespread and serious diseases of roses. The powdery covering consists of fungal strands and spores. The spores are spread by the wind to healthy plants. The fungus saps plant nutrients, causing distortion, discoloration, and often death of the leaves and canes. Powdery mildew may occur on roses any time during the growing season when rainfall is low or absent, temperatures are between 70° and 80°F, nighttime relative humidity is high, and daytime relative humidity is low. In areas where there is high rainfall in spring and summer, control may not be needed until the drier months of late summer. Rose varieties differ in their susceptibility to powdery mildew.

Solution: Apply ORTHO RosePride Funginex Rose & Shrub Disease Control or ORTHO RosePride Orthenex Insect & Disease Control at the first sign of mildew. Repeat the spray at intervals of 7 to 10 days if mildew reappears. Rake up and destroy leaves in the fall. Plant resistant varieties.

ROSA (Rose) (continued)

Rust	Spider mites	Viruses

Rust

Rust on rose foliage.

Problem: Yellow to brown spots, up to ¼ inch in diameter, appear on the upper surfaces of leaves, starting in the spring or late fall. The lower leaves are affected first. On the undersides of leaves are spots or blotches containing a red, orange, or black powdery material that can be scraped off. Infected leaves may become twisted and dry and drop off the plant, or they may remain attached. Twigs may also be infected. Severely infected plants lack vigor.

Analysis: Rose rust is caused by any of several species of fungi (*Phragmidium* species) that infest only rose plants. Rose varieties differ in their susceptibility to rust. Wind spreads the orange fungal spores to rose leaves. With moisture (rain, dew, or fog) and moderate temperatures (55° to 75°F), the spores enter the tissue on the undersides of leaves. Spots develop directly above, on the upper surfaces. In the fall, black spores develop in the spots. These spores can survive the winter on dead leaves. In spring, the fungus produces the spores that cause new infections. Rust may also infect and damage young twigs.

Solution: At the first sign of rust, pick off and destroy the infected leaves and spray with ORTHO RosePride Funginex Rose & Shrub Disease Control or ORTHO RosePride Orthenex Insect & Disease Control. Repeat at intervals of 7 to 14 days for as long as conditions remain favorable for infection. Rake up and destroy infected leaves in the fall. Prune off and destroy infected twigs. Apply ORTHO Dormant Disease Control Lime-Sulfur Spray during the dormant season. Plant resistant varieties.

Spider mites

Spider mite damage and webbing.

Problem: Leaves are stippled, bronzed, and dirty. A silken webbing may be found on the lower surfaces of the leaves or on new growth. Infested leaves often turn brown, curl, and drop off. New leaves may be distorted. Plants are usually weak and appear spindly. To determine if a plant is infested with mites, examine the bottoms of the leaves with a hand lens. Or hold a sheet of white paper underneath an affected leaf and tap the leaf sharply. Minute specks the size of pepper grains will drop to the paper and begin to crawl around. The pests are easily seen against the white background.

Analysis: Spider mites, related to spiders, are major pests of many garden and greenhouse plants. They cause damage by sucking sap from the undersides of leaves. As a result of their feeding, the plant's green leaf pigment disappears, producing the stippled appearance. Spider mite webbing traps cast-off skins and debris, making the plant messy. Many leaves may drop off. Severely infested plants produce few flowers. Mites are active throughout the growing season but they thrive in hot, dry weather (70°F and up). By midsummer, they can build to tremendous numbers.

Solution: Spray with ORTHO RosePride Orthenex Insect & Disease Control, ORTHO Bug-B-Gon Ready-Spray, or ORTHO Isotox Insect Killer when damage is first noticed. Cover the undersides of the leaves thoroughly. Repeat the application two more times at intervals of 7 to 10 days.

Viruses

Virus disease in rose.

Problem: Yellow or brown rings, or yellow splotches of various sizes, appear on the leaves. The uninfected portions remain dark green. New leaves may be puckered and curling; flower buds may be malformed. Sometimes there are brown rings on the canes. The plants are usually stunted.

Analysis: Several viruses infect roses. The viruses are transmitted when an infected plant is grafted or budded to a healthy one. This generally occurs in the nursery where the plant is grown. Some plants may show symptoms in only a few leaves. The virus lives throughout the plant, however, and further symptoms may appear later. Most rose viruses are fairly harmless unless there is extensive yellowing or browning. The virus suppresses the development of chlorophyll, causing the splotches or rings. Food production is reduced, which may result in stunted plant growth.

Solution: No cure is available for virus-infected plants. Rose viruses rarely spread naturally; therefore, only weak plants need to be removed. When purchasing rose bushes, buy only healthy plants from a reputable dealer.

Rose aphid

Rose aphids (8× life size).

Problem: Tiny (⅛-inch), green or pink, soft-bodied insects cluster on leaves, stems, and developing buds. When insects are numerous, flower buds are usually deformed and may fail to open properly. A shiny, sticky substance often coats the leaves. A black, sooty mold may grow on the sticky substance. Ants may be present.

Analysis: Rose aphids (*Macrosiphum rosae*) do little damage in small numbers. Plants can tolerate fairly high populations without much effect. The aphids are extremely prolific, however, and populations can rapidly build to damaging numbers during the growing season. Damage occurs when the aphid sucks the juices from the rose stems and buds. The aphid is unable to digest fully all the sugar in the plant sap and excretes the excess in a fluid called *honeydew*, which often drops onto the leaves below. A sooty mold may develop on the honeydew, causing the rose plants to appear black and dirty. Ants feed on the sticky substance and are often present where there is an aphid infestation. When aphid populations are high, flower quality and quantity are reduced.

Solution: Spray with ORTHO Isotox Insect Killer, ORTHO Bug-B-Gon Ready-Spray, ORTHO Rose & Flower Insect Killer, or an insecticidal soap when clusters of aphids are noticed. Repeat the treatment if the plant becomes reinfested.

Beetles

Fuller rose beetle (4× life size).

Problem: Holes appear in the flowers and flower buds; open flowers may be entirely eaten. Often affected buds fail to open, or they open deformed. Stem tips may be chewed, or the leaves may be notched or riddled with holes. Red, green-spotted, brownish, or metallic green beetles up to ½ inch long are sometimes seen on the flowers or foliage.

Analysis: Several types of beetles infest roses. They may destroy the ornamental value of the plant by seriously damaging the flowers and foliage. The insects usually spend the winter as larvae in the soil or as adults in plant debris on the ground. In late spring or summer, mature beetles fly to roses and feed on the flowers, buds, and sometimes leaves. Punctured flower buds usually fail to open, and flowers that do open are often devoured. Many beetles feed at night, so their damage may be all that is noticed. Female beetles lay their eggs in the soil or in flowers in late summer or fall. The emerging larvae crawl down into the soil to spend the winter, or they mature and pass the winter as adults. The larvae of some beetles feed on plant roots before maturing in the fall or spring.

Solution: Spray with ORTHO Isotox Insect Killer or ORTHO Orthene Systemic Insect Control when damage is first noticed. Repeat the spray if the rose becomes reinfested.

Crown gall

Crown gall appearing at the base of a plant.

Problem: Large, corky galls up to several inches in diameter appear at the base of the plant and on the stems and roots. The galls are rounded, with rough, irregular surfaces, and may be dark and cracked. Plants with numerous galls are weak; growth is slowed and leaves turn yellow. Branches or the entire plant may die back. Plants with only a few galls often show no other symptoms, however.

Analysis: Crown gall is caused by a soil-inhabiting bacterium (*Agrobacterium tumefaciens*) that infects many ornamentals and fruit trees in the garden. The bacteria are often brought to a garden initially on the stems or roots of an infected plant and are spread with the soil and contaminated pruning tools. The bacteria enter the plant through wounds in the roots or the stem. They produce a compound that stimulates rapid cell growth in the plant, causing gall formation on the roots, crown, and sometimes branches. The galls may disrupt the flow of water and nutrients up the roots and stems, weakening and stunting the top of the plant. Galls do not usually cause the death of the plant.

Solution: Crown gall cannot be eliminated from a plant. An infected plant may survive for many years, however. To improve its appearance, prune out and destroy galled stems. Disinfect pruning shears after each cut by dipping in a solution of 1 part chlorine bleach and 9 parts water. Destroy severely infected plants. The bacteria will remain in the soil for 2 to 3 years. If you wish to replace the infected roses soon, select other plants that are resistant to crown gall.

ROSA (Rose) *(continued)*

Roseslug

Roseslugs (life size).

Problem: The upper or lower surfaces of leaves are eaten between the veins; the lacy, translucent layer of tissue that remains turns brown. Later, large holes or the entire leaf, except the main vein, may be chewed. Pale green to metallic green sluglike worms, up to ¾ inch long, with large brown heads, may be found feeding on the leaves. Some have hairs covering their bodies, and others appear wet and slimy.

Analysis: Roseslugs are the larvae of black-and-yellow wasps called *sawflies*. The adult wasps appear in spring. They lay their eggs between the upper and lower surfaces of leaves along the leaf edges, with a sawlike egg-laying organ. Depending on the species of sawfly, some of the larvae that emerge exude a slimy substance, giving them a sluglike appearance. Others are hairy. The roseslugs begin feeding on one surface of the leaf tissue, skeletonizing it. Later, several species chew holes in the leaf or devour it entirely. When they are mature, the larvae drop to the ground, burrow into the soil, and construct cells in which to pass the winter. Some roseslugs pupate, emerge as sawflies, and repeat the cycle two to six times during the growing season. Severely infested roses may be greatly weakened and produce fewer blooms.

Solution: Spray with an insecticide containing *diazinon* when damage is first noticed. Repeat as necessary if the rose becomes reinfested.

SALIX (Willow)

Willow borers

Damaged stem. Inset: Adult (2× life size).

Problem: Swollen areas with holes in their centers develop on twigs, branches, or the trunk. Many side shoots may grow from below the swellings, destroying the tree's natural form. The leaves on infested twigs and branches turn yellow and may be chewed on the edges. Sawdust and many broken twigs are often found beneath the tree.

Analysis: At least five species of willow borers (*Saperda* species) feed on willows, causing galls, or swollen areas, to form. The 1-inch, striped and spotted, brownish or gray beetles with long antennae appear in late spring or early summer. Females lay their eggs in small holes that they gnaw in the bark of twigs and branches that are more than ½ inch in diameter. The emerging legless, whitish grubs tunnel into the wood. Excess tissue accumulates around the wound, resulting in a gall. When infestations are severe, nearly all twigs and branches more than ½ inch in diameter have one or more galls. The galls weaken the twigs, causing them to break and litter the ground during stormy weather. The grubs remain in the wood 1 to 2 years before maturing into adults.

Solution: Remove and destroy severely damaged trees. Spray the bark of less severely damaged trees with an insecticide containing *chlorpyrifos* in late May or early June. Repeat the spray 2 weeks later.

Poplar-and-willow borer

Poplar-and-willow borer (life size).

Problem: Leaves turn yellow and holes appear in the twigs. Large quantities of sawdust cling to the bark just below the holes. Sap often oozes from the holes. Young trees may be killed, and older trees may lose their natural form due to the production of numerous side shoots. Small (⅜-inch), black or dark brown weevils with pale yellow spots (scales) and long snouts may be seen around the tree from midsummer until fall.

Analysis: All willows and most species of poplar may be attacked by the poplar-and-willow borer (*Cryptorhynchus lapathi*), a weevil also called the *mottled willow borer*. The adult weevils cause minor injury by chewing holes in the bark of twigs. The major damage is caused by the C-shaped larvae, which are white with brown heads. During mid- to late summer, the larvae hatch from eggs laid in holes chewed by the female weevils. The larvae burrow into and feed on the inner bark. In the spring, large quantities of frass (sawdust and excrement) are expelled from the holes as the larvae tunnel into the center of the twigs to pupate. The feeding and tunneling cause branches to break easily and disrupt nutrient and water movement through the tree. The leaves turn yellow, and the tree often becomes bushy from the growth of numerous side shoots. The larvae pupate in June and emerge as adults in midsummer. Several other borers may also infest the trunk and branches of willows.

Solution: Remove and destroy severely infested trees or branches before early summer. Spray the bark with an insecticide containing *chlorpyrifos* in late July or early August.

SORBUS (Mountain ash)   SYRINGA (Lilac)

Fire blight

Fire blight on mountain ash.

Problem: The blossoms and leaves of infected twigs suddenly wilt and turn black as if scorched by fire. The leaves curl and hang downward. The bark at the base of the blighted twig becomes water-soaked, then dark, sunken, and dry; cracks may develop at the edge of the sunken area. In warm, moist spring weather drops of brown ooze appear on the sunken bark. Young trees may die.

Analysis: Fire blight is caused by a bacterium (*Erwinia amylovora*) that is destructive to many trees and shrubs. (For a list of susceptible plants, see page 341.) The bacteria spend the winter in the sunken cankers on the branches. In the spring, the bacteria ooze out of the cankers. Bees, flies, and other insects are attracted to the sticky ooze and become smeared with it. When the insects visit a mountain ash flower for nectar, they infect it with the bacteria. The bacteria spread rapidly through the plant tissue in warm (65°F or higher), humid weather. Insects visiting these infected blossoms later carry bacteria-laden nectar to healthy blossoms. Rain, wind, and tools may also spread the bacteria. Tender or damaged leaves may be infected in midsummer.

Solution: Prune out infected branches 12 to 15 inches beyond any visible discoloration and destroy them. Disinfect pruning shears after each cut by dipping in a solution of 1 part chlorine bleach and 9 part water. A protective spray of a bactericide containing basic copper sulfate applied before bud-break in the spring will help prevent infection. Spray with a bactericide containing *streptomycin* during bloom. Repeat at intervals of 5 to 7 days until the end of bloom.

Powdery mildew

Powdery mildew on lilac.

Problem: The leaves are covered with a thin layer or irregular patches of a grayish white powdery material. Infected leaves may turn yellow and drop off. New growth is often stunted. In late summer, tiny black dots (spore-producing bodies) are scattered over the white patches like ground pepper.

Analysis: Powdery mildew is caused by a fungus (*Microsphaera alni*) that thrives in both humid and dry weather. The powdery patches consist of fungal strands and spores. The fungus saps plant nutrients, causing yellowing and sometimes the death of the leaf. Since this mildew attacks many kinds of trees and shrubs, the fungus from a diseased plant may infect other plants in the garden.

Solution: Spray plants with ORTHO RosePride Funginex Rose & Shrub Disease Control or ORTHO RosePride Orthenex Insect & Disease Control when the plant shows the first sign of powdery mildew. Cover the upper and lower surfaces of leaves thoroughly. Repeat the treatment at intervals of 7 to 10 days until the mildew disappears.

Lilac borer

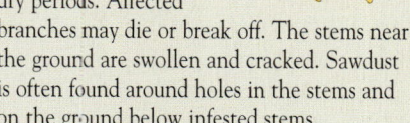

Borer damage in lilac.

Problem: Branch tips wilt in late summer, especially during warm, dry periods. Affected branches may die or break off. The stems near the ground are swollen and cracked. Sawdust is often found around holes in the stems and on the ground below infested stems.

Analysis: The lilac borer (*Podosesia syringae*), also called the *ash borer*, is the larva of a brownish, clear-winged moth that resembles a wasp. Moths may be seen flying around the plant in late spring. The moths lay their eggs in cracks or bark wounds at the base of the stems. The cream-colored larvae bore into the wood and feed on the sapwood and heartwood. The stems become swollen and may break where the larvae are feeding. Their feeding also cuts off the flow of nutrients and water through the stems, causing the shoots to wilt and die. The larvae spend the winter in the stems. In the spring, they feed for a few weeks before maturing into moths.

Solution: Before the moths emerge in the spring (April to May), cut out infested stems to ground level and destroy them. In late April, spray or paint the trunks and stems with ORTHO Borer & Leaf Miner Spray. Repeat the treatment two more times at intervals of 7 to 10 days. For more precise spray timing, use pheromone traps to sample pests. Spray 10 days after the first male is trapped. Kill borers by inserting a flexible wire into the borer hole in early summer. Avoid pruning during the spring when moths are present. Avoid wounding the shrubs with lawn mowers.

TAXUS (Yew)

Poor soil

Yew in poorly drained soil.

Problem: Young leaves turn yellow. Eventually the entire plant may turn yellow, wilt, and die. The plant is growing in heavy, poorly drained, acid or alkaline soil.

Analysis: Yews are particularly sensitive to improper growing conditions. When planted in soil that is heavy, poorly drained, very acid (between pH 4.5 and 5.5), or very alkaline (above pH 7.5), the plants do not usually survive. The bark on the roots decays and sloughs off, and the roots die. The roots can no longer supply sufficient amounts of nutrients and water to the leaves, resulting in leaf yellowing and wilting. The plant usually dies within several months.

Solution: Improve the soil drainage, or, if the plant is small, move it to an area with better drainage. Check the acidity of your soil. If the pH is below 6.0, add ground limestone around the base of the plant. Add aluminum sulfate to the soil if the pH is above 7.0. The optimum pH for yews is between 6.0 and 6.5. Do not water yews heavily.

ULMUS (Elm)

GROWING GUIDE

Adaptation: Zones 3 through 10. To determine your zone, see the map on page 336. Not adapted to desert areas.

Light: Full sun.

Soil: Tolerates a wide variety of soils.

Fertilizer: Fertilize with Scotts Evergreen, Shrub and Tree Food according to label directions.

Water: Water during periods of extended drought.

Pruning: Prune off dead wood and destroy or strip off the bark to eliminate breeding sites for elm bark beetles, which transmit Dutch elm disease.

Elm leaf beetle

Larva (life size). Inset: Adult and eggs (3× life size).

Problem: The lower surface of the leaf is eaten between the veins, giving the leaves a lacy appearance. Small holes may be in the leaves. Severely infested leaves turn brown; the entire tree may appear scorched. Many leaves drop off by midsummer. Small (½-inch) yellow-and-black larvae or ¼-inch yellowish green-and-black striped beetles may be found on the undersides of leaves.

Analysis: Elm leaf beetles (*Pyrrhalta luteola*) attack all species of elm, but the beetles may have local preferences. The beetles spend the winter as adults in buildings or in protected places outside. In the fall, when the beetles are looking for shelter, they often become a nuisance inside homes. The adults fly back to elm trees in the spring. They eat small holes in the developing leaves, mate, and lay clusters of yellow eggs. The emerging black larvae feed on the lower surface of the leaf between the veins. As the larvae mature, they turn a dull yellow with black stripes. After feeding for several weeks, the larvae pupate. Bright yellow pupae may be seen around the base of the tree in late June or early July. Adults emerge, and one or two more generations may follow. Trees that lose many of their leaves early in the season may grow new ones, which also may be eaten.

Solution: Spray with an insecticide containing *diazinon*, *methoxychlor*, or *chlorpyrifos* when damage is first noticed. To prevent severe leaf damage, apply sprays just as the leaves grow to full size. Repeat the spray if the tree becomes reinfested.

Dutch elm disease

Dutch elm disease.

Diseased stem on left.

Failure to bloom

Failure to bloom.

Problem: Leaves wilt, curl, and turn yellow on one or more branches in the top of the tree; many leaves drop off. Trees may die slowly over a period of a year or longer. Or trees wilt and die within a few weeks, often in the spring soon after they have leafed out. Sections of infected branches may show a ring of brown dots or brown streaks in the wood just underneath the bark. Small holes may be found in the bark of infected branches.

Analysis: Dutch elm disease is caused by a fungus (*Ophiostoma ulmi*) that invades and plugs the water-conducting vessels in the tree. The fungus enters the wood through feeding wounds made by elm bark beetles. In the spring, the adult beetles emerge from holes in the bark of elm trees where they have spent the winter. If the trees are infected with the Dutch elm fungus, the beetles have sticky spores of the fungus on and inside their bodies. The beetles fly to healthy elm trees, where they feed in crotches of small twigs, usually high in the tree, and deposit the fungus spores in the wounds. The fungus then spreads downward through the tree. The infected elm usually develops the disease that summer. The fungus produces a toxin that interferes with the water-conducting vessels in the wood, reducing the amount of water available to the leaves. The foliage on the infected branch wilts, turns yellow, and drops. Surrounding branches, and eventually the entire tree, become infected, and the tree dies. When elms are closely planted (50 feet or less between trees), the fungus may spread through natural root grafts between trees.

Solution: Curing a tree of Dutch elm disease is usually not possible. Severely infected trees should be removed promptly to prevent the spread of the disease. Disease development may be delayed on lightly infected trees (where less than 5 percent of the foliage and branches show evidence of the disease) that were initially infected by bark beetles rather than by root grafts. Removal of early infections by tree surgery may save a tree for a number of years. Remove yellowing branches at least 10 feet below the point where brown streaks are visible in the wood. Spray the wound with an insecticide containing *methoxychlor*. Systemic fungicides (fungicides that are carried throughout the tree) injected into the tree by a trained arborist may increase the life span of a lightly infected tree. Contact an arborist when yellowing is first noticed. Prevention involves four different measures that should be carried out on a community-wide basis. **1.** A good sanitation program will slow the spread of the fungus. Because bark beetles carrying the fungus breed in dead or dying elm wood, all dead or dying trees, damaged limbs, and prunings should be removed and burned or buried. The bark of all stumps should be peeled to just below ground level. **2.** Controlling bark beetles with insecticides is practical only on valuable specimens, when used in conjunction with a sanitation program. Spray with an insecticide containing *methoxychlor* in early spring before leaves come out. Use high-pressure spray equipment, and cover the entire tree thoroughly. **3.** To prevent transmission of the fungus through the roots of closely planted trees, grafted roots should be severed either by mechanical trenching or by soil injection of a chemical. This is important if a diseased tree is nearby. Contact a professional arborist. **4.** An annual scouting program should be established to detect and deal with infected trees. When replanting, select resistant elms. (For a list of resistant elms, see page 356.)

Problem: No flowers appear in the spring, but the vine is healthy and vigorous.

Analysis: Most wisterias purchased from nurseries bloom well after 2 or 3 years. These plants are usually asexually propagated (started from cuttings or by some method other than from seed). Vines started from seed often do not bloom for 10 or more years. Wisterias may also fail to bloom because of improper growing conditions, poor pruning practices, or freeze damage. Young plants should be well fed and watered. Plants old enough to bloom flower best with less food and water. Too much nitrogen fertilizer during the growing season causes lush, overly vigorous, green growth and poor flower bud production (flower buds for the following season's bloom are produced in early summer). Heavy pruning also may produce lush, overly vigorous growth, or flower buds may be mistakenly removed.

Solution: Do not grow wisteria from seed. Buy nursery-grown vines. If your old wisteria did not bloom in spring and is lush and growing vigorously, do not use nitrogen fertilizer for an entire season. Fertilize in early summer with 0–10–10 fertilizer to promote flower bud formation for the following season. Prune back vigorous shoots in summer. In winter, cut back or thin out side shoots from the main stems. Spurs (short fat stems bearing flower buds) develop on these side shoots. Cut back spurs to two or three buds. Do not drastically prune side shoots. Drastic pruning will eliminate all the spurs so no flowers are produced.

Apple 'Summer Red'

The desire to taste old-fashioned fruit flavor has led rural gardeners to experiment with antique varieties and suburban gardeners to plant their first dwarf apple or apricot trees. The savor of a handful of homegrown walnuts has inspired many gardeners to raise their own nut crops. A fresh ripe peach, plum, or pear is one of the delights of summer. Growing healthy trees and a satisfying crop requires thoughtful consideration throughout the plants' life cycle and in all seasons.

PRODUCING A CROP

To tend your fruit and nut trees well, you must be aware of natural production cycles; take an interest in the mysteries of pollination; learn the practical art of thinning; and, as always, remain alert for problems. Most gardeners agree unhesitatingly that the reward of a bumper harvest is worth the trouble.

Pollination: If your fruit tree is festooned with blossoms but yields pea-sized fruit that drops off instead of growing, poor pollination may be the cause. Some fruit trees, including apricots, nectarines, peaches, pomegranates, persimmons, sour cherries, and most citrus, are self-fruitful. They can be successfully fertilized by their own pollen. But most sweet cherries, plums, apples, papayas, pears, and nut trees must be grown within 100 feet of a different variety of the same fruit or nut tree that flowers at the same time.

An early-blooming tree cannot cross-pollinate a late-blooming one. If your garden lacks space for cross-pollinating trees and there are none in the immediate vicinity, talk to your local nursery or a friend with fruit trees. Sometimes setting a bucket of compatible blossom-filled branches under your target tree will do the trick.

You may see labels declaring plums to be self-fruitful. Some are, to a limited extent. But fruiting will be markedly better if pollen comes from another plum tree. Plums are quite fussy about which plum tree variety provides this pollen, so read specifics when you purchase to prevent problems.

Apples and pears have some of the same fussiness about their mates. If the nearby variety blossoms irregularly or has a different genetic base, fruit may not set. If trees were present when you moved onto the property, or the label has been misplaced, you may

have to do some investigation to find out which varieties you have in order to improve fruiting. Your local nursery, county agricultural extension service, or old-time apple association can usually help.

Insect pollination: Insect pollinators are necessary to transfer pollen within a tree, or from one fruit or nut tree to another. Though birds, animals, and breezes transfer some pollen, insects move most of the pollen within a tree or from one tree to another. In its search for food, an insect may land on the male part of a flower, or anther. The anther contains pollen, which sticks to the insect's body. Later, if the insect visits the female part of a flower, the stigma, pollen may brush off. If it does, pollination occurs. Pollination is the necessary first step to producing a crop.

But bad weather limits insect movements. If there is continued heavy rain, unseasonal cold, or heavy wind during prime flowering season, not only are some of the blossoms knocked off, but pollinators don't get out to do their job.

Honeybees perform most fruit and nut tree pollination. Experts have estimated that they are crucial to over $20 billion in food crops each year. Honeybees move from flower to flower, collecting both nectar and pollen. Nectar is the sweet juice from inside

a flower; bees transform it into honey. Bees need both nectar and pollen to survive.

In today's world of rapid construction and decreasing open space the honeybee population seems to be declining. If you want bees for fruit and nut tree pollination, you should take steps to encourage their presence. Provide them with a continuous supply of clean water. A water supply for bees must be shallow, as they may drown when trying to drink from a deeper container. Place some gravel in a bucket, then fill the bucket with just enough water to cover the stones. The gravel provides a place for the bees to settle as they refresh themselves.

Bees are extremely sensitive to controls for other insects. The best course is not to use pesticides on any plant in an area where bees are feeding. If you must use a pesticide, do so in the late afternoon, when fewer bees are visiting flowers.

Insects known as syrphids, hover flies, or flower flies are second only to bees as pollinators. These ¼- to ½-inch-long insects are often confused with honeybees or yellowjacket wasps. Unlike bees and wasps, however, syrphids do not sting. They hover in the air, seemingly motionless, over blossoms, occasionally dipping down to feed on both nectar and pollen. In addition to their work as pollinators, syrphids help control pests.

Growing a fruit tree as an espalier is a good way to enjoy a crop in limited space.

Overgrown with a dense jumble of branches, this 'Mutsu' apple is ready for pruning.

Proper thinning lets in light and air. Branching is reduced by about one-half while the productive spurs are preserved.

Thinning reduces the number of fruits, but because the remaining fruits receive more nutrients they are bigger and healthier.

Young syrphids, ⅛- to ½-inch-long green or brown worms, eat an average of one aphid per minute and also feed on mealybugs, leafhoppers, and the like.

Wasps do a share of pollination and pest control. Wasps destroy flies, beetle larvae, and caterpillars. Treat wasps as beneficial insects unless their presence is potentially harmful to people. Some wasp species, most notably the yellowjacket, can be irritated by swatting and other disturbances. The result could be a series of painful stings.

Hand pollination: What do you do if rain, wind, cold, or other environmental factors keep insect pollinators from doing their job? You must literally put your hand to the task. The best time to pollinate by hand is when the weather has been warm and dry for at least 2 days. Shaking blossoms of one variety over those of another is a reasonably effective means of cross-pollination. To be more precise, move your fingertip over an anther. If yellow grains come off on your finger, pollen is available. Use a small natural-hair artist's brush or a cotton swab to transfer pollen from anthers of one variety to stigmas of another. Whichever method you use, transfer pollen every day until the trees are finished flowering. Hand-pollination may not result in a bumper crop, but in the absence of insect pollinators it is the only alternative if you want fruit or nuts.

OTHER CROP PROBLEMS

Fruit and nut yields are subject to natural cycles as well as weather conditions and the vicissitudes of pollination.

Failure to bear: Before worrying about a tree that is not bearing fruit or nuts, make sure it is mature enough to produce a crop. Peaches bear after three years, dwarf apples at two to three years, and plum trees at about four years. Sweet-cherry trees bear at five to seven years.

If your tree is mature yet produces little or no fruit despite adequate pollinators, again you may have no cause for concern. Some tree varieties, particularly apple, pear and citrus, tend toward light crops in alternate years, even when they are healthy. Heavy fruiting takes energy away from flower production for the following year. If the number of flowers is low, the yield is small. Thorough thinning helps trees produce consistent yields from year to year.

Poor weather: Poor fruiting and premature fruit drop occur if temperatures are too low or if hot and cold weather alternate in late fall and early spring. In either case the tree is fooled into thinking spring has arrived. It sets new leaves and blossoms, which are particularly susceptible to damage from temperature extremes.

Alternate freezing and thawing can cause soil movement that damages roots. If the damage is severe, the nutrient flow to the trunk is disrupted.

To protect fruit and nut trees from extreme weather, mulch them with a 6-inch layer of straw, evergreen branches, chopped leaves, wood chips, or pine needles. If you expect the temperature to drop below freezing during the night, temporarily cover trees with fabric or plastic sheets. Make sure the covers are loose and remove them in the morning; left in place they block sunlight

and impede air circulation. For prized fruit or nut trees, consider installing a heat source. Sometimes the heat from a simple 60-watt bulb turned on and set under a tree draped with clear plastic is enough to prevent freezing. Ask local nursery professionals or the county extension agent what heat source works best in your microclimate.

Thinning: By natural dieback and leaf drop, trees normally do some of their own thinning so the remaining fruit will receive the nutrients to reach proper size. It takes 30 healthy leaves to ripen one full-size orange or apple. If all the fruit a tree set stayed on the branches, each fruit would probably be undersized because of nutrient shortage.

Even so, in most cases you must take an active role in ensuring large fruit or nuts by removing some immature crop yourself. This process is known as thinning. In addition to providing each remaining fruit or nut with a bigger share of nutrients, thinning opens each fruit to more sun and air, making it less susceptible to disease.

Each type of fruit or nut tree has its own thinning requirements. For example, thin apples and pears by cutting the stems with sharp scissors or pinching the stems between thumb and forefinger. Leave the stem behind when you cut. Thin pears after natural spring fruit drop, when the fruits turn downward. Leave 2 pears per cluster. With apples, leave one per cluster. When thinning, you may notice that the "crown" apple (the fruit in the middle of a cluster) is malformed. If so, be certain to remove it. But if not, leave it, as it can be larger and quite good. The clusters should be about 6 inches apart.

CORRECTING GENERAL PROBLEMS

In addition to solving problems relative to crops, gardeners must often solve problems such as drought, wind stress, and sucker and seedling growth.

Drought: A lack of water can be caused by a lack of rainfall, of course, but it can also result from inadequate saturation.

Make the most of rain and irrigation by digging a shallow basin around your fruit or nut tree. Extend the basin about a foot beyond the branch tips, and keep enlarging it as the tree grows. Construct an earthen barrier about a foot from the trunk; it should be high enough to keep the water in the basin from touching the trunk.

Lawn watering does not suit fruit or nut trees. In most cases, lawn watering only soaks the top 2 inches of soil. Sprinkler systems keep trunk and top roots wet, encouraging plant diseases. Trees that are only watered along with the lawn may grow slowly and have small yellowing leaves that drop early.

To lessen the problems of lawn watering, aim sprinklers so they do not reach the trunk or base of the tree. If possible, leave a 2-foot space between tree trunk and grass. To ensure soil saturation, irrigate deeply to deliver water 2 to 3 feet below the surface.

Wind stress: Constant wind causes rapid water evaporation. If the roots cannot take in and move as much water as the leaves are losing, the result is wilted leaves. Wilting is most common in young trees, which may not have root systems substantial enough to counter the evaporation. Constant wind stress slows growth.

Keeping the stressed plant well-watered at all times can mitigate wind damage. Another solution is a windbreak, wall, or shrub that blocks the wind. If you plant a living windbreak, make certain it does not compete with the stressed tree for water or sun. In serious cases of wind stress, consider moving the plant to another location.

Suckers: A sucker is a shoot that grows from the roots or the lower part of the trunk. Sucker leaves may look different than those on higher branches. If suckers appear at the base of a new tree and there is no tree top growth, the cause may be root injury during planting or cultivation around the trunk. Cut suckers at the base as soon as they emerge from the ground. Suckers weaken a tree by using nutrients that should go into tree growth. Provide sufficient water and wind protection to prevent stress while roots recover.

Seedlings: Fruits and nuts often fall from the trees that bore them; seedlings often sprout from the fallen crop. Though the seedlings may bear edible fruit, this second generation is usually inferior. In many cases, the seedlings do not develop true to type. If you want to experiment, move the seedling to another site. Otherwise, it may grow larger than the parent tree and shade it out.

PURCHASING A NEW TREE

A tree buyer has two goals: acquiring a healthy tree and doing so at the right time of year.

The area of a tree just under the bark is called the cambium. This is the area in which the tree creates new cells; therefore, an assessment of the health of the cambium provides a test of whether the tree you are thinking about buying has the potential to grow, given supportive conditions. The cambium should be bright green. The bark that covers it should not be shriveled.

Check containerized trees for girdling, or roots that circle around inside or outside the container. As girdling roots grow, they wrap ever more tightly around each other, cutting off nutrient flow. The result can be the death of the tree. To check for girdling roots inside the container, poke your finger 2 to 3 inches into the soil near the trunk. Brush away a bit of topsoil. If you see or feel a root that is damaged or constricted by another root, select a different tree.

To ease transplant shock, purchase your tree as soon as the weather is warm enough to permit safe planting. In early spring the tree is still dormant; it has no leaves or buds to support. After you plant it, the tree can devote its energy to repairing the damage that even the most careful installation causes. By the time the tree is ready to produce spring growth, it has recovered from the trauma of planting.

PLANTING THE TREE

To get a new tree off to the best possible start, provide trunk support and take care to minimize transplant shock.

Trunk support: Too much trunk movement prevents roots from getting a good grip into the soil. If not anchored solidly by roots, the tree trunk may snap in a strong wind. To prevent this, install support posts before installing the tree. A standard-sized tree requires 8-foot posts, semi-dwarfs

A regular spray program ensures the highest quality home-grown fruit.

Cylindrical wire cages that reach 2 feet above the normal snowline protect vulnerable bark from rabbits and mice.

201

'Eureka' lemon

Clockwise from left: English walnuts, black walnuts, and Persian walnuts.

6-foot posts, and dwarf trees 4-foot posts. Oak stakes stand up well to all types of weather. To avoid damage to the roots of the new tree, place the post in the hole before lowering the tree into the hole. Place the stake on the side of incoming winds. Use rubber ties, plastic tape, or wire shielded with lengths of garden hose to connect the tree to the stake. Provide enough slack to allow the trunk to sway slightly. There must be some leeway, for this encourages a strong trunk. Check the ties regularly and loosen or replace them as necessary to prevent constriction and damage to the trunk.

Transplant shock: In a nursery, a bare-root, burlapped, or container tree is carefully watered, shaded, and nurtured. Newly arrived in the garden, the same tree may be propped against a heat-reflecting wall, causing it to bake. If the tree is unwrapped or removed from its container, the plant is even more susceptible to hot sun and drying wind. Left unprotected for even a few hours, a fruit or nut tree can be in shock and dying even before being placed in the planting hole. If possible, plant the tree immediately and water it well. When transplanting a tree from a container to the ground, soak the root ball in water while it is still in the container; then place the tree into the planting hole.

If you must wait a few days before planting the tree, place it in a shallow holding trench you have dug in a shady location before purchasing the plant. Lean the trunk and root ball against one side of the trench. Thoroughly cover the roots with soil and water. Place the tree into the planting hole as soon as possible.

Growing in containers: If you want a fruit crop but have limited space or poor drainage, a dwarf tree in a container may be a solution. A dwarf tree also tends to bear fruit earlier in its life than a standard-sized tree, so a dwarf may also be the answer if you want a quick crop. Unfortunately, containerized dwarf trees are quickly affected by poor care. The results are yellowing, drooping, curling leaves, and poor fruit production.

Purchase only healthy plants. Leaves should be large and green. Choose a tree with few flowers or fruit rather than one full of blossoms or fruit. The less ornamented tree may not be as appealing now, but will put more energy into valuable first-year root development. Avoid root-bound plants. Circling or matted roots must be cut back to encourage them to grow outward, and cutting encourages transplant shock.

The container for a dwarf tree should be about 3 inches wider than the roots when they are spread out. Do not use soil directly from the garden unless you have good loam. Avoid clay; it holds water and fruit trees do not like soggy soil.

Fruit trees in containers cannot expand out into surrounding soil to obtain moisture or nutrients. What you provide in the container is what they attempt to live with. Lack of water is a major problem.

Water when the top of the soil feels dry to the touch. Do not allow the leaves to dry to the point of wilting; repeated wilting can cause death. Do not overwater. Containerized soil that stays soggy invites crown and root rot.

Apply fruit-tree fertilizer once a month to correct or prevent iron chlorosis or zinc deficiencies. If leaf edges turn brown and dry, excess fertilizer may be accumulating in the container. To leach out the fertilizer, put a garden hose into the container and turn the water on low. Let water run through the container for about 20 minutes.

HARVESTING

Picking fruit at just the right time ensures peak flavor. But it's often difficult to tell exactly when a fruit is ready to be taken off the tree. Color is a good indicator on trees such as pears. Citrus fruit, however, can look ripe for several months before they are actually edible. Your only course is to learn the harvest time for each type of fruit. The list that follows presents guidelines for harvesting a few common trees.

■ Taste citrus fruit to determine picking time. Leaving them on too long will do no harm; ripe citrus can remain on a tree for more than 2 months.

■ Harvest pears when they reach full size and are starting to lose their green color. Do not allow them to soften or turn yellow on the tree. Overripe pears are mealy, mushy, or gritty.

■ Pick plums, astringent persimmons, figs, and apricots when they are fully colored and slightly soft.

■ Remove cherries from the tree when they are fully colored and soft or firm. Soft cherries tend to be sweeter, but the birds may get to them first.

DISEASES OF FRUIT AND NUT TREES

Anthracnose	Serious fungal diseases affecting mango, papaya, and walnut trees. Brown spots appear on leaves and fruit. Infected areas decay rapidly. Fruit or nuts drop prematurely. Clear and destroy debris regularly. At 2-week intervals, spray with fungicide containing benomyl, maneb, or zineb.
Bacterial canker	Affects mainly cherry, fig, nectarine, peach, plum, and almond trees. Fruit is not usually infected, but entire branches may die back. Difficult to control; spraying in fall with basic copper sulfate may help. Plant resistant varieties. See pages 207, 219, 222.
Bacterial leaf spot	One of the more destructive diseases of stone-fruit trees, infecting apricot, nectarine, peach, and plum. Fruit may be ruined. Spraying with basic copper sulfate when buds open may help combat disease, but no total control exists. Plant resistant varieties. See pages 219, 222, 227, 234.
Black rot	Infects mainly apple and pear trees. Fruit may be ruined. Remove all rotted fruit from tree and ground. Prune as advised (see page 216). Spray with captan as soon as disease is noticed; respray the following spring, when growth begins.
Brown rot	Fungal disease destructive to stone-fruit trees—mainly apricot, cherry, nectarine, peach, and plum. Sometimes infects citrus. Fruit is often ruined. Spray uninfected blossoms and maturing fruit with captan or benomyl. Plant resistant varieties. See pages 218, 228, 235.
Crown and root rot	May infect nearly any fruit or nut tree. The whole tree generally loses vigor and appears ill. Only the upper branches may bloom and fruit. No controls are available. Infected trees usually die; replace them with resistant varieties. See page 213.
Fire blight	Severe on pear trees; can also infect apple. Fruit may be ruined. Spraying with basic copper sulfate at intervals before and throughout blooming helps combat the disease. Plant resistant varieties. See page 230.
Leaf curl	Infects peach and nectarine trees. Fruit crop is decreased, but fruit is edible. Leaf curl cannot be treated once it appears, but spraying in fall and spring with lime sulfur or a dormant disease control may prevent recurrence. See page 227.
Powdery mildew	Affects apple, cherry, papaya, and almond trees. Infected apples and papaya may be edible if peeled, but infected cherries are generally ruined. Spray at 2-week intervals with benomyl, dinocap, or sulfur. See page 210.
Scab	Serious problem on apple trees but can also affect apricot, avocado, citrus, mango, nectarine, peach, pear, and pecan. Infected fruit is edible if peeled. Scab cannot be treated once it appears, but spraying with captan in spring, after blossoms drop, may prevent recurrence. Plant resistant varieties. See pages 216, 218, 226, 230.
Shot-hole fungus	Infects apple, apricot, nectarine, peach, and almond trees. Yield is reduced, but fruit is edible if peeled. Shothole fungus cannot be treated once it appears, but spraying in spring with captan, lime sulfur, or a dormant disease control should help prevent recurrence. See page 227.

POOR FRUITING

Fruit too small

Small cherries.

Problem: The tree is healthy and produces many small fruits. No signs of pests or diseases exist.

Analysis: When a tree overbears, it distributes smaller quantities of nutrients to each maturing fruit, resulting in large numbers of small fruits. Certain fruit trees, including peaches, nectarines, Japanese plums, and apples, tend to produce large quantities of small fruits when the trees are not pruned or thinned adequately. If the fruit-bearing wood is not pruned during the dormant season, the tree will set much more fruit than it is able to bear to full size. Even when properly pruned, certain fruit trees have a tendency to overbear. A tree has only a limited amount of nutrients that can be supplied to the fruit.

Solution: Prune trees properly during the dormant season, and thin the young fruits when they are thumbnail-sized (4 to 8 weeks after bloom). For pruning and thinning details, look up your tree in the alphabetical section beginning on page 214.

Poor-tasting fruits and nuts

Poor-quality pear.

Poor-quality apples.

Problem: Fruits and nuts are not flavorful. Fruits may be dry, watery, pulpy, or grainy and may taste sour or tart.

Analysis: Fruits and nuts may be poor in flavor for several reasons.

1. Lack of nutrients: Trees planted in infertile soil grow poorly and often produce small crops of inferior-tasting fruits and nuts.

2. Environmental stress: Trees may be stressed by too much or too little soil moisture, excessively high or low temperatures, or rapid, unseasonable weather changes. Under environmental stress, fruits or nuts of many types of trees may fail to ripen properly, producing dry, pulpy, or otherwise poor-tasting crops.

3. Disease or insect damage: Diseases and pests often slow root, shoot, and leaf growth and prevent fruits and nuts from ripening properly. Fruits and nuts themselves may also be infected or infested, resulting in poor flavor.

4. Untimely harvest: When fruits are prematurely harvested, they may taste tart, flavorless, dry, or starchy. Pears taste gritty or mealy if they are allowed to ripen on the tree.

5. Fruit naturally unflavorful: Tree varieties vary considerably in the quality of their fruits and nuts. No matter how healthy and vigorous your tree is, the variety may just naturally produce a flavorless crop. Seedling trees often produce insipid fruits.

Solution: The numbered solutions below refer to the numbered items in the analysis.

1. Fertilize trees regularly according to label directions with Scotts Evergreen, Shrub and Tree Food.

2. To reduce stress caused by too much or too little soil moisture, avoid overwatering or underwatering your tree. Minimize tree stress resulting from weather and temperature fluctuations by maintaining the tree in good health. For specific cultural information, look up your fruit tree in the alphabetical section beginning on page 214.

3. To help ensure high-quality fruit and nut crops, keep your trees as free of insect pests and diseases as possible. To determine what is infecting or infesting your tree, look up your fruit tree in the alphabetical section beginning on page 214.

4. As a general rule, fruits are ripe when they are fully colored, slightly soft, and easy to separate from the branch when gently lifted. For details on harvesting, see page 203 or look up your fruit tree in the alphabetical section beginning on page 214.

5. If your tree appears to be healthy but has continued to produce poor-quality fruit over several years, plant a variety that bears more flavorful fruit or graft the tree with a good variety. Check with your local county extension office for a list of flavorful fruit and nut trees adapted to your area.

ANIMAL DAMAGE

Bark-feeding animals

Ground squirrel damage to almond trees.

Problem: Leaves and buds or shoots are chewed from trees. Bark may be gnawed from trunk or lower branches. In some cases, the trunk is girdled. Deer, rabbits, or mice may be seen in the yard, or their tracks may be seen on the ground or in snow. Damage is usually most severe during winter months.

Analysis: Several animals chew on bark.
1. Deer damage trees by feeding on the leaves, shoots, buds, and bark. They feed by pulling or twisting the bark or twigs, leaving ragged or twisted twig ends or patches of bark. The males may also damage trees by rubbing their antlers on trunk and branches.
2. Rabbits damage fruit trees by chewing on the bark at the base of the trunk and clipping off tender shoots. They chew bark and twigs off cleanly, leaving a sharp break. The damaged trunk is often marked with paired gouges. Rabbits feed no more than 2 feet above the ground or snow level. They damage young or dwarf trees most severely.
3. Field or meadow mice damage fruit trees by chewing off the bark at the base of the trunk, just at or slightly above or below ground or snow level. They may girdle the trunk, killing the tree. Mice leave tiny scratches in the exposed wood.

Solution: Protect individual trees with wire-mesh cages (1- to 2-inch mesh) or fencing. Cage or fencing need to be at least 8 feet high to keep out deer and 2 feet above the average snow level to keep out rabbits. Hardware cloth cylinders that extend 12 inches above the average snow level will keep out mice. Extend rabbit and mice barriers 3 to 4 inches below soil level.

Animals eating fruits and nuts

Bird-damaged apple.

Squirrel eating green cherry.

Problem: Ripened fruits and nuts have holes in them and may be partially eaten. Fruits and nuts may disappear from the tree or may have been knocked to the ground. Birds, tree squirrels, or raccoons may be seen feeding in the trees.

Analysis: Some birds and animals, especially tree squirrels and raccoons, feed on tree fruits and nuts.

1. Birds: Birds are notorious pests of many tree fruits, especially cherries, figs, persimmons, and other soft, sweet fruits. They peck at the ripening fruit, leaving holes in the flesh. The wounded fruit may decay, becoming inedible. Some birds also feed on fruit blossoms and tiny developing fruits, greatly reducing the overall fruit yield.

2. Squirrels: Tree squirrels feed on a large variety of foods, including bark, leaves, insects, and eggs. They prefer maturing nuts and fruits, however. They can strip entire trees of nuts, many of which they store for later use. They often leave partially eaten nuts on the ground around the tree. Tree squirrels are especially fond of filberts.

3. Raccoons: Raccoons are usually found in wooded areas near a source of water. They feed on a wide variety of foods, including ripening fruits and nuts. Raccoons may strip off fruits and nuts and carry them away or feed on them in the tree. In the process of feeding, they may knock many fruits and nuts to the ground.

Solution: The numbered solutions below correspond to the numbered items in the analysis.

1. The most effective way to control birds is to throw nets over the trees and secure the nets tightly around the trunk. Birds are most likely to damage ripening fruits. Check the trees every morning, and harvest fruits and nuts that have ripened. Bright, shiny objects hung in trees frighten birds and will repel them for a while.

2. Prevent tree squirrels from climbing fruit and nut trees by wrapping 2-foot-wide bands of metal (made from materials like aluminum roof flashing) snugly around tree trunks at least 6 feet above ground level. Prune trees so that all branches are at least 6 feet above the ground and 6 feet away from other trees and structures. If necessary, completely enclose dwarf trees or shrubs in a chicken-wire cage. If permissible in your area, you can also trap tree squirrels.

3. Raccoons are intelligent, inquisitive animals that can be difficult to control. To discourage raccoons from climbing between trees or from a building to a tree, keep limbs pruned so that they do not touch each other and do not make contact with the roof. Wrap metal guards at least 18 inches wide around tree trunks at least 3 feet above the ground.

BARK OR WOOD PROBLEMS

Limb breakage

Fruit overload on apple branch.

Problem: Branches laden with large quantities of ripening fruit break off.

Analysis: Limb breakage as fruit reaches full size can be a problem with trees that produce large fruits, such as peaches, nectarines, apples, and pears. Branches that have not been pruned or thinned properly are most likely to break. Branches with narrow crotch angles may break where they join the tree. Trees that produce fruit on thin, year-old wood—for example, peaches and nectarines—are most susceptible to limb breakage.

Solution: Prune off stubs where branches have broken. Prop up any branches that are bent or appear to be ready to break by cutting a notch at one end of a board and placing the board about one-third of the way in from the tip of the sagging branch. Set the branch into the notch, and then push the board into the ground. If the soil is not soft, you may need to dig a hole in the ground first. The board should be pushed in at a slight angle (20 degrees), leaning toward the center of the tree. During the dormant season, prune the tree properly and thin young fruit. For more pruning and thinning details, look up your fruit tree in the alphabetical section beginning on page 214.

Gummosis

Gummosis on peach.

Gummosis on apricot.

Problem: Beads of sticky, amber-colored sap appear on healthy bark, cankers, wounds, or pruning cuts.

Analysis: Oozing sap (gummosis) occurs in all trees to some extent and is caused by one or several of the following factors.

Solution: The numbered solutions below correspond to the numbered items in the analysis.

1. Natural tendency: Certain species of fruit trees, especially cherries, apricots, peaches, and plums, have a natural tendency to ooze sap. Small beads of sap often form on the bark of these trees.

1. As long as the bark appears healthy, you do not need to worry.

2. Environmental stress: Trees under stress because they are growing in wet, poorly drained, or very dry soil may produce large quantities of sap, even though they are not diseased. Also, many fruit trees respond to rapid changes in weather conditions or soil moisture by gumming profusely.

2. If your tree is growing in wet, poorly drained soil, allow the soil to dry out between waterings. Provide for drainage of water away from tree trunks and roots. To help prevent crown rot, carefully remove enough soil around the base of the trunk to expose the first major roots.

3. Mechanical injury: Most trees ooze sap when the bark or wood is wounded. Wounding results from limb breakage, lawnmower injury, pruning, improper staking, tying, or guying techniques—and other practices that damage the bark and wood.

3. Avoid unnecessary mechanical injuries to the tree. Stake, tie, and prune properly.

4. Disease: Fruit trees respond to certain fungal and bacterial infections by forming cankers that gum profusely. Gummosis is often one of the initial signs of infection.

4. Remove badly infected branches and cut out cankers. Keep the tree healthy. For cultural information, look up your fruit tree in the alphabetical section beginning on page 214.

5. Insect damage: Several species of insects bore into tree bark, causing sap to ooze from the damaged areas. The tunnels they form in the wood often become infected with decay.

5. Borers are difficult to control once they have burrowed into the wood. Keep trees healthy and vigorous. When pruning, do not leave stubs that can attract or harbor borers. Avoid injuries to roots and trunks.

Cankers

Cytospora canker on peach.

Problem: Sunken, oval, or elongated dark lesions (cankers) develop on the trunk or branches. The bark at the edge of the canker may thicken and roll inward. Sticky, amber-colored sap may ooze from the canker. The foliage on infected branches may be stunted and yellowing; some of the leaves may turn brown and drop off. Twigs and branches may die back, and the tree may eventually die.

Analysis: Several species of fungi and bacteria cause cankers on fruit and nut trees. These organisms may be spread by wind, splashing water, or contaminated tools. Infection usually occurs through injured or wounded tissue. Bark that has been damaged by sunscald, cold, pruning wounds, or mechanical injury is especially susceptible. The decay organisms sometimes infect the leaves directly, then spread down into healthy twigs. Cankers form as the decay progresses. Many fruit trees produce a sticky sap that oozes from the cankers. The portion of the branch or stem above the canker may die from decay or from clogging of the water and nutrient-conducting vessels in the branch. Cankers that form on the trunk are the most serious and may kill the tree. The tree may halt the development of a canker by producing callus tissue, a growth of barklike cells, to wall off the infection.

Solution: Remove badly infected branches and cut out cankers. Keep trees healthy, and avoid wounding. For cultural information, look up your fruit tree in the alphabetical section beginning on page 214.

Sunscald

Sunscald may kill young trees.

Problem: Patches of bark on the trunk or branches darken and die. These patches appear on the sunny side of the tree. Cracks and sunken lesions (cankers) may eventually develop in the dead bark. Damaged trees have recently been transplanted or heavily pruned.

Analysis: When a tree is shaded by other trees or structures or is covered with dense foliage, the bark on the trunk and branches remains relatively thin. If the tree is suddenly exposed to intense sunlight, the newly exposed bark and the wood just beneath the bark may be injured by the sun's heat. This frequently happens when young trees are moved from a shaded nursery to an open area and when trees are heavily pruned during periods of intense sunlight. The problem also occurs on cold, clear days in winter, as cold bark is quickly warmed by the sun. The damaged bark usually splits open, forming long cracks or cankers. Decay fungi may invade the exposed wood. Sunscald is most severe when the soil is dry. Young trees may die from sunscald.

Solution: Unless the tree is very young or extremely damaged, it will usually recover with proper care. Water the tree and fertilize it with Miracle-Gro Tree & Shrub Fertilizer Spikes to stimulate new growth. To prevent further damage, wrap the trunks and main branches of recently pruned or newly transplanted trees with tree-wrapping paper. Or paint the exposed bark with a white interior latex paint or whitewash. The tree will eventually adapt to increased exposure by growing more foliage and producing thicker bark.

INSECTS

Borers

Borer holes.

Problem: Foliage on a branch or at the top of the tree is sparse; eventually the twigs and branches die. Holes or tunnels are apparent in the trunk or branches. Sap or sawdust may be present near the holes. The bark over the tunnels may die or slough off, or knotlike swellings may be found on the trunk and limbs. Weakened branches break during wind- or snowstorms. Weak, young, or newly transplanted trees may be killed.

Analysis: Borers are the larvae of beetles or moths. Many kinds of borers attack fruit and nut trees. Females lay their eggs in bark crevices throughout the summer. The larvae feed by tunneling through the bark or wood. Borer tunnels stop the flow of nutrients and water through the area by damaging the conducting vessels; branch and twig dieback result. Sap flow may act as a defense against borers if the tree is healthy. When the borer burrows into the wood, tree sap fills the hole and drowns the insect. Trees that are weakened by mechanical injuries, disease, poor growing conditions, or insect infestation are more susceptible to borer attack.

Solution: Cut out and destroy all dead and dying branches, and remove severely infested young trees. Spray or paint the trunk and branches with an insecticide containing *chlorpyrifos* to kill young larvae before they burrow into the wood. Make sure that your tree is listed on the product label. Maintain tree health by watering and fertilizing regularly and controlling insects and disease-producing organisms.

INSECTS (continued)

Scales

Cherry scale. Inset: Lecanium scale (life size).

Problem: Dark crusty bumps; thick, white, waxy bumps; or clusters of somewhat flattened scaly bumps cover the stems or undersides of leaves. The bumps can be scraped or picked off; the undersides are usually soft. Leaves turn yellow and may drop. In some cases, a shiny or sticky substance coats the leaves. A black, sooty mold often grows on the sticky substance.

Analysis: Several types of scales infest fruit trees. They lay their eggs on leaves or bark, and in spring to midsummer the young scales, called *crawlers*, settle on leaves and twigs. The small (1/10-inch), soft-bodied young feed by sucking sap from the plant. The legs usually atrophy, and a hard crusty or waxy shell develops over the body. Mature female scales lay their eggs underneath their shell. Some species of scales are unable to digest fully all the sugar in the plant sap, and they excrete the excess in a sticky fluid called *honeydew*. An uncontrolled infestation of scales may kill the plant after 2 or 3 seasons.

Solution: Spray with an insecticide containing *malathion* when the young are active. Make sure that your fruit tree is listed on the product label. Early the following spring, before new growth begins, spray the trunk and branches with ORTHO Volck Oil Spray to control overwintering insects.

Aphids

Aphids (life size).

Problem: The youngest leaves are curled, twisted, discolored, and stunted. Leaves may drop, and in severe cases the tree may defoliate. Developing fruit may be small and misshapen. A shiny or sticky substance may coat the leaves. A black, sooty mold often grows on the sticky substance. Tiny (1/8-inch) yellow, green, purplish, or black soft-bodied insects cluster on the young shoots and on the undersides of leaves.

Analysis: Many species of aphids infest fruit trees. Aphids do little damage in small numbers. They are extremely prolific, however, and populations can rapidly build up during the growing season. Damage occurs when the aphid sucks the juices from the leaves and immature fruit. The aphid is unable to digest fully all the sugar in the sap, and it excretes the excess in a sticky fluid called *honeydew*, which often drops onto the leaves below. A sooty mold fungus may develop on the honeydew, causing the leaves to appear black and dirty. At harvest time, the fruit may be small, misshapen, and pitted due to aphid damage earlier in the season.

Solution: Spray with ORTHO Diazinon Ultra Insect Spray, ORTHO Malathion 50 Plus Insect Spray, or ORTHO Home Orchard Spray. Make sure that your fruit is listed on the product label. Repeat the spray if the tree becomes reinfested.

Cottony cushion scales, mealybugs, and woolly aphids

Citrus mealybugs.

Problem: The stems, branch crotches, trunk, or undersides of leaves are covered with white, cottony masses. Leaves may be curled, distorted, and yellowing, and knotlike galls may form on the stems, trunk, or roots. Sometimes a shiny or sticky substance coats the leaves. Twigs and branches may die.

Analysis: Cottony cushion scales, mealybugs, and woolly aphids produce white, waxy secretions that cover their bodies. Their similarity makes separate identification difficult. When the insects are young, they are usually inconspicuous on the host plant. Their bodies range in color from yellowish green to brown, blending in with the leaves or bark. As the insects mature, they exude filaments of white wax, giving them a cottony appearance. Mealybugs and scales generally deposit their eggs in white, fluffy masses. Damage is caused by the withdrawal of plant sap from the leaves, branches, or trunk. The insects are unable to digest fully all the sugar in the plant sap, and they excrete the excess in a sticky fluid called *honeydew*, which often drops onto the leaves or plants below.

Solution: Spray the branches, trunk, and foliage with ORTHO Diazinon Ultra Insect Spray, ORTHO Malathion 50 Plus Insect Spray, or ORTHO Volck Oil Spray. Make sure that your fruit tree is listed on the product label. During the dormant season, spray the trunk and branches with ORTHO Volck Oil Spray.

Ants

Ants feeding on papaya (½ life size).

Problem: Ants crawl on the trunk, branches, and fruit. In many cases, the trees are also infested with aphids, leafhoppers, mealybugs, scales, and whiteflies.

Analysis: Most ants do not directly damage plants. They may be present for any of several reasons. Many ants feed on *honeydew,* a sweet, sticky substance excreted by several species of insects, including aphids, leafhoppers, mealybugs, scales, and whiteflies. Ants are attracted to plants infested with these pests. Their presence may discourage predators that would otherwise control pests. Ants may also feed on flower nectar, fruit, or tree sap or on fruit that has had its skin broken or is rotting. Ants usually live in underground nests. Some species make colonies in trees and building foundations.

Solution: Destroy ant nests by treating anthills with ORTHO Diazinon Granules or by spraying the nest and surrounding soil with ORTHO Diazinon Ultra Insect Spray. Control aphids, leafhoppers, mealybugs, scales, and whiteflies by spraying the infested plants with an insecticide containing *malathion* or *diazinon.* Make sure that your fruit tree is listed on the product label. To prevent ants from crawling up the trunk, apply a ring of sticky barrier substance to the trunk.

Tent caterpillars and fall webworm

Tent caterpillar nest (life size).

Problem: In spring or summer, silk nests appear in branch crotches or on ends of branches. Leaves are chewed; branches or the entire tree may be defoliated. Groups of bluish, black, tan, or greenish hairy caterpillars with spots or stripes are feeding in or around the nests.

Analysis: These insects feed on many fruit and ornamental trees. Several species of tent caterpillars (*Malacosoma* species) and the fall webworm (*Hyphantria cunea*) are distributed throughout the United States. In the summer, adult tent caterpillar moths lay masses of eggs in a cementing substance around twigs. The eggs hatch in early spring as the leaves unfold, and the young caterpillars immediately begin to construct their nests. On warm, sunny days, they devour the surrounding foliage. In mid- to late summer, brownish or reddish adult moths appear. The fall webworm moth lays many eggs on the undersides of leaves in the spring. In early summer, the young caterpillars begin feeding and surrounding themselves with silk nests. The caterpillars drop to the soil to pupate. Up to four generations occur between June and September.

Solution: Spray with ORTHO Diazinon Ultra Insect Spray, ORTHO Malathion 50 Plus Insect Spray, ORTHO Home Orchard Spray, or a compound containing the bacterial insecticide *Bacillus thuringiensis* (Bt). Make sure that your plant is listed on the product label. For the best results, use *Bacillus thuringiensis* while the caterpillars are small. Remove egg masses found in the winter. Prune out and destroy branches with tents and worms.

Honeydew

Yellowjacket feeding on honeydew.

Problem: A shiny or sticky substance coats the leaves, fruit, and sometimes twigs. A black, sooty mold often grows on the sticky substance. Insects may be found on the leaves, and ants, flies, or bees may be present.

Analysis: Honeydew is a sweet, sticky substance secreted by aphids, mealybugs, psyllids, whiteflies, and some scales. These sucking insects cannot digest all the sugar in the plant sap, and they excrete the excess in this sticky fluid. The honeydew drops onto the leaves directly below and onto anything beneath the tree. Ants, flies, and bees feed on honeydew and may be found around the plant. A sooty mold often develops on the honeydew, causing the leaves, fruit, and twigs to appear black and dirty. The fungus does not infect the leaf but grows on the honeydew. Extremely heavy infestations may prevent light from reaching the leaf, reducing food production.

Solution: Honeydew will eventually be washed off by rain, or it may be hosed off. Prevent more honeydew by controlling the insect that is producing it. Inspect the foliage to determine what type of insect is present. For control instructions for aphids, mealybugs, and scales, see page 208.

POWDERY MATERIAL ON LEAVES

DISCOLORED OR MOTTLED LEAVES

Powdery mildew

Powdery mildew on crabapple.

Problem: Grayish white, powdery patches appear on the leaves. New growth is often stunted, curled, and distorted. Infected buds may open later than usual, and infected flowers and leaves often turn brittle and die. The fruit is sometimes small and misshapen. It may be russeted in a network pattern or in patches, or it may be covered with white, powdery patches.

Analysis: Powdery mildew is caused by a fungus that thrives in both humid and dry weather. The fungus spends the winter in leaf and flower buds. In the spring, spores are blown to the new leaves, which are very susceptible to infection. The fungus saps plant nutrients, causing distortion and often death of the tender foliage. Powdery mildew thrives in warm days and cool nights, reduced light, and lack of rainfall.

Solution: Spray infected trees with a fungicide containing *myclobutanil* or sulfur. (Do not use sulfur on apricots.) Make sure that your fruit tree is listed on the spray label. Most fruit trees should be sprayed at regular intervals of 10 to 14 days from bud-break until 3 to 4 weeks after the petals have fallen from the blossoms. Resume spraying whenever the mildew recurs. For more details about spraying, look up your fruit tree in the alphabetical section beginning on page 214.

Sooty mold

Sooty mold on citrus.

Problem: A black, sooty mold grows on the leaves, fruit, and twigs. It can be completely wiped off the surfaces of the leaves. Cool, moist weather hastens the growth of this substance.

Analysis: Sooty mold, a common black mold, is found on a wide variety of plants in the garden. Sooty mold is unsightly but is fairly harmless because it does not attack the leaf directly. The presence of sooty mold indicates that the tree is infested with insects. It is caused by any of several species of fungi that grow on the sugary material left on plants by aphids, mealybugs, scales, whiteflies, and other insects that suck sap from the plant. The insects are unable to digest fully all the sugar in the sap, and they excrete the excess in a sticky fluid called *honeydew*, which drops onto the leaves and fruit below. The sooty mold develops on the honeydew, causing the leaves to appear black and dirty. Extremely heavy infestations prevent light from reaching the leaf, so the leaf produces fewer nutrients and may turn yellow.

Solution: Rain will eventually wash off sooty mold. Prevent more sooty mold from growing by controlling the insect that is producing the honeydew. Inspect the foliage to determine what type of insect is present. For control instructions for aphids, mealybugs, and scales, see page 208.

Leaf scorch

Leaf scorch on hickory.

Problem: During hot weather, leaves turn brown around the edges and between the veins. Sometimes the leaves die. Many leaves may drop during late summer. This problem is most severe on the youngest branches. Trees rarely die, but growth is impaired.

Analysis: Leaf scorch occurs when water evaporates from the leaves faster than it can be replenished. In hot weather, water evaporates rapidly from the leaves. If the roots cannot absorb and convey water fast enough to replenish this loss, the leaves turn brown and wither. This usually occurs in dry soil, but leaves can also become scorched when the soil is moist and temperatures are very high for extended periods. Drying winds, severed roots, limited soil area, salt buildup, low temperatures, or weed killers can also cause scorch.

Solution: To prevent further scorch, water trees during periods of hot weather to wet the entire root space. Newly transplanted trees should be watered whenever the root ball is dry 2 inches below the surface. Controls are not available for scorch on trees in moist soil. Plant trees adapted to your climate.

Iron deficiency

Iron-deficient apple leaves.

Nitrogen deficiency

Nitrogen-deficient peach.

Nitrogen-deficient citrus.

Problem: Leaves turn pale green or yellow. The newest leaves (those at the tips of the branches) are most severely affected. Except in extreme cases, the veins of affected leaves remain green. Older leaves may remain green. Fruit production may be reduced, and fruit flavor may be poor.

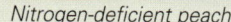

Analysis: Plants frequently suffer from deficiencies of iron and other minor nutrients such as manganese and zinc, elements essential to normal tree growth and development. Often these minor nutrients are present in the soil, but alkaline soil with a pH of 7.5 or higher or wet soil conditions cause them to form compounds that cannot be used by the tree. An alkaline condition can result from overliming, from lime leached from cement or mortar, or from calcium occurring naturally in the soil. Regions where soil is derived from limestone or where rainfall is low usually have alkaline soils.

Solution: To correct the iron deficiency, spray a chelated iron fertilizer on the foliage, then apply Miracid Plant Food, which supplies needed iron as well as acidifies the soil. Check the soil pH. If necessary, take additional steps to correct the pH of the soil by treating it with ferrous sulfate and watering it well or by adding organic matter to the soil.

Problem: Older leaves turn yellow. Eventually the rest of the leaves turn yellow-green and then yellow. Yellow leaves usually die and drop off. New leaves are small, and growth is slow. Fruit production is poor.

Analysis: Nitrogen, which is essential in the formation of plant tissue, green leaf pigment, and many other compounds necessary for plant growth, is deficient or unavailable in most soils. When a tree cannot obtain enough nitrogen from the soil, it uses nitrogen from its older leaves for new growth. The older leaves become deficient in nitrogen and turn yellow. A continuing shortage of nitrogen causes overall yellowing, stunting, death of the older leaves, and a reduced fruit yield. Fast-growing, young fruit trees usually require large amounts of nitrogen. Nitrogen naturally present in the soil is made available to trees as organic matter decomposes. Soils that are low in organic matter, such as sandy and readily leached soils, are often infertile, and nitrogen is leached from the soil more quickly when rainfall or irrigation is heavy. Poorly drained, overwatered, and compacted soils lack oxygen, which is necessary for the utilization of nitrogen. Trees growing in these soils often exhibit symptoms of nitrogen deficiency. In addition, trees growing in cold (50°F and below), hot (90°F and above), acid (pH of 5.5 or lower), or alkaline (pH of 7.8 or higher) soils are often low in nitrogen.

Solution: Fertilize trees regularly with a balanced fertilizer, such as Scotts Evergreen, Shrub and Tree Food. Add organic matter to compacted soils and soils low in organic substances. Improve soil drainage. Do not keep the soil constantly wet. Adjust pH in soils that are acid or alkaline.

DISCOLORED OR MOTTLED LEAVES (continued)

TREE STUNTED OR DECLINING

Mites

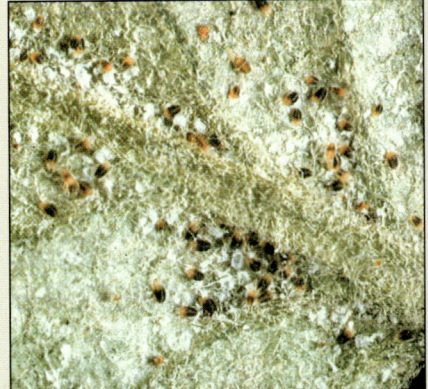

Spider mites (2× life size).

Problem: Leaves are stippled, yellowing, silvered, or bronzed. There may be webbing over flower buds, between leaves, or on the lower surfaces of leaves. The fruit may be roughened or russet colored. To check for certain species of mites, examine the bottoms of the leaves with a hand lens. Or hold a sheet of white paper underneath an affected leaf and tap the leaf sharply. Minute specks the size of pepper grains will drop to the paper and begin to crawl around.

Analysis: Mites, related to spiders, commonly attack fruit trees and other garden plants. Certain mites, such as the twospotted spider mite, are large enough to be detected on white paper. Smaller mites, such as plum and pear rust mites, are microscopic and cannot be seen without the aid of a strong hand lens or microscope. Mites cause damage by sucking sap from the fruit surface and the undersides of leaves. As a result of their feeding, the plant's green leaf pigment disappears, producing the stippled or silvered appearance. Mites are active throughout the growing season but thrive in hot, dry weather (70°F and up). By midsummer, they can build up to tremendous numbers.

Solution: Spray infested trees with ORTHO Diazinon Ultra Insect Spray, ORTHO Malathion 50 Plus Insect Spray, or ORTHO Volck Oil Spray. Repeat the spray at least two more times at intervals of 7 to 10 days. Make sure that your tree is listed on the spray label. Predatory mites, which feed on other mites, may be available for control.

Nematodes

Root knot nematode damage.

Problem: Leaves are bronzed and yellowing. They may wilt on hot, dry days but recover at night. The tree is generally weak. After several years, the tree is noticeably stunted, and branches may die. Some nematodes produce galls; others produce lesions on roots.

Analysis: Most nematodes are microscopic worms unrelated to earthworms that live in the soil. They feed on tree roots, damaging and stunting them. Damaged roots can't supply sufficient water and nutrients to the branches and leaves, and the tree is stunted or slowly dies. Nematodes are found over North America but are most severe in the southeastern United States. They prefer moist, sandy loam soils. They move only a few inches each year on their own, but may be carried long distances by soil, water, tools, or infested plants. Testing roots and soil is the only positive method for confirming the presence of nematodes. Contact your county extension office for sampling instructions and addresses of testing laboratories. Soil and root problems such as poor soil structure, drought stress, nutrient deficiency, and root rots can produce symptoms similar to those caused by nematodes. Eliminate these problems as causes before soil and root samples are sent for testing.

Solution: No available chemicals kill nematodes in planted soil. They can be controlled before planting, however, by soil fumigation or solarization. To solarize soil, cultivate and thoroughly wet an area before covering it with clear polyethylene film during the hottest time of the year. Allow the film to remain in place for 4 to 6 weeks.

Crown gall

Crown gall on roots.

Problem: Large corky galls up to several inches in diameter appear at the base of the tree and on the roots. The galls are rounded, with a rough, irregular surface. Trees with numerous galls are weak; growth may be slowed, and the foliage may turn yellow.

Analysis: Crown gall is caused by a soil-inhabiting bacterium (*Agrobacterium tumefaciens*) that infects many ornamentals, fruits, and vegetables in the garden. The bacteria are often brought to a garden initially on the roots of an infected plant and are spread in the soil. The bacteria enter the tree through wounds in the roots or the base of the trunk (the crown). They produce a substance that stimulates rapid cell growth in the plant, causing galls to form. The galls disrupt the flow of water and nutrients up the roots and trunk, weakening and stunting the top growth. They do not usually cause the tree to die.

Solution: Infected trees cannot be cured. They often survive for many years, however. Although the disease cannot be eliminated from the tree, individual galls can be removed by professionals. If you wish to remove galls from valued trees, consult a professional horticulturist or landscape contractor. The bacteria will remain in the soil for as long as 3 years after an infected tree has been removed. Replant with a resistant tree.

Root and crown rots

Crown rot on cherry.

Phytophthora *rot on apple.*

Tree neglect

Neglected fruit tree.

Problem: Normal leaf color dulls, and the plant loses vigor. Leaves may wilt or turn yellow or light brown. Major branches or the entire tree may die. The tree sometimes lives for many months in a weakened condition, or it may die quickly. Fruit growing only on the upper branches is a common sign of a tree's decline. The roots and cambium (thin layer of tissue just beneath the bark) of the lower trunk are brownish, and the roots may be decayed. In the Southwest only, fine, woolly, brown strands may form on the roots and white, powdery spores on the soil surface. Or fan-shaped plaques of white strands may appear between the bark and wood of the roots and lower stems. Mushrooms may appear at the base of the plant in the fall.

Analysis: Root and crown rots are caused by any of several fungi that live in the soil and on roots. They are spread by water, soil, and transplants.

1. Phytophthora species: These fungi cause browning and decay of the roots and browning of the cambium and wood of the lower trunk. The tree usually dies slowly, but young trees may wilt and die rapidly. The disease is most prevalent in heavy, waterlogged soil.

2. Phymatotrichum omnivorum: This fungus, commonly known as *cotton root rot* or *Texas root rot*, is a serious problem on many plants in the Southwest. Young trees may suddenly wilt and die. Brown strands form on the roots, and white powdery spores on the soil. The disease is most severe in heavy, alkaline soils.

3. Armillaria mellea: This disease is commonly known as *shoestring root rot, mushroom root rot,* or *oak root fungus.* It is identified by fan-shaped plaques of white fungal strands between the bark and the wood of the roots and lower trunk. Honey-colored mushrooms may appear at the base of the plant.

Solution: The numbered solutions below correspond to the numbered items in the analysis.

1. Remove dead and dying trees. When replanting, use plants that are resistant to phytophthora. (For a list, see page 342.) Improve soil drainage. Avoid overwatering plants.

2. Remove dead and dying trees. When replanting, use only resistant varieties. (For a list, see page 344.) Before planting, in areas with neutral or alkaline soil increase the soil acidity by adding 1 pound of ammonium sulfate for every 10 square feet of soil. Make a circular ridge around the planting area, and fill the basin with 4 inches of water. Repeat treatment in 5 to 10 days. Improve soil drainage.

3. Remove dead trees and as much of the root systems as possible. If the plant is newly infected, expose its base to air for several months by removing 3 to 4 inches of soil. Prune off and destroy diseased roots. Fertilize with Scotts Evergreen, Shrub and Tree Food to stimulate growth. When replanting, use resistant plants (see page 345).

Problem: Overall growth is slow. Foliage may be sparse, discolored, or stunted. Branches are dense and intertwined. Twigs and branches may be dying or dead, and fruit is small and of poor quality. Twigs and branches may be covered with moss or lichens. The tree may be diseased or infested with insects.

Analysis: Most fruit trees decline in vigor and fruitfulness if neglected for several years. Fruit trees are susceptible to fungal and bacterial diseases that can affect the roots, crown (where the trunk meets the roots), branches, foliage, and fruit. Many insect pests also weaken fruit trees by feeding on, boring into, or otherwise damaging them. Most fruit trees need to be pruned annually to stimulate the production of new fruiting wood or to eliminate excess or old fruiting wood. Proper watering and fertilizing are also important for maintaining tree health and fruitfulness.

Solution: Control pests and plant diseases. To determine what types of insects and diseases are affecting your tree and how to control them, and for specific details on pruning, watering, and fertilizing, look up your fruit tree in the alphabetical section beginning on page 214. Thin out weak, diseased, intertwined, and dying twigs and branches. If the tree needs to be heavily pruned or restructured, gradually prune it into proper shape over a period of 3 years. Remove all invasive growth, such as grass or ground cover, from around the base of the tree. Water and fertilize the tree properly.

APPLE

GROWING GUIDE

Adaptation and Pollination:
See variety chart on page 358.

Soil: Any good, deep, well-drained soil. pH 5.5 to 7.5.

Fertilizer: Fertilize with Scotts All Purpose Plant Food or Miracle-Gro Plant Food according to label directions.

Water: How much: Apply enough water to wet the soil 3 to 4 feet deep.
How often: Water when the soil 6 inches below the surface is just barely moist.

Pruning: Apples are borne on short fruiting branches (spurs) that grow on 2-year-old wood. The spurs continue to fruit for about 10 years. Prune lightly, since the removal of many spurs will reduce the apple yield. Thin out weak, crossing, or dead twigs and branches. When apples are thumbnail size, thin them to 6 inches apart, with 1 fruit per spur.

Harvest: Apples are ripe when their seeds turn dark brown to black, the flesh turns creamy white, and the apple stem separates easily from the spur when the fruit is gently lifted. Harvest earlier if you prefer tart apples.

Codling moth

Codling moth damage. Inset: Exit hole.

Problem: The fruit is blemished by small holes surrounded by dead tissue. Brown, crumbly material resembling sawdust may surround holes. Brown-headed, pinkish white worms up to 1 inch long may be in the fruit. Fruit interior is often dark and rotted. Affected apples drop prematurely.

Analysis: This worm is the larva of the codling moth (*Cydia pomonella*), one of the most serious apple pests in the United States. The ½-inch-wide, gray-brown moths appear in spring when apple trees are blooming, usually flying at twilight. They lay eggs on leaves, twigs, and developing fruit. When the eggs hatch, the larvae tunnel into fruit. They feed for several weeks, then emerge to pupate, often leaving dark excrement on the skin and inside the fruit. After pupating, another generation of moths emerges in midsummer. Apples may be damaged throughout summer. In fall, larvae spin cocoons in protected places on or around the tree, such as under loose bark or in tree crevices. They spend winter in the cocoons, emerging as moths in spring. They may also overwinter on other plants under the trees.

Solution: Once worms have penetrated the apples, it is impossible to kill them. To protect uninfested apples, spray with ORTHO Home Orchard Spray or ORTHO Malathion 50 Plus Insect Spray every 10 days until a week before harvest. Remove and destroy all fallen apples, and clean up debris. The following spring, begin spraying 2 to 3 weeks after petal fall. For more precise spray timing, use pheromone traps to monitor moth flights. Hang two per tree after the first two sprayings. Spray 1 week after catches reach five moths per week.

Apple maggot

Apple maggots (4× life size). Inset: Adult.

Problem: The fruit may be dimpled and pitted, with brown trails winding through the flesh. White, tapered, legless maggots about ⅜ inch long may be present in the fruit. Severely infested apples are brown and pulpy inside. Frequently, many apples drop prematurely.

Analysis: Apple maggots, (*Rhagoletis pomonella*), also known as *railroad worms* or *apple fruit flies*, are the larvae of flies that resemble the common housefly. Apple maggots infest plums, cherries, and pears in addition to apples. Adult flies emerge from pupae between late June and the beginning of September. They lay eggs in the fruit through holes they puncture in the skin. The maggots that emerge from the eggs make brown trails through the flesh as they feed. Infested apples usually drop to the ground. Mature maggots emerge from the apple and burrow in the soil to pupate. They remain in the soil throughout the winter and emerge as adult flies the following June.

Solution: Maggots cannot be killed after apples are infested. Protect healthy apples from adult flies by spraying at intervals of 7 to 10 days from the end of June until the beginning of September with ORTHO Home Orchard Spray. Pick up and destroy fallen apples every week throughout the summer. For more precise spray timing, use apple maggot traps. These red balls look like ripe apples but are coated with a sticky material that traps adult flies. Pheromone attractants are also available for use with red ball traps. Spray as soon as adult flies appear. Red ball traps can also be used to control apple maggots without spraying. Hang at least two in each tree.

Sooty blotch and fly speck

Sooty blotch and fly speck.

Problem: Clusters of a few to 100 sharply defined, shiny black spots appear on the fruit. Sooty or cloudy blotches may also appear. Both appear only on the skin of the fruit without affecting the flesh.

Analysis: Although sooty blotch and fly speck are not caused by the same fungus, they are so commonly found in association that they are usually described together. Sooty blotch can be caused by any of several related fungi. Fly speck is caused by the fungus *Zygophiala jamaicensis*. These fungi spend the winter on the twigs of apples and many other woody plants. During mild, wet weather, the fungi produce spores that are blown to and infect the developing apples. Infection can occur anytime after petal fall but is most prevalent in mid- to late summer. About a month after the initial infection, specks and blotches appear on the maturing fruit. Although these diseases are unsightly, they are external and do not generally affect the taste of the apples.

Solution: You cannot clear infected apples, but the fruit is edible. To prevent the diseases the following year, improve air circulation by pruning to open dense trees and by thinning apple clusters. Remove wild brambles around the trees to eliminate sources of infection. The following spring, spray with ORTHO Home Orchard Spray every 10 to 14 days, from 1 or 2 weeks after petal fall until 3 weeks before harvest. Trees in the Southeast need annual spraying.

Bitter rot

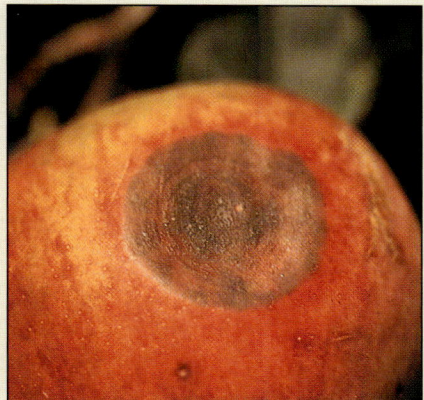

Bitter rot affects both apples and pears.

Problem: Sunken, light brown, circular spots appear on half-grown to mature fruit.  These spots gradually enlarge to 1 inch in diameter. Concentric rings and sticky pink masses of spores may appear on the spotted areas during moist weather. The rotted apple flesh tastes bitter. Sunken lesions may form on the branches.

Analysis: Bitter rot, which also affects pears, is caused by two closely related fungi (*Glomerella cingulata* and *Colletotrichum gloeosporioides*). The fungi spend the winter in rotted apples left on the tree and on the ground, and in sunken lesions on the branches. Spores are spread by splashing rain to healthy apples in the spring. Infection can occur throughout the fruiting season. In areas where hot, humid conditions last for long periods of time, bitter rot can quickly destroy an entire apple crop. This disease primarily attacks the fruit and does not severely damage the health of the tree.

Solution: Spray infested trees with ORTHO Home Orchard Spray. To prevent recurrence the following year, prune out and destroy deadwood and branches with lesions, and remove and destroy rotted apples. The following spring, spray the trees with ORTHO Home Orchard Spray every 10 to 14 days from petal fall until 7 days before harvest. Trees in the red zone on the accompanying map need annual treatment. Plant resistant varieties.

San Jose scale

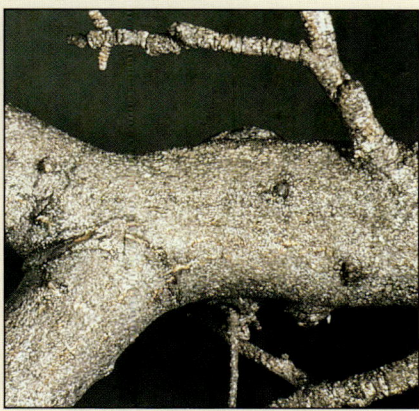

San Jose scales (½ life size).

Problem: Some leaves are pale green to yellow and may drop prematurely from weakened limbs. The bark is encrusted with small (1/16-inch), hard, circular, slightly raised bumps with dull yellow centers. If the hard cover is scraped off, the insect underneath is found to be yellow or olive. Limbs may die back severely, and entire branches may be killed. Red spots with white centers mar the apple skins.

Analysis: San Jose scale (*Quadraspidiotus perniciosus*) is an insect that infests the bark, leaves, and fruit of many fruit trees. The scales bear live young in the spring. In late spring to midsummer the young scales, called *crawlers*, settle on leaves and twigs. The soft-bodied young feed by inserting their mouthparts and sucking sap from the plant. The legs atrophy, and a hard, crusty shell develops over the body. An uncontrolled infestation of San Jose scales may kill large branches after 2 or 3 seasons.

Solution: During the dormant season, just prior to growth in early spring, spray the trunk and branches with ORTHO Volck Oil Spray. Begin checking for crawlers about a month after full bloom. Continue monitoring through the summer to catch later generations. Crawlers may be killed with ORTHO Malathion 50 Plus Insect Spray or an insecticide containing *diazinon*.

215

APPLE (continued)

Fruit tree leafroller

Fruit tree leafrollers (life size).

Problem: Irregular holes appear in the leaves and fruit. Some of the leaves are rolled and held together with a web. Inside these rolled leaves are pale green worms, up to ¾ inch long, with black heads. The maturing apples are scarred and misshapen.

Analysis: The fruit tree leafroller, the larva of a brown moth (*Archips argyrospilus*), is a common pest of many fruit and ornamental trees. The moths lay their eggs on branches or twigs in June or July. The eggs hatch the following spring, and the emerging larvae feed on the blossoms and developing fruit and foliage. Leafrollers often wrap leaves around ripening fruit, then feed on the fruit inside. After about a month, the mature larvae pupate within rolled leaves, to emerge as moths in June or July.

Solution: If practical, pick off and destroy rolled leaves to reduce the numbers of moths that will emerge later in the season. The following spring, spray the tree with ORTHO Home Orchard Spray, ORTHO Malathion 50 Plus Insect Spray, or an insecticide containing *diazinon* when 75 percent of the petals have fallen from the blossoms. Repeat the spray according to the directions on the label. For more precise timing, use pheromone traps. Begin spraying when moth catches peak and begin to decline. Eggs are hatching at that time.

Scab

Scab, a fungal disease.

Problem: Olive brown velvety spots, ¼ inch or more in diameter, appear on the leaves and fruit. Spots develop first on the undersides of leaves, then on both sides. As the infected apples mature, the spots become light to dark brown corky lesions. The fruit is often cracked and malformed and may drop prematurely. Severely infected trees may completely defoliate.

Analysis: Scab is caused by a fungus (*Venturia inaequalis*). It is one of the most serious diseases of apples in areas where spring weather is mild (60° to 70°F) and wet. The fungus spends the winter in infected leaf debris on the ground. Beginning at bud-break, spores are ejected into the air when the leaf debris becomes wet. Air currents carry them to emerging leaves. If a film of water is present on the leaf, the spores germinate and infect the leaf. The infected tissues produce more spores, which in turn infect other leaf and fruit surfaces. If the fruit stays wet for 2 or 3 days at a time in late summer or early fall, spores can infect the fruit, but symptoms do not develop until the fruit has been stored, sometimes for several months.

Solution: Unless severely infected, the apples are edible. To prevent recurrence of the disease the following year, remove and destroy leaf debris and infected fruit in the fall. The following spring, spray with ORTHO Home Orchard Spray or with a fungicide containing *captan* or *myclobutanil*.

Black rot

Black rot is a fungus that also attacks pears.

Problem: A firm spot composed of concentric light and dark brown rings appears on the apple. This spot gradually turns dark brown or black and enlarges, rotting part or all of the fruit. Spots on the leaves, first appearing from 1 to 3 weeks after petal fall, are also formed of light and dark concentric rings. Reddish brown, slightly sunken lesions up to several feet in length often appear on the branches or trunk of the tree.

Analysis: Black rot, also known as *frog-eye leaf spot*, is caused by a fungus (*Botryosphaeria* species) that also attacks pears. The fungus spends the winter in rotted apples, deadwood, and old fire blight cankers. When temperatures reach 60°F and higher in the spring, spores are produced on infected tissues and splashed by water to the foliage and fruit. Branches and trunks may be infected with black rot, especially when the bark has been infected by fire blight or weakened by sunscald, cold, or heavy shade. As the fungus decays the wood, cankers form, weakening the branches and reducing the overall vigor of the tree. Fruit infection is generally more severe in warm, moist areas, and canker formation is more prevalent in cooler climates.

Solution: Spray with ORTHO Home Orchard Spray according to label directions as soon as infestation is noticed. Remove all rotted apples from the tree and ground and destroy them. Prune out deadwood and branches infected with fire blight. Destroy the prunings. Thin densely branched trees to provide adequate light and air circulation. The following spring, begin spraying when buds break. Spray every 10 to 14 days.

Cedar-apple rust

Cedar-apple rust alternates with junipers.

Problem: Pale yellow spots appear on the upper surfaces of leaves and on fruit in mid- to  late spring. These spots gradually enlarge, turn orange, and develop minute black dots. Small (¹⁄₁₆-inch) cups with fringed edges form on the lower surfaces of leaves. Infected leaves and fruit may drop prematurely; the fruit is often small and deformed.

Analysis: Cedar-apple rust is caused by a fungus (*Gymnosporangium juniperi-virginianae*) that affects both apples and certain species of juniper and red cedar. This fungus cannot spread from apple to apple, or from juniper to juniper, but must alternate between the two. In the spring, spores from brown and orange galls on juniper or cedar are blown up to 3 miles to apple trees. During mild, wet weather, the spores germinate and infect the leaves and fruit, causing spotting and, eventually, premature leaf and apple drop. During the summer, spores are produced in the small cups on the undersides of leaves. These spores are blown back to junipers and cedars, causing new infections and starting the cycle again.

Solution: Cedar-apple rust cannot be controlled on the current season's apples and leaves. The following spring, spray apple trees with a fungicide when the flower buds turn pink, again when 75 percent of the petals have fallen from the blossoms, and once more 10 days later. When practical, do not plant apples within several hundred yards of junipers or red cedar.

Flatheaded borers

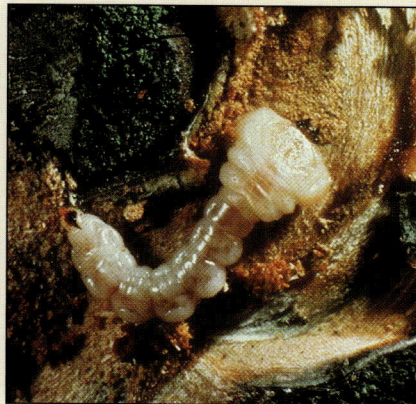

Pacific flatheaded borer (2× life size).

Problem: Leaves wilt and turn brown. Patches of bark on the trunk are sunken and discolored  and may be soaked with sap. Holes about ³⁄₈ inch in diameter may appear in the affected bark. Sawdust-filled tunnels in the wood may contain yellowish white, flat-headed grubs about ¾ inch long. In late spring to midsummer, bronze or copper-colored beetles ½ to ¾ inch long may be seen feeding on the foliage. Newly planted or weak trees are most severely affected.

Analysis: In addition to damaging apples, flatheaded borers (*Chrysobothris* species) attack many other trees and shrubs. In late spring to midsummer, the females begin to lay eggs in crevices in the bark. The emerging larvae bore through the bark into the outer layer of wood, creating winding tunnels. These tunnels damage the nutrient- and water-conducting vessels in the tree, causing twig and branch dieback and sometimes killing the tree. The mature larvae bore deep into the heartwood to pupate; adult beetles emerge the following spring. Newly transplanted, weakened, and diseased trees are most susceptible to borer infestation. Borers often invade sunburned areas on the trunk of newly planted trees.

Solution: Apply an insecticide containing *chlorpyrifos* according to label directions. Keep your tree healthy and vigorous by watering, fertilizing, and pruning it properly. Discourage borer infestation by wrapping the trunk soon after bloom with tree wrapping paper or burlap.

APRICOT

GROWING GUIDE

Adaptation and Pollination: See variety chart on page 359.

Soil: Any good, deep, well-drained garden soil. pH 5.5 to 8.0.

Fertilizer: Fertilize with Scotts All-Purpose Plant Food or Miracle-Gro Plant Food according to label directions.

Water: How much: Apply enough water to wet the soil 3 to 4 feet deep.
How often: Water when the soil 6 inches below the surface is just barely moist.

Pruning: Apricots are borne on short fruiting branches (spurs), which grow on 2-year-old wood. The spurs continue to fruit for 2 to 4 years. Prune back last year's growth to half its length, and thin out spurred branches that have stopped fruiting (4 years or older). Thin out weak, crossing, or dead twigs and branches. For large apricots, thin fruits when they are large enough to handle to 3 to 5 inches apart, with no more than 2 fruits per spur.

Harvest: Harvest when the fruits are fully colored and slightly soft. When they are ripe, the apricot stem separates easily from the spur when the fruit is gently lifted.

APRICOT (continued)

Brown rot

Brown rot mummies.

Brown rot on stems, leaves, and fruit.

Scab

Scab, a fungal disease.

Problem: Blossoms and young leaves wilt, decay, and turn brown during the first 2 weeks of the bloom period. Often the decayed blossoms and leaves fail to drop, and they may hang on the tree throughout the growing season. In humid conditions, masses of gray spores may appear on the infected flower parts. Extensive twig dieback often occurs. (Twig dieback is more common in the West than the East.) Sunken lesions (cankers) may develop on the twigs and branches as the season progresses; a thick, gummy material often oozes from the cankers.

On the fruit: Small circular brown spots appear on the young apricots. Later in the season, as the apricots start to mature, these spots may enlarge to rot part or all of the fruit. During moist weather, the rotted apricots are covered with tufts of gray spores. When the infected fruit is sliced open, the flesh inside is found to be brown, firm, and fairly dry. Infected apricots either drop prematurely or dry out, turn dark brown, and remain on the tree past the normal harvest period.

Analysis: Brown rot, caused by either of two closely related fungi (*Monilinia laxa* or *M. fructicola*), is destructive to apricots and all other stone fruits. The fungi spend the winter in twig cankers or in rotted apricots (mummies) on the tree or on the ground. In the spring, spores are blown or splashed from cankers or mummies to healthy flower buds. After penetrating and decaying the flowers, the fungus grows down into the twigs, producing brown, sunken cankers. During moist weather a thick, gummy sap sometimes oozes from the cankers, and tufts of gray spores may form on the infected areas. Spores from cankers and infected blossoms or mummies are splashed and blown to maturing fruit. Young apricots are fairly resistant to infection, but maturing apricots are vulnerable. Brown rot develops most rapidly in mild, moist conditions.

Solution: If uninfected blossoms remain on your tree, spray with ORTHO Home Orchard Spray or ORTHO Multi-Purpose Fungicide Daconil 2787® Plant Disease Control to protect them from further infection. Repeat the spray 10 days later. The following spring, spray trees when the first flowers begin to open and continue to spray according to label directions. To protect maturing apricots from infection, spray them with ORTHO Home Orchard Spray about 3 weeks before they are to be harvested. Remove and destroy all infected apricots and mummies. Prune out cankers and blighted twigs. Clean up and destroy all debris around the tree. Plant resistant varieties.

Problem: Small olive green spots appear on half-grown fruit. These spots are usually centered around the stem end of the apricot. The spots eventually turn brown and velvety. The fruit is often dwarfed, deformed, or cracked. The leaves may have small brown spots and holes in them, and many twigs die back.

Analysis: Scab is caused by a fungus (*Cladosporium carpophilum*) that attacks peaches, nectarines, cherries, and plums as well as apricots. The fungus spends the winter on twig lesions. In the spring, spores are splashed and blown to the developing foliage and fruit. The young fruit does not show scab lesions for at least a month after it is infected. Spores produced on the infected leaves, twigs, and fruit continue to infect healthy apricots throughout the fruiting season. The disease resembles shot-hole fungus and is often confused with it.

Solution: You cannot do anything about the spots on the current year's fruit, but it is edible if peeled. The following year, spray a fungicide containing *captan* on the tree when the petals have fallen from the blossoms. If scab is serious in your area, continue to spray at intervals of 10 to 14 days until about a month before the apricots are harvested.

Plum curculio

Plum curculio (5× life size).

Bacterial leaf spot

Bacterial leaf spot also attacks peaches.

Cytospora canker

Cytospora canker, showing lesions on branch.

Problem: The ripening fruit is misshapen and rotten and often drops prematurely. Holes about ⅛ inch in diameter and deep, crescent-shaped scars appear on the fruit. When cut open, the damaged fruit may contain crescent-shaped grayish white grubs about ⅓ inch long.

Analysis: Plum curculios (*Conotrachelus nenuphar*), insects found east of the Rocky Mountains, attack other stone fruits, apples, and pears, as well as apricots. The adult insects are mottled brown beetles, about ¼ inch long, with long, curved snouts. They hibernate in debris and other protected places during the winter. The beetles emerge in the spring when new growth starts and begin feeding on young leaves, blossoms, and developing fruit. After 5 to 6 weeks, the female beetles start to lay eggs in the young fruit. The grubs that hatch from the eggs feed for several weeks in the fruit. Usually the infested apricots drop to the ground. The grubs eventually leave the fruit and bore into the soil, where they pupate. The emerging beetles feed on fruit for a few weeks, then go into hibernation. Or, in the South, they lay eggs, producing a second generation of grubs in the late summer.

Solution: Once fruit is infested, you cannot kill the grubs inside. Spray with ORTHO Home Orchard Spray or ORTHO Malathion 50 Plus Insect Spray to kill beetles that may be feeding on fruit or laying eggs. Pick up and destroy all fallen fruit. The following spring, spray the trees when the petals are falling from the blossoms; repeat applications according to directions on the label.

Problem: Water-soaked spots on the undersides of leaves turn brown or black; often the centers of the spots fall out. The tips of the leaves may die, and eventually the leaves drop. When the fruit sets, the surface may be dotted with spots that later turn into deep, sunken brown pits, often surrounded by yellow rings. Sunken lesions can often be seen at the joints of the twigs.

Analysis: Bacterial leaf spot is caused by a bacterium (*Xanthomonas arboricola* pv. *pruni*) that also attacks peaches, nectarines, and plums. The disease is common east of the Rocky Mountains and is one of the more destructive stone fruit diseases. The bacteria spend the winter in the lesions on the twigs, oozing out in the spring to be carried by splashing raindrops to the young leaves and shoots, which they infect and decay. Periods of frequent rainfall foster the infection.

Solution: This disease is difficult to control adequately. Remove and destroy infected twigs in the fall. Spray with basic copper sulfate as soon as the leaves shed in the fall and again in the spring as buds begin to fatten.

Problem: Oval or oblong sunken lesions on the bark enlarge gradually. A sticky gum may ooze from the lesion and is sometimes followed by the emergence of curly orange threads. Later, small black freckles appear on the bark along the edge of the lesions. Leaves on affected branches may turn brown and die, or the entire branch may die.

Analysis: Cytospora canker, also known as *perennial canker,* is a plant disease caused by two related fungi (*Cytospora cincta* and *C. leucostoma*) that also attack peach, plum, and cherry. The fungi spend the winter in cankers or on deadwood. In the spring, black fungal bodies develop in the bark, and curly orange fungal chains form. These chains release spores that are spread by wind and splashing rain to healthy trees. Infection usually occurs through injured tissues. Bark that is damaged by sunscald, cold, pruning wounds, or mechanical injury is especially susceptible. Depressed cankers form and may ooze a sticky gum. The branch or stem above the canker may die because the water-conducting tissue of the branch decays or clogs. Mild, wet weather (70° to 85°F) enhances development of this disease.

Solution: No fully adequate control is available; a combination of procedures must be used. Remove and destroy badly infected branches and cut out cankers. Avoid mechanical injuries to the tree, and paint the trunk with white latex paint to protect against cold injury and sunburn. Avoid fertilizing in late summer or fall.

APRICOT (continued)

Peach twig borer

Peach twig borer (2× life size).

Problem: New growth at the tips of the twigs wilts and dies. When the affected twigs are sliced open lengthwise, worms about ½ inch long are found inside. The reddish brown color of these worms distinguishes them from oriental fruit moth worms, which cause similar damage. Later in the season, these worms may be found in some cut-open maturing fruit. During the summer, cocoons may be attached to the branches or tree crotches.

Analysis: The peach twig borer (*Anarsia lineatella*) attacks all of the stone fruits and is particularly damaging along the Pacific coast. The young larvae hibernate during the winter in burrows under loose bark or in other protected places on the tree. When the tree blooms in the spring, the larvae emerge and bore into the young buds, shoots, and tender twigs, causing twig and leaf death. When mature, they leave the twigs and pupate in cocoons attached to branches. After several weeks, gray moths emerge and lay eggs on the twigs, leaves, and fruit. Egg laying and larval damage can occur all through the growing season. Later in the summer, larvae feed primarily on the maturing fruit. In addition to ruining the fruit, these pests may cause abnormal branching on young trees.

Solution: Worms in the twigs and fruit cannot be killed with pesticides. To prevent future worm damage, kill the overwintering larvae by spraying during the dormant season with a dormant oil spray combined with *diazinon*. To control adults, spray again at petal fall with ORTHO Bug-B-Gon Insect Killer. Repeat according to label instructions.

CHERRY

GROWING GUIDE

Adaptation and Pollination: See the variety chart on page 359.

Soil: Any good, deep, very well-drained soil. pH 5.5 to 8.0.

Fertilizer: Fertilize with Scotts All-Purpose Plant Food or Miracle-Gro Plant Food according to label directions.

Water: How much: Apply enough water to wet the soil 3 to 4 feet deep.
How often: Water when the soil 6 inches below the surface is just barely moist.

Pruning: Cherries are borne on short fruiting branches (spurs) that grow on 2-year-old wood. The spurs are long-lived and often continue to bear for 10 years or more. Prune lightly, thinning out weak, crossing, or dead twigs and branches.

Harvest: Harvest when the cherries are fully colored. With ripe cherries, the stem will easily separate from the spur when the fruit is gently lifted or pulled. When picking cherries, avoid damaging the spurs.

Cherry fruit flies

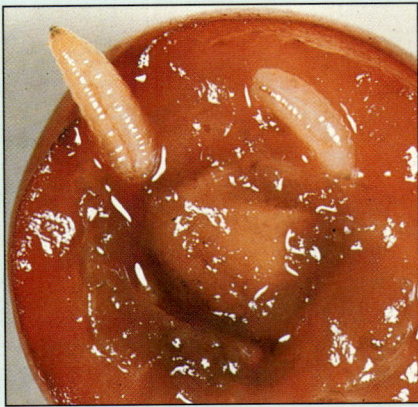

Cherry fruit fly maggots (3× life size).

Problem: The fruit is malformed, shrunken, or shriveled. Often the cherries are rotten and pulpy, with one side turning red before maturity. Holes may appear in the fruit. Tapered, yellow-white, legless worms up to ¼ inch long may be found in the cherries.

Analysis: These worms are the larvae of several closely related cherry fruit flies (*Rhagoletis* species). The adult flies, about half the size of the common housefly, are black with dark bands on the clear wings. Adults appear in the late spring for a period of about a month. They lay eggs in the cherries through holes they puncture in the skin, beginning when the fruit has turned yellow. After several days, the eggs hatch into maggots that tunnel through the cherry flesh. Mature maggots leave the fruit and drop to the ground, where they burrow into the soil to pupate. They remain in the soil through winter and emerge as adults the following spring. Damaged cherries have an exit hole. They may fall or remain on the tree.

Solution: You cannot control the worms in the current year's fruit, but you can probably prevent the problem from happening again the following year. Look for the adult flies in late spring as the fruit is coloring. Monitor their presence with yellow sticky traps. As soon as fruit flies appear, spray with ORTHO Bug-B-Gon Insect Killer or ORTHO Diazinon Ultra Insect Spray. Repeat the application two more times at intervals of 7 to 10 days. Prevent maggots from burrowing into the ground to pupate by spreading black plastic under the tree.

Pearslug

Pearslug larva (2× life size).

Problem: The upper surfaces of leaves are chewed between the veins, leaving a lacy, translucent layer of tissue that turns brown. Olive green to blackish, slimy, sluglike worms with heads wider than the bodies, up to ½ inch long, may be feeding on the leaves. Severely infested trees may be defoliated.

Analysis: Although pearslugs (*Caliroa cerasi*), also called *cherry slugs*, closely resemble slugs, they are actually the larvae of black-and-yellow flies that infest pears, plums, and some ornamental trees in addition to cherries. Adult flies, which appear in the late spring, lay their eggs in the leaves. The young larvae that hatch from these eggs exude a slimy, olive green substance, giving them a sluglike appearance. They feed on the foliage for about a month, then drop to the ground, burrow into the soil, and pupate. Young trees severely infested by pearslugs may be greatly weakened and will produce fewer, poor-quality cherries.

Solution: If only a few worms are found, wash them from the tree with a strong spray of water. For heavier infestations, spray the tree with ORTHO Diazinon Ultra Insect Spray. Repeat the spray if infestation recurs.

Oriental fruit moth

Oriental fruit moth damage.

Problem: New growth at the tips of the twigs wilts and dies. When the affected twigs are sliced open lengthwise, pinkish white worms up to ½ inch long with brown heads are found inside. Later in the season, some of the maturing fruit may also be found to contain these worms.

Analysis:
The larvae of the night-flying oriental fruit moth (*Grapholita molesta*) damage stone fruits, apples, and pears. The larvae hibernate in cocoons on tree bark or in branch crotches. In the spring they pupate, emerge as brown adult moths, and lay eggs on the young cherry twigs and leaves. The larvae bore into the young buds, shoots, and tender twigs, causing twig and leaf death. When mature, they leave the twigs, spin cocoons, and pupate in the tree or in debris on the ground. After several weeks, moths emerge to lay their eggs. Egg laying and larval damage can occur all through the growing season. Later in the summer, larvae feed primarily on the maturing cherries. They leave gum-filled holes in the cherries when they exit to pupate. In addition to ruining the fruit, these pests may cause abnormal branching patterns on young trees when large numbers of twigs are infested.

Solution: Worms in the twigs and fruit cannot be killed with pesticides. To prevent future worm damage, kill the moths by spraying infested trees with an insecticide containing *carbaryl*. Apply as directed on the label.

Shot-hole borer

Shot-hole borer beetle.

Problem: Many small holes (¹⁄₁₆ to ⅛ inch in diameter) are bored in the twigs, the branches, and sometimes the trunk of the tree. A sticky gum often oozes from the holes. The holes may be plugged with the bodies of dead insects. When sliced open, the branches reveal sawdust-filled tunnels in the wood. Pinkish white, slightly curved grubs may be found in the tunnels. Brownish black beetles ¹⁄₁₀ inch long are often present on the bark. The foliage on damaged branches sometimes wilts and turns brown. Twigs and buds may be killed, and entire branches may die.

Analysis: The shot-hole borer (*Scolytus rugulosus*), a beetle sometimes known as the *fruit tree bark beetle*, also attacks other stone fruits and many ornamental trees. The adult beetles that emerge in the late spring or early summer feed at the base of buds and small twigs, often killing them. Female beetles bore into the wood, creating tunnels in which they lay their eggs. The grubs that hatch from the eggs bore into the inner wood, creating sawdust-filled burrows 2 to 4 inches long. The grubs pupate just under the bark, then emerge as adult beetles. The last generation of grubs spends the winter in the tunnels, emerging the following spring. Weakened, diseased, and dying trees are most susceptible to infestation.

Solution: Keep the tree healthy by watering and fertilizing it properly. Remove and destroy infested weakened trees. Protect sour cherries from borer damage by spraying the trunk and larger branches with ORTHO Borer & Leaf Miner Spray while adults are present. Chemical treatment is not recommended on sweet cherries.

CHERRY (continued)

Bacterial canker

Bacterial canker, showing oozing lesions.

Problem: Sunken, elliptical lesions appear on the trunk or branches. Throughout the fall, winter, and spring, thick, amber, sour-smelling gum oozes from these lesions. In the spring, especially when the weather is very wet and cold, blossoms may die and fail to open, or they may open, then turn brown and wither. Individual branches may die back. Angular holes may appear in the leaves.

Analysis: Bacterial canker, also known as *bacterial gummosis* or *bacterial blast,* is found on various fruit and nut trees but is most severe on cherries. This disease is caused by bacteria (*Pseudomonas syringae*). Splashing rain spreads the bacteria to dormant buds, twigs, and branches. Infection occurs through wounds in the twigs and branches, and bacterial decay causes cankers to form. During the fall, winter, and early spring, large quantities of bacteria-containing gum ooze from the cankers. Bacterial activity decreases in the summer. Slowly developing cankers may encircle a branch, however, and by midsummer, affected branches and limbs start to die back. With the onset of cool, wet fall weather, bacterial activity increases again. This disease is most serious on young trees.

Solution: Bacterial canker is difficult to control. Prune out diseased branches. Disinfect pruning shears after each cut by dipping in a solution of 1 part chlorine bleach and 9 parts water. In the fall, spray with a fungicide containing basic copper sulfate. Prune in late winter or early spring rather than early in the dormant season. Keep the tree healthy. For care of cherries, see page 220.

San Jose scale

San Jose scales can kill large branches.

Problem: Some of the leaves are pale green to yellow and may drop prematurely from weakened limbs. Or they may turn brown and wither but remain attached to the tree into the winter. The bark is encrusted with small (1/10-inch), gray, hard, circular, slightly raised bumps with dull yellow centers. If the hard cover is scraped off, the insect underneath is found to be yellow or olive. Limbs may die back severely, and entire branches may be killed. The fruit may be marred by specks.

Analysis: San Jose scales (*Quadraspidiotus perniciosus*) infest the bark, leaves, and fruit of many fruit trees. The scales bear live young in the spring. In late spring to midsummer the young scales, called *crawlers,* move about and settle on leaves and twigs. The small (1/16-inch), bright yellow, soft-bodied young feed by inserting their mouthparts and sucking sap from the plant. The legs atrophy, and a hard, crusty shell develops over the body. An uncontrolled infestation of San Jose scales may kill large branches after 2 or 3 seasons.

Solution: During the dormant season, just prior to growth in early spring, spray the trunk and branches with ORTHO Volck Oil Spray. Begin checking for crawlers about a month after full bloom. Continue monitoring through the summer to catch later generations. Crawlers may be killed with ORTHO Bug-B-Gon Insect Killer, ORTHO Malathion 50 Plus Insect Spray, or ORTHO Diazinon Ultra Insect Spray.

Cherry leaf spot

Cherry leaf spot results in poor-quality fruit.

Problem: Purple spots appear on the upper surfaces of leaves. The centers of the spots may fall out, leaving holes. Many of the spotted leaves are yellow and dying. The undersides of the leaves may be dotted with cream-colored masses of spores. In severe cases, the fruit is also spotted. The tree defoliates prematurely, and cherry yield is reduced and of poor quality. The fruit is often soft and watery.

Analysis: Cherry leaf spot, also known as *yellow leaf spot,* is caused by a fungus (*Coccomyces hiemalis*). The fungus spends the winter in fallen leaves. About the time the cherry trees are finished blooming, large numbers of spores are splashed and blown from leaves on the ground to the emerging leaves. The infection and premature death of the leaves greatly reduces the amount of food the tree can make and store. This results in weakened trees, reduced yields, and poor-quality fruit. Infected trees are much more susceptible to cold injury during the following winter. Cherry leaf spot is most severe during mild (60° to 70°F), wet weather.

Solution: Spray with ORTHO Multi-Purpose Fungicide Daconil 2787® Plant Disease Control or ORTHO Home Orchard Spray. In the fall, remove and destroy all leaf debris around the trees. The following spring, spray when the petals fall from the tree; repeat the spray at least two more times at intervals of 10 to 14 days. If the problem is severe, continue spraying until 7 days before harvest.

GROWING GUIDE

Adaptation: See the variety chart on page 358.

Soil: Any good, well-drained soil. pH 5.5 to 8.0.

Fertilizer: Fertilize with Scotts Citrus Food according to label directions.

Water: How much: Apply enough water to wet the soil 3 to 4 feet deep.
How often: Water when the soil 6 to 12 inches below the surface is just barely moist.

Pruning: Prune only to shape the tree, and to remove suckers and dead twigs and branches. Lightly thin the inside of the tree if the growth is dense.

Harvest: When the fruits are fully colored, taste one to determine if it is ripe. If the flavor is not yet sweet enough for the variety, allow the fruits to ripen for another few weeks, then try again. To harvest, clip the fruits off with pruning shears rather than pulling them off.

Citrus thrips

Citrus thrips damage.

Problem: Leaf buds shrivel and turn brown. Some of the leaves are silvery gray, leathery, curled, and distorted. The fruit may be silvery, scabbed, or streaked, often in a distinct ring around the stem.

Analysis: Citrus thrips (*Scirtothrips citri*), minute pale yellow or orange insects, are pests of citrus and various ornamental trees. Thrips cause their damage by rasping the tissue of young leaves and immature fruits. They feed on the plant sap that exudes from the injured tissue. Adult thrips lay their eggs in leaves and stems in the fall. The following spring, the young thrips that emerge from these eggs begin feeding on the new growth. These pests are often found in protected areas such as the insides of leaf buds. Thrip damage occurs throughout the growing season and is especially severe during hot, dry weather.

Solution: Damage to the fruit is only cosmetic; fruit may be eaten. Spray infested plants with ORTHO Malathion 50 Plus Insect Spray. Follow directions carefully, and repeat at regular intervals of 7 to 10 days as long as new damage is seen. Keep plants well irrigated and vigorous. Vigorous plants often outgrow thrips damage.

Mealybugs and cottony cushion scale

Cottony cushion scale (2× life size).

Problem: White, cottony masses cluster on the leaves, the stems, the branches, and possibly the trunk. Some of the foliage may wither and turn yellow; leaves and fruit may drop. A shiny or sticky substance often coats the leaves. A sooty mold may grow on the sticky substance.

Analysis: Mealybugs and cottony cushion scale (*Planococcus citri* and *Icerya purchasi*), insects frequently found together on citrus plants, look so much alike that separate identification is difficult. In late spring and summer, females are covered with white, cottony masses containing up to eight hundred eggs. Females lay their conspicuous egg masses on leaves, twigs, and branches. The inconspicuous young insects that hatch from these eggs are yellow to red. They feed by sucking sap from the plant tissues. They are unable to digest fully all the sugar in the plant sap, and they excrete the excess in a fluid called *honeydew*. A sooty mold may develop on the honeydew, causing the citrus leaves to appear black and dirty. Mealybugs and scales can be spread in any of several ways. The wind can blow egg masses and insects from plant to plant, or active young insects can crawl to new locations.

Solution: Natural enemies usually control these pests. Don't spray unless leaves begin to turn yellow. Then spray infested trees with ORTHO Malathion 50 Plus Insect Spray, covering both surfaces of the leaves. Repeat the spray 5 to 7 days later. Do not spray when the plant is in full bloom. Do not apply within 7 days of harvest.

CITRUS (continued)

Mites	Lack of nitrogen	Iron deficiency

Damaged leaves. Inset: Citrus red mite (10x life size).	*Nitrogen-deficient citrus leaves.*	*Iron deficiency on orange leaves.*

Mites

Problem: Leaves are stippled, yellowing, or scratched in appearance. Webbing may be seen over flower buds, between leaves, or on the lower surfaces of leaves. The fruit is often brown or russet colored, leathery, or silvery and may drop prematurely. To determine if a plant is infested with mites, examine the bottoms of the leaves with a hand lens. Or hold a sheet of white paper underneath an affected branch and tap the branch sharply. Minute green, red, or yellow specks will drop to the paper and begin to crawl around.

Analysis: Mites, related to spiders, are damaging to all types of citrus. Several species of mites attack citrus, including citrus red mites, citrus bud mites, purple mites, and citrus rust mites. Mites cause damage by sucking sap from the leaves and young fruit. As a result of their feeding, the plant's green leaf pigment disappears, producing a yellow, stippled appearance. Mite webbing traps cast-off skins and debris, making the plant messy. Feeding damage also causes tissue death, resulting in browning and silvering of the fruit and foliage. Mites are active throughout the growing season but thrive in hot, dry weather (70°F and up). By midsummer, they have built up to tremendous numbers. A severe mite infestation weakens the plant and can seriously reduce the size and quality of the fruit.

Solution: Spray infested trees with a miticide containing *dicofol* or with ORTHO Volck Oil Spray. Be sure to cover both the upper and lower surfaces of the leaves. Repeat the spray at least two more times at intervals of 7 days.

Lack of nitrogen

Problem: Foliage turns pale green, and the older leaves gradually turn yellow and often fall off. Overall growth is stunted. The tree may flower profusely but fail to set much fruit.

Analysis: Most soils are naturally nitrogen-deficient. Nitrogen is essential in the formation of plant protein, fiber, enzymes, chlorophyll (green leaf pigment), and many other compounds. When a plant lacks nitrogen, it breaks down chlorophyll and other compounds in its older leaves to recover nitrogen, which it reuses for new growth. The loss of chlorophyll causes the older leaves to turn yellow. Soils that are low in organic matter or that are sandy and porous are frequently deficient in nitrogen. These kinds of soils in particular need to be supplemented with fertilizers. Poor drainage, cold (50°F and below), and acidity or alkalinity can also make soil nitrogen less available for plant use.

Solution: Fertilize plants with Scotts Citrus Food as directed on the label.

Iron deficiency

Problem: Some leaves turn pale green or yellow. The newest leaves (those at the tips of the branches) are most severely affected. Except in extreme cases, the veins of affected leaves remain green. In extreme cases, the newest leaves are small and all-white or yellow. Older leaves may remain green.

Analysis: Citrus trees frequently suffer from deficiencies of iron and other minor nutrients such as manganese and zinc, elements essential to normal plant growth and development. Deficiencies can occur when one or more of these elements are depleted in the soil. Often these minor nutrients are present in the soil, but alkaline soil with a pH of 7.5 or higher or wet soil conditions cause them to form compounds that cannot be used by the tree. Alkalinity can result from overliming or from lime leached from cement or mortar. Regions where soil is derived from limestone or where rainfall is low usually have alkaline soil. Some citrus trees turn yellow naturally in cold weather, but if iron is available, the foliage will turn green again when the weather warms.

Solution: For a quick response, apply a chelated iron fertilizer to the foliage. Then use Scotts Citrus Food to keep the problem from recurring. Improve soil drainage.

Cold damage

Cold-damaged Valencia oranges.

Cold damage to foliage and flowers.

Bird damage

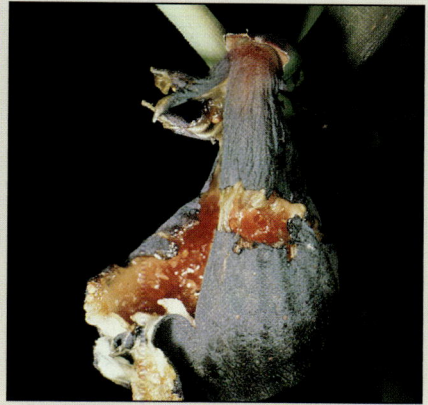

Bird damage to mission fig.

Problem: The tree has been exposed to freezing temperatures. Tender shoots may blacken and die; older foliage often turns yellowish brown and leathery and eventually withers and falls. The fruit rind may be scarred with brown or green sunken lesions. Or, even though the fruit rind appears normal, when the fruit is cut open, the flesh is dry. Severely damaged trees suffer twig and branch dieback. The bark along the branches or trunk may split open.

Analysis: Citrus plants are frost-tender and easily damaged by temperatures below 32°F. Although many citrus plants can recover from a light frost, they cannot tolerate long periods of freezing weather. Damage to fruit occurs when the juice-filled cells freeze and rupture. The released fluid evaporates through the rind, leaving the flesh dry and pulpy. In addition to causing leaf and twig dieback, temperatures of 20°F and lower promote bark splitting, which may not become apparent for several weeks or months.

Solution: Don't prune back damaged branches immediately. If danger of frost continues, drive four stakes into the ground around the tree and cover the tree with fabric, cardboard, or plastic. If possible, place a lamp under the cover. Turn the light on during cold nights. Remove this cover when the weather warms up. To protect the trunks and main limbs of young trees, wrap them with cornstalks, palm fronds, or fiberglass building insulation. Do not shade the foliage. Keep the soil moist during a freeze, but be careful not to overwater. Limit fertilizer. Damaged fruit can be removed immediately following the freeze. Always wait for new growth to appear before pruning. As soon as the danger of frost is past, you can prune blackened shoots and withered foliage. If the tree has suffered serious injury, you may not be able to determine the extent of the damage to the trunk and main limbs for up to 6 months. You can then prune out the deadwood.

Problem: Ripened figs have holes in them and may be partially eaten. They may have been knocked to the ground. Birds may be seen feeding on ripening figs.

Analysis: Some birds feed heavily on ripening figs. When the fruit is fully ripe, birds peck at the soft flesh, leaving holes in the fruit. The wounded figs may decay, becoming inedible.

Solution: You can save many of your figs by harvesting daily. Check the tree every morning and harvest those figs that have ripened. Nets thrown over the tree are also effective in reducing bird damage. Buy nets at your local nursery or hardware store.

PEACH AND NECTARINE

GROWING GUIDE

Adaptation and Pollination: See the variety chart on page 360.

Soil: Any good, deep, well-drained soil. pH 5.5 to 7.5.

Fertilizer: Fertilize with Scotts All-Purpose Plant Food or Miracle-Gro Plant Food according to label directions.

Water: How much: Apply enough water to wet the soil 3 to 4 feet deep.
How often: Water when the soil 6 inches below the surface is just barely moist.

Pruning: Peaches and nectarines bear on 1-year-old wood. Prune half of last year's growth annually. Thin out weak, crossing, or dead twigs and branches. Thin fruits to 6 inches apart when they are thumbnail size.

Harvest: Harvest when the fruits are fully colored and slightly soft. With a ripe peach or nectarine, the stem will easily separate from the branch when the fruit is gently lifted.

Plum curculio

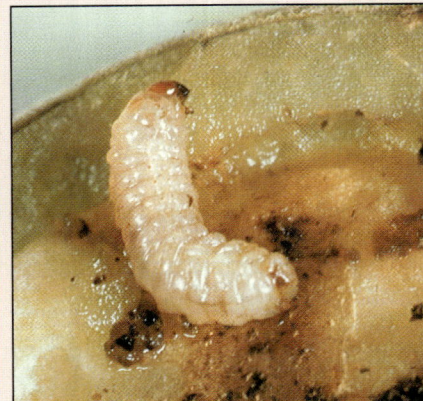

Plum curculio larva (4× life size).

Problem: The ripening fruit is misshapen and rotten and often drops prematurely. Holes about ⅛ inch in diameter and deep, crescent-shaped scars appear on the fruit. When cut open, damaged fruit may be found to contain crescent-shaped yellow-gray grubs with brown heads.

Analysis: Plum curculios (*Conotrachelus nenuphar*) are insects that attack stone fruits, apples, and pears. The adults are mottled gray-and-brown beetles with long, curved snouts. They hibernate in debris and other protected places during the winter. The beetles emerge in the spring when new growth starts and begin feeding on young leaves, blossoms, and developing fruit. After 5 to 6 weeks, the female beetles start to lay their eggs in the young fruit, cutting distinctive, crescent-shaped slits in the peaches and nectarines. The grubs that hatch feed for several weeks in the fruit. Infested fruits usually drop to the ground. The grubs eventually leave the fruit and bore into the soil, where they pupate. The emerging beetles feed on fruit for a few weeks and then go into hibernation. In the South, they lay eggs, producing a second generation of grubs in late summer.

Solution: You cannot kill the grubs inside the fruit. Spray with ORTHO Home Orchard Spray or ORTHO Malathion 50 Plus Insect Spray to kill beetles that are feeding on fruit or laying eggs. Pick up and destroy all fallen fruit. The following spring, spray the trees when the petals are falling from the blossoms. Repeat applications according to directions on the label.

Scab

Scab lesions concentrate near the stem end.

Problem: Small olive green spots appear on the half-grown fruit. These spots are generally centered around the stem end of the peach or nectarine. The spots eventually turn brown and velvety. The fruit is often dwarfed, deformed, or cracked. The leaves may have small brown spots and holes in them. Oval lesions with raised purple margins occasionally appear on shoots and twigs.

Analysis: Peach scab is caused by a fungus (*Cladosporium carpophilum*) that attacks all the stone fruits. In the spring, spores are splashed and blown from lesions on the twigs to the developing foliage and fruit. The young fruit does not show scab lesions for at least a month after it is infected. Spores that are produced on the infected leaves, twigs, and fruit continue to infect healthy peaches and nectarines throughout the growing season.

Solution: You cannot do anything about the spots on the current year's fruit, but it is edible if peeled. The following year, spray with ORTHO Multi-Purpose Fungicide Daconil 2787® Plant Disease Control, ORTHO Home Orchard Spray, or a fungicide containing *captan* when the petals have fallen from the blossoms. If scab is a serious problem in your area, continue to spray at intervals of 10 to 14 days until about a month before the fruit is harvested.

Shot-hole fungus

Shot-hole fungus. Inset: Infected fruit.

Problem: Small purplish spots appear on the young twigs, leaves, and developing fruit in early spring and eventually turn brown. These leaf spots often drop out, leaving shot holes in the leaf. Infected buds, shoots, and leaves may die. The spots on the maturing peaches and nectarines turn scablike, drop off, and leave rough, corky lesions.

Analysis: Shot-hole fungus, also called *coryneum blight* or *peach blight*, is caused by a fungus (*Wilsonomyces carpophilus*) that attacks peaches, nectarines, apricots, plums, and almonds. The fungus spends the winter in lesions on the twigs and buds. In spring, the spores are splashed by rain to the developing buds, leaves, and fruit, causing spotting and tissue death. Infection causes the leaf tissue to produce a layer of cells that walls off the damaged area. The center of the spot then drops out. Severe infection may cause extensive twig and bud blighting and possibly premature defoliation, reducing peach and nectarine yield. The disease thrives in wet spring weather.

Solution: To prevent twig and leaf bud infection, prune out infected twigs and branches as soon as they are discovered. Spray the tree with ORTHO Dormant Disease Control Lime-Sulfur Spray or ORTHO Multi-Purpose Fungicide Daconil 2787® Plant Disease Control in the fall immediately after the leaves have dropped. To reduce or prevent fruit infection, apply ORTHO Multi-Purpose Fungicide Daconil 2787® Plant Disease Control 1 to 2 weeks after petals have fallen.

Leaf curl

Leaf curl on peach.

Problem: Leaves are puckered, thickened, and curled from the time they first appear in the spring. Emerging shoots are swollen and stunted. Initially the infected foliage may be pink or red, but frequently it is pale green to yellow. As the season progresses, a grayish white powdery material develops on the leaves. Eventually these leaves shrivel and drop. Fruiting is poor, and the fruit that is present may be covered with raised, wrinkled, irregular lesions.

Analysis: Leaf curl is caused by a fungus (*Taphrina deformans*) that attacks peaches and nectarines wherever they are grown. Infection occurs as soon as the buds begin to swell in early spring. Fungal spores are splashed from the bark to the buds by spring rains. Later in the season, the infected leaves develop a grayish white covering of spores that are blown onto the bark. Infected trees are greatly weakened by the premature loss of foliage in early summer. Leaf curl is most severe when spring weather is cool and wet.

Solution: Infected leaves cannot be cured. To prevent recurrence of the disease the following year, spray trees with ORTHO Dormant Disease Control Lime-Sulfur Spray or ORTHO Multi-Purpose Fungicide Daconil 2787® Plant Disease Control in the fall immediately after the leaves have dropped or in the spring before the buds begin to swell. If the disease has been severe in past years, spray in both fall and spring.

Bacterial leaf spot

Bacterial leaf spot on nectarine.

Problem: Brown or black, angular spots appear on the leaves. The centers of the spots often fall out. The tips of the leaves may die, and eventually the leaves turn yellow and drop. The surface of the fruit may be dotted with brown to black spots and become pitted and cracked. Sunken lesions may form on the twigs. Severely infected trees may drop all their leaves by harvest time.

Analysis: Bacterial leaf spot is caused by a bacterium (*Xanthomonas campestris* pv. *pruni*) that also attacks apricots and plums. This is one of the more destructive diseases of stone fruits east of the Rocky Mountains. In the spring, bacteria ooze from lesions on the twigs to be carried by splashing rain to young leaves, shoots, and developing fruits. Frequent rainfall fosters the infection. Trees that defoliate early in the summer are weakened and produce small crops of poor-quality peaches and nectarines.

Solution: This disease cannot be adequately controlled. Spraying with basic copper sulfate when the flower buds open in the spring may help suppress the disease but will not eliminate it. Plant resistant varieties.

PEACH AND NECTARINE *(continued)*

Peach twig borer

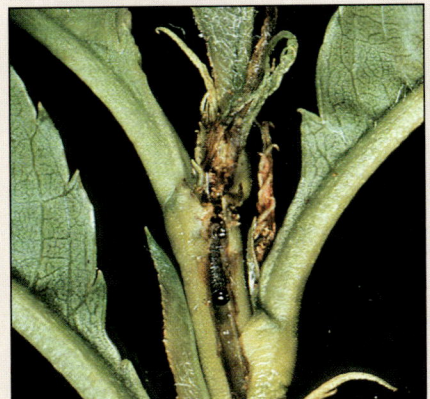

Peach twig borer (life size).

Problem: New growth at the tips of the twigs wilts and dies. When affected twigs are sliced open lengthwise, worms about ½ inch long are discovered inside. The reddish brown color and alternating light and dark bands and dark head of these worms distinguishes them from oriental fruit moth worms, which cause similar damage. Later in the season, some of the maturing fruit also contains these worms. During the summer, cocoons may be attached to the branches or tree crotches.

Analysis: The peach twig borer (*Anarsia lineatella*) attacks all of the stone fruits; it is particularly damaging along the Pacific coast. The young larvae hibernate during the winter in silk-lined burrows under loose bark or in other protected places on the tree. When the tree starts to bloom in the spring, the larvae bore into young buds and shoots. They feed on the tender twigs, killing twigs and leaves. When mature, they leave the twigs and pupate in cocoons attached to branches. After several weeks, gray moths emerge and lay their eggs on the twigs, leaves, and fruit. Egg laying and larval damage can occur all through the growing season. Later in the summer, larvae feed primarily on the maturing fruit.

Solution: Worms in the twigs and fruit cannot be killed with pesticides. To prevent future worm damage, kill the moths by spraying infested trees with ORTHO Bug-B-Gon Insect Killer or ORTHO Home Orchard Spray. The following spring, spray again just before blossoms open. Repeat the treatment according to label directions.

Brown rot

Blighted shoots.

Fruit rot.

Problem: Blossoms and young leaves wilt, decay, and turn brown during the first 2 weeks of the bloom period. The decayed blossoms may fail to drop and may hang on the tree throughout the growing season. In humid conditions, masses of gray spores may appear on infected flower parts. Extensive twig dieback often occurs. Sunken lesions (cankers) develop on the twigs and branches as the season progresses. A second wave of twig dieback may develop around harvest time as the fungus grows from the infected fruit into spurs and small branches.

On the fruit: Small circular brown spots appear on the young fruit. Later in the season, as the peaches or nectarines start to mature, these spots may enlarge to rot part or all of the fruit. During moist weather, the rotted fruit is covered with tufts of gray spores. When the infected peaches or nectarines are sliced open, the flesh inside is found to be brown, firm, and fairly dry. Infected fruit either drops prematurely or dries out, turns dark brown, and remains on the tree past the normal harvest period. Healthy fruit may rot when it contacts infected fruit in storage.

Analysis: Brown rot, caused by either of two closely related fungi (*Monilinia laxa* or M. *fructicola*), is very destructive to all of the stone fruits. The fungi spend the winter in twig cankers or in rotted fruit (mummies) in the tree or on the ground. In the spring, spores are blown or splashed from cankers or mummies to healthy flower buds. After penetrating and decaying the flowers, the fungus grows down into the twigs, producing brown, sunken cankers. During moist weather, a thick, gummy sap oozes from the lesions, and tufts of gray spores may form on the infected areas. Spores from cankers and infected blossoms or mummies are splashed and blown to the maturing fruit. Young peaches and nectarines are fairly resistant to infection, but maturing fruit is vulnerable. Brown rot develops most rapidly in mild, moist conditions.

Solution: If uninfected blossoms remain on your tree, spray with ORTHO Multi-Purpose Fungicide Daconil 2787® Plant Disease Control or a fungicide containing *captan* to protect them from further infection. Repeat the spray 10 days later. To protect maturing peaches and nectarines from infection, spray them with ORTHO Home Orchard Spray or a fungicide containing *captan* about 3 weeks before harvest. Remove and destroy all infected fruit and mummies. Prune out cankers and blighted twigs. Clean up and destroy all debris around the tree. The following spring, spray trees when most flowers show pink color and the first flowers are beginning to open. Continue to spray according to label directions.

Catfacing

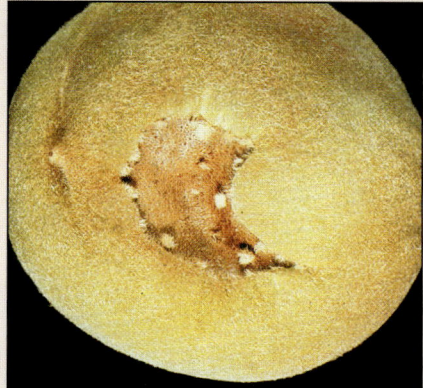

Catfacing on peach.

Problem: Sunken, corky areas mar the fruit surface. Blossoms may drop without setting fruit. Many of the young fruits drop prematurely. Some of the developing leaves and twigs are deformed. Brown, green, or rust-colored bugs ¼ to ½ inch long may be feeding on the buds and fruit.

Analysis: Catfacing, the sunken, corky "catface" disfigurations that appear on the fruit, are usually caused by the tarnished plant bug (*Lygus lineolaris*) and various species of stinkbugs. The insects hibernate in vetch or other broadleaf weeds during the winter. When the trees start to bloom in the spring, these bugs feed on the young buds, blooms, and fruits, causing bud and fruit drop, twig malformation, and catfacing. Most of the damage occurs early in the season, although the bugs may occasionally feed on the fruit up until harvest. Hail or cold weather may also damage the tender blooms and fruit surfaces, causing catface injuries.

Solution: To control plant bugs, spray with ORTHO Home Orchard Spray or ORTHO Malathion 50 Plus Insect Spray when the buds turn pink. Repeat the spray when the petals have dropped from most of the blossoms and whenever bugs are seen in the trees. The following fall, clean up weeds and plant debris to eliminate hibernating locations for the overwintering bugs.

Aphids

Aphid damage.

Problem: New leaves are curled and twisted. Leaves may turn yellow and drop. Developing fruit may be small and misshapen. A shiny or sticky substance may coat the leaves. A black, sooty mold often grows on the sticky substance. Tiny (⅛-inch) yellow, light green, or black soft-bodied insects cluster on the young shoots and on the undersides of leaves. Ants may be present.

Analysis: Several species of aphids, including the green peach aphid (*Myzus persicae*), infest peaches and nectarines. Aphids do little damage in small numbers. They are extremely prolific, however, and populations can rapidly build up to damaging numbers during the growing season. Damage occurs when the aphid sucks the juices from the young peach and nectarine leaves. The aphid is unable to digest fully all the sugar in the sap, and it excretes the excess in a fluid called *honeydew*, which often drops onto the leaves below. A sooty mold may develop on the honeydew, causing the leaves to appear black and dirty. Ants feed on this sticky substance and are often present where there is an aphid infestation.

Solution: Spray with ORTHO Home Orchard Spray, ORTHO Malathion 50 Plus Insect Spray, or ORTHO Bug-B-Gon Insect Killer as soon as the insects appear. Repeat the spray if the tree becomes reinfested. To avoid killing bees, do not spray during bloom.

PEAR

GROWING GUIDE

Adaptation and Pollination: See the variety chart on page 359.

Soil: Any good, deep, well-drained soil. pH 5.5 to 8.0.

Fertilizer: Fertilize with Scotts All-Purpose Plant Food or Miracle-Gro Plant Food according to label directions.

Water: How much: Apply enough water to wet the soil 3 to 4 feet deep.
How often: Water when the soil 6 inches below the surface is barely moist.

Pruning: Pears are borne on short fruiting branches (spurs) that grow on 2-year-old wood. The spurs continue to fruit for 5 to 8 years. Prune lightly, because removal of many spurs will reduce the pear yield. Thin out weak, crossing, or dead twigs and branches.

Harvest: Unlike most other fruit, pears should not be allowed to ripen on the tree. Pick them when they have reached their mature size and are starting to lose their green color. Don't let them soften or turn entirely yellow before harvesting. Most pears may be safely harvested during late summer. After harvesting, place the pears in a plastic bag and refrigerate them for at least 2 weeks. To ripen them, remove them from the refrigerator and keep at room temperature. After 5 to 10 days, they should be fully ripe.

PEAR *(continued)*

Fire blight

Blighted twig.

Infected blossoms.

Problem: Blossoms turn black and die. Young leafy twigs wilt from the tips down, turn black, and die. Leaves remain attached. A bend often develops at the tips of the infected shoots. On the branches, and at the base of the blighted twigs, the bark becomes water-soaked in appearance, then dark, sunken, and dry. Cracks may develop at the edge of the sunken area. In warm, moist spring weather, drops of brown ooze appear on the surface of these lesions. During the summer, shoots or branches may wilt and turn dark brown to black. Infected fruit shrivels, turns black, and remains on the tree.

Analysis: Fire blight is caused by a bacterium (*Erwinia amylovora*) that is very severe on pears and also affects apples and several ornamental plants in the rose family. (For a list of susceptible plants, see pages 358 and 359.) The bacteria spend the winter in cankers on the branches and twigs. In the spring, the bacteria ooze out of the cankers and are carried by insects to the pear blossoms. Once a few of the blossoms have been contaminated, splashing rain, honeybees, and other insects continue to spread the bacteria to healthy blossoms. The bacteria spread down through the flowers into the twigs and branches, where cankers develop. Often developing cankers encircle a shoot or branch by midsummer, causing conspicuous branch and twig dieback. Although fire blight is spread primarily through flower infection, leaves and twigs damaged by hail or wounded in some other manner are also susceptible to infection, as are tender, succulent shoots and sprouts. Although severely diseased trees may be killed, more commonly only the fruiting stems (spurs) are killed, resulting in greatly reduced fruit yields. Fire blight is most severe during warm (65° to 85°F), wet weather.

Solution: After the infection has stopped spreading in the summer or fall, prune out and destroy infected twigs and branches at least 12 inches beyond visible decay. Disinfect pruning shears after each cut by dipping them in a solution of one part chlorine bleach and 9 parts water. The following spring, apply a protective spray of basic copper sulfate soon after bud-break, when about ¼ inch of green tip is showing. To prevent blossom infections, apply a bactericide containing *streptomycin* when about 20 percent of the blossoms have opened, and repeat at intervals of 3 to 5 days until the end of the blooming period. To prevent excess growth of shoots and suckers, avoid fertilizing with high-nitrogen fertilizers. Plant varieties that are less susceptible to fire blight.

Scab

Scab on young fruit, twigs, and foliage.

Problem: Olive brown velvety spots, ¼ inch or more in diameter, appear on the leaves and young fruit. As the infected pears mature, the spots develop into light to dark brown corky lesions. The fruit is often cracked and malformed and may drop prematurely. Many of the green twigs have small, blisterlike pustules on them.

Analysis: Scab is caused by a fungus (*Venturia pirina*) that commonly infects pears. The fungus spends the winter in infected plant debris and twig lesions. In the spring, spores are produced and discharged into the air. They are blown to the developing leaves, flowers, twigs, and young pears. If the leaves and fruit are wet, the fungus infects them and spots develop. The infected tissues produce more spores, which further spread the fungus. As temperatures increase in the summer, the fungus becomes less active.

Solution: Unless they are severely infected, the pears are edible if the scabby areas are removed. To prevent recurrence of the disease the following year, remove and destroy leaf debris and infected fruit in the fall. The following spring, spray with ORTHO Home Orchard Spray or with a fungicide containing *captan* when buds are just opening and green tips are about ¼ inch long. Repeat every 10 days until 3 or 4 weeks after petal fall.

Codling moth

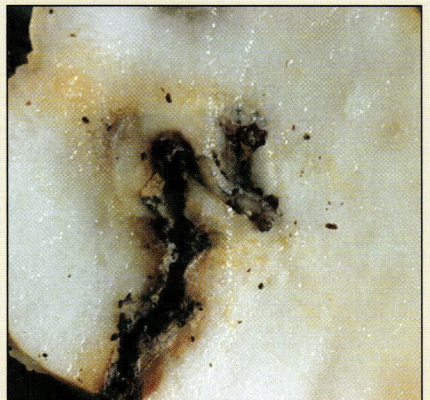

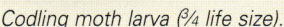

Codling moth larva (¾ life size).

Adult (3× life size)

Mites

Spider mite damage.

Problem: Fruit is blemished by small holes surrounded by dead tissue. A brown, crumbly material that resembles sawdust may surround the holes. Brown-headed, pinkish worms up to 1 inch long may be found in the fruit. The interior of the fruit may be dark and rotted. Many pears drop prematurely.

Analysis: This worm, the larva of the small (½-inch) gray brown codling moth (*Cydia pomonella*), attacks apples, quinces, and several other fruit and nut trees in addition to pears. The moths appear in the spring and lay their eggs on the leaves, twigs, and developing fruit, flying at sunset when temperatures are above 65°F. The eggs hatch in about a week, and the larvae that emerge tunnel into the fruit. They feed for several weeks, then emerge from the pears, often leaving a mass of dark excrement on the skin and inside the fruit. After pupating in sheltered locations on or around the tree, another generation of moths emerges in midsummer. Pears may be damaged by worms continuously throughout the summer. In fall the mature larvae spin cocoons in protected places, such as under loose bark or in tree crevices. They spend the winter in these cocoons and, with the warming temperatures of spring, pupate and emerge as moths.

Solution: Once worms have penetrated the pears, it is impossible to kill them. To protect uninfested pears, spray with ORTHO Home Orchard Spray at intervals of 10 to 14 days as directed on the label. Remove and destroy all fallen pears, and clean up debris around the trees. The following spring, begin spraying 10 to 14 days after petal fall. This may control moths for the rest of the summer. If codling moths have been a serious problem in the past, continue spraying according to label directions. To more accurately predict when to make the first treatment, record temperatures at sunset each day. Spray 1 week after sunset temperatures have been above 65°F for 2 or more days.

Problem: Leaves are stippled, yellowing, or bronzed. There may be webbing over flower buds, between leaves, or on the lower surfaces of leaves. Fruit may be russeted. To determine if a tree is infested with mites, examine the bottoms of the leaves with a hand lens. Or hold a sheet of white paper underneath an affected leaf and tap the leaf sharply. Minute specks the size of pepper grains will drop to the paper and begin to crawl around.

Analysis: Several species of mites, including the twospotted spider mite (*Tetranychus urticae*) and the pear rust mite (*Epitrimerus pyri*), attack pears. Twospotted spider mites, which cause leaf stippling and webbing, may be detected on white paper. Pear rust mites, which cause fruit russeting and leaf stippling and bronzing, cannot be seen without the aid of a microscope or strong hand lens. These pests, related to spiders, cause damage by sucking plant sap. As a result of their feeding, the tree's green leaf pigment disappears, producing the stippled or bronzed appearance. Mites are active throughout the growing season but thrive in hot, dry weather (70°F and up).

Solution: Spray infested trees with ORTHO Malathion 50 Plus Insect Spray. Repeat the spray two more times at intervals of 7 to 10 days. After the leaves have dropped the following fall, spray the tree with a mix of ORTHO Dormant Disease Control Lime-Sulfur Spray and ORTHO Volck Oil Spray. If you are not sure whether your trees are infested with pear rust mites, bring an infested fruit spur to your county extension service for confirmation.

231

PEAR *(continued)*

PECAN

San Jose scale

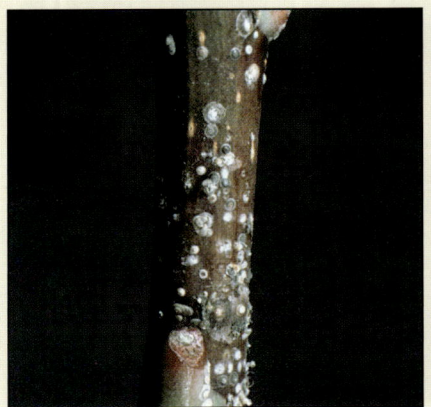

San Jose scales (life size).

Problem: Some of the leaves are pale green to yellow and may drop prematurely on weakened limbs. The bark is encrusted with small (1/16-inch), gray, hard, circular, slightly raised bumps with dull yellow centers. If the hard cover is scraped off, the insect underneath is found to be yellow or olive. Entire branches may be killed. Reddish-purple spots mar some of the infested fruit and shoots. An uncontrolled infestation of San Jose scales may kill large branches or entire trees after 2 or 3 seasons.

Analysis: San Jose scale (*Quadraspidiotus perniciosus*) infests the bark, leaves, and fruit of many fruit trees. The scales bear live young in the spring. In late spring to midsummer the young, bright yellow scales, called *crawlers*, move about and then settle on leaves, twigs, and fruit. The small (1/16-inch), soft-bodied young feed by inserting their mouthparts and sucking sap from the plant. The legs atrophy, and a hard, crusty shell develops over the body.

Solution: During the dormant season, just prior to growth in early spring, spray the trunk and branches with ORTHO Volck Oil Spray. Begin checking for crawlers about a month after full bloom. Continue to monitor through the summer to catch later generations. When crawlers are present, spray with ORTHO Malathion 50 Plus Insect Spray or an insecticide containing *diazinon*.

Pecan weevil

Pecan weevil (2× life size).

Problem: Immature pecans that drop to the ground during August are marked with dark patches and tobacco-like stains. Later in the season, some of the ripe nuts have 1/8-inch holes in them. When cut open, the kernels are found to be destroyed and may contain creamy white curved grubs up to 1/2 inch long. Reddish brown to gray, long-beaked beetles 1/2 inch long may be seen in the tree. If the limbs are shaken, these beetles drop to the ground.

Analysis: Both the immature and adult stages of the pecan weevil (*Curculio caryae*) are very damaging to pecans and hickories. Adult weevils emerge from the soil in late summer and feed on immature pecans. The injured nuts drop from the tree. As soon as the kernels harden, female weevils drill holes through the shucks and shells and lay their eggs in the kernels. The grubs that hatch from these eggs feed on the kernels for several weeks, then chew a hole in the shell about 1/8 inch in diameter, leave the nut, drop to the ground, and burrow into the soil. They emerge as adult weevils after 2 to 3 years.

Solution: Spray severely infested trees with an insecticide containing *chlorpyrifos*. If nut drop is excessive, spray prior to shell hardening. Otherwise, beginning at shell hardening, spray at intervals of 7 to 10 days until the shucks split from the shells. Weevils may also be partially controlled by shaking them from lightly infested trees. Place sheets under the tree, then lightly jar the limbs. Collect and kill the dislodged weevils that fall onto the sheet. Repeat every 3 or 4 days until weevils are no longer present.

Hickory shuckworm

Hickory shuckworm (2× life size).

Problem: Cream-colored worms up to 1/2 inch long, with reddish brown heads and black spots, are feeding in the immature nuts, many of which fall to the ground prematurely. Later in the season, after the shells have hardened, the worms may be found in the green shucks.

Analysis: The hickory shuckworm, the larvae of a small, dark gray moth (*Laspeyresia caryana*), is also known as *pecan shuckworm*. It is a pest of pecan and hickory trees wherever they are grown. The larvae spend the winter in shucks (the outer covering of the nut) on the ground or in the tree. The shuckworms pupate and emerge as adult moths in the spring to lay their eggs on pecan leaves and nuts. The larvae that hatch tunnel into the soft green pecan shells and feed on the developing kernels. Infested nuts usually drop. Later in the season, after the nutshells have hardened, the larvae tunnel into the shucks. Their feeding damage interferes with the development of the kernels. Shuckworm damage may occur throughout the spring and summer.

Solution: Chemical controls are not practical for the home gardener. Clean up and destroy all dropped nuts and shucks to eliminate many of the overwintering larvae.

Pecan nut casebearer

Pecan nut casebearer damage.

Problem: Olive green worms up to ½ inch long with yellow brown heads are feeding on the twigs, foliage, and developing nuts. Some of the young shoots are wilting. Nut clusters may be webbed together. Some nuts may have holes in them, and many kernels have been destroyed. Many nuts drop prematurely. Some may contain worms or pupae, either in the kernel or in the shuck.

Analysis: The pecan nut casebearer, the larva of a small, dark gray moth (*Acrobasis nuxvorella*), is damaging to pecans. The larvae come out of hibernation when buds open in the spring. They feed on the developing buds for a short time, then tunnel into the new shoots to pupate. Adult moths emerge just as the nuts start to form and lay their eggs on the young pecans. This second generation of worms binds clusters of nuts together with silken webs, then bores into and feeds on the nuts. This generation usually damages many pecans, because each worm eats three or four of the immature nuts during its larval stage. After reaching mature size, the larvae pupate inside the nuts and become moths. Damage by larvae continues throughout the summer but lessens in severity as the nuts enlarge.

Solution: Spray with an insecticide containing *diazinon, malathion,* or *chlorpyrifos.* The following spring, contact your county extension service to determine when moths are laying eggs in your area. Spray during this period and again 6 weeks later. Destroy all infested nuts that fall to the ground.

Sunburn

Sunburn.

Problem: During hot weather, usually in August or September, dark brown or black patches appear on the developing fruit. Leaves may turn brown around the edges and between the veins.

Analysis: Sunburn is caused by excessive evaporation of moisture from the leaves and fruit. In hot weather, water evaporates rapidly from the fruit and foliage. If the roots can't absorb and convey water fast enough to replenish this loss, fruit surfaces exposed to the sun overheat and burn; in severe cases, the leaves turn brown and wither. This usually occurs in dry soil, but fruit and leaves can also burn when the soil is moist and temperatures are around 100°F. Drying winds, severed roots, and a limited soil area can also cause sunburn.

Solution: You cannot do anything about damaged fruit, but it is still edible. To help prevent further sunburn, deep-water plants during periods of hot weather to wet down the entire root space. Water newly transplanted trees whenever the rootball is dry 2 inches below the surface.

Fruit drop

Fruit drop.

Problem: Fruit drops prematurely. The tree appears to be healthy; no signs of insects, pests, or diseases exist.

Analysis: Persimmons have a natural tendency to drop their fruit prematurely. Large quantities of fruit may drop when the tree is under stress. Stress may be caused by conditions such as excessive heat, drought, cold, or overwatering. Excessive fruit drop may also occur on trees that are growing vigorously because of heavy nitrogen fertilization.

Solution: Although fruit drop cannot be eliminated, it can be reduced. Avoid overfertilizing and overwatering or underwatering the tree.

PLUM

GROWING GUIDE

Adaptation and Pollination: See the variety chart on page 360.

Soil: Any good, deep, well-drained soil. pH 5.5 to 8.0.

Fertilizer: Fertilize with Scotts All-Purpose Plant Food or Miracle-Gro Plant Food according to label directions.

Water: How much: Apply enough water to wet the soil 3 to 4 feet deep. **How often:** Water when the soil 6 inches below the surface is just barely moist.

Pruning: European and American plums are borne on short fruiting branches (spurs) that continue to fruit for many years. Prune lightly, because removal of many spurs will reduce the plum yield. Thin out weak, crossing, and dead twigs and branches. Japanese plums are borne on 1-year-old wood and on spurs that grow on 2-year-old wood. The spurs continue to fruit for 2 to 4 years. Prune back last year's growth to half its length, and thin out spurred branches that have stopped fruiting. Thin out weak, crossing, or dead twigs and branches. For large plums, thin fruit to 4 to 6 inches apart.

Harvest: Harvest when the fruit are fully colored and slightly soft. When they are ripe, the plum stem will easily separate from the spur or branch when the fruit is gently lifted off.

Plum curculio

Plum curculio damage.

Problem: Ripening fruit is misshapen and rotten and often drops prematurely. Holes about ⅛ inch in diameter and deep, crescent-shaped scars appear on the fruit. When cut open, such fruit may be found to contain crescent-shaped, yellow-gray, legless grubs with brown heads.

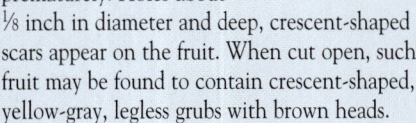

Analysis: The plum curculio (*Conotrachelus nenuphar*), an insect found east of the Rocky Mountains, commonly attacks other stone fruits, apples, and pears, as well as plums. The adult insects are brown beetles with long, curved snouts. They hibernate in debris and other protected places during the winter. The beetles emerge in the spring when new growth starts and begin feeding on young leaves, blossoms, and developing fruit. After 5 to 6 weeks, the female beetles start to lay eggs in the young fruit. During this process they cut distinctive, crescent-shaped slits into the plums. The grubs that hatch from the eggs feed for several weeks in the fruit. The infested plums usually drop to the ground. The grubs eventually leave the fruit and bore into the soil, where they pupate. The emerging beetles feed on fruit for a few weeks, then go into hibernation. Or, in the South, they lay eggs, producing a second generation of grubs in the late summer.

Solution: Once the fruit is infested, you cannot kill the grubs inside the fruit. Spray with an insecticide containing *malathion* or *phosmet* to kill beetles that may be feeding on fruit or laying eggs. Pick up and destroy all fallen fruit. The following spring, spray the trees when the petals are falling from the blossoms; repeat applications according to directions on the label.

Bacterial leaf spot

Bacterial leaf spot on Stanley prunes.

Problem: Brown or black angular spots develop on the leaves. The centers of the spots often fall out. The tips of the leaves may die, and severely infected leaves turn yellow and drop. When the fruit sets, the surface may be dotted with brown to black spots. The surface of the fruit becomes pitted and cracked. Sunken lesions may form on the twigs. Severely infected trees may defoliate.

Analysis: Bacterial leaf spot, caused by bacteria (*Xanthomonas campestris* pv. *pruni*), also attacks apricots and peaches. The disease is common east of the Rocky Mountains. The bacteria spend the winter in lesions on the twigs and in buds, oozing out in the spring to be carried by splashing rain to young leaves, shoots, and developing fruit. Periods of frequent rainfall foster the infection. Trees that defoliate early in the summer become weakened and produce few, poor-quality plums.

Solution: This disease cannot be adequately controlled. Spraying when the flower buds open in the spring with basic copper sulfate may help suppress the disease but will not eliminate it.

Brown rot

Blossom blight.

Fruit rot.

Aphids

Aphids (2× life size).

Problem: Blossoms and young leaves wilt, decay, and turn brown during the first 2 weeks of the bloom period. Often the decayed blossoms fail to drop, and they may hang on the tree through the growing season. In humid conditions, masses of gray spores may appear on infected flower parts and twigs. Extensive twig dieback often occurs. Cankers sometimes develop on the twigs and branches as the season progresses. These cankers usually exude a sticky ooze.

On the fruit: Small circular brown spots appear on the young fruit. Later in the season, as the plums start to mature, these spots may enlarge to rot part or all of the fruit. During moist weather the rotted fruit is covered with tufts of gray spores. When the infected plums are sliced open, the flesh inside is found to be brown, usually firm, and fairly dry. Infected plums either drop prematurely or dry out, turn dark brown, and remain on the tree past the normal harvest period.

Analysis: Brown rot, caused by either of two closely related fungi (*Monilinia laxa* or *M. fructicola*), is destructive to all of the stone fruits. The fungi spend the winter in twig cankers or in rotted fruit (mummies) in the tree or on the ground. In the spring, spores are blown or splashed from cankers or mummies to healthy flower buds. After penetrating and decaying the flowers, the fungus grows down into the twigs, producing brown, sunken cankers. During moist weather, a thick, gummy sap oozes from the lesions, and tufts of gray spores may form on the infected areas. Spores from cankers and infected blossoms or mummies are splashed and blown to maturing fruit. Young fruit is fairly resistant to infection, but maturing fruit is vulnerable. Brown rot develops most rapidly in mild, moist conditions.

Solution: If uninfected blossoms remain on the tree, protect them from further infection by spraying with a fungicide containing *captan* or with ORTHO Multi-Purpose Fungicide Daconil 2787® Plant Disease Control. Repeat the spray 10 days later. To protect maturing plums from infection, spray them with a fungicide containing *captan* about 3 weeks before harvest. Remove and destroy all infected fruit and mummies. Prune out cankers and blighted twigs. Clean up and destroy all debris around the tree. The following spring, spray trees when the first flowers begin to bloom. Continue to spray according to label directions.

Problem: The youngest leaves are curled, twisted, discolored, and stunted. Leaves may drop; in severe cases, the tree may defoliate. Developing plums may be small and misshapen. A shiny or sticky substance may coat the leaves. A black, sooty mold often grows on the sticky substance. Tiny (⅛-inch) green, yellow, purplish, or black soft-bodied insects cluster on the young shoots and on the undersides of leaves.

Analysis: Several species of aphids infest plums. Aphids do little damage in small numbers, but they are extremely prolific, and populations can rapidly build up to damaging numbers during the growing season. Damage occurs when the aphid sucks the juices from plum leaves. The aphid is unable to digest fully all the sugar in the plant sap, and it excretes the excess in a fluid called *honeydew,* which often drops onto the leaves and fruit below. A sooty mold may develop on the honeydew, causing the plum leaves to appear black and dirty.

Solution: Spray with ORTHO Bug-B-Gon Insect Killer or ORTHO Diazinon Ultra Insect Spray as soon as the insects appear. Repeat if the tree becomes reinfested. To kill overwintering insects or eggs, spray with ORTHO Volck Oil Spray just prior to bud-break in the spring while the tree is still dormant.

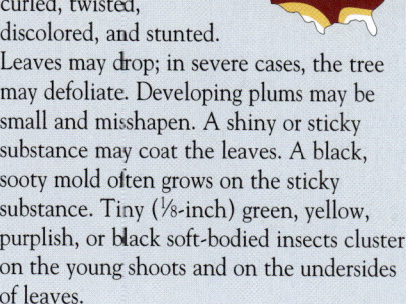

235

PLUM (continued)

Oriental fruit moth

Oriental fruit moth damage.

Problem: New growth at the tips of the twigs wilts and dies. When the affected twigs are sliced open lengthwise, worms about ½ inch long with brown heads are found inside. The pinkish white color of these worms distinguishes them from peach twig borers (see pages 220 and 228), which cause similar damage. Later in the season, some of the maturing fruit also contain these worms. Often the plums have holes in them filled with a sticky gum.

Analysis: The larvae of the night-flying oriental fruit moth (*Grapholita molesta*) damage stone fruits, apples, and pears. The larvae hibernate in cocoons on tree bark or buried in branch crotches. In the spring they pupate, emerge as brown adult moths, and lay eggs on the young plum twigs and leaves. The larvae bore into the young buds, shoots, and tender twigs, causing twig and leaf death. When mature, they leave the twigs, spin cocoons, and pupate in the tree or in debris on the ground. After several weeks, moths emerge to lay eggs. Egg laying and larval damage continue throughout the growing season. Later in the summer, the larvae feed mainly on the maturing fruit. They leave gum-filled holes in the plums when they exit to pupate. In addition to ruining the fruit, these pests may cause abnormal branching on young trees.

Solution: Worms in the twigs and fruit cannot be killed with pesticides. To prevent additional worm damage, kill the moths by spraying infested trees with a pesticide containing *malathion*. The following spring, spray again when the petals have fallen from the blossoms.

Scales

Terrapin scale (3× life size).

Problem: Brownish, crusty bumps; thick, white, waxy bumps; or clusters of flattened yellowish, gray, or brownish scaly bumps cover the stems, small branches, or undersides of leaves. The bumps can be scraped or picked off; the undersides are usually soft. Leaves turn yellow and may drop. In some cases, a shiny or sticky substance coats the leaves. A black, sooty mold often grows on the sticky substance.

Analysis: Several different types of scales infest plum trees. They lay their eggs on leaves or bark, and in spring to midsummer the young scales, called *crawlers*, move about and then settle on leaves and twigs. The small (¹⁄₁₀-inch), soft-bodied young feed by inserting their mouthparts and sucking sap from the plant. The legs usually atrophy, and a hard crusty or waxy shell develops over the body. Mature female scales bear live young or lay their eggs underneath their shell. Some species of scales are unable to digest fully all the sugar in the plant sap, and they excrete the excess in a fluid called *honeydew*, which often drops onto the leaves below. A sooty mold may develop on the honeydew, causing the plum leaves to appear black and dirty.

Solution: Spray with ORTHO Diazinon Ultra Insect Spray, ORTHO Bug-B-Gon Insect Killer, or ORTHO Volck Oil Spray in midsummer when the crawlers are active. Early the following spring, while the tree is still dormant, spray the trunk and branches with ORTHO Volck Oil Spray to control overwintering insects.

Black knot

Black knot forms corky galls on branches.

Problem: Soft greenish knots or elongated swellings form on twigs and branches. These knots develop into black, corky, cylindrical galls that range from ½ to 1½ inches in diameter and may be more than 12 inches in length. Twigs and branches beyond the galls are usually stunted and eventually die.

Analysis: Black knot is caused by a fungus (*Apiosporina morbosa*) that is severe on plums and occasionally attacks cherries. Fungal spores form during wet weather in the spring. Galls appear 6 months to 1 year after infection. The galls slowly enlarge and elongate. They eventually cut off the flow of water and nutrients to the branches, causing stunting, wilting, and dieback. Black knot spreads most rapidly during warm (55° to 75°F), wet, spring weather.

Solution: Prune out and destroy infected twigs and branches during the fall and winter. When pruning, cut at least 4 inches below visible signs of infection. Cut out knots on the trunk or large limbs down to the wood and at least ½ inch outward past the diseased tissue. The following spring, spray the tree with a fungicide containing *captan* just before the buds open. Repeat the spray two more times at intervals of 7 to 10 days. Plant resistant varieties.

WALNUT

Walnut husk flies

Walnut husk fly damage.

Problem: Soft, blackened, decayed areas cover part or all of the walnut husk. Cream- to yellow-colored maggots up to ⅜ inch long are feeding inside the husk. The walnut shells are stained dirty black; sometimes the husks stick to the shells. Kernels are often stained and may be shriveled.

Analysis: These maggots are the larvae of several closely related flies. The adult walnut husk flies (*Rhagoletis* species), slightly smaller than a housefly, are yellow brown with banded wings. The flies begin laying their eggs in developing walnut husks in late July and early August. The maggots that hatch from these eggs feed in the husks for about a month, then drop to the ground and pupate in the soil until the following summer. As a result of maggot feeding, the husks become black and decayed, and the shells and kernels are stained. The maggots never feed on the kernels, and although the walnut meats may be shriveled or discolored, they usually taste normal.

Solution: Walnut husk flies are very difficult to control. Partial control may be obtained by killing the adult flies before they lay their eggs. Spray with an insecticide containing *malathion*, covering all of the foliage thoroughly. Respray according to label directions. Do not apply after the husks split. Contact your county extension service for information regarding husk fly emergence dates in your area. Keep the area under the tree clear of fallen walnuts. Dispose of husks in a tightly sealed container.

Walnut blight

Walnut blight attacks twigs, foliage and nuts.

Problem: Buds turn dark brown to black and die. Brown spots are seen on the leaves and dead, sunken lesions on the shoots. A shiny black fluid may exude from these lesions. Some of the leaves may be deformed. Black, sunken, hard areas develop on the nuts. Infected nuts have stained shells or shriveled kernels. Nut yield may be reduced.

Analysis: Walnut blight is caused by a bacterium (*Xanthomonas campestris* pv. *juglandis*) that is common on walnuts. The bacteria spend the winter in diseased buds, twig lesions, and old infected nuts attached to the tree. A thick, shiny fluid containing millions of bacteria exudes from the infected plant parts in the spring. Spring rains splash the bacteria to the buds, shoots, flowers, and developing nuts, starting new infections. Bacterial infection reduces nut set and can continue to spread to healthy nuts and foliage throughout the summer during periods of wet weather. If the nuts are infected before their shells harden (when the nuts are three-quarters grown), the bacteria may spread into and decay the kernels. Because the wet, rainy conditions of spring faster the rapid spread of walnut blight, early-blooming walnut varieties are most susceptible to this disease.

Solution: The following spring, spray with basic copper sulfate when catkins (flower spikes resembling cats' tails) start to shed pollen; spray again when the small nutlets start to appear. If the weather remains wet, spray once more after 2 weeks. Plant late-blooming varieties.

Crown rot

Crown rot is caused by soil-inhabiting fungi.

Problem: The bark on part or all of the trunk just above or below the soil line is darkened. Sap may ooze from the affected bark. When the diseased bark is scraped away, the underlying sapwood is found to be discolored tan to black. Foliage may be sparse and yellowing. Little new growth occurs, and the tree may be stunted.

Analysis: Crown rot is caused by soil-inhabiting fungi (*Phytophthora* species) that infect many trees and shrubs. The fungi penetrate the bark of the lower trunk or upper roots, forming lesions. As the fungi progress inward, they rot the nutrient-conducting tissue under the bark, interfering with the flow of nutrients to the roots. If left unchecked, the fungi will encircle the trunk, eventually causing the death of the tree. Crown rot is greatly fostered by wet soil. Trees planted in lawns, flower beds, or other moist areas are highly susceptible.

Solution: If the fungus has not completely encircled the tree, the tree might be saved. To let the crown dry out, remove the soil within 4 feet of the base of the tree, exposing the major roots. Keep this area dry. During the rainy season, slope the remaining soil away from the hole to keep it from filling with water. After the tree shows signs of recovering (look for healthy new growth), replace the soil around the base of the tree. Or, if possible, fill the hole with stones instead of soil to help water to drain away rapidly. Avoid planting flowers, shrubs, or other vegetation immediately around the tree.

A bountiful vegetable harvest is the reward of careful planning and good gardening techniques.

Raising your own vegetables can be a source of enormous satisfaction. It can also be an extremely frustrating enterprise if your crop is attacked by insects, overrun by weeds, eaten by birds and other animals, or just simply dies. Planning and regular maintenance, however, should ensure a bumper crop.

MINIMIZING TRAFFIC DAMAGE

Pathways between vegetable rows are a way to reach produce without trampling on anything tasty. A narrow, uncomfortable path or a muddy one takes much of the pleasure from picking, however.

Gardeners usually make paths wide enough to allow easy foot traffic, but problems occur when the access is too narrow for wheelbarrows, carts, or other large equipment. What looks like a sufficient path in spring can become annoyingly tight by summer, when vegetable foliage spreads over furrows and into the passageways. In gardens in which space is at a premium, plan 12- to 18-inch-wide pathways—the minimum width for comfort. If possible, make pathways 2 feet wide.

Some gardeners like to establish grass pathways, which are appealing to the eye and comfortable to walk on. These pathways require mowing, however, and grass may escape into the vegetable beds if path edges aren't trimmed faithfully.

Permanent pathways can be created from brick, concrete, or stones. They absorb sunlight during the day, then warm the surrounding soil at night.

Mulches make effective pathways, too, and many varieties of mulch are available. Path mulches include leaves, sawdust, wood chips, grass clippings, and aged hay. Unfortunately, these materials provide shelter for pests such as snails, slugs, sowbugs, rodents, and earwigs. And, since organic mulches decompose, you will have to renew them regularly. Try to keep path mulches at least 3 inches thick—a thin layer of organic mulch can permit weeds to emerge. The disadvantages of mulch pathways are balanced by the benefits the soil receives from mulch decomposition.

The list of path mulches includes newspapers. Being lightweight, they must be held down by stones or other material. Try putting a layer of organic mulch over a layer of newspaper to create a solid, renewable walkway. Black plastic or weed block rolled out along pathways does the job, too. Though not the most attractive path material, plastic coverings do not need the constant renewal that degradable mulches do. In addition, black plastic prevents almost all weeds in the covered area. Clear plastic can be used as a covering, but it does not control weeds as well.

After two to three years, plastic becomes brittle and may crack. Though still usable as a walkway, cracked plastic allows weed emergence. Weeds that spread into vegetable beds can prove an enormous problem because applying an herbicide in already planted areas requires much care. Check plastic mulches at the beginning of each planting season, and replace them when they begin to show signs of wear.

CONTROLLING WEEDS

Weeds in the vegetable garden compete for water, nutrients, and sunlight. If prolific, weeds post a fire hazard, give shelter to furry as well as insect pests, and make harvesting difficult.

Some crops—onions, carrots, strawberries, lettuce, spinach, and celery—need intensive weeding because their tops don't grow large enough to shade sunlight-reaching weeds. Brussels sprouts, pole tomatoes, and corn require comparatively minimal weeding once past the seedling stage. They provide enough shade to reduce weed competition.

Methods of weed control in the vegetable garden include hand-weeding, cultivation, mulching, solarization, and herbicide use. These methods are often used in combination, depending on the garden crop and land.

Annual weeds germinate, flower, and die back in one season. However, their seeds are extremely plentiful and are spread by gardeners, mulches, manure, compost, birds, animals, wind, and rain.

Common annual vegetable-garden weeds are chickweed, cheeseweed, wild oats, wild barley, mustards, shepherd's purse, sow thistle, annual bluegrass, bur clover, groundsel, nettle, crabgrass, nightshade, horseweed, purslane, fleabane, lamb's-quarter, prickly lettuce, milk thistle, sweet clovers, bristly oxtongue, and barnyard grass. A single barnyard grass weed can produce over 1 million seeds in its short life.

Perennial weeds live on through winter, although they may die back. They reproduce from underground bulbs, rhizomes, or crowns on taproots. Common perennial weeds interfering with crop growth include nutsedge, witchgrass, bermudagrass, dallisgrass, milkweed, field bindweed, johnsongrass, and oxalis. If some of these names seem familiar, it is because what's considered a weed in the vegetable garden may be considered grass in the front yard.

Unfortunately, grass seeds germinate as well in a strawberry bed or tomato patch as they do in a lawn. Sometimes better, if you keep garden soil in top condition. All that open space provides much less competition than that offered by a crowded front yard.

The cardinal precept of weed control is keep weeds out. Hand-pull all weeds as soon as they appear. Never let them go to seed in the vegetable garden or in a surrounding area. Weeding a thriving berry patch is difficult but necessary; consider the size of the weed crop of tomorrow if berry-eating birds spread the weed seeds through your garden today.

Make certain all commercially purchased soil additives, manures, or mulches are certified as weed-free. Even then, be vigilant—weed seeds may survive sterilization. Fresh manure is loaded with weed seeds because most animals feed outdoors. Compost fresh manure, and make certain the compost pile is hot enough to kill weed seeds.

Another weed source is seed set out for wild birds. Keep it in containers and away from planting areas. Weeds may also be spread by dumping leftover food of caged birds, guinea pigs, and hamsters, as well as the animals' waste products, outdoors.

Living sunlight blockers: Some vegetable plants can be a definite help in weed control. For example, tall corn shades out weeds. It is also easy to cultivate between the rows. Other weed shaders include cauliflower, broccoli, and tomato plants. Potato plants also shade the ground, but don't use them in any area infested with nutsedge—this particular weed encourages potato rot by piercing tubers. After two successive years of weed-shading vegetables, plant garlic, lettuce, carrots, strawberries, and other minimal-shade plants to reinvigorate the soil.

Mulching with translucent polyethylene film discourages weeds, reduces water loss due to evaporation, and helps to warm the soil in the spring.

Raised beds make it easier to reach weeds with less stooping.

Soil solarization: An increasingly popular way to decrease weeds as well as insect pests that overwinter or pupate in the ground is through soil solarization. Solarization involves using the rays of the sun to bake out problems. You need steady sun to accomplish this, because the temperature of the treated soil should stay at 80°F or above. Therefore, solarization works best in the hottest months of summer or fall.

Before beginning solarization, level and smooth the soil to be treated. Then wet the soil to encourage maximal heat penetration. Cover the entire area with a sheet of clear polyethylene plastic. Plastic that is 1 mil thick will do for most areas. If your garden receives a lot of wind, however, 2-mil-thick plastic has greater resistance to tearing. Smooth out the plastic as you place it; close contact with the wet soil encourages transmission of sunlight into the earth. Leave the plastic in place for 4 to 6 weeks. After that, unless you are using plastic that contains an ultraviolet inhibitor, you will have to remove the solarization covering. Left on too long, it may shred and be difficult to remove completely.

Not all weeds and weed seeds are destroyed by soil solarization. Nutsedge, sweet clover, purslane, crabgrass, and field bindweed are particularly resistant.

Hand-weeding: Pulling weeds manually doesn't always do the trick, since even a portion of some weeds, such as dandelion, will quickly regrow into an entire plant.

Various implements are available to make weeding more effective, including hand tines, weeding hoes, chopping hoes, and push-pull hoes. You must remove all tuber and rhizome segments, and this may mean digging deep. Some gardeners use rotary tilling as a means of weed control. Although it benefits soil with aeration, this kind of tilling may make a weed problem worse by chopping and spreading the still-viable weed roots.

Do not leave chopped or dug-out weed parts in the garden; put them in the compost heap or in a closed container. Weed seeds can spread from your gleanings, and weeds such as purslane can actually reroot where thrown if water is available.

Mulches: Weeds tend to be opportunists. They come up earlier than vegetables and grow much faster, crowding out desirable crops. By applying weed-blocking mulch early, you can give vegetable seedlings a head start on the weeds. Mulch keeps soil cool and may deter weed sprouting as well as early growth. Remove all weed parts before applying the covering. If weeds sprout through the mulch, dig them out immediately by hand. Be careful not to damage adjacent vegetable roots when hand-pulling or hoeing.

New synthetic weed-blocking fabrics avoid many of the disadvantages of organic mulches, such as pest shelter and insufficient coverage. Your local garden center may offer photodegradable plastic film, nonwoven polypropylene fabric, or heavy pressed fibrous paper. Each has advantages you may want to investigate.

Herbicides: Using herbicides is an option if weeds cannot be satisfactorily controlled by other means. Preemergent herbicides eliminate annual weeds as they sprout. They may be used among perennial crops such as cane berries and asparagus. Preemergent herbicides are not always selective about which annual seeds they eliminate, however. Check container instructions for a list of seeds that will not be killed by the herbicide. Be sure the seeds you intend to sow are included in that list, or do not use the herbicide.

Herbicide use in the vegetable garden always requires caution. Some weeds have susceptible crop cousins, so read the list of herbicide-resistant vegetables and berries carefully. Follow label instructions about when to apply the control, and learn what weeds are present, so you can select a postemergent herbicide with the appropriate ingredients. Avoid long-life herbicides. They might protect your summer crop but remain in soil past harvest and harm the winter crop.

Weeder geese: Some gardeners use Chinese geese to control weeds in the vegetable and berry garden. The birds eat slugs, also. If you wish to try this option, read several books on goose care before beginning. Geese must be kept in a fenced area away from domestic and wild animals. Use 3-foot-high stakes and chicken wire of similar height to create a movable fence line. Like any other animal, geese must be provided with clean water and protection from rain, heat, and cold. In the vegetable garden, some supplemental food will be necessary.

Geese eat young tender weeds. If there is not a sufficient supply of weeds, they also eat crop seedlings. Move the fence to varying sites to provide an ongoing food supply. All geese feed primarily in the early morning and late evening. If you must set them out in the garden at specific times, work according to their schedule. Never use pesticides or herbicides in any area immediately before introducing weeder geese or when they are present. The herbicide label will cite the length of time that must pass before you can reintroduce animals in the treated area; follow the instructions.

A sturdy weeding hoe is an essential garden tool. When mechanically pulling weeds, be sure to remove all plant parts.

Growing vegetables in containers reduces the need for weeding. These potatoes enjoy quickly-warmed soil in black plastic trash

bags that make harveting a snap. A short length of pvc pipe facilitates deep watering.

CONTROLLING EROSION

If your vegetable garden is on a sloping hillside, erosion may present a problem. With water and wind action, soil tends to move from high areas to low ones. If the topsoil is depleted, bolster the soil with fresh topsoil and fertilizer before attempting to plant anything.

Also develop some type of erosion control. You may find examples of terracing and retaining structures in pictures of hillside gardening in South America, Japan, India, Asia, and Italy, among many other countries where agricultural space is at a premium.

The basic concept of terracing is to provide a retaining structure that prevents soil loss. Heavy rain can wash soil past even the most effective retainers, however, so a sturdy wall of some type must be at the bottom of the terraced garden to hold the topsoil that makes it down the hill.

Construct retaining walls from railroad ties, concrete blocks, tires held with concrete and wire mesh, logs, bricks, or stones. You can terrace the entire area or separate sections, depending on need. Always follow land contour, rather than altering it.

PROVIDING FERTILIZER

Lack of available soil nutrients can ruin a vegetable or berry crop. Plants may grow but be stunted or distorted, or produce a limited yield. Since plants in poor health tend to suffer more from insect attack than healthy plants, other problems may follow nutrient deficiency.

If symptoms of nutrient deficiency continue after the correct fertilizer is applied, pH imbalance may be the cause. A soil pH over 8.0 may make certain nutrients, such as manganese and iron, unavailable to plants.

Treatments for deficiencies:
Nutrients required in soil for plant growth include phosphorus (see page 250), nitrogen (see page 254), iron, boron, potassium, magnesium, calcium, molybdenum, and zinc. The sections that follow describe how various deficiencies affect specific plants.

■ **Boron deficiency:** Most common in dry weather and in alkaline sandy soil, boron deficiency in beets causes a condition known as heart rot. Black areas appear on the skin of the beet and inside the root. The root may be wrinkled and cracked. Most leaves die, leaving only a few deformed leaves. Plant growth slows markedly. Boron deficiency in celery causes brown horizontal cracks to appear across stalks. Leaves turn yellow and may die. Growth slows.

As soon as you identify boron deficiency in your garden soil, add borax, which is available in several forms. Before sowing, rake sodium tetraborate into the soil at the rate of 1 ounce for every 20 square yards of soil; mix in enough light sand to provide even distribution. Or, if plants are already growing, use 1 tablespoon of household borax dissolved in 12 quarts of water. This solution is sufficient for a 100-foot row. Repeat the treatment in 2 to 3 weeks. If necessary, correct soil pH to bring it between 6.0 and 7.0.

Water as needed to prevent the soil from drying out.

■ **Iron deficiency:** Also called lime-induced chlorosis, iron deficiency causes leaves to turn pale green or yellow. New leaves are most severely affected, and they may be undersized. Leaf edges and leaftips may scorch. In severe cases, new leaves are all yellow or all white, though leaf veins may remain green. Iron deficiency almost always occurs when soil pH is above 7.5. (The condition affects raspberries at pH 8.0.) Reduce soil pH by adding acidic materials such as peat, or use an aluminum sulfate or ferrous sulfate additive. After treatment, wait until the pH reaches at least 7.0 before planting.

■ **Magnesium deficiency:** A shortage of magnesium shows up as yellowing (chlorosis) of older leaves and upward curling of leaf edges. The edges may yellow, leaving a green area in the center of the leaf. Symptoms begin in the lower plant and proceed upward. Magnesium deficiency is common in all types of berries and vegetables because magnesium is easily washed out of soil during heavy rains. In addition, magnesium becomes unavailable to plants in the presence of potassium.

Vegetable plants are notoriously heavy feeders; adequate soil fertility is essential throughout the growing season for maximum yields.

Magnesium deficiency of potatoes appears as yellowing of leaf areas between veins. The leaf then browns and becomes brittle. Plant growth slows. Magnesium deficiency of beets causes older leaves to turn pale between veins, then brown. Some beet varieties may develop bright red leaf tints. Magnesium deficiency of tomato plants is common in plants receiving high-potassium fertilizers. Yellow-orange bands appear between leaves. A similar coloring may appear on raspberry leaves. Lower leaves are affected first. Symptoms spread upward, turning older leaves brown. In grapevines, magnesium deficiency may appear as purple blotches rather than the typical yellow-orange discoloration between veins.

To treat magnesium deficiency, spray with a product containing magnesium sulfate; apply it at 7- to 10-day intervals throughout the growing season. Some leaf yellowing may persist on tomato plants, but the crop will be unaffected.

■ **Manganese deficiency:** The symptoms of manganese deficiency resemble those of magnesium deficiency and often occur in conjunction with iron deficiency. In general, leaf yellowing occurs. In beet and spinach plants, leaves also roll inward. Beet leaves may assume a triangular appearance. Yellow blotches appear between leaf veins. In severe cases, entire leaves may turn pale yellow.

Manganese deficiency in peas, which is called marsh spot, causes a dark rust-red spot or cavity in pea centers. Pods appear normal, but leaves may be slightly yellow between veins. Manganese deficiency usually occurs in sand, alluvial silt, and clay soils. It is more common where pH is over 7.5. Symptoms may appear suddenly after a heavy rainfall, since soggy soil may impede manganese release to plants. Prevention includes adding manganese to soil before planting, or spraying with a solution of 2 ounces of manganese sulfate in 3 gallons of water. Add an agent to help the solution stick to the plants. Repeat 2 or 3 times at 2-week intervals.

■ **Molybdenum deficiency:** Called whiptail, molybdenum deficiency is a problem only on brassicas such as broccoli and cauliflower. Heads of affected plants fail to develop. Leaf blades become thin, straplike, and rippled. Molybdenum deficiency occurs in acidic soils only. Add lime before sowing or planting.

■ **Potassium deficiency:** Also known as potash deficiency, a lack of potassium causes small brown spots to appear along leaf edges. Leaf edges turn yellow, then gradually turn brown, curl downward, and die. Vegetable and berry yields are small. The crops that do appear may be distorted and ripen poorly. Vegetables and berries with potassium deficiency are susceptible to fungus and viral diseases.

Potassium aids in moving food supplies from leaves to roots and stems. Without it, crops cannot grow properly. Light, sandy soil and peat and chalk soil are often low in potassium, as are soils in high-rainfall areas. Some crops, notably tomatoes, beans, and raspberries, demand more potassium than others. To remedy potassium deficiency, spray plant foliage with a liquid plant food, or water with a liquid plant-food additive. Apply a general-purpose fertilizer formulated for tomatoes—such fertilizers are usually high in potassium. For raspberries, apply a fertilizer that contains sulfate of potash.

Fertilizer burn: The condition called fertilizer burn may occur when gardeners use too much fertilizer or too strong a concentration, do not water fertilizer in properly, or allow undissolved granules to remain on leaves. In these cases, the accumulation of salts from both spray and granular fertilizers interferes with water use by the plant. Salts may accumulate in leaf edges, causing a burnt or scorched appearance. In strawberries, leaf edges and areas between veins turn dark brown and die. After fertilizer burn occurs in any plant, no amount of watering will restore the green to browned edges. The burned leaves may drop. Plant growth may slow or stop. In severe cases of fertilizer burn, plants die.

To prevent fertilizer burn, follow application instructions carefully. In areas with poor drainage, be especially careful because fertilizer will remain in the standing water. If possible, improve drainage before applying the treatment. Add fertilizer to moist soil only. Water fertilizer well; do not permit it to rest on leaves or on the soil

Trenching next to rows is a good way to deliver water and fertilizer to roots, as these healthy beans show.

surface. If you apply excess fertilizer by mistake, water the area thoroughly to leach salts out.

If you plan to add a dry fertilizer at seeding time, make furrows for fertilizer 3 inches from the empty seed rows and 2 inches deeper than seed depth. Put fertilizer in the furrows, and refill them with soil. Then plant the seeds.

Seedlings are especially sensitive to fertilizer burn. Excess fertilizer can cause seedling root damage or prevent seedlings from emerging. The key to correct application is following label directions.

Like commercial fertilizers, fresh manure from cattle and poultry can cause fertilizer burn because the manure is high in salts. Do not use fresh manure of any kind. In addition to causing fertilizer burn, fresh manure may contain large numbers of weed seeds, depending on what the animal ingested. Fresh manure may also contain insect eggs and larvae, if these were on the grasses eaten. Though fresh manure is often touted for its high nitrogen content, the amount varies considerably—again, according to the producer's diet. Before applying fresh manure to the soil or using manure as a mulch, thoroughly compost it. Add just a bit to the compost heap at a time or the pile may overheat, destroying beneficial microorganisms and earthworms.

PREVENTING COMPETITION

Competition from trees can prove fatal to many vegetables. Feeder roots, which reach out underneath the soil at least as far as the branches reach overhead, can divert nutrients and water from crops. If possible, establish your garden far from shallow-rooted trees such as elms, maples, poplars, and willows. If limited space makes proximity to shallow-rooted trees unavoidable, dig a 3-foot-deep trench around the garden. If feeder roots are visible, cut them; limited root pruning will not harm the trees. Line the dividing trench with heavy-duty plastic or sheet metal, then backfill with soil. The feeder roots will take a while to penetrate this barrier.

Black walnut trees present special problems for tomato plants. Black walnut roots give off a substance that causes wilting and dwarfing of many tomato varieties. (Falling walnut leaves will not cause problems.) Digging a dividing trench may not be sufficient to save the plants. If you have no other growing space, plant tomatoes in containers.

WATERING

Without water to soften and break seed coats, seeds will not germinate. Once germinated, seedlings are especially sensitive to water stress. Their delicate roots may reach only into the top inch of soil. If this dries out, the seedling may die despite diligent rescue efforts. In hot weather in which soil becomes dry to the touch, seedlings may need water once a day. Even mature plants may not have a root system deep enough to allow them to survive temporary dry periods.

Effects of drought: Plant wilting, followed by recovery after watering, is usually the first sign of water stress. Reliance on a plant's recuperative powers during drought is seldom wise. Repeated wilting results in poor plant growth. Fruit on stressed plants may not mature properly. It may crack open and have poor flavor. Cucumbers without sufficient water grow in odd shapes and have a bitter taste. Overcompensating for water stress by overwatering can also be harmful. Under

these conditions, sweet-potato tubers may develop cracks and black spots within roots.

Berry bushes can tolerate water stress better than other garden plants. Once berry bushes are established, they have an extensive root system. Unfortunately, their outstretched feeder roots are also pathways for fungi. The natural susceptibility of berry bushes to fungus diseases is enhanced by overwatering and high humidity.

Irrigation schedules and methods: When establishing a watering schedule for your garden, plan on including one thorough soaking per week during the growing season. Remember that most common vegetables are almost 90 percent water and must retain this amount to mature into tasty table food.

The amount of watering you need to do depends greatly on what type of soil you have. Sandy soil retains less water than clay, and requires more frequent irrigation. Use water effectively by adding organic materials, such as compost, to the soil. Compost greatly improves the moisture retention and water distribution in any soil.

■ **Drip irrigation:** When watering vegetables, avoid overhead sprinkling if possible. Leaves that do not dry quickly support fungi and water-transmitted viruses. Some form of drip irrigation—whether a drip, trickle, or soaker system—provides water without encouraging runoff. Drip systems are becoming increasingly popular because they are economical as well as effective. They provide water at a slow rate, allowing it to seep to where it is most needed, at the root zone. If you want to give your tomato seedlings an even bigger boost than a drip system alone can provide, make shallow watering basins around each plant and fill them at each watering.

■ **Water conservation through mulching:** Using mulch helps slow water evaporation, cutting watering needs by as much as one third. Mulching may be a necessity for gardeners in warm climates where water is scarce. Put mulch down early for weed control, but do not apply it around crops until late spring, giving soil a chance to warm up. Use mulch around transplants and any seedling that is at least 6 inches tall.

Effective organic mulches include peat moss, seaweed, sawdust, dry composted manure, bark chips, and straw. Do not use fresh hay as a mulch. If you have an ample

Regular watering is essential for vegetables and fruits grown in containers.

hay supply, let it rot outdoors for a year before applying in the garden.

Mulch depth should be about 3 inches; piled up too high, mulch may encourage rot.

Black plastic mulch discourages weed growth and lessens evaporation caused by wind. However, black plastic allows water to reach the ground only through whatever holes you make in it. During and after rain, puddles may form and stay on top of the covering. A new development, weed-blocking fabric, avoids this disadvantage. This fabric effectively prevents weeds but allows sunlight and water to pass through to the ground.

Application instructions vary slightly with each product. Basically, however, installation involves covering the vegetable patch with the plastic or fabric, then cutting holes for areas to be seeded or for transplanted seedlings. Seed holes will be slightly larger than seedling holes but, to prevent weed infiltration, make openings only as large as necessary. Hold the material in place with earth, wood chips, or rocks at the edges. Black plastic or woven plastic mulch is valuable for vine crops, such as squash and melon. The vines are free to spread without competition from weeds, and the crop is kept off of damp earth.

Clear plastic is not effective for water conservation or weed prevention. Sunlight penetrates clear plastic, warming soil. This added warmth can encourage heat-loving seeds, but it can also aid weed germination or produce enough heat to kill everything.

WEATHERPROOFING

The two most common weather problems for home gardeners are wind and extremes in temperature.

Windbreaks: If wind is a constant or intermittent problem, create a windbreak for the vegetable garden. Windbreaks serve multiple purposes. Not only do they cut the effect of drying summer winds and destructive winter winds, but they help retain heat—of special importance to spring seedlings. Also, windbreaks encourage bees to visit and pollinate—bees don't like strong winds any more than plants do.

Placing the garden where natural features or buildings shield it from the wind is the first step toward solving a wind problem. Building a low fence around the garden is another solution.

Young transplants need protection from hot sun and wind; a few shingles will do.

They also need protection from late frosts; these are getting an early start.

Hedges that do not block sunlight work well as living windbreaks. Protective hedges that also attract birds include barberry, yew, euonymus, and arborvitae. Trees can also be effective windbreaks if their branches are not allowed to grow dense and shade the garden. Trees or large shrubs may be inappropriate, however, if space is limited, because in a small area they compete with garden plants for nutrients, water, and light. Plan living windbreaks carefully, with growth patterns in mind, and prune hedge or tree foliage as necessary.

Since it takes time and labor to establish a fence or hedge windbreak, consider planting a flower windbreak. Some annuals grow high enough in a season to block a light breeze, although they are ineffective against strong winds. To protect garden seedlings against strong winds, shield the plants by placing cans with the ends removed around the stems.

Protection from heat and cold: If you have access to old tires and do not mind their appearance, they make fine windbreaks for warmth-loving vegetables such as tomatoes and eggplant. Tires hold heat, particularly if you put stones where the inner tube used to be. Some gardeners create raised beds throughout their gardens from tires of various sizes that are filled with fertile soil. The elevation of the plants in the tires alleviates drainage problems. However, just as a tire holds heat during cold spells, it can produce a warmer

environment during hot weather. Plants inside tires may need extra watering.

During extremely hot weather, drape newspaper tents or cheesecloth over susceptible crops. Anchor the covering with stones or dirt.

Frost is a continual concern in some areas and an intermittent concern in others. Among the possible solutions is covering the garden with plastic, blankets, or sheets on nights when frost is expected. Another emergency measure is covering plants with temporary terrariums made from clear plastic drink bottles, bottoms removed. If frost threatens frequently, you may want to place heat-absorbing objects in the garden such as large, dark-colored boulders—objects that, after sunset, release stored heat and protect the vegetables from frost.

If you must plant warmth-loving vegetables in frost pockets, keep in mind that cold air tends to collect in low places. Block cold air with a hedge, stone fence, or embankment. However, since cold air, like water collecting behind a dam, can overflow, provide an outlet. Make a pathway through the embankment, allowing the air to flow harmlessly away from susceptible plants.

Since cold air and the frost that accompanies it settles in ditches and ground hollows, you might want to go one step further and dig out a low-lying catch basin at a site of your choosing, away from your vegetable garden. Raising the vegetable beds is another means of frost protection.

Trellises can be an attractive part of the garden.

PROVIDING SUPPORT

Staking vegetable and berry plants can be a nuisance. The alternative, however—leaving plants to trail on the ground—can produce as much rot as food. Some plants, such as cane berries, turn into impenetrable thickets unless controlled by staking.

Always use rot-resistant wood when staking vegetables. A metal stake or uninsulated wire will heat up in summer, possibly burning any plant that touches it.

Gardeners sometimes underestimate the growth potential of plants. If you decide to stake them, provide 8-foot-high stakes at planting time rather than trying to lash stakes together to rig a higher support late

This A-frame trellis is ideal for supporting cucumbers.

in the season. Lashed stakes do not balance properly and are prone to toppling. Dig stakes 1 foot into the soil to anchor them against plant weight and wind. Keep stakes about 4 inches away from plants.

To stake red and yellow raspberries, use sturdy 3-foot-high posts that are 3 inches thick. Anchor the posts well into the ground; superficially anchored posts will topple in the wind or fall over from the weight of the canes or shifting wet soil. String two smooth 10- or 11-gauge wires between the posts, one wire near the tops and the other halfway up from the ground.

Climbing vegetables, such as beans and peas, do best off the ground, away from insects. One type of support structure for climbers consists of wood supports tilted in an inverted V shape. To create this structure, place 8-foot-high supports in 2 rows. Place the rows 24 inches apart, and set the stakes within each row 12 inches apart. Tilt the tops inward and secure them. Make certain that supports are extremely well anchored. If they fall over while supporting a heavy crop, they will be difficult to replace without damaging crops.

Mesh netting with large openings is a space-effective support alternative for pea and bean vines. Attach the netting to any vertical support. To avoid wind damage to the support system, encourage the vines to grow through the netting.

Cucumbers left to lie on the ground can take up quite a bit of space in a garden. An easy way to let them climb is to provide nylon netting, which can be hooked over a sunny fence. Garden-

supply stores and catalogs offer green nylon netting, which is easily camouflaged by growing foliage.

For tomato plants, staking continues to be controversial. Some gardeners prefer to let them sprawl, keeping foliage open and fruit exposed to the sun. You will need at least 15 square feet per plant if you choose this option. Slugs and hornworms go after lower tomatoes first, and ground rot may occur. Watch low-lying plants carefully, and take corrective action if necessary.

One compromise between staking tomato plants and letting them sprawl is placing them on a 6-inch-thick mulch. Setting plants through slits in protective black plastic is another alternative. Tomatoes also receive ground protection from multiple layers of newspaper.

Cages to support tomato plants are available at most garden-supply stores. Place a cage around a plant when it is small. Placing a cage over a maturing bush or vine often damages foliage. Anchor the cage securely into the ground.

Tomato vines damage easily. Tie them to stakes or supports with soft cloth strips, bits of old nylon stocking, or commercially available tomato ties. Never bend vines at a sharp angle when handling. If you do accidentally angle a stem and it breaks partially, you may be able to do some garden doctoring. Join, then tape, the edges together. If the leaves do not wilt in a few days, the vine will survive.

HARVESTING

■ Asparagus should not be harvested the first year. The second year, take no more than 2 spears from each plant, and remove them from mid-April to mid-June only. After this, the spears become spindly. Allow them to form feathery foliage to help build plant strength for the following season.
■ Harvest all beets before severe frost. Beets are especially delicious when small, about the 1-inch size. They become less tender as they get larger.
■ Pick broad beans when the seeds are about fingernail size. Left overlong on the vine, the pods develop black streaks, which indicate that the seeds have ripened and their skins have become tough.
■ Harvest brussels sprouts from the bottom of the plant upward. Snap or cut them off cleanly and as close to the stem as possible.
■ Dig up carrots when they reach baby

length, about 3 inches long. Although they can remain in soil for quite a while when mature, carrots longer than 3 inches tend toward woodiness. Remove all carrots before winter. Left in damp ground they are easy targets for wireworms and slugs, and heavy rains may cause split roots.

■ Inspect corn carefully to determine picking time. Turn back the corn sheath until just a few kernels are visible. Pierce a kernel with your fingernail. If the juice is watery, the corn isn't ready yet. Test again a few days later. If the juice is milky white, the corn is ready. Remove the ear by giving it a quick twist. Cook corn as soon as possible.

■ Harvest the cucumber crop regularly, and do not allow the cucumbers to get too large. Pickling size is 2 inches long; slicing cucumbers range from 6 to 8 inches long. If the vine is overburdened, production will be limited.

■ Butterhead lettuce is ready for the table when a loose head forms. Tight-growing crisp lettuce varieties are best when heads are firm.

■ Pick melons when they pass the smell test—that is, when the stem end exudes a strong pleasant aroma. The exception is cantaloupe, which is ready when fruit comes off the stem easily.

■ Harvest onions when leaftips turn yellow and start bending over. Be sure that these are indeed signs of maturity, not of damage by thrips. Don't jerk the onions from the ground; ease them out with a hand spade. Let bulbs dry in the sun. If you have many onions and limited drying space, make an onion rope from heavy twine. Attach onions to the dangling cord, beginning at the bottom of the twine.

■ Harvest summer squash when the skin is not yet firm. The converse is true of winter squash, which should have firm skin when harvested.

■ Pick tomatoes at their peak, which comes about 6 days after first color appears. At their ripest, they are brightly colored, full-fleshed, and shiny. Three days after ripening, they begin to lose flavor. Even refrigerator storage does not maintain prime taste. If you must pick full-sized tomatoes while they are still green, ripen them indoors between 60° and 70°F. Shield them from direct sunlight. Keep fruit from touching in the ripening container.

ANIMALPROOFING

As housing moves into former forests and fields, many native animals become garden pests. Birds, rabbits, gophers and moles, raccoons, and others can do a lot of damage. So can birds, both exotic species and common city and suburban residents.

Birds: Starlings, cowbirds, grackles, blackbirds, and crows are particularly voracious feeders. They and other species are as fond of berries as are people and insects. Birds are attracted to the high sugar level in ripe fruit and often get to a juicy berry right before you're ready to pick it. Many gardeners like birds so much that they place a few extra berry plants in the yard to provide enough fruit for all. But if birds are eating into the family food supply, several remedies are available.

For raspberry protection, create a wood or metal frame around plants. Over this, drape netting with ¾-inch mesh. This size lets in air, water, and sunlight. Secure netting at the bottom so berry predators cannot slip underneath.

To protect blueberries, set 8-foot posts around the growing area to form a frame; keep the frame at least a foot away from the blueberry bushes. Nail netting with ¾-inch mesh around sides and tops of posts. In large growing areas, place the screening and posts in rows. This prevents birds that have somehow gotten into one area from feeding in all areas.

For strawberry protection, cover plants with wire cages or netting with ¾-inch mesh stretched over a frame.

To protect grapes, loosely tie paper bags over ripening fruit clusters. Allow just enough space around the vine to permit air circulation. Cut off the bottom corners of bags for additional necessary circulation. Provide bagged clusters with some afternoon shade, or grapes within may cook.

To protect vegetable seedlings from bird feeding, use floating row covers. Sold in rolls at most local garden centers, these plastic coverings resemble fabric and were created especially for the vegetable garden. They are lightweight and permit sun, water, and air to pass through.

Cover the seedling rows with the plastic. Secure the edges by weighting them down or burying them so animal pests cannot crawl underneath. Hefty rocks are effective weights, as are 2×4's. If you bury the edges, add weights at corners and at intervals along the sides. The drawback to row covers is

The best way to protect blueberries from birds is to cover the bushes with bird netting. Be sure to gather it at the base of the plants to keep birds out.

247

that they can create temperatures underneath that are up to 30° F higher than the air temperature. Remove row covers in summer to avoid baked plants.

In some areas, crows are a particular problem. With great skill, these large noisy birds pick up coverings, shred material, and find their way around obstacles. Ordinary row covers or nylon netting may not keep them away. Try using removable chicken wire cages over vegetable seedlings. Anchor the cages.

Scarecrows seldom work. Birds become accustomed to them quickly and learn that they are harmless. Shiny aluminum foil streamers fluttering from rope placed across the garden may discourage bird feeding. However, birds soon get accustomed to this, too. If netting or row covering is not an option, an active cat may do the job. A cat may be too effective, however, in that birds will flee the garden entirely, leaving insect pests unchecked.

Rabbits: In small numbers, rabbits just nibble on row ends; large populations go after the greenery as it comes from the ground. Dogs can sometimes keep them away. Some rabbits, however, just seem to run circles around canine protectors.

The first thing to do if you see signs of rabbit feeding is get rid of potential hiding places. These might include a woodpile or tall grass and weeds.

One method to deter rabbit feeding is to create a 24-inch fence. Around the garden periphery, pound in green plastic stakes of the appropriate height. Use metal clips to attach green plastic-coated wire fencing

with 1- by 2-inch mesh to the stakes. The green mesh will blend with garden foliage, and the clips will allow you to adjust the fence when necessary.

Since rabbits can burrow underneath fences, you may want to extend the wire mesh at least 6 inches below ground. Determined rabbits will eventually tunnel underneath, but the below-ground shield may delay them long enough to allow crops to mature and be harvested. An alternative, particularly if you want mobile fencing, is to buy 36-inch-high wire mesh and fold out the bottom 12 inches around the entire periphery. Securely staked or weighted down, this horizontal portion will make it more difficult for rabbits to dig under the fence edge.

If fencing your garden is impractical, try protecting seedlings from rabbits by shielding the plants with large tin cans from which the metal tops and bottoms have been removed. In early spring, when the temperature needs a boost anyhow, leave on the plastic top that comes with many cans. It allows sunlight to come through and provides protection from wind and cold. Make certain that sufficient water condenses inside the container to provide moisture. If necessary, remove the cans to water the seedlings, then replace the cans. Using cans as protection becomes ineffective when the plant tops peek over the metal edges.

Gophers and moles: Signs of gopher or mole invasion include wilted plants that have no roots and mounds of fresh pulverized earth. These animals eat underground vegetable parts, often

eliminating them altogether in just one evening. They also pull entire plants down into their burrows.

Although gophers and moles are not communal animals, there can be as many as 20 per acre, each with a separate series of tunnels. If your garden is severely infested, you may have to dig the entire vegetable bed to a depth of 2 feet, line it with ½-inch–mesh chicken wire, and replace the soil.

If you cannot line the entire bed, consider making or purchasing individual wire mesh baskets for each plant. Again, mesh must be ½ inch or less and protect all below-ground portions of the plant.

Trapping is the most thorough method of gopher and mole control. Using just one trap works only if there is only one hole. Gophers and moles make many exit and entry holes, so they have many escape routes. Purchase at least 2 traps. Box traps are easier to use than Maccabee traps, but both are effective when employed according to directions.

Dig down to the main horizontal runway that connects with the surface hole. The runway can be as deep as 18 inches. Place traps on either side of your excavation. Since commercial traps vary, read and follow instructions. Bait the area with carrot tops or other fresh greens. Cover your excavation with a board to block out all light. Check the traps frequently.

Many gardeners do not consider poison an option because of the potential danger to cats, dogs, and even curious youngsters. Though accidental poisoning is rare, it can happen. If you do choose to use poison,

Grapes protected in paper bags.

Strawberries protected under bird netting.

Blueberries protected by bird netting.

place bait in the deep tunnels close to living quarters rather than in surface tunnel runways, which may be used intermittently. Make certain to close your digging hole with boards or earth after placing the bait.

Gopher and mole elimination may require persistence as well as multiple methods.

Raccoons: Raccoons are becoming much more common in urban and suburban vegetable gardens as home construction cuts a swath into their natural habitats. Raccoons are adaptable creatures, handy at removing garbage-can lids in search of food and curious enough to crawl down the chimney. Cute as they may be, raccoons are wild animals and may bite. Never handle a raccoon under any circumstances. If a raccoon is trapped under the house or by a dog, call animal control for help.

Because raccoons feed on pest insects as well as fruit and seeds, many homeowners tolerate their presence. Once raccoons have discovered a food source, however, they keep returning. In the process, they may discover your vegetable garden and berry patch. Therefore, protecting your garden means keeping the whole area free of raccoon-attracting food. Secure garbage-can lids thoroughly; locking devices are available. A spotlight on the garbage-can area is another raccoon deterrent. Bring pet food inside at night. Keep your compost pile turned, and hot enough to decompose fruit and other kitchen remainders.

Traps are most effective before your crop ripens. A ripe crop provides foods that compete successfully with baits. Raccoons like sweets, so peanut butter, marshmallows, or bread and honey are effective attractants. Raccoons may be protected by law in your vicinity. Ask at the county extension service or local animal control agency about regulations.

Deer: Like raccoons, deer are becoming increasingly problematic as suburbia intrudes into forested areas. Deer can easily jump 5-foot fences to get at vegetable greens. They tend to feed in the early morning or late evening, when few gardeners are around to chase them off. A large watchdog may scare them away, although barking can annoy neighbors.

Fencing is perhaps the best deerproofing. To exclude deer, a traditional fence must be at least 8 feet high. In some area,s a fence of this height is prohibited by building codes or is infeasible. Since deer jump high or wide, but usually not both ways at the same time, a deer fence may be the answer. This is 6 feet high with a 3-foot-wide section across the top, parallel to the ground.

Some gardeners in deer areas have tried electrical fencing. The strands of such a fence should be 10 inches apart and the bottom one not more than 8 inches from the ground. Consult the county extension service for rules concerning electrified fencing in your area.

Rats and mice: Rodents take up residence where the food pickings are easy, such as in a garden with a lot of debris or in an inefficient compost heap. Organic mulches provide rats and mice with hiding places. To keep the animals' winter homes away from your food supply, thin out strawberry plants in early fall and wait until the first frost to mulch.

In all seasons, keep your yard clear of debris. This includes trash piles, leaves, firewood, newspapers, boxes, pipes, logs, old doghouses, and tree cuttings. Clean out and close off the area under steps. If you must store construction materials outdoors, put them on platforms 12 to 18 inches off the ground.

Despite what you do, rats and mice may enter your garden from a neighboring property. If the problem is severe, you may have to press local officials to enforce trash abatement rules. This procedure can take a while. In the meantime, a pair of feisty cats may persuade rodents that living is easier elsewhere.

Ground squirrels: Vegetable gardens that abut fields or open space are often plagued by ground squirrels. They feed on roots and tubers as well as aboveground plants. These squirrels store excess food in burrows, which may be as deep as 4 feet below the surface.

In some areas, regulations protect ground squirrels, so consult the county extension service before using traps or poison. Place baited box traps outside burrows. Do not use any type of poison bait if pets or children are present. Peanut butter is an effective bait for a live trap. Be sure to release squirrels in more appropriate surroundings, where they will not do more damage.

Enclosed cages protect vegetables from rabbits.

Larger cages protect plants from deer.

Wire cages protect lettuce from rabbits.

POOR PRODUCE AND SLOW GROWTH

Poor-quality produce

Poor-quality tomatoes.

Slow growth

Phosphorus deficiency on corn.

Slow growth from too acid soil.

Problem: Fruit has poor or strong flavor and is smaller than normal. Fruit yield is low.

Analysis: Vegetables and berries may yield poor-quality produce for many reasons.
1. Hot weather affects the flavor of some fruits, especially cool-season crops such as lettuce and members of the cabbage family. Their taste becomes strong and bitter.
2. Fluctuations in soil moisture affect the flavor of many fruits, especially cucumbers. Sweet and white potato tubers crack and appear un-appetizing. Alternating wet and dry soil results in erratic growth with poor-quality fruit.
3. Poor soil fertility causes plants to grow poorly and yield little if any fruit. Any fruit produced is frequently off-flavor and may not develop fully.
4. Overmature fruit deteriorates rapidly and loses its flavor. Allowing fruit to overripen on the plant reduces future yields by taking energy the plant could have used to produce more fruit.

Solution: Follow these guidelines for better-quality produce. (For more information on each of these recommendations, see the entry for your plant in the alphabetical section beginning on page 261.) The numbered solutions below correspond to the numbered items in the analysis.
1. Plant vegetables and small fruits at the times of year suggested on page 361.
2. Follow irrigation guidelines for your plant.
3. Fertilize as instructed.
4. Pick fruit as it matures, so more fruit will be produced. Follow the harvesting guidelines specific to your plant.

Problem: Plants grow slowly. Leaves are pale green to yellow, or darker than normal, and dull. Few flowers and fruit are produced. Fruit that is produced matures slowly. Plants pulled from the soil may have small, black, rotted root systems. Plants may die.

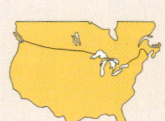

Analysis: Plants may grow slowly and produce less fruit for any of several reasons.

1. Excess water: If the soil is constantly wet, either from frequent watering or poor drainage, lack of available oxygen causes the roots to become shallow, stunting and sometimes killing the plant. Wet soil also encourages root-rotting fungi that destroy the root system and kill the plant.

2. Phosphorus deficiency: Phosphorus is a major nutrient needed by plants for root formation, flower and fruit production, and overall cell growth. Phosphorus-deficient plants grow slowly and have dark leaves that may be tinted with purple or have purple veins. This nutrient may be lacking either because it is not present in the soil or because it is in a form that is unavailable to plants.

3. Incorrect pH: The soil pH limits the amounts and kinds of nutrients available to plants. Vegetables grow best with a pH of 6.0 to 7.0. Required nutrients are usually available in adequate amounts within this range.

4. Cool weather: Warm-season vegetables, such as tomatoes, beans, and okra, require temperatures above 70°F for best growth and fruit production. If they are planted too early in the season, the cool weather slows their growth. Affected plants may take several months to recover from this setback and still may not produce abundantly.

Solution: The numbered solutions below refer to the numbered items in the analysis.

1. Allow the soil around plants to dry out. Remove and destroy any plants with rotting roots. Avoid future root rot problems by planting in well-drained soil.

2. Spray the leaves with a liquid or soluble fertilizer for a quick response. Fertilize with a balanced fertilizer. Follow the rates on the label or those given for each vegetable in the alphabetical section beginning on page 261.

3. Test the soil pH and correct to 6.0 to 7.0.

4. Plant vegetables at the correct time of year. For planting times in your area, see page 361. Discard warm-season vegetable plants that have been set back by cool temperatures. Replace with healthy transplants.

Insufficient light

Slow growth due to insufficient light.

Problem: Plants grow slowly or not at all. Leaves are light green, and few or no flowers or fruit are produced. Plants are shaded for much of the day. Lower leaves may turn yellow and drop.

Analysis: Plants need sunlight to manufacture food and produce fruit. Without adequate sunlight, they grow slowly. Vegetable and berry plants that yield fruit such as tomatoes, brambles, strawberries, beans, and peppers require at least 6 hours of sunlight per day. Some leafy and root vegetables, however, will tolerate light shade.

Solution: Prune surrounding trees to allow more sunlight. Plant vegetables that require less sunlight in somewhat shady areas of the garden. If possible, move your garden site, or grow vegetables in containers on a sunny patio or porch.

Seedlings eaten

Pea seedlings eaten by rabbits.

Bean seedlings eaten by slugs.

Problem: Seedling leaves are chewed ragged. Seedlings may completely disappear, or short stubs of stems may remain.

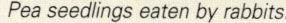

Analysis: Several animal and insect pests feed on seedlings.

Solution: The numbered solutions below correspond to the numbered items in the analysis.

1. Birds: Many kinds of birds, especially grackles, blackbirds, and crows, eat entire seedlings. The plants are most susceptible to bird attack when they have just emerged.

1. Cover seedlings with a tent made of cheesecloth or netting stretched over a wooden frame. Dangling aluminum pie plates may help to deter birds.

2. Snails and slugs: These pests feed at night and on cloudy days, chewing holes in leaves or devouring entire seedlings. Silvery winding trails are evidence of their presence. During the day they hide in damp places, such as under rocks and flower pots or in debris.

2. Control snails and slugs with ORTHO Bug-Geta Snail & Slug Killer. Lightly wet the area before application to activate the pellets. The moisture will also attract snails and slugs. Treat along the seeded rows and in hiding places, such as under rocks, boards, and flower pots or around compost piles. For more information on snails and slugs, see page 256.

3. Earwigs: These dark brown insects with pincers projecting from the rear of their bodies feed at night and chew holes in leaves and stems.

3. To control earwigs, treat with ORTHO Bug-Geta Plus Snail, Slug & Insect Killer along the seeded rows and in hiding places, such as under rocks and in debris. For more information on earwigs, see page 260.

4. Rabbits: These animals may eat entire seedlings or may leave only short stubs of the stems standing in the soil.

4. Keep rabbits out of the garden by erecting a 2-foot-high fence of small-gauge fencing wire around the garden. Anchor the bottom of the fence with boards 1 to 2 inches deep into the soil to prevent rabbits from digging underneath.

5. Grasshoppers: These insects are present throughout the growing season and migrate from area to area as their food source is depleted. They eat entire seedlings. Grasshoppers are prevalent in hot, dry weather.

5. Control grasshoppers and protect uneaten seedlings with ORTHO Bug-Geta Plus Snail, Slug & Insect Killer. Repeat at weekly intervals as long as grasshoppers are present. For more information on grasshoppers, see page 258.

SEEDLING PROBLEMS *(continued)*

Damping-off

Damping-off of radish seedlings.

Cutworms

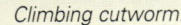

Climbing cutworm.

Cutworms on celery (life size).

Problem: Seeds rot. Seedlings fail to emerge or fall over soon after they emerge. Areas on the stem at the soil line are water-soaked and discolored. The base of the stem is soft and thin.

Analysis: Damping-off is a common problem caused by several fungi and aggravated by wet soil with a high nitrogen level. Wet, rich soil promotes damping-off in two ways: the fungi that cause it are more active under these conditions, and the seedlings are more susceptible to attack. Damping-off is often a problem when the weather remains cloudy and wet and when seedlings are heavily shaded or crowded. Seedlings started indoors in a wet, unsterilized medium are also susceptible.

Solution: To prevent fungi from attacking seedlings, take the following precautions.
1. Allow the surface of the soil to dry slightly between waterings.
2. Do not start seeds in soil that has a high nitrogen level. Add nitrogen fertilizer only after the seedlings have produced their first true leaves.
3. Protect the seeds during germination with a fungicide containing *captan* or *thiram*. Add a pinch of fungicide to a packet of seeds (or ½ teaspoon per pound) and shake well to coat the seeds.
4. For starting seeds indoors, use a sterilized medium, such as a pasteurized potting soil or a mixture of peat moss, perlite, and vermiculite.
5. Thin seedlings to allow good air circulation through the planting.

Problem: Stems of young plants are chewed or cut off near the ground. Gray, brown, or black worms, up to 2 inches long, may be found in the top 2 inches of the soil near the base of the damaged plants. The worms coil when disturbed.

Analysis: Several species of cutworms attack plants in the vegetable garden. Surface-feeding cutworms are common pests of young vegetables planted early in the season. A single surface-feeding cutworm can sever the stems of many young plants in one night. It eats through the stems just above ground level. Tomatoes, peppers, peas, beans, and members of the cabbage family are particularly susceptible. Some cutworms can climb up the stem or trunk of grapes, blueberries, brambles, tomatoes, and other garden crops to feed on young leaves, buds, and fruit. Cutworms hide in the soil during the day and feed only at night. Adult cutworms are dark, night-flying moths with bands or stripes on their forewings.

Solution: Apply ORTHO Dursban® Lawn & Garden Insect Control, ORTHO Diazinon Soil & Turf Insect Control, or ORTHO Bug-Geta Plus Snail, Slug & Insect Killer around the base of undamaged plants where cutworm damage has been observed. Make sure that your plant is listed on the product label. Because cutworms are difficult to control, applications may need to be repeated at weekly intervals. Before transplanting new plants into the area, apply a preventive treatment of ORTHO Dursban® Lawn & Garden Insect Control or ORTHO Diazinon Soil & Turf Insect Control, and work it into the soil. Cultivate the soil thoroughly in late summer and fall to expose and destroy eggs, larvae, and pupae. Further reduce damage with a cutworm collar around the stem of each plant. These collars should be at least 2 inches high, surround the plant stem fairly closely, and be pressed into the soil. Make collars out of stiff paper or aluminum foil bent into a cylinder or out of tin cans or paper cups with the bottoms removed. To reduce injury from climbing cutworms, inspect your plants at night with a flashlight and pick off and destroy any cutworms you find.

Wilt diseases

Fusarium wilt on tomato.

Problem: Leaves wilt, turn yellow, and may turn brown. Little or no fruit is produced. Growth slows and plants may be stunted. When the stem is sliced open lengthwise, the tissue just under the surface is usually found to be brown.

Analysis: Wilt diseases infect many vegetables and berries. They may be caused by any of several fungi that live in the soil. The fungi are spread by contaminated plants, soil, and equipment. They enter the plant through the roots and spread through water-conducting vessels in the stems. The vessels become discolored and plugged. This plugging cuts off the flow of water and nutrients to the leaves, resulting in leaf yellowing and wilting. The plugging also results in a reduced yield and poor-quality fruit. Severely infected plants die.

Solution: No available chemicals will cure infected plants. Wilt fungi can be removed from the soil only by fumigation or solarization techniques. To solarize soil, cultivate an area and cover it with a layer of clear polyethylene film at the hottest time of the year. Leave the film in place for 4 to 6 weeks. The best solution is usually to avoid plants that are susceptible to wilt diseases. Don't put the same kind of plant back where a wilt-infected plant grew. For a list of vegetables and berries susceptible to wilt diseases, see page 362.

Root and stem rot

Root rot on broccoli.

Problem: Leaves wilt and turn yellow. The lower leaves are affected first, then the upper ones. Little or no fruit is produced. Plants do not recover when watered, and they usually die.

Analysis: Root and stem rot is caused by any of several fungi known as water molds. These fungi thrive in waterlogged, heavy soils. The fungi attack the plant roots or the stems at the soil level. Infection causes the roots and stems to decay, resulting in wilting, then yellowing leaves, and eventually the death of the plant. Many of these fungi also cause *damping-off* of seedlings. For more information on damping-off, see page 252.

Solution: Allow the soil around the plants to dry out, minimizing drought stress by placing container plants in shade or providing shade for plants in the ground and then placing clear plastic tents over plants. Begin watering again when the plant shows signs of drought stress, such as heavy wilting or yellowing and dropping of leaves. Remove and discard severely infected plants. Avoid future root rot problems by planting in well-drained soil. Avoid overwatering by following the watering guidelines for your plant found in the alphabetical section beginning on page 261.

Nematodes

Nematode damage on bean leaves.

Problem: Plants wilt in hot, dry weather and recover at night. They are stunted and yellow. Round and elongated nodules may occur on the roots. Plants may die.

Analysis: Nematodes are microscopic worms (unrelated to earthworms) that live in the soil. They feed on plant roots, damaging and stunting them. The damaged roots can't supply sufficient water and nutrients to the aboveground plant parts, and the plant is stunted or slowly dies. Nematodes may also transmit certain viruses. Nematodes are found throughout the United States but are most severe in southeastern areas. They prefer moist, sandy loam soil. They can move only a few inches each year on their own, but they may be carried long distances by soil, water, tools, or infested plants. Testing roots and soil is the only positive method for confirming the presence of nematodes. Contact your local county extension service for sampling instructions and addresses of testing laboratories. Soil and root problems, such as poor soil structure, drought stress, nutrient deficiency, and root rots, can also produce similar symptoms of decline. These problems should be eliminated as causes before soil and root samples are sent for testing.

Solution: No chemicals available to homeowners kill nematodes in planted soil. The worms can be controlled before planting, however, by soil solarization: At the hottest time of the year, cultivate soil and cover the area with a layer of clear polyethylene film. Leave the film in place for 4 to 6 weeks.

DISCOLORED OR MOTTLED LEAVES

Root rot

Root rot on bush bean.

Problem: Leaves turn yellow, starting with the older, lower leaves and progressing to the younger ones. Plants grow very little. Flowers yellow and drop. Fruit shrivels and does not ripen. When the plant is pulled up, the roots appear black, soft, and rotted. The soil has frequently been very moist.

Analysis: Any of several fungi present in most soils cause root rot. Some of these fungi normally do little damage but can cause root rot in wet or waterlogged soil. Waterlogged soil may result from overwatering or from poor soil drainage. Infection causes the roots to decay, resulting in wilting, yellowing leaves, flower and fruit drop, reduced fruit yield, and eventually the death of the plant.

Solution: To avoid root rot problems, do not overwater. Follow the watering guidelines under the entry for your plant in the alphabetical section beginning on page 261. Remove and destroy severely infected plants. Avoid future root rot problems by planting in well-drained soil. Do not plant the same crop in a part of the garden where root rot was a problem in the past 3 to 5 years.

Nitrogen deficiency

Nitrogen-deficient tomato.

Problem: The bottom leaves, including the veins, turn light green to pale yellow and may die or drop. Growth slows, and new leaves are small. Flowers turn yellow and drop. Fruit is small and may be distorted and discolored.

Analysis: Plants need nitrogen to make chlorophyll, the essential green pigment in their leaves, and for overall healthy growth and high-quality fruit production. When plants lack nitrogen, growth slows, blossoms drop, and fruit yield is reduced. Nitrogen may either be lacking in the soil or be present in a form that is not available to the plant. In cool, rainy, early spring weather, little or no nitrogen is available. Later in the season, as vegetables grow rapidly and mature, plants may deplete the supply of nitrogen in the soil. During a drought, nitrogen is carried to the soil surface as the water it is dissolved in evaporates. Once the nitrogen is above the plant's root zone, the roots are unable to absorb it. Rain or irrigation water carries the nitrogen back to the root zone, where it is available to the plant.

Solution: For a quick response, spray the foliage with a liquid or soluble fertilizer. Water the plant with the same solution. Remove fruit from severely affected plants to allow the plants time to resume normal growth. Fertilize according to the recommendations for your plant in the alphabetical section beginning on page 261.

Spider mites

Spider mite damage on string bean.

Problem: Leaves are stippled, yellowing, and dirty. They may dry out and drop. There may be webbing over flower buds, between leaves, or on the lower surfaces of leaves. To determine if a plant is infested with mites, examine the bottoms of the leaves with a hand lens. Or hold a sheet of white paper underneath an affected leaf or stem and tap the leaf or stem sharply. Tiny specks the size of pepper grains will drop to the paper and move about. They are easily seen against the white background.

Analysis: Spider mites, related to spiders, are major pests of many garden and greenhouse plants. They cause damage by sucking sap from the undersides of leaves. As a result of their feeding, the plant's green leaf pigment disappears, producing the stippled appearance. Spider mite webbing traps cast-off skins and debris, making the plant messy. Although mites do not attack fruit directly, they cause leaf drop, which weakens the plant and reduces fruit yield. If plants are severely infected, flowers do not form, bloom, or produce fruit. Mites are active throughout the growing season but thrive in dry weather with temperatures of 70°F and up. By midsummer (or as early as March in Florida), they may have built up to tremendous numbers.

Solution: Treat infested plants with ORTHO Bug-B-Gon Insect Killer, ORTHO Malathion 50 Plus Insect Spray, or an insecticidal soap at the first sign of damage. Repeat at weekly intervals until no further damage occurs. Make sure that your vegetable or berry is listed on the label. Hose plants frequently with a strong stream of water to wash off mites and webs.

Salt damage

Salt damage on radishes.

Problem: Leaf edges and areas between the veins turn dark brown and die. Burned leaves  may drop. Growth slows or stops. A white or dark crust may be seen on the soil.

Analysis: Salt damage occurs when salt accumulates in the soil to damaging levels. This can happen in either of two ways: (1) the garden does not receive enough water from rainfall or irrigation to wash the salts from the soil or (2) the drainage is so poor that water does not pass through the soil. In either case, as water evaporates from the soil and plant leaves, the salts that were dissolved in the water accumulate near the soil surface. In some cases, a white or dark brown crust of salts forms on the surface. Salts can originate in the soil, in irrigation water, or in applied fertilizers.

Solution: The only way to eliminate salt problems is to wash the salts through the soil with water. If the damage is only at a low spot in the garden, fill in the spot to level the area. If the entire garden drains poorly, improve the drainage. If the soil drains well, increase the water applied at each watering by 50 percent or more, so that excess water will leach salts below the root zone of the plants. Fertilize according to instructions for your plant in the alphabetical section beginning on page 261.

Pesticide burn

Grape with spray injury.

Problem: Irregular spots occur on leaves. Young leaves and blossoms may be distorted and brown. Plants may have recently been sprayed with a pesticide.

Analysis: Insecticides, fungicides, and herbicides damage plants when used improperly. Damage usually occurs within 24 hours of the time the plants were sprayed. In some crops, such as grapes, however, damage may not be apparent for several days after spraying. Pesticides may drift from other areas on windy days and damage plants. Pesticides may burn plants when the temperature is above 90°F at the time of spraying or if it rises above 90°F within a few hours of spraying. Damage is also common during damp, humid weather when spray is slow to dry on the plant. Pesticides not mixed and applied according to label directions may also burn plants. Because traces of herbicides are difficult to remove from sprayers, other pesticides used in tanks once used for herbicides may be contaminated.

Solution: Once plants are damaged, you cannot do anything except try to keep the plants healthy with regular watering and fertilizing. Do not spray when temperatures are above 90°F. Avoid spraying on windy days, when sprays may drift. Mix according to label directions. Do not increase the dosage. When mixing two or more chemicals, be sure they are compatible. Always keep the solution well mixed by shaking the tank periodically while spraying. This is especially important when using wettable powder formulations. Purchase fresh pesticides each year. Keep a separate sprayer for herbicides.

Leaf spots

Leaf spot on strawberry.

Problem: Spots and blotches appear on the leaves. The spots may be yellow, red, tan, gray, or brown. They range in size from barely visible to ¼ inch in diameter. Several spots may join to form blotches. Leaves may be yellow and dying. Leaf spotting is most severe in warm, humid weather. Fruit may be spotted like the leaves or be discolored brown or yellow. A fine gray or white mold sometimes covers the infected leaf or fruit tissue in damp conditions.

Analysis: Several fungi cause leaf spots. Some of these may eventually kill the plant or weaken it so that it becomes susceptible to attack by other organisms. Others merely cause spotting that is unsightly but not harmful. Infected fruit is less appetizing, some is inedible, and the yield is reduced. Fruit infected only slightly is still edible if the discolored tissue is cut away. Leaf-spotting fungi are spread by splashing water or wind. They generally survive the winter in diseased plant debris. Most of the fungi do their greatest damage in mild weather (50° to 85°F).

Solution: Treat infected plants with ORTHO Multi-Purpose Fungicide Daconil 2787® Plant Disease Control or a fungicide containing captan at the first sign of the disease. Repeat the treatment at intervals of 7 to 10 days until weather conditions favorable to the spread of the disease no longer occur. Make sure that your plant is listed on the product label. Remove all plant debris from the garden after harvest.

DISCOLORED OR MOTTLED LEAVES (continued)

LEAVES AND FLOWERS CHEWED

Powdery mildew

Powdery mildew on cucumber.

Problem: A white powdery growth covers the upper surfaces of the leaves and sometimes the stems. Areas of the leaves turn brown and dry. Older leaves are affected first, with the disease progressing to younger leaves. Leaves may become cupped, showing a silvery underside. Fruit may also be covered with the white powdery growth.

Analysis: Powdery mildew is caused by fungi that thrive in both humid and dry weather. The powdery patches consist of fungal strands and spores. The spores are spread by the wind to healthy plants. The fungus saps plant nutrients, causing yellowing and sometimes death of the leaf. Fruit yield may also be reduced. A severe infection may kill the plant. Since powdery mildews attack many different kinds of plants, the fungus from a diseased plant may infect other types of plants in the garden. For a list of powdery mildews and the plants they attack, see page 343. Under favorable conditions, powdery mildew can spread rapidly through a closely spaced planting.

Solution: Control powdery mildew on vegetable and berry plants with ORTHO Multi-Purpose Fungicide Daconil 2787® Plant Disease Control or with a sulfur dust at the first sign of the disease. Continue spraying at intervals of 7 days as long as the disease is a problem. Make sure that your plant is listed on the label. Clean up and destroy plant debris after harvest. When available, grow resistant varieties.

Leafminers

Leafminer trails on tomato.

Problem: Light-colored, irregular blotches, blisters, or tunnels appear on the leaves. The tan areas peel apart easily like facial tissue. Tiny black specks are found inside the tunnels.

Analysis: Leafminers belong to a family of leafmining flies. The tiny black or yellow adult fly lays its white eggs on the undersides of leaves or in the leaves. The maggots that hatch from these eggs tunnel between the upper and lower surfaces of the leaves, feeding on the inner tissue. The tunnels and blotches are called *mines*. The black specks inside are the maggots' droppings. Damaged portions of leaves are no longer edible. The yield is usually not affected on fruit-producing vegetables and berries unless many leaves are damaged. Several overlapping generations occur during the growing season, so larvae are present continually from spring until fall.

Solution: Control leafminers with ORTHO Diazinon Ultra Insect Spray when the egg clusters are first seen under the leaves. Repeat two times at weekly intervals to control succeeding generations. Once leafminers enter the leaves, sprays are ineffective. Spraying after the miners first appear in the leaves will control only those leafminers that attack after the application. Make sure that your plant is listed on the insecticide label. Clean all plant debris from the garden after harvest to reduce overwintering spots for the pupae. Adult flies can be kept from laying eggs with floating row covers.

Snails and slugs

Slug damage.

Problem: Irregular holes with smooth edges are chewed in the leaves. Some leaves may be sheared off entirely. Silvery trails wind around the plant and soil nearby. At night, check with a flashlight for slimy creatures—with or without hard brown shells—feeding on the leaves.

Analysis: Snails and slugs are mollusks and are related to clams, oysters, and other shellfish. They feed on the leaves of a wide variety of garden plants and may completely devour a young seedling. Snails and slugs may attack ripe and unripe fruit lying on the ground, especially if the fruit is shaded by the foliage, as with strawberries and unstaked tomatoes. Like other mollusks, snails and slugs need to be moist all the time. For this reason, they avoid direct sun and dry places and hide during the day in damp places, such as under flowerpots or in thick ground covers. They emerge at night or on cloudy days to feed. Slugs lay white eggs encased in a slimy mass in protected places. Snails bury their eggs in the soil, also in a slimy mass. The young look like miniature versions of their parents.

Solution: Apply ORTHO Bug-Geta Snail & Slug Killer in the areas you wish to protect. Also apply in areas where snails and slugs might be hiding, such as in dense ground covers, weedy areas, compost piles, or pot-storage areas. Wet down the treated areas to encourage snail and slug activity that night. Repeat every 2 weeks as needed.

Nocturnal pests

Cabbage plants damaged by nocturnal pests.

Problem: Young plants are chewed or cut off near the ground. Some of the leaves, stems, flowers, and fruit are chewed. When affected plants are inspected at night with a flashlight, insects may be seen feeding on them.

Analysis: Several kinds of insects, including some beetles, weevils, and caterpillars as well as all earwigs and cutworms, feed on plants only at night. Beetles are hard-bodied insects with tough, leathery wing covers. Weevils are a type of beetle with a long snout. Earwigs are reddish brown, flat, elongated insects up to 1 inch long with pincers projecting from the rear of the body. Caterpillars and cutworms are smooth or hairy, soft-bodied worms. All of these nocturnal pests usually hide in the soil, in debris, or in other protected locations during the day.

Solution: Control nocturnal pests by treating with ORTHO Bug-Geta Plus Snail, Slug & Insect Killer. For more information on earwigs, see page 260. For more information on caterpillars, see the next column on this page.

Caterpillars

Cabbage looper (2× life size).

Problem: Irregular or round holes appear in the leaves and buds. Leaves, buds, and flowers may be entirely chewed off. Worms or caterpillars are feeding on the plants.

Analysis: Several species of these moth or butterfly larvae feed on many vegetable and berry plants. Some common caterpillars are budworms, hornworms, and loopers. Most moths or butterflies start to lay their eggs on garden plants with the onset of warm weather in the spring. The larvae that emerge from these eggs feed on the leaves, flowers, and buds for 2 to 6 weeks, depending on weather conditions and species. Mature caterpillars pupate in cocoons attached to leaves or structures or buried in the soil. There may be one generation or more overlapping generations during the growing season. The last generation of caterpillars in the fall survives the winter as pupae. Moths and butterflies emerge from the pupae the following spring.

Solution: Spray infested plants with ORTHO Tomato & Vegetable Insect Killer or an insecticide containing *diazinon*, *malathion*, or *methoxychlor*. The bacterial insecticide *Bacillus thuringiensis* (Bt) is effective against the early stages of caterpillars. Make sure that your plant is listed on the product label. Repeat the treatment if reinfestation occurs, allowing at least 7 days between applications.

Japanese beetle

Japanese beetles on grape (⅓ life size).

Problem: Leaf tissue has been chewed between the veins, giving the leaf a lacy appearance. Metallic green-and-bronze winged beetles, ½ inch long, feed in clusters on the foliage, especially on the tender new leaves.

Analysis: As its name suggests, the Japanese beetle (*Popillia japonica*) is native to Japan. It was first seen in New Jersey in 1916 and has since become a major pest in the eastern United States. It feeds on hundreds of plant species. The adult beetles are present from the beginning of summer to early fall. They feed only in the daytime, rapidly defoliating plants. Leaves exposed to direct sun are the most severely attacked. Badly damaged leaves drop. Any reduction in leaf tissue ultimately affects the overall vigor and production of fruit. The larva of the Japanese beetle, a white grub, feeds on grass roots, frequently killing entire lawns.

Solution: Treat infested plants with ORTHO Malathion 50 Plus Insect Spray or ORTHO Home Orchard Spray. Japanese beetles can fly up to 5 miles and travel from garden to garden. Consequently, repeated sprays are necessary to control them. Make sure that your plants are listed on the product label. Treat for grubs in the lawn as outlined on page 49. Use resistant plants.

INSECTS ON THE PLANT *(continued)*

Cucumber beetles

Spotted cucumber beetle (½ life size).

Problem: Holes are chewed in the leaves, leafstalks, and stems by yellow-green beetles with black spots or stripes.

Analysis: Both striped cucumber beetles (*Acalymma* species) and spotted cucumber beetles (*Diabrotica* species) are common pests of vegetable plants. It is important to control these beetles, because they may infect plants with two serious diseases that damage and may kill cucurbits (such plants as squash, melons, and cucumbers): squash mosaic virus (see page 282) and bacterial wilt (see page 281). Adult beetles survive the winter in plant debris and weeds. As soon as vegetable plants are set in the garden in the spring, the beetles attack the leaves and stems and may destroy the plants. Adults lay their yellow-orange eggs in the soil at the base of the plants. The grubs that hatch from these eggs eat the roots and the stems below the soil line, causing the plants to be stunted or to die prematurely. Severely infested plants produce little fruit. The slender white grubs feed for several weeks, pupate in the soil, and emerge as adults to repeat the cycle. One generation occurs per year in northern parts of the United States and two or more in southern areas.

Solution: Treat plants with ORTHO Tomato & Vegetable Insect Killer or ORTHO Diazinon Ultra Insect Spray at the first sign of the beetles. Make sure that your plant is listed on the product label.

Flea beetles

Flea beetles on turnip leaf (2× life size).

Problem: Leaves are riddled with shot holes about ⅛ inch in diameter. Tiny (¹⁄₁₆ inch) black beetles jump like fleas when disturbed. Leaves of seedlings and eventually whole plants may wilt and die.

Analysis: Flea beetles jump like fleas but are not related to fleas. Both adult and immature flea beetles feed on a wide variety of garden vegetables and berries. The immature beetle, a legless gray grub, injures plants by feeding on the roots and the lower surfaces of leaves. Adults chew holes in leaves. Flea beetles damage seedlings and young plants most. Leaves of seedlings riddled with holes dry out quickly and die. Adult beetles survive the winter in soil and garden debris. They emerge in early spring to feed on weeds until vegetables sprout or plants are set in the garden. Grubs hatch from eggs laid in the soil and feed for 2 to 3 weeks. After pupating in the soil, they emerge as adults to repeat the cycle. There are one to four generations per year. Adults may feed for up to 2 months.

Solution: Control flea beetles on vegetable and berry plants with ORTHO Diazinon Ultra Insect Spray, ORTHO Tomato & Vegetable Insect Killer, or an insecticide containing *methoxychlor* when the leaves first show damage. Watch new growth for evidence of further damage, and repeat the treatment at weekly intervals as needed. Make sure that your plant is listed on the product label. Clean up and destroy plant debris after harvest to reduce the number of overwintering spots for adults.

Grasshoppers

Grasshopper on blueberry (2× life size).

Problem: Large holes are chewed in the margins of leaves. Greenish yellow to brown jumping insects, ½ to 1½ inches long with long hind legs, are eating the plants. Some fruit and corn ears may be chewed.

Analysis: Grasshoppers attack a wide variety of plants. They eat leaves and occasionally fruit, migrating as they deplete their food sources. In vegetable gardens, they are most numerous in the rows near weedy areas. In late summer, adult grasshoppers lay their eggs in pods in the soil. The adults continue to feed until cold weather kills them. The eggs hatch the following spring. Grasshopper problems are most severe during hot, dry weather. Grasshoppers migrate into green gardens and yards as surrounding areas dry up in the summer heat. Periods of cool, wet weather help keep their numbers under control.

Solution: Treat plants with ORTHO Bug-Geta Plus Snail, Slug & Insect Killer or an insecticide containing *diazinon* as soon as grasshoppers appear. Repeat at weekly intervals if plants become reinfested. Make sure that your plant is listed on the product label.

Leafhoppers

Leafhopper (3× life size).

Problem: Spotted, pale green insects up to ⅛ inch long hop, move sideways, or fly away quickly when a plant is touched. The leaves are stippled. Cast-off skins may be found on the undersides of leaves.

Analysis: Leafhoppers feed on many vegetables and small fruits. They generally feed on the undersides of leaves, sucking the sap, which causes stippling. Severely infested vegetable and small fruit plants may become weak and produce little edible fruit. One leafhopper, the aster leafhopper (*Macrosteles fascifrons*), transmits aster yellows, a plant disease that can be quite damaging. Leafhoppers at all stages of maturity are active during the growing season. They hatch in the spring from eggs laid on perennial weeds and ornamental plants. Even areas where the winters are so cold that the eggs cannot survive are not free from infestation, because leafhoppers migrate in the spring from warmer regions.

Solution: Spray infested plants with ORTHO Diazinon Ultra Insect Spray, ORTHO Malathion 50 Plus Insect Spray, ORTHO Tomato & Vegetable Insect Killer, or an insecticidal soap. Be sure to cover the lower surfaces of the leaves. Repeat the spray as often as necessary to keep the insects under control. Allow at least 10 days between applications. Make sure that your plants are listed on the product label. Eradicate nearby weeds—especially thistles, plantains, and dandelions—that may harbor leafhopper eggs.

Whiteflies

Whiteflies (life size).

Problem: Tiny white-winged insects feed mainly on the undersides of leaves. Nonflying, scalelike larvae covered with white, waxy powder may also be present on the undersides of leaves. When the plant is touched, insects flutter rapidly around it. Leaves may be mottled and yellowing.

Analysis: Whiteflies are a common pest of many garden and greenhouse plants. The four-winged adult lays eggs on the undersides of leaves. The larvae are the size of a pinhead, flat, oval-shaped, and semitransparent. Both immature and adult forms suck sap from the leaves. The nymphs are more damaging because they feed more heavily. Adults and nymphs cannot fully digest all the sugar in the plant sap. They excrete the excess in a fluid called *honeydew*, which often drops onto the leaves or plants below. A sooty mold may develop on the honeydew, causing the leaves to appear black and dirty. Whiteflies are unable to live through extended periods of freezing weather. One species, the silverleaf whitefly (*Bemisia argentifolii*), began causing serious damage to vegetable crops in the Southwest in 1990.

Solution: There is no completely effective remedy for whiteflies on vegetables. You can gain some control by spraying with ORTHO Malathion 50 Plus Insect Spray or a horticultural oil spray every 7 to 10 days as necessary. Spray the foliage thoroughly, covering the upper and lower surfaces of the leaves. Be sure your plant is listed on the label. Whiteflies may also be partially controlled with yellow sticky traps.

Aphids

Bean aphids (4× life size).

Problem: Pale green, yellow, purple, or black soft-bodied insects cluster on the undersides of leaves. Leaves turn yellow and may be curled, distorted, and puckered. Plants may be stunted and produce little fruit.

Analysis: Aphids are one of the most common pests in the garden. They do little damage in small numbers, but they are extremely prolific, and populations can rapidly build up to damaging numbers during the growing season. Damage occurs when the aphids suck the juices from vegetable or small fruit leaves and flower buds. Aphids usually prefer young, tender leaves. Severely infested plants may be stunted and weak, producing little fruit. Fruit yield is also reduced when the aphids spread viral plant diseases. Aphids feed on nearly every plant in the garden and are spread from plant to plant by wind, water, and people. Aphids are unable to digest fully all the sugar in the plant sap, and they excrete the excess in a fluid called *honeydew*, which often drops onto lower leaves. Ants feed on this sticky substance and are often present where there is an aphid infestation.

Solution: Control aphids on vegetables and berries with ORTHO Malathion 50 Plus Insect Spray, ORTHO Tomato & Vegetable Insect Killer, an insecticidal soap, or ORTHO Diazinon Ultra Insect Spray as soon as the insects appear. Repeat the spray if plants become reinfested. Make sure that your plant is listed on the product label.

INSECTS ON THE PLANT
(continued)

ANIMAL PESTS

Earwigs

Earwig (2× life size).

Wildlife

Snapbeans eaten by rabbits.

Raccoon.

Problem: Dark, reddish brown insects up to ¾ inch long with pincers projecting from the rear of the body are found under objects in the garden. They scurry for cover when disturbed. Holes may be chewed in leaves and blossoms.

Analysis: Although seen in most gardens, earwigs are only minor pests of vegetables and berries unless populations are high, when they can become major pests. They feed predominantly on decaying plant material and other insects. Earwigs feed at night and hide under stones, debris, and bark chips in the daytime. They have wings but seldom fly, preferring to run instead. Adult earwigs lay eggs in the soil in late winter to early spring. The young that hatch from these eggs may feed on green shoots and eat holes in leaves. As the earwigs mature, they feed occasionally on blossoms and ripening fruit. Earwigs are beneficial when they feed on other insect larvae and on snails. They sometimes invade homes.

Solution: Because earwigs typically cause only minor damage, insecticide control is seldom needed in the vegetable garden. If they are numerous and troublesome, however, treat the soil with ORTHO Diazinon Soil & Turf Insect Control or ORTHO Bug-Geta Plus Snail, Slug & Insect Killer. Because earwigs are most active at night, treat in the late afternoon or evening. Make sure that your plant is listed on the product label. Pick fruit as it ripens. (For information on controlling earwigs indoors, see page 312.)

Problem: Plants are chewed or completely eaten. Ripening fruit, pods, and ears are partially or completely eaten. Deer, raccoons, squirrels, rabbits, woodchucks, or mice may be seen in the garden.

Analysis: Various forms of wildlife feed in the garden.

Solution: Wildlife can be kept from your garden in several ways. The numbered solutions below correspond to the numbered items in the analysis.

1. **Deer** feed on leaves and fruit. They are most active at dawn and dusk.

1. The only sure way to exclude deer is with a woven wire fence 8 feet tall (see page 249. Deer are sometimes repelled by cotton drawstring bags filled with bloodmeal fertilizer or human hair.

2. **Raccoons** knock over cornstalks to feed on maturing ears. They may also feed on other ripening fruit. Raccoons feed at night.

2. Exclude raccoons with a 6-foot-tall fence. Electric fencing above the fence may also be needed. Protect corn by interplanting with members of the cucurbit family. Raccoons will not walk on the prickly vines. Sprinkle ripening corn with cayenne pepper.

3. **Squirrels** feed on ripening fruit and climb cornstalks to feed on maturing ears.

3. Protect ears of corn from squirrels by sprinkling corn silks with cayenne pepper.

4. **Rabbits** feed on young bean, pea, lettuce, and cabbage plants, eating the young leaves and frequently leaving short stubs of the stems standing in the soil.

4. To exclude rabbits, erect a fence 18 inches high and anchor it by burying the edges 5 to 6 inches deep in the soil.

5. **Woodchucks,** also called *groundhogs*, feed in the afternoons but avoid tomatoes, eggplants, red and green peppers, chives, and onions.

5. Deter woodchucks with a wire fence buried 1 foot deep and extended horizontally underground 10 to 12 inches.

6. **Mice** may bite into ripening tomatoes, cucumbers, and beans that are close to the ground to eat the seeds inside. Mice frequently travel underground in mole tunnels.

6. Stake, trellis, or cage plants to keep fruit off the ground and away from mice.

ASPARAGUS

GROWING GUIDE

Adaptation: Throughout the United States, except the Southeast.

Planting Time: See page 361.

Planting Method: Set crowns, buds upward, 12 inches apart in a trench 6 to 8 inches deep. As the plants grow, gradually fill in the trench with soil until it is even with the surrounding soil.

Soil: Any good garden soil. pH 6.5 to 8.0.

Fertilizer: At planting time, use 2 pounds of Scotts Vegetable Food per 100 square feet. Side-dress in late July to early August with ¾ pound per 100-foot row. After the first year, fertilize twice a year, once in the spring before the shoots emerge, and again just after harvest.

Water: How much: Apply enough water at each irrigation to wet the soil 1 to 1½ feet deep.
How often: Water when the soil is just barely moist.

Harvest: Plants must grow for 2 years before harvesting to build strong roots to ensure crops for at least 10 years. In the spring of the third year, pick when the spears are 7 to 10 inches tall with tight heads. To avoid wounding the crown, snap off the spears, do not tear or cut them. Harvest for 2 weeks. The fourth year, pick for 4 weeks, and the fifth and following years for 8 weeks.

Small spears

Small spears can have several causes.

Problem: Spears are small and skinny.

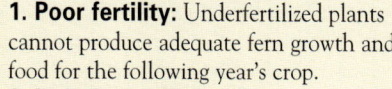

Analysis: Asparagus spears may be small for several reasons:
1. Poor fertility: Underfertilized plants cannot produce adequate fern growth and food for the following year's crop.
2. Immature plants: Asparagus crowns produce small spears for the first 2 or 3 years following planting.
3. Poor drainage: Asparagus plants do not produce well in poorly drained soil.
4. Overharvested plants: When harvest continues late in the season, the plants are unable to produce enough foliage and do not store enough food for the following year's crop.

Solution: Follow the growing and harvesting guidelines at left. Improve soil drainage and aeration.

Crooked spears

Crooked spears are caused by insect damage.

Problem: Spears are crooked or misshapen.

Analysis: Asparagus spears grow crooked when the growing shoot is damaged by insects (especially the asparagus beetle—see page 262), cultivation wounds are on the crown, or the tender shoots have been pelted by windblown sand. The injured areas grow more slowly than the uninjured side, causing the stem to curve. Although the spears are misshapen, they are still edible.

Solution: Control asparagus beetles. To avoid wounding the crown, do not cultivate closer or deeper than 2 inches. Tall plants or fencing can be used as windbreaks to prevent damage from windblown sand.

ASPARAGUS (continued)

Freeze injury

Damaged spear.

Problem: Spears turn brown and may be soft, or they may dry and wither.

Analysis: Asparagus is the earliest vegetable in the spring garden and is sometimes damaged by spring frosts. The damage may not be noticed in the early morning following the frost, but as the temperature warms through the day, the tissue discolors and softens. Damaged spears then dry rapidly as temperatures rise and humidity drops. A slight freeze injury may result in a crooked spear or slightly damaged tip of the spear. For information on other causes of crooked spears, see page 261.

Solution: Harvest and discard frost-damaged spears. When night temperatures below 32°F are predicted, protect the spears with a mulch of straw, newspaper, or leaves. Remove the mulch in the morning.

Fusarium wilt

Fusarium wilt.

Problem: The growing shoot turns yellow to dingy brown and wilts. During wet weather, white or pink cottonlike strands often appear under the leaf scales. Wilting is most severe among full-grown plants during the months of July and August. When the infected plant is dug up, all or part of the root system is seen to be a reddish color. The plant eventually dies.

Analysis: Fusarium wilt is caused by a fungus (*Fusarium* species). Asparagus plants under stress from poor growing conditions (such as drought, poor drainage, or insect or disease injury) are more severely affected. The fungus lives on organic matter in the soil. The disease is spread by contaminated seeds, plants, soil, and equipment; it often enters a garden on the roots of a transplant. The fungus enters the plant through the roots and spreads up into the stems and leaves through water-conducting vessels in the stems. The vessels become discolored and plugged, cutting off the flow of water to the leaves. Because *Fusarium* thrives in wet weather and warm temperatures (70° to 85°F), the fungi build up in warm soils.

Solution: No chemical control is available. It is best to destroy infected plants. The following year, plant healthy crowns where asparagus has not been planted for 2 to 4 years. The fungi can be eradicated from the soil by fumigation or solarization techniques. To solarize soil, cultivate the area and cover it with a layer of clear polyethylene film during the hottest time of the year. Leave the film in place for 4 to 6 weeks.

Asparagus beetle and spotted asparagus beetle

Larvae (4× life size). Inset: Adults (life size).

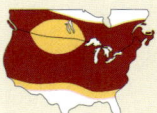

Problem: The tips of young asparagus spears are chewed and scarred. Later, when the spears develop into asparagus ferns, the ferns are also chewed. Small (¼-inch), metallic blue or black beetles with yellow markings and a narrow red head may feed on the tips of the spears and later on the ferns and stems. Reddish orange beetles with black spots may be present. Shiny black specks are found on the spear tips. Humpbacked orange or slate gray grubs may also be seen.

Analysis: The asparagus beetle (*Crioceris asparagi*) and spotted asparagus beetle (*C. duodecimpunctata*) injure asparagus plants throughout the growing season. The blue-black asparagus beetle is found throughout the United States. The orange and black spotted asparagus beetle is found east of the Mississippi River. The beetles are a particular problem when young asparagus shoots emerge in the spring. Both adults and grubs injure the plants by feeding on shoots, ferns, and stems. This feeding robs the root system of food manufactured in the foliage and necessary for healthy growth the following year.

Solution: Apply ORTHO Tomato & Vegetable Insect Killer or an insecticide containing *rotenone* when the beetles are first noticed. Repeat the applications as long as the beetles or grubs are feeding. Also treat fern growth in late summer or early fall to prevent adults from overwintering on the ferns and reinfesting the following year's crop. Handpick adults where practical.

Rust

Rust can seriously weaken asparagus plants.

Problem: Fronds turn yellow, then brown, and die back in early- to midsummer. Reddish brown, orange, or black blisters appear on the ferns and stems but not on newly emerged spears. Spears mature earlier than usual.

Analysis: Rust is caused by a fungus (*Puccinia asparagi*). The current year's rust weakens the plant by reducing the amount of food manufactured in the leaves to be stored in the roots. Because this stored food supplies the energy for the following year's crop, fewer shoots are produced the following year. Damage is worst when the tops are attacked several years in a row. In severe cases, the plants die. Warm temperatures and high humidity from fog, heavy dew, or overhead watering foster rust infection. Rust spores from diseased tops that have been left in the garden infect new shoots as they emerge in the spring. Wind spreads the spores from plant to plant.

Solution: After the harvest, spray the ferns and stems with a fungicide containing *maneb* or *mancozeb*. Repeat the treatment at intervals of 10 days when weather conditions promote rust infection. Do not spray while spears are still being harvested. Cut the tops close to the ground after they die in the fall, and destroy them. Don't add them to the compost pile or leave them lying around the garden. If you plant more asparagus, select rust-resistant varieties.

GROWING GUIDE

Adaptation: Throughout North America.

Planting Time: See page 361.

Planting Method: Sow seeds 1 to 1½ inches deep and 1 to 2 inches apart. Thin to 2 to 4 inches apart when the seedlings have their first set of leaves.

Soil: Any good garden soil, somewhat friable for easy seedling emergence. pH 6.0 to 8.0.

Fertilizer: At planting time, use 1½ pounds of Scotts Vegetable Food per 100 square feet, or ½ pound per 25-foot row. Side-dress every 3 to 4 weeks with ½ pound per 25-foot row.

Water: How much: Apply enough water to wet the soil 8 to 10 inches deep. **How often:** Water when the soil is just barely moist.

Harvest: Break the beans from the vines, being careful not to jerk or tear the vine. Pick beans when they are ready. If over-mature beans remain on the vines, production is greatly reduced. Cool beans in the refrigerator until use.

Harvesting times:
Snap (green) beans: When the seed in the pod begins to swell.
Lima beans: When the pods are well-filled, but still bright green.

Mexican bean beetle

Mexican bean beetle (2× life size).

Problem: The tissue between the leaf veins is eaten, giving the leaves a lacelike appearance. Copper-colored beetles about ¼ inch long feed on the undersides of the skeletonized leaves. The beetles have sixteen black spots on their backs. Orange to yellow soft-bodied grubs about ⅓ inch long with black-tipped spines on their backs may also be present. Leaves dry up, and the plant may die.

Analysis: The Mexican bean beetle (*Epilachna varivestis*) is found throughout the United States. It prefers lima beans but also feeds on pole and bush beans and cowpeas. Feeding damage by both adults and larvae can reduce pod production. Adult beetles spend the winter in plant debris in the garden and emerge in late spring and early summer. They lay yellow eggs on the undersides of leaves. Larvae that hatch from these eggs, in early to midsummer, are green at first and gradually turn yellow. One to four generations occur per year. Frequently, all stages of beetles appear at the same time throughout the season. Hot, dry summers and cold winters reduce the beetle population.

Solution: Apply ORTHO Diazinon Ultra Insect Spray when the adults first appear. Be sure to spray the undersides of leaves where the insects feed. Early treatments to control the adults may save extra applications later to control the larvae, which are more damaging and harder to control. Remove and destroy all plant debris after the harvest to reduce overwintering spots for adults.

BEANS (continued)

Bean leaf beetle

Bean leaf beetle (life size).

Problem: Round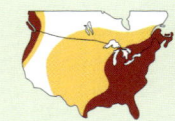
holes are chewed in the
leaves. Yellow to red
beetles with black spots
and a black band around the outer edge of
the body are feeding on the undersides of
leaves. The plant may later turn yellow and
wilt. If you pull the plant up, you may see
slender white grubs up to ⅓ inch long
feeding on the roots.

Analysis: The bean leaf beetle (*Cerotoma
trifurcata*), a widely distributed insect,
attacks all beans, as well as peas. Adult
beetles feed on the undersides of leaves,
blossoms, and pods throughout the growing
season. Grubs feed on the roots and stems
below the soil line. Females lay clusters of
orange eggs on the soil at the base of plants.
The grubs that hatch from these eggs attack
the plant below the soil, feeding on the
roots and sometimes girdling the stem at soil
level. This feeding can kill the plant. Both
adults and grubs cause serious damage to
young plants. One to three generations
occur per year.

Solution: Apply ORTHO Diazinon Ultra
Insect Spray at the first sign of damage.
Be sure to spray the undersides of the leaves
where the beetles feed. Repeat at intervals
of 7 to 10 days whenever damage occurs.
Clean all debris from the garden at the end
of the season to eliminate overwintering
spots for adults.

Bean anthracnose

Anthracnose spots.

Problem: Small
brown specks on pods
enlarge to black,
circular, sunken spots. In
wet weather, a salmon-colored ooze appears
in the infected spots. Elongated dark reddish
brown spots appear on the stems and veins
on the undersides of leaves. If seedlings are
attacked, the stems may rot, or the first
young leaves may be spotted. In either case,
the seedlings die.

Analysis: Bean anthracnose is caused
by a fungus (*Colletotrichum lindemuthianum*)
that affects all kinds of beans but is most
destructive on lima beans. It occurs in the
eastern and central states, rarely west of the
Rocky Mountains. The fungus thrives in
warm, wet weather. The salmon-colored
ooze that often appears in the infected
spots during wet weather is a mass of spores.
The spores are carried by splashing water,
animals, people, or tools to healthy plants.
The fungus survives from one season to the
next on diseased bean seeds and on plant
debris that has been left in the garden.

Solution: Remove and destroy any
diseased plants. The disease spreads rapidly
on moist foliage, so do not work in the
garden when the plants are wet. To avoid
reintroducing this fungus into your garden,
buy seeds produced in the western United
States. Do not plant beans in the infected
area for 2 or 3 years. Rotate your bean
planting site every year.

Bacterial blights

Halo blight.

Problem: Small,
water-soaked spots
appear on the leaves.
These spots enlarge,
turn brown, and may kill the leaf. In cool
weather, narrow greenish yellow halos may
border the infected spots. The leaves either
turn yellow and die slowly or turn brown
rapidly and drop off. Long reddish lesions
may girdle the stem. In moist conditions, a
tan or yellow ooze is produced in spots on
pods.

Analysis: Two widespread bacterial
blights on beans are *common blight* (caused
by *Xanthomonas phaseoli*) and *halo blight*
(caused by *Pseudomonas phaseolicola*). These
bacteria attack all kinds of beans. Common
blight is more severe in warm, moist
weather; halo blight prefers cool
temperatures. The bacteria are usually
introduced into a garden on infected seed
and can live on infected plant debris in the
soil for as long as 2 years. They are spread by
rain, splashing water, and contaminated
tools. The bacteria multiply rapidly in
humid weather. If water-conducting tissue is
invaded, bacteria and dead cells eventually
clog the veins, causing leaf discoloration.
Often bacteria ooze from infected spots in
a cream-colored mass.

Solution: No chemical controls bacterial
blights. Avoid overhead watering. Do not
work with the beans when the plants are
wet. Do not plant beans in the same area
more often than every third year. Buy new
seed each year from a reputable company.

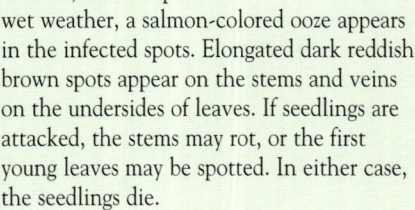

Bean rust

Rust on underside of bean leaf.

Problem: Rust-colored spots form, mostly on the undersides of leaves. Angular yellow spots develop above each rust spot on the top side of the leaf. Severely infected leaves turn yellow, wilt, dry, and fall off. Stems and pods may also have spots.

Analysis: Bean rust is caused by a fungus (*Uromyces phaseoli*) that affects only bean plants. It is most common on mature plants and is most damaging to pole beans and lima beans. Scarlet runner beans are sometimes mildly affected. Each rust spot develops thousands of spores that are spread by wind and splashing water. The disease develops rapidly during periods of cool nights and warm days. High humidity from rain, dew, or watering also encourages rust. Heavy vine growth that shades the ground and prevents air circulation produces ideal conditions for the disease. At the end of the summer, when the nights are longer, the fungus produces another type of spore. This thick-walled black spore spends the winter in infected bean plant debris.

Solution: Apply ORTHO Multi-Purpose Fungicide Daconil 2787® Plant Disease Control to beans at the first sign of the disease. Weekly applications may be necessary if conditions favorable to the disease continue. Avoid overhead irrigation. Water in the morning rather than evening to allow wet foliage to dry quickly. Thin seedlings, and space plants far enough apart to allow air to circulate freely. In the fall, remove and destroy all infected plants to prevent the fungus from surviving and reinfecting in the spring. Do not plant beans in the same area more often than every third year. Select resistant varieties.

Mold

White mold. Inset: Gray mold.

Problem: Soft, watery spots appear on the stems, leaves, or pods. Under moist conditions, these spots enlarge rapidly. A fuzzy gray, gray-brown, or white mold forms on the infected tissue. Small, hard, black, seedlike structures may be embedded in the white mold. Bean plants may yellow, wilt, and die. Rotted pods are soft and mushy.

Analysis: Mold on beans is caused by two related fungi. One fungus, *Sclerotinia sclerotiorum*, is responsible for white mold, also known as *watery soft rot*. The white mold fungus forms dark, seedlike structures that can drop to the soil and survive through adverse conditions to infect bean crops for the next few years. The fungus *Botrytis cinerea* is responsible for gray mold. This fungus produces tan to gray-brown fungal strands. Both fungi often attack weak or dead plant parts, such as old blossoms. Once established on a plant, these diseases can be spread to healthy plants by wind or splashing water or when infected plant parts touch healthy ones. Mold spreads quickly in cool, wet weather.

Solution: Remove and destroy all diseased plants as soon as symptoms appear. Water healthy plants early in the day so they will dry quickly. Avoid wetting plant foliage. Do not plant beans in the affected area for 3 to 4 years. Until then, plant other resistant vegetables, such as corn, beets, Swiss chard, or spinach. Always plant beans in well-drained soil, and to improve air flow, avoid overcrowding. Spray the foliage with ORTHO Multi-Purpose Fungicide Daconil 2787® Plant Disease Control.

Hopperburn

Hopperburn. Inset: Leafhopper (3× life size).

Problem: Leaves are stippled. Some are scorched, with a green midrib and brown edges curled under. Spotted pale green winged insects up to ⅛ inch long hop, run, or fly away quickly when the plant is touched.

Analysis: Hopperburn is caused by the potato leafhopper (*Empoasca fabae*), which injects a toxin into the leaves as it feeds. Bean yields may be drastically reduced from hopperburn. Leafhoppers are active throughout the growing season. They hatch in the spring from eggs laid on perennial weeds and ornamental plants. Even areas that have winters so cold that the eggs cannot survive are not free from infestation, because leafhoppers migrate in the spring from warmer regions.

Solution: Treat infested plants with ORTHO Malathion 50 Plus Insect Spray, ORTHO Diazinon Ultra Insect Spray, ORTHO Tomato & Vegetable Insect Killer, or an insecticidal soap at the first sign of damage. Be sure to cover the lower surfaces of leaves, where the leafhoppers feed. Repeat the spray as often as necessary to keep the insects under control. Allow at least 10 days between applications.

BEANS *(continued)*

BEETS

Root rot

Fusarium root rot.

Problem: Leaves turn yellow and fall prematurely. Overall growth is slow, and the plant may be dwarfed. In hot, dry weather, the plant suddenly wilts and sometimes dies. Few pods are produced. Red spots or streaks may be seen on stems and roots. Underground stems may be streaked dark brown or black.

Analysis: Root rot is caused by several soil-inhabiting fungi (*Fusarium* species, *Pythium* species, and *Rhizoctonia solani*) that attack beans and many other vegetables. The fungi live in the soil, invading the plant through the roots and underground stem. Plants are usually attacked when they are in a weakened state. As the disease progresses, the roots decay and shrivel. The leaves turn yellow, and growth slows. The plant becomes dwarfed, wilts, and sometimes dies. Under favorable growing conditions, bean plants may grow new side roots to replace the rotted ones. These plants will survive, but the yield will be reduced. Root rots develop most rapidly at soil temperatures between 60° and 85°F.

Solution: No chemical completely and effectively controls this problem. Pull out and discard wilted plants. Rotate your bean planting site yearly. Plant in well-drained soil, and let the soil surface dry out between waterings. Keep plants growing vigorously.

Bean mosaic

Bean yellow mosaic.

Problem: Leaves are mottled yellow and green and may be longer and narrower than usual and puckered. Raised dark areas develop along the central vein, and the leaf margins curl downward. The whole plant is stunted. Pods on affected plants may be faded, rough, and few in number. The seeds inside are shriveled and small.

Analysis: Bean common mosaic and bean yellow mosaic are widespread viral diseases that are difficult to distinguish from one another. Common mosaic virus affects only French and snap beans; yellow mosaic virus affects lima beans, peas, summer squash, clover, gladiolus, and other perennial flowers. Both diseases are spread by aphids, which transmit the virus as they feed. In warmer parts of the country with large aphid populations, the disease spreads rapidly. Common bean mosaic is also spread in infected seed. If the infection occurs early in the season when the plants are young, the plants may not bear pods. Infection later in the season does not affect pod production as severely.

Solution: No chemical controls plant viruses. Remove all infected plants and all clover plants in the vicinity of the garden. Plant virus-resistant varieties.

Flea beetles

Flea beetle (5× life size).

Problem: Leaves are riddled with shot holes about 1/8 inch in diameter. Tiny (1/16-inch) black beetles jump like fleas when disturbed. Leaves of seedlings and eventually whole plants may wilt and die.

Analysis: Flea beetles jump like fleas but are not related to fleas. Both adult and immature flea beetles feed on a wide variety of garden vegetables. The immature beetle, a legless gray grub, injures plants by feeding on the roots and the lower surface of leaves. Adults chew holes in leaves. Flea beetles damage seedlings and young plants most. Leaves of seedlings riddled with holes dry out quickly and die. Adult beetles spend the winter in soil and garden debris. They emerge in early spring to feed on weeds until vegetables sprout or plants are set in the garden. (Damage is often worse in weedy areas.) Grubs hatch from eggs laid in the soil and feed for 2 to 3 weeks. After pupating in the soil, they emerge as adults to repeat the cycle. One to four generations occur per year. Adults may feed for up to 2 months.

Solution: Control flea beetles with an insecticide containing *carbaryl* or *rotenone* or with an insecticidal soap when the leaves first show damage. Watch new growth for further evidence of feeding, and repeat treatments at weekly intervals as needed. Remove all plant debris from the garden after the harvest to eliminate overwintering areas for the adult beetle. Remove weeds in and around the garden.

Beet leafminer

Leafminer damage.

Problem: Light-colored, irregular blotches, blisters, or tunnels appear in the leaves. Brown areas peel apart easily like facial tissue. Tiny black specks are found inside the tunnels.

Analysis: The beet leafminer (*Pegomya hyoscyami*) belongs to a family of leafmining flies. The tiny black or yellow adult fly lays its white eggs on the undersides of leaves. When the eggs hatch, the cream-colored maggots bore into the leaf. They tunnel between the upper and lower surfaces of the leaf, feeding on the inner tissue. The tunnels and blotches are called *mines*. The black specks inside are the maggot's droppings. The beet leaves are no longer edible, but the root is. Several overlapping generations occur during the growing season, so larvae are present continually from spring until fall.

Solution: Spray with ORTHO Malathion 50 Plus Insect Spray or an insecticide containing *diazinon* when the white egg clusters are first seen under the leaves. Repeat two more times at weekly intervals to control succeeding generations. Once the miners enter the leaves, sprays are ineffective. Spraying after the mines first appear will control only those leafminers that attack after the application. If practical, destroy infested leaves. Adult flies can be kept from laying eggs with floating row covers.

Leaf spot

Cercospora leaf spot.

Problem: The leaves have small, circular, distinct spots with dark borders. These spots may run together to form blotches or dead areas. The leaves often turn yellow and die. Leaf spotting is most severe in warm, humid weather, when a fine gray mold may cover the infected tissue. Older leaves are more severely affected than younger ones.

Analysis: Leaf spot, a common, destructive fungus (*Cercospora beticola*), attacks beets, spinach, and Swiss chard. The fungus invades the leaves but not the beet root. Severe infection damages many leaves and hinders the development of the root, however. The spotting makes the leaves unappetizing. The fungus is spread by wind, contaminated tools, and splashing water and is fostered by moist conditions and high temperatures. Leaf spot is most common during the summer months and in warm areas. This fungus survives the winter on plant debris not cleaned out of the garden.

Solution: Picking off and destroying the first spotted leaves retards the spread of leaf spot. Leaf spots on beets seldom cause enough damage to warrant fungicide sprays. If possible, avoid overhead watering; use drip or furrow irrigation instead. Use a mulch to reduce the need for watering. Clean all plant debris from the garden after the harvest to reduce the number of overwintering spores.

Curly top

Curly top, a virus spread by beet leafhoppers.

Problem: The leaf margins roll upward and feel brittle. The undersides of leaves are rough, with puckering along the veins. The leaves and roots are stunted; the plant may die.

Analysis: Curly top, a virus disease, affects many vegetables, including beets, beans, tomatoes, squash, and melons. The virus is transmitted from plant to plant by the beet leafhopper (*Circulifer tenellus*) and is common only in the West, where this insect lives. The beet leafhopper is a pale greenish yellow insect about $\frac{1}{8}$ inch long that feeds from early May through June. It sucks the sap and virus from the infested leaves and then injects the virus into healthy plants at its next feeding stop. Curly top symptoms vary in their severity, depending on the variety and age of the plant. Young infected plants usually die. Older plants may turn yellow and die.

Solution: Nothing directly controls curly top. To reduce the chance of infection, control the beet leafhopper with ORTHO Malathion 50 Plus Insect Spray, beginning when the insect swarms first appear. Destroy infected plants.

BLUEBERRIES

BRAMBLES

Cherry fruitworm or cranberry fruitworm

Cranberry fruitworms (³⁄₄ life size).

Problem: Clusters of berries are webbed together with silk. Berries may be shriveled and full of sawdustlike material. Inside the berries are smooth, pink-red or pale yellow green caterpillars about ³⁄₈ to 1 inch long.

Analysis: The cherry fruitworm (*Grapholita packardi*) and cranberry fruitworm (*Acrobasis vaccinii*) are serious pests of blueberries and cranberries. They also attack cherries and apples. Each caterpillar destroys two to six berries. The caterpillars do not damage the leaves. In midspring, adult moths lay eggs on developing berries and leaves. The caterpillars that hatch from these eggs bore into the berries at the junction of the stem and berry. When they are about half-grown, the fruitworms move to another berry, usually one touching the infested berry. This way the fruitworm can move to a new food source without exposing itself. About mid-June, when the caterpillar is full grown, it crawls to the soil, garden debris, or a pruning stub, where it remains through the rest of the growing season and the winter. Adult moths emerge in midspring to start the cycle again.

Solution: Once the fruit is infested, the worms inside cannot be killed. Handpick and destroy all infested fruit. If your planting is large, or if the fruitworm infestation the previous year was severe, spray the plants immediately after bloom, before the berries are ¼ inch in diameter, with an insecticide containing diazinon. Repeat 10 days later.

Blueberry maggot

Blueberry maggots and adult (2× life size).

Problem: Ripening blueberries leak juice and are soft and mushy. White, tapered maggots about ³⁄₈ inch long are feeding inside the berries.

Analysis: Blueberry maggot (*Rhagoletis mendax*), also called the *blueberry fruit fly*, is the most important pest of blueberries in the Midwest and on the East Coast. Maggots attack both green and ripe fruit and often ruin an entire crop. They are most severe after unusually cold winters and when the weather is very wet at harvest time and frequent pickings are not possible. The maggots spend the winter as pupae in the soil. From late June through August, the adult flies lay eggs just under the skin of the fruit. The maggots that hatch from the eggs feed in the berry for about 20 days, then drop to the soil, where they pupate for 1 to 2 years. Only one generation occurs per year.

Solution: Harvest frequently, destroying any infested berries. The following year, beginning in early July, treat the plants with an insecticide containing *diazinon*. Repeat the treatment at intervals of 10 days throughout the harvest. Treatments the following year may be necessary to control any pupae that mature after 2 years.

Cane blight

Cane blight on raspberry.

Problem: Branches wilt and die in midsummer. Brownish purple areas occur on pruned ends or wounded areas of canes and extend downward, sometimes encircling the stem. Infected canes turn gray in late summer. Symptoms are associated with wounds and are usually not visible on first-year canes.

Analysis: Cane blight is caused by a fungus (*Leptosphaeria coniothyrium*). It is more prevalent on black raspberries than on red and purple raspberries and blackberries. The fungus spends the winter on diseased canes. The spores are spread from plant to plant by the wind. In wet weather in late spring and early summer, the spores enter canes through cracks in the bark, broken fruit stems, and wounds from pruning and insects. The spores don't infect young tissue directly. Canes weakened from cane blight are more subject to winter injury.

Solution: Apply ORTHO Dormant Disease Control Lime-Sulfur Spray in late fall and early spring. Follow with two applications of a fungicide containing *ferbam*: one just before bloom, when the new canes are 1½ to 2 feet tall; and the other just after harvest. Remove infected canes before growth starts in the spring. Prune just above a bud in dry weather at least 3 days before rain so the wounds have time to dry. Follow the summer topping of black raspberries with an application of a fungicide containing *benomyl* or *captan*. Avoid overhead irrigation. Avoid unnecessary wounding of canes.

Cane and crown gall

Crown gall on blackberry.

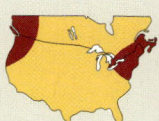

Problem: Plants are stunted, break and fall over easily, and produce dry, seedy berries. Irregular, wartlike growths (galls) may appear on canes, especially near soil level. The canes may dry out and crack. Galls also occur just below soil level on roots and crowns. They range from pinhead-sized to several inches in diameter and are white or grayish brown.

Analysis: Cane and crown gall are caused by bacteria (*Agrobacterium rubi* and *A. tumefaciens*) that occur on blackberries, raspberries, boysenberries, loganberries, and youngberries throughout the United States. The plants become weak and produce fewer berries. The bacteria are often brought to a garden initially on the roots of an infected plant and are spread with the soil and by contaminated pruning tools. The bacteria enter the plant through wounds in the roots or the base of the stem (the crown). They produce a substance that stimulates rapid cell growth in the plant, causing gall formation on the roots, crown, and canes. The galls disrupt the flow of water and nutrients up the roots and stems, weakening and stunting the top of the plant. Galls do not usually cause the plant to die.

Solution: Crown gall cannot be eliminated from the plant. Although infected plants may survive for many years, they produce few berries. Dig up and discard diseased plants and the soil within 6 inches of a gall. Wait at least 3 years before replanting brambles in areas where the disease has occurred. Plant gall-free plants in clean soil. Prune and cultivate carefully to avoid wounding plants. Select resistant varieties.

Powdery mildew

Powdery mildew on raspberry.

Problem: A whitish gray powder covers the leaves, fruit, and young growing tips of the canes. Canes may be dwarfed or distorted. Fruit may be covered with white powder and wither.

Analysis: Powdery mildew seriously affects red raspberries and occasionally attacks purple and black raspberries and blackberries. It is caused by a fungus (*Sphaerotheca macularis*) that thrives in both humid and dry weather. The powdery patches consist of fungal strands and spores. The spores are spread by the wind to healthy plants. The fungus saps plant nutrients, causing yellowing and sometimes the death of the leaf. A severe infection may kill the plant. Since this powdery mildew attacks many different kinds of plants, the fungus from a diseased plant may infect other types of plants in the garden. (For a list of susceptible plants, see page 343.) Under conditions favorable to it, powdery mildew can spread through a closely spaced planting of berries in a matter of days or weeks.

Solution: Spray the plants with flowable sulfur when the blossoms first open. Repeat at weekly intervals if the plants become reinfected. The following fall, and again in the spring as the buds begin to swell, spray the plants with ORTHO Dormant Disease Control Lime-Sulfur Spray. Keep the row width narrow or the canes far enough apart to allow good air circulation so they can dry rapidly after rain or watering.

Orange rust

Orange rust on blackberry.

Problem: Leaves are dwarfed, misshapen, and yellowish. Young shoots are spindly and clustered. Blisterlike pustules and bright orange dust cover the undersides of leaves. Plants remain infected throughout their lives and produce no blossoms.

Analysis: Orange rust is caused by the fungi *Arthuriomyces peckianus* and *Gymnoconia nitens*. It is severe on wild and cultivated blackberries. It sometimes attacks black raspberries but does not affect red and purple raspberries. The disease spreads throughout the entire plant. Infected plants never recover and never bloom. Spores are produced on the plant each year after it is infected. The fungus spreads from plant to plant on the wind. In the spring, orange spores land on leaves. After infecting them, the fungus spreads throughout the plant into the canes, crown, and roots. The disease spreads into new shoots as the plant grows.

Solution: Fungicide sprays and pruning are not effective. Remove and destroy infected plants as soon as they are noticed. Remove any wild blackberries growing nearby. Thin the plants for good air circulation. Pull out all weeds. Grow rust-resistant varieties.

CABBAGE

GROWING GUIDE

Adaptation: Throughout North America.

Planting Time: See page 361.

Planting Method:
Spacing: Cabbage, cauliflower, and broccoli—15 to 24 inches apart; brussels sprouts—30 to 36 inches apart; kohlrabi—4 to 6 inches apart; kale and collards—8 to 12 inches apart.

Soil: Any good garden soil. pH 6.0 to 7.0.

Fertilizer: At planting time use 1 to 2 pounds of Scotts Vegetable Food per 100 square feet, or 1 pound per 25-foot row. Side-dress every 3 to 4 weeks with ½ pound per 25-foot row.

Water: How much: Apply enough water at each irrigation to wet the soil 8 to 10 inches deep.
How often: Water when the soil 1 inch deep is barely moist.

Harvest: Broccoli: When the buds are still tight, before they open into yellow flowers. Smaller heads will grow along the stem after the first head is harvested.
Brussels sprouts: Pick sprouts from the bottom up when they are about 1 inch in diameter and still tight. Remove the leaves as you pick.
Cabbage: Cut just below the head when firm and before it cracks.
Chinese cabbage: Cut the entire head when it is 12 to 16 inches tall.
Cauliflower: Cut when the heads are 6 to 8 inches in diameter with tight curds.
Collards and kale: Pick the leaves when the plant is 1 foot tall.
Kohlrabi: Harvest when the bulb is 2 to 3 inches in diameter.

Bolting

Bolting broccoli.

Problem: In hot weather, an elongated stalk with flowers grows from the main stem of broccoli, cauliflower, and brussels sprouts plants. On cabbage, the head splits open, and the stalk emerges from within.

Analysis: Bolting, or seed stalk formation, results from exposure of the plants to cold temperatures early in their lives. Once a plant has matured to the point where its leaves are about 2 inches wide, exposure to cold (40° to 50°F) temperatures for several days in a row causes flower buds to form within the growing point. These buds remain dormant until hot weather arrives. With the arrival of hot weather, the buds develop into tall flower stalks. As the plant bolts, its flavor deteriorates and bitterness develops.

Solution: Discard plants that have bolted. Avoid setting out plants too early in the winter or spring. If cabbage plants are set out in the fall or winter, harvest them before hot weather causes them to bolt. Cutting the flower stalk will not prevent poor flavor from developing.

Diamondback moth

Diamondback moth larva (life size).

Problem: Small holes and transparent holes appear in the leaves. Small (¼-inch) green worms feed on the undersides of leaves. When disturbed, the worms wriggle rapidly and often drop from the plant on a silk thread.

Analysis: This worm is the larva of a gray or brown diamondback moth (*Plutella xylostella*) that flies in the evening. The adult moths don't damage plants. The pale green larvae feed on the undersides of leaves of members of the cabbage family, chewing holes and eating the lower surfaces of leaves. The feeding damage often allows soft rots to enter the leaves. After feeding for about 2 weeks, the larvae pupate in transparent silken cocoons attached to the undersides of leaves. Adult moths spend the winter hidden under plant debris. Two to seven generations per year damage both spring and fall plantings.

Solution: Apply ORTHO Tomato & Vegetable Insect Killer, ORTHO Diazinon Ultra Insect Spray, or the bacterial insecticide *Bacillus thuringiensis* (Bt) to the foliage when the young worms first appear or when the leaves show feeding damage. The insecticide must reach the undersides of the leaves in order to be most effective. Repeat the treatment each week as long as the caterpillars are found. Adults can be kept from laying eggs with row covers. Clean up plant debris in the fall and cultivate the soil thoroughly to expose and destroy overwintering moths. Kill wild mustard and other weeds of the cabbage family in the vicinity of the garden several weeks before planting in the spring.

Cabbageworms

Imported cabbageworm. Inset: Adult (life size).

Problem: Round or irregular holes appear in leaves. Green worms with light stripes down their backs, up to 1½ inches long, feed on the leaves or heads. Masses of green or brown pellets may be found between the leaves. Cabbage and cauliflower heads may be tunneled.

Analysis: Cabbageworms are either the *cabbage looper* (*Trichoplusia ni*) or the *imported cabbageworm* (*Pieris rapae*). Both worms attack all members of the cabbage family, as well as lettuce. Adults lay eggs throughout the growing season. Adults of the imported cabbageworm attach yellow, bullet-shaped eggs to the undersides of leaves. These white butterflies are frequently seen around cabbage plants in the daytime. The brownish cabbage looper moth lays pale green eggs on the upper surfaces of leaves in the evening. Worms may be present from early spring until late fall. In the South, they may be present all year. Worms spend the winter as pupae attached to a plant or nearby object.

Solution: Control cabbageworms with ORTHO Diazinon Ultra Insect Spray, ORTHO Tomato & Vegetable Insect Killer, or an insecticidal soap. Spray as soon as damage is seen. Cabbageworms can also be killed with the bacterial insecticide *Bacillus thuringiensis* (Bt) while they are small. Repeat treatments weekly as long as worms are found, but stop 3 days before harvest. Adults can be kept from laying eggs with row covers. Remove plant debris after harvest to destroy the pupae.

Cutworms

Cutworm (2× life size).

Problem: Seedlings and young transplants are chewed or cut off near the ground. If you dig near the freshly damaged plant about 2 inches down, you may find dull gray, brown, or black worms. They have spots or stripes on their smooth bodies, are about 1½ to 2 inches long, and coil when disturbed.

Analysis: Several species of these moth larvae are pests in the vegetable garden. The most likely pests of young cabbage plants set out early in the season are surface-feeding cutworms. They spend the days hidden in the soil and feed at night. A single cutworm can sever the stems of several young plants in a single night. Adult cutworms are dark, night-flying moths with stripes on their forewings.

Solution: Apply ORTHO Diazinon Soil & Turf Insect Control or ORTHO Bug-Geta Plus Snail, Slug & Insect Killer around the base of undamaged plants when stem cutting is observed. Treat in late evening just before cutworms become active. Because cutworms are difficult to control, applications may need to be repeated at weekly intervals. Clean up any weedy growth in the vicinity of the garden. Before transplanting in an area previously full of weeds, apply a preventive treatment of ORTHO Diazinon Soil & Turf Insect Control or the bacterial insecticide *Bacillus thuringiensis* (Bt), and work it into the soil. Further reduce damage with a cutworm collar around the stem of each plant (see page 252). Cultivate the soil in late summer and fall to expose and destroy eggs, pupae, and larvae.

Cabbage maggot

Cabbage maggots (6× life size).

Problem: Young plants wilt in the heat of the day. They may later turn yellow and die. Soft-bodied, white maggots about ⅓ inch long are feeding in the roots. The roots are honeycombed with slimy channels and scarred by brown grooves. Damage is particularly severe during cool, moist weather in the spring, early summer, and fall.

Analysis: The cabbage maggot (*Delia radicum*) is an important pest in the northern United States. Early maggots attack the roots and stems of cabbage, cauliflower, broccoli, radishes, and turnips in the spring and early summer. Later insects injure cabbage, turnips, and radishes in the fall. The adult is a gray fly, somewhat smaller than a housefly, with black stripes and bristles down its back. It lays eggs on stems and nearby soil. The maggots hatch in 2 to 3 days and tunnel into the stems and roots of plants, sometimes to a depth of 6 inches. The cabbage maggot causes feeding damage and also spreads black rot bacteria (see page 272).

Solution: Once the growing plant wilts and turns yellow, nothing can be done. To control maggots in the next planting, mix ORTHO Diazinon Soil & Turf Insect Control (granules) or ORTHO Dursban® Lawn & Garden Insect Control 4 to 6 inches into the soil before seeding or transplanting. Control lasts about 1 month. Since control depends on the maggots coming into direct contact with the insecticide, correct timing and thorough mixing are important. Plant as late as possible to avoid the period of maximum maggot damage in late spring. Adult flies can be kept from laying eggs with floating row covers.

CABBAGE (continued)

Flea beetles	Aphids	Black rot

Flea beetles (2× life size).

Aphids on cabbage (2× life size).

Black rot can be spread by insects.

Flea beetles

Problem: Leaves are riddled with shot holes, about ⅛ inch in diameter. Tiny (1⁄16 inch) black or striped beetles jump like fleas when disturbed. Seedlings may wilt and die.

Analysis: Flea beetles jump like fleas but are not related to fleas. Both adult and immature flea beetles feed on a wide variety of garden vegetables. The immature beetle, a legless gray grub, injures plants by feeding on the roots. Adults chew holes in leaves. Flea beetles damage seedlings and young plants most. Adult beetles spend the winter in soil and garden debris. They emerge in early spring to feed on weeds until vegetables sprout or plants are set in the garden. Grubs hatch from eggs laid in the soil and feed for 2 to 3 weeks. After pupating in the soil, they emerge as adults to repeat the cycle. One to four generations occur per year. Adults may feed for up to 2 months.

Solution: Control flea beetles with ORTHO Bug-B-Gon Insect Killer, an insecticidal soap, or ORTHO Tomato & Vegetable Insect Killer when the leaves first show damage. Watch new growth for evidence of further damage, and repeat the treatment at weekly intervals as needed. Clean all debris from the garden after harvest to eliminate overwintering spots for adult beetles. Remove weedy growth in the vicinity of the garden to eliminate alternate hosts of the beetles. Use row covers to exclude the beetles.

Aphids

Problem: Some leaves are yellowed and cupped downward. Tiny (⅛-inch) pale green or gray soft-bodied insects cluster under leaves, on stems, and on heads. A shiny, sticky substance may coat the leaves. Ants may be present.

Analysis: Aphids do little damage in small numbers. They are extremely prolific, however, and populations can rapidly build up to damaging numbers during the growing season. Damage occurs when the aphid sucks the juices from the plant. The aphid is unable to digest fully all the sugar in the plant sap, and it excretes the excess in a fluid called *honeydew*. The honeydew often drops onto the leaves below. Ants feed on the sticky substance and are often present where there is an aphid infestation. The most common aphid attacking the cabbage family is the cabbage aphid (*Brevicoryne brassicae*). Cabbage aphids spend the winter as eggs on plant debris in the garden. In the South, they are active year-round.

Solution: Treat with ORTHO Bug-B-Gon Insect Killer, ORTHO Malathion 50 Plus Insect Spray, ORTHO Tomato & Vegetable Insect Killer, or an insecticidal soap as soon as the insects appear. Repeat at intervals of 1 week if the plant is reinfested. Clean all plant debris from the garden after harvest to reduce the number of over-wintering eggs.

Black rot

Problem: Young plants turn yellow, then brown, and die. On older plants, yellow areas develop along the leaf margins, then progress into the leaf in a V shape; these areas later turn brown and die. Lower (older) leaves wilt and drop off. The veins running from the infected leaf margins to the center stem are black. When the stem is cut across, a black ring and sometimes yellow ooze are seen in the cross section.

Analysis: Black rot is caused by a bacterium (*Xanthomonas campestris*) that affects all members of the cabbage family at any stage in their growth. The bacteria live in or on seeds or in infected plant debris for as long as 2 years. Insects, splashing water, and garden tools carry bacteria to other leaves and plants. The bacteria enter the plant through natural openings or wounds and spread in the water- and nutrient-conducting vessels of the plant. The infected tissue may form pockets where dead cells and bacteria accumulate as a yellow ooze. Warm, humid weather promotes the spread and development of the bacteria. Black rot can kill seedlings rapidly. The diseased heads are edible but unappetizing.

Solution: No chemical controls black rot. Discard infected plants. To avoid further infection, plant only disease-free seed and healthy plants. Place plants far enough apart to allow good air circulation. Avoid overhead irrigation. Plant in soil that has not grown cabbage for at least 2 years.

Clubroot

Clubroot on kohlrabi.

Problem: The plant wilts on hot, sunny days and recovers at night. The older, outer leaves turn yellow and drop. The roots are swollen and misshapen. The largest swellings are just below the soil surface. Growth slows, and the plant eventually dies.

Analysis: Clubroot is caused by a fungus (*Plasmodiophora brassicae*) that persists in the soil for many years. Warm, moist weather conditions, together with an acid soil, encourage the infection. The fungus causes cells to grow and divide within the root tissue. This causes swelling and a general weakening of the plant, allowing other fungi and bacteria to invade the roots and cause root rot. As the roots decompose, they liberate millions of spores into the soil, where they are spread by shoes or tools or in runoff water. Most members of the cabbage family are susceptible to clubroot, but most rutabaga varieties and many turnip varieties are resistant. Weeds in this family, such as mustard, pennycress, and shepherd's purse, can also harbor the fungus.

Solution: Once a plant is infected, it cannot be cured. To reduce the severity of the disease the following year, do not grow susceptible plants where any member of the cabbage family has grown for the past 7 years. To discourage infection, lime the soil with garden lime to a pH of 7.2.

Carrot weevil

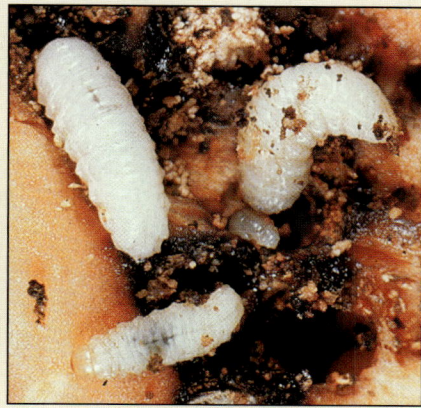

Carrot weevil larvae (3× life size).

Problem: The upper part of the root is scarred with zigzag tunnels. White, curved, legless grubs, about ⅓ inch long, may be found in the root and soil.

Analysis: The carrot weevil (*Listronotus oregonensis*) feeds on carrots, dill, celery, parsley, parsnips, and Queen Anne's lace. The adult is a dark brown, nonflying beetle that lays eggs in the carrot tops. In May and June the eggs hatch into white grubs, which travel down to the developing root. There the grub tunnels into and feeds on the upper tissue, scarring the root. After a short resting period in the soil, the pest emerges as an adult and lays eggs that hatch into a second generation in August. The adults of this second generation spend the winter in debris in and around the garden. The following spring, the cycle repeats itself.

Solution: No insecticides are currently registered for home use on this insect pest. Some control may be obtained when spraying for leafhoppers (see page 259). Keep adults from laying eggs by using row covers, especially when the young carrots have five or six leaves. Remove Queen Anne's lace from the garden area. Clean up garden debris in the fall to eliminate overwintering spots for adults. Do not add infested debris to the compost pile. The following year, rotate carrots to an area previously planted with a nonsusceptible crop.

Wireworms

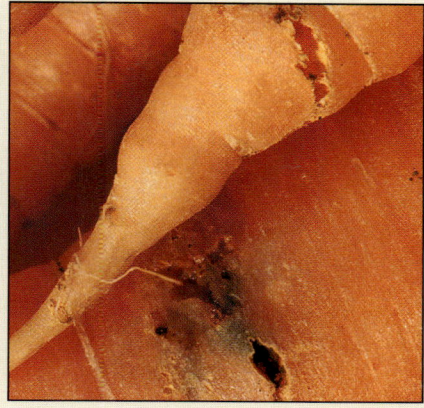

Wireworm damage.

Problem: Plants are stunted and grow slowly. Roots are poorly formed. Tunnels wind through the roots. Shiny, hard, jointed, creamy yellow, dark brown, orange, or gray worms up ⅝ inch long are found in the roots and soil.

Analysis: Wireworms attack carrots, corn, potatoes, beets, peas, beans, lettuce, and many other plants. They feed only on underground plant parts, devouring seeds, underground stems, tubers, and roots. Infestations are most extensive in soil where turfgrass has previously grown. The adult is known as a *click beetle* because it makes a clicking sound when turning from its back to its feet. The adult lays eggs in the spring. After the eggs hatch, the wireworms feed for 2 to 6 years before maturing into adult beetles. All sizes and ages of wireworms may be found in soil at the same time.

Solution: Treat with ORTHO Diazinon Soil & Turf Insect Control just before planting, working the insecticide into the top 6 to 8 inches of soil. Avoid planting carrots in areas that have held clover, grass, pasture, or grassy weeds.

CARROTS *(continued)*

Aster yellows

Aster yellows. Inset: Infected root.

Problem: Inner leaves are yellow and stunted and grow in tight bunches. Outer leaves turn rusty red to reddish purple. The roots are stunted and deformed and have a bitter taste. Tiny hairlike roots grow in great profusion out of the main root. Numerous tiny leaves grow from the top of the root.

Analysis: Aster yellows is a plant disease caused by phytoplasmas, microscopic organisms similar to bacteria. The phytoplasmas are transmitted from plant to plant primarily by leafhoppers. (For information on leafhoppers, see page 259.) The symptoms of aster yellows are more severe and appear more quickly in warm weather. Even when the disease is present in the plant, aster yellows may not manifest its symptoms in temperatures of 55°F or less. The disease infects many other vegetables, ornamental plants, and weeds. For a list of plants susceptible to aster yellows, see page 352.

Solution: Aster yellows cannot be eliminated entirely, but it can be controlled. Move and destroy infected plants. To remove sources of infection, eradicate nearby weeds that may harbor aster yellows and leafhopper eggs. Treat leafhopper-infested plants with ORTHO Bug-B-Gon Insect Killer or ORTHO Malathion 50 Plus Insect Spray. Repeat the treatment whenever leafhoppers are seen. Keep leafhoppers from the plants with floating row covers.

CORN

GROWING GUIDE

Adaptation: Throughout North America.

Planting Time: Mid- to late spring.

Planting Method: Sow seeds 1 to 2 inches deep and 12 to 16 inches apart. Or set transplants so that the top of their rootball is even with the garden soil. Space plants 12 to 16 inches apart.

Soil: Any good garden soil that is high in organic matter. pH 6.0 to 7.5.

Fertilizer: At planting time, use 2 pounds of Scotts Vegetable Food per 100-foot row. Side-dress with the same amount when the plants are 8 inches tall, and again when they are 18 inches tall.

Water: How much: Apply enough water at each irrigation to wet the soil 8 to 10 inches deep.
How often: Keep the soil moist, never allowing it to dry out. Corn needs constant moisture all season.

Harvest: Corn is generally ready to be picked 3 weeks after the silks appear. When ready, the silks become dark brown and dry. Test ripeness by pressing a kernel with your finger. If it spurts milky juice, it is ready. To pick, grab the ear at its base, bend it downward, and twist. Do not damage the main stalk. Pick corn just before cooking, as its flavor deteriorates rapidly when ears are stored.

Armyworms

Armyworm (½ life size).

Problem: Leaf edges are chewed, and some leaves may be completely eaten. Light tan to dark brown caterpillars 1½ to 2 inches long with yellow, orange, or dark brown stripes are feeding on the leaves and may be boring into the ears.

Analysis: Armyworms attack corn, grains, grasses, and other garden crops. These pests do not overwinter in cold-winter areas, but the moths migrate great distances in the spring in search of places to lay their eggs. The tan to gray adult moths lay eggs on the blades of grasses and grains. The caterpillars that hatch from these eggs feed on corn leaves, ears, and ear stalks. They get their name from their feeding habits. After they have eaten everything in one area, they crawl in droves to another area in search of more food. After several weeks of feeding, they pupate in the soil, then emerge as adult moths to repeat the cycle. Several generations occur each year, beginning in mid-May. These pests are most numerous after cold, wet spring weather that slows the development of the natural parasites and diseases that help keep the population in check.

Solution: When the worms are first seen, spray with ORTHO Bug-B-Gon Ready-Spray. Results may not be seen for 4 or 5 days. Repeat at weekly intervals if the plants become reinfested. To control worms that bore into the ears, spray when 10 percent of the ears show silk. Repeat three or four more times at intervals of 3 days.

Corn earworm

Corn earworm (¾ life size).

Adult (2× life size).

Problem: Striped yellow, brown, or green worms are feeding on the tip of the ear inside the husk. The worms range in size from ¾ to 2 inches long. Leaves may be chewed and ragged.

Analysis: The corn earworm (*Helicoverpa zea,* formerly *Heliothis zea*) is the most serious pest of corn. It attacks many other garden vegetables and flowers as well and is also known as the *tomato fruitworm* and the *cotton bollworm.* The worm is the larva of a light gray-brown moth with dark lines on its wings. In the spring, the moth lays yellow eggs singly on corn silks and on the undersides of leaves. The worms that hatch from these eggs feed on the new leaves in the whorls. This feeding doesn't reduce the corn yield, but the leaves that develop are ragged and plant growth may be stunted. More serious damage is caused when worms feed on the silks, causing poor pollination, and when they feed on the developing kernels. Worms enter the ear at the silk end, or they may bore through the husk. Several generations occur per year. In the South, where these pests survive the winter, early and late plantings suffer the most damage. Adult moths migrate into northern areas, where late plantings are severely damaged. Undamaged parts of infested ears are still edible.

Solution: Once the worms are in the ears, insecticides are ineffective. In the future, spray or dust plants with an insecticide containing *pyrethrins, endosulfan,* or *permethrin* when 10 percent of the ears show silk. Repeat the treatment three or four times at intervals of 3 days. If infestation continues, repeat as necessary until harvest. To avoid unnecessary spraying, monitor for adult moths with pheromone traps. Put out traps when the ears begin to silk. Begin spraying when moths are trapped. Physical means that make it difficult for the young worm to enter the ear are somewhat effective: Select varieties with long, tight husks, or pinch the husk closed with a clothespin where the silk exits the husk.

Poor pollination

Poor pollination results in poorly filled ears.

Problem: Corn ears are not completely filled with kernels. Plants are healthy.

Analysis: Poorly filled ears result from ineffective or incomplete pollination. Pollen grains produced on the tassels must fall on the sticky silks for complete pollination. Each strand of silk is attached to a kernel, so for each silk that is pollinated, one kernel develops on the ear. Corn pollen is spread by the wind; if the wind blows across a single row of corn, the pollen on the tassels is carried away from the silks, resulting in poorly filled ears. Poor pollination can also result from dry soil during pollination and from hot, dry winds. Prolonged periods of rain reduce the amount of pollen shed from the tassels. Damage to the silks from corn earworms, rootworm adults (see page 276), armyworms (see page 274), and grasshoppers (see page 277) may also result in incomplete ears.

Solution: To help ensure pollination, grow corn in blocks of at least three short rows rather than in one long row. Plant seeds 12 to 16 inches apart in rows 30 to 36 inches apart. Keep the soil moist, letting the surface dry slightly between waterings. Control corn insect pests. Corn can be hand-pollinated by shaking the tassels onto the silks.

CORN *(continued)*

Corn smut

Corn smut galls full of black powdery spores.

Problem: Puffballs or galls appear on the stalk, leaves, ears, or tassels. Galls are white and may be smooth, or they may be covered with a black, greasy or powdery material. They range from pea-sized to 5 inches in diameter.

Analysis: Corn smut is caused by a fungus (*Ustilago maydis*) that attacks any aboveground part of corn. Germinating seedlings are not affected. Galls are full of black powdery spores that survive the winter in soil and corn debris. They are spread from plant to plant by wind and water. A gall forms only where a spore lands. The disease does not spread throughout the plant. Younger plants are more susceptible; most plants are infected when they are 1 to 3 feet tall. Corn is less susceptible after the ears have formed. Corn smut is most prevalent in warm temperatures (80° to 95°F) and when dry weather early in the season is followed by moderate rainfall as the corn matures. Smut doesn't reduce the corn yield directly but rather saps the plant's energy, reducing ear development.

Solution: No chemical controls this disease. Cut off smuts before they break open and release the black powdery spores. Grow varieties tolerant of corn smut. Clean all plant debris from the garden after harvest.

Flea beetles

Flea beetle (12× life size).

Problem: Leaves are riddled with shot holes about ⅛ inch in diameter. Tiny (1⁄16-inch) black beetles jump like fleas when disturbed. Leaves of seedlings and eventually whole plants may wilt and die.

Analysis: Flea beetles jump like fleas but are not related to fleas. Both adult and immature flea beetles feed on a wide variety of garden vegetables. Some flea beetles are responsible for spreading the bacterial wilts that kill corn plants. The immature beetle, a legless gray grub, injures plants by feeding on the roots and the lower surfaces of leaves. Adults chew holes in leaves. Flea beetles damage seedlings and young plants most. Leaves of seedlings riddled with holes dry out quickly and die. Adult beetles survive the winter in soil and garden debris. They emerge in early spring to feed on weeds until vegetables sprout or plants are set in the garden. Grubs hatch from eggs laid in the soil and feed for 2 to 3 weeks. After pupating in the soil, they emerge as adults to repeat the cycle. One to four generations occur per year. Adults may feed for up to 2 months.

Solution: Control flea beetles on corn with ORTHO Bug-B-Gon Insect Killer or an insecticide containing *pyrethrins* when the leaves first show damage. Watch new growth for evidence of further damage, and repeat the treatment at weekly intervals as needed. Remove all plant debris from the garden after harvest to eliminate overwintering spots for adult beetles.

Corn rootworms

Rootworm damage. Inset: Adult (3× life size).

Problem: Yellow, pale green, or brownish red beetles with very long antennae crawl on the plants. Some may have black spots or stripes. The ears are malformed, with undeveloped or partially developed kernels; leaves may be chewed.

Analysis: Both adult beetles and young worms of corn rootworms (*Diabrotica* species) attack corn. For information on the larva, see page 278. The adult beetles feed on the pollen, silks, and tassels, resulting in malformed ears and undeveloped kernels from improper or incomplete pollination. Some beetles may also feed on the leaves. The adults lay yellow-orange eggs in the soil at the base of the corn plants. The young worms that hatch from these eggs feed on the corn roots for several weeks, then pupate in the soil to emerge as beetles in late July and August. These pests are most common where corn has grown consecutively for 2 or more years. Late-planted corn and corn under drought stress are the most susceptible.

Solution: Treat with an insecticide containing *carbaryl* when the beetles first appear on the plants. Repeat at weekly intervals if the plants become reinfested. Control the rootworm larvae as suggested on page 278.

Grasshoppers

Grasshopper (½ life size).

Problem: Large holes are chewed in the margins of leaves. Greenish yellow to brown jumping insects, ½ to 1½ inches long with long hind legs, infest corn plants. Kernels on ears may be chewed or undeveloped.

Analysis: Grasshoppers attack a wide variety of plants, including corn, grains, and grasses. They eat corn leaves and silk, migrating as they mature and as they deplete their food sources. They are most numerous in the rows adjacent to weedy areas. In the late summer, adult grasshoppers lay eggs in pods in the soil. The adults continue to feed until cold weather kills them. The eggs hatch the following spring. Grasshopper populations are most severe during hot, dry weather. The insects migrate into green gardens and yards as surrounding areas dry up in the summer heat. Periods of cool, wet weather help keep their numbers under control. The loss of a small amount of leaf tissue to a small population of grasshoppers doesn't reduce the corn yield significantly.

Solution: Apply ORTHO Bug-Geta Plus Snail, Slug & Insect Killer or ORTHO Bug-B-Gon Insect Killer as soon as grasshoppers appear. Repeat at weekly intervals if the plants become reinfested. Clean up weedy areas near the garden, and destroy plant debris after harvest.

European corn borer

European corn borer (2× life size).

Problem: Leaves are riddled with tiny shot holes. Tassels may be broken and ear stalks bent. Holes filled with sawdust are bored into the main stalks. Pinkish caterpillars with dark brown heads, two rows of brown dots, and indistinct reddish longitudinal stripes are found inside the stalks or within the ears.

Analysis: The European corn borer (*Ostrinia nubilalis*) is one of the most destructive pests of corn. It also feeds on tomatoes, potatoes, and peppers. Early plantings are most affected, although late plantings can also be severely damaged in areas where more than one generation occurs in a season. The borer survives the winter in corn plants, pupates in the spring, and emerges as an adult moth in early summer. The moth, which is tan with dark wavy lines on the wings, lays clusters of about 20 cream-colored eggs on the undersides of the lower corn leaves. The borers that hatch from these eggs feed first in the whorl of leaves, riddling the leaves with shot holes. Later they bore into stalks and the bases of ears. This feeding results in broken stalks and tassels, poor ear development, and dropped ears. The borers continue feeding for a month, pupate, and emerge as moths to repeat the cycle.

Solution: At the first sign of borers or when 10 percent of the ears show silk, treat ear shoots and centers of leaf whorls with an insecticide containing *carbaryl* or *rotenone* or with the bacterial insecticide *Bacillus thuringiensis* (Bt). Repeat at weekly intervals until borers are no longer seen. Destroy the plants at the end of the season. Avoid early planting.

Common stalk borer

Stalk borer damage. Inset: Larva (life size).

Problem: Young plants are stunted and may die. Leaves are chewed and ragged. Stalks don't produce ears and may be distorted and curled. Dark brown to purple caterpillars 1 inch long with white stripes and bands may be found inside the stalks.

Analysis: Common stalk borer (*Papaipema nebris*) is a serious pest of corn east of the Rocky Mountains. These borers feed on a variety of plants but prefer corn. They spend the winter as eggs on grasses and weeds, especially on giant ragweed. After hatching in the early spring, the worms feed in the leaf whorls and then bore into the side of the stalks and burrow upward. After pupating in the soil, the adult moths emerge in late summer and early fall. These grayish brown moths lay eggs on grasses for the following year's generation. Only one generation occurs per year.

Solution: Once the damage is noticed, it is too late for any controls. Destroy all infested plants. Clean all plant debris from the garden after harvest. Eliminate nearby grasses and weeds, especially giant ragweed. If stalk borers are serious, treat the plants the following year with an insecticide containing *diazinon* or *carbaryl* in early to midspring.

CORN *(continued)*

Corn rootworm larvae	Cutworms	Seedcorn maggot

Corn rootworm larvae (life size).

Black cutworm (⅓ life size).

Seedcorn maggot damage.

Problem: Corn plants are dwarfed and yellow and fall over easily. The base of the stalks may have a crook-necked shape. The remaining roots have a stubby appearance. In the soil, white worms, from ½ to ¾ inch long with brown heads, are eating the roots.

Analysis: Several species of the corn rootworm larvae (*Diabrotica* species) attack corn. The beetle larva, or worm, feeds on corn roots, completely devouring small ones and tunneling into larger ones. The worms hatch from eggs in early summer to midsummer and migrate through the soil, feeding on corn roots. This feeding damage can be so serious to young plants that some gardens may need to be replanted. When populations are large, all the roots may be destroyed. The worms pupate in the soil and emerge as greenish yellow or light brown beetles, some with black spots or stripes, to repeat the cycle. One to three generations occur per year. Adult beetles feed on leaves and silks. (For information on adult corn rootworms, see page 276.) Corn plants are affected most severely when they are growing in dry soil, where root regrowth is minimal. Rot diseases may enter the damaged roots, injuring them further. Several parasitic insects and diseases help keep beetle populations under control most years.

Solution: Discard damaged plants and clean up weedy areas where the insects lay eggs and spend the winter. Apply an insecticide containing *diazinon* or *chlorpyrifos* or the bacterial insecticide *Bacillus thuringiensis* (Bt) to the soil at planting time. Do not grow corn in the same soil for more than 2 consecutive years.

Problem: Young plants are chewed or cut off near the ground. Gray, brown, or black worms, 1½ to 2 inches long, may be found about 2 inches deep in the soil near the base of the damaged plants. The worms coil when disturbed.

Analysis: Several species of cutworms attack plants in the vegetable garden. The most likely pests of corn seedlings are surface-feeding cutworms. The two most common on corn are the black cutworm (*Agrotis ipsilon*) and the dingy cutworm (*Feltia ducens*). A single surface-feeding cutworm can sever the stems of many young plants in one night. Cutworms hide in the soil during the day and feed only at night. All adult cutworms are dark, night-flying moths with bands or stripes on their forewings. For more information on cutworms, see page 252.

Solution: Apply ORTHO Diazinon Soil & Turf Insect Control or ORTHO Bug-Geta Plus Snail, Slug & Insect Killer around the base of undamaged plants when stem cutting is observed. Because cutworms are difficult to control, applications may need to be repeated at weekly intervals. Before planting more corn in the same area, apply a preventive treatment of the bacterial insecticide *Bacillus thuringiensis* (Bt) or ORTHO Dursban® Lawn & Garden Insect Control and work it into the soil. Cultivate the soil thoroughly in late summer and fall to expose and destroy eggs, larvae, and pupae.

Problem: Seeds don't sprout, or the seedlings are weak and don't develop leaves. Pearly white worms, ¼ inch long, are feeding in the seeds. The seeds are hollow.

Analysis: Seedcorn maggots (*Delia platura*) feed on seeds and seedlings. They are attracted to large-seeded vegetables such as peas, beans, cucumbers, watermelon, and corn. The maggots are most numerous in cool periods in the spring and fall and in cold soil that is high in organic matter. The black, hairy adult flies are attracted to the organic matter and lay eggs in the soil. The maggots that hatch from these eggs burrow into the seeds and eat the inner tissue, leaving a hollow shell. Rot fungi may enter a damaged seed and further destroy it. After 1 to 2 weeks, the maggots burrow deep into the soil and pupate. The adult flies that emerge feed on nectar and plant juices before laying more eggs. Several generations occur per year. In warm-winter areas, these pests are active year-round. In cold-winter areas, they survive the winter as pupae and emerge as adults in the early spring.

Solution: Treat seed with an insecticide powder that contains *diazinon*. Mix ¼ teaspoon of insecticide with each packet of seed. Shake off the excess, and plant the seeds 1 to 2 inches deep. To speed germination, plant seeds when the soil has reached a temperature of 55°F or higher. Since adult flies are attracted to organic matter, don't add manure to the soil in the spring when planting beans, peas, cucumbers, watermelon, or corn.

CUCURBITS

GROWING GUIDE

Adaptation: Throughout North America.

Planting Time: See page 361.

Planting Method: Start seeds indoors or directly in the garden. Plant cucumber, muskmelon, and watermelon seeds 1 to 2 inches deep. Plant squash and pumpkins 2 to 3 inches deep. Set transplants in the garden so that the top of their rootball is even with the surrounding garden soil. Space transplants or thin seedlings as follows: Cucumbers and muskmelons—12 inches apart, or 24 to 36 inches apart if planted in hills; pumpkins and vining squash—36 to 40 inches apart; and bush squash and watermelon—24 to 36 inches apart.

Soil: Any good garden soil high in organic matter. pH 5.5 to 8.0.

Fertilizer: At planting time, use 1 pound of Scotts Vegetable Food per 50-foot row, or 1 tablespoon per plant. Side-dress when the runners are 12 to 18 inches long and again when the first fruits have set.

Water: How much: Apply enough water at each irrigation to wet the soil 1 to 1½ feet deep.
How often: Water when the soil 2 inches deep is barely moist.

Harvest:
Cantaloupe: When netting becomes pronounced and skin color turns yellow-tan; stem should separate or slip easily from the fruit.
Pumpkins and gourds: After the vines die in the fall but before a hard frost.
Summer squash: Continuous picking ensures a steady supply. Zucchini and crookneck: 1½ to 2 inches in diameter. Bush scallop: 3 to 4 inches.
Watermelon: When the spot where the melon touches the ground turns from white to creamy yellow; green skin is dull, not shiny; and the melon makes a dull thud when hit with the palm of the hand.
Winter squash: After the vines die in the fall, but before a hard frost; when your fingernail doesn't scratch the hardened skin.

Storage: Most cucurbits are used soon after havesting. Cool in refrigerator until use. Winter squash and pumpkins can be stored over the winter. Pick only mature fruit, and cure them at temperatures between 80° and 85°F for 10 days. Then move the squash or pumpkins to a well-ventilated place with temperatures between 50° and 60°F. Store in a single layer; they rot easily if stored in piles.

Blossom-end rot

Blossom-end rot on squash.

Problem: A water-soaked, sunken spot develops on the blossom end (opposite the stem end) of squash and watermelon. The spot enlarges and turns brown to black. Mold may grow on the spot.

Analysis: Blossom-end rot, a disorder of squash, watermelon, tomatoes, and peppers, is caused by a lack of calcium in the developing fruit. This lack of calcium is the result of slowed growth and damaged roots caused by any of the following factors:
1. Extreme fluctuations in soil moisture, either very wet or very dry.
2. Rapid plant growth early in the season, followed by extended dry weather.
3. Excessive rain that smothers root hairs.
4. Excess soil salts.
The first fruits of the season are the most severely affected. As the name implies, the disorder always starts at the blossom end, and it may enlarge to affect half of the fruit. Moldy growths on the rotted area are caused by fungi or bacteria.

Solution: The numbered solutions below correspond to the numbered items in the analysis.
1. Maintain uniform soil moisture by mulching and by watering adequately.
2. Avoid high-ammonia fertilizers and large quantities of fresh manure. Water regularly during dry periods.
3. Plant in well-drained soil.
4. If your soil or water is salty, provide more water at each watering to help leach salts through the soil. Avoid using high-ammonia fertilizers and fresh manure.

CUCURBITS *(continued)*

Beet leafhopper

Beet leafhopper damage to zucchini.

Problem: Spotted, pale green, winged insects up to ⅛ inch long hop and fly away quickly when a plant is touched. The leaves are stippled and may be curled, puckered, and brittle.

Analysis: The beet leafhopper (*Circulifer tenellus*) is a western insect found only as far east as Missouri and Illinois. It attacks all members of the cucurbit family and frequently infects them with the virus that causes curly top (see page 267). Plants infected with curly top are stunted and brittle and sometimes die. The beet leafhopper feeds from early May through June. It sucks the sap and virus from infected leaves and then injects the virus into healthy plants at its next feeding stop.

Solution: Treat infested plants with ORTHO Bug-B-Gon Insect Killer or an insecticidal soap. Be sure to cover the lower surfaces of the leaves. Repeat the treatment at intervals of 7 to 10 days if the plants become reinfested.

Cucumber beetles

Striped cucumber beetle (4× life size).

Problem: Holes are chewed in the leaves, leafstalks, and stems by yellow-green beetles with black stripes or spots. Plants may wilt and die.

Analysis: Cucumber beetles, both striped (*Acalymma* species) and spotted (*Diabrotica* species), are common pests of cucumbers, melons, squash, and pumpkins. It is important to control these beetles, because they carry two serious diseases that damage and may kill cucurbits: mosaic (see page 282) and bacterial wilt (see page 281). The adults survive the winter in plant debris and weeds. They emerge in the early spring and feed on a variety of plants. As soon as cucurbits are planted in the garden, the beetles attack the leaves and stems and may totally destroy the plant. They lay their yellow-orange eggs in the soil at the base of the plants. The grubs that hatch from these eggs eat the roots and the stems below the soil line, causing the plant to be stunted or to wilt. The slender white grubs feed for several weeks, pupate in the soil, and emerge as adults to repeat the cycle. One generation occurs per year in northern parts of the United States and two or more in southern areas.

Solution: Treat plants with ORTHO Tomato & Vegetable Insect Killer or ORTHO Bug-B-Gon Insect Killer at the first sign of the beetles. Repeat at weekly intervals as the plants become reinfested. Control early in the season helps prevent susceptible young seedlings and plants from becoming infected with bacterial wilt.

Squash bug

Squash bugs (⅓ life size).

Problem: Squash and pumpkin leaves wilt and may become black and crisp. Bright green to dark gray or brown, flat-backed bugs, about ½ inch long, cluster on the plants.

Analysis: Both the young (nymphs) and adult squash bugs (*Anasa tristis*) attack cucurbits and are most serious on squash and pumpkins. They injure and kill the plants by sucking the sap from leaves and stems. Seedlings are especially susceptible. The dark brown adults, which are sometimes incorrectly called *stinkbugs*, emit a disagreeable odor when crushed. They lay brick red egg clusters on the leaves in the spring. Although only one generation occurs per year, all stages are found throughout the summer.

Solution: Squash bugs are elusive and difficult to control. Handpick the bugs and crush egg masses. Trap adults by laying boards on the ground and destroying bugs that congregate under them during the day. Treat the plants and soil with an insecticide containing *methoxychlor* when the bugs first appear. Repeat the treatment every 7 days until the bugs are controlled. Plant varieties that are resistant to squash bugs.

Squash vine borer

Borer damage.

Problem: Squash vines suddenly wilt. Holes in the stems are filled with a green or tan sawdustlike material, which spills out of the holes. Fat white worms up to 1 inch long are found in the affected vines when the stems are slit open lengthwise with a knife.

Analysis: The squash vine borer (*Melittia satyriniformis*) primarily attacks squash, pumpkins, and gourds and only rarely attacks cucumbers and melons. Hubbard squash is especially susceptible. The larvae damage and kill the plants by tunneling in the stems, preventing the rest of the vine from receiving the water and nutrients it needs. The metallic green adult moth lays eggs on the vines in early summer (April and May in the South and in June and July in the North). When the eggs hatch, the white larvae bore into the stems and feed for 4 to 5 weeks. They then crawl out of the stem and into the soil to pupate.

Solution: Applying insecticides after the borer is inside the stem is not effective. Instead, slit the affected stems with a knife and destroy the borer. If the plant has not died, cover the damaged portion of the stem with soil. Keep the soil moist to encourage new roots to grow. The vine may recover. Destroy plant residues in the fall. The following year, dust the plant with an insecticide containing *methoxychlor*, or spray with ORTHO Malathion 50 Plus Insect Spray during the egg-laying period, at about the time the vines start to run. Row covers will keep adults from laying eggs.

Bacterial wilt

Bacterial wilt. Inset: Bacterial ooze.

Problem: A few leaves wilt and dry and may be chewed. Wilted leaves often recover at night but then wilt again on sunny days and finally die. Fruit shrivels. To test for bacteria, cut a wilted stem near the base of the plant and squeeze out the sap, looking for a milky white substance. Touch a knife to the sap and withdraw it slowly. Look for a white ooze that strings out in a fine thread as you withdraw the knife.

Analysis: Bacterial wilt, caused by a bacterium (*Erwinia tracheiphila*), is more prevalent on cucumbers and muskmelons than on pumpkins and squash. Watermelons are not affected. The bacteria spend the winter in striped or spotted cucumber beetles and are spread to plants when the beetles feed. (For more information on cucumber beetles, see page 280.) An entire plant may become infected within 15 days. The disease is most prevalent in cool weather in areas with moderate rainfall.

Solution: No chemicals control bacterial wilt. Remove and discard all infected plants promptly. Control cucumber beetles with ORTHO Diazinon Ultra Insect Spray. Repeat the treatments every 7 days if the plants become reinfected. Grow varieties resistant to this disease.

Anthracnose

Anthracnose on melon.

Problem: Yellow, water-soaked areas spot melon and cucumber leaves, enlarge rapidly, and turn brown and dry. These spots then shatter, leaving a ragged hole in the spot. On watermelon leaves, the spots turn black. Elongated dark spots with light centers may appear on the stems. Whole leaves and vines die. Large fruit is spotted with sunken, dark brown, circular spots. Pinkish ooze may emerge from the spots. Young fruit darkens, shrivels, and dies.

Analysis: Anthracnose is caused by a fungus (*Colletotrichum lagenarium*) and is the most destructive disease of melons and cucumbers in the East. It rarely attacks squash and pumpkins. The disease affects all aboveground parts of the plant and is most prevalent in humid weather with frequent rain and a temperature range of 70° to 80°F. The spores overwinter in seeds and plant debris not removed from the garden. They are spread by splashing water, cucumber beetles, and tools.

Solution: Treat plants with ORTHO Multi-Purpose Fungicide Daconil 2787® Plant Disease Control at the first appearance of the disease. Repeat every 7 days or more frequently if warm, humid weather occurs. Grow varieties resistant to this disease (see page 363). Avoid overhead watering. Clean up crop residue or turn it under at the end of the season.

CUCURBITS (continued)

Mosaic virus

Mosaic on squash. Inset: Mosaic on cucumbers.

Problem: Leaves are  mottled yellow and green and are distorted, stunted, and curled. Cucumber fruits are mottled with dark green and pale green to white blotches and covered with warts. Sometimes the skin is smooth and completely white. Summer squash fruit may also be covered with warts.

Analysis: Mosaic virus is caused by several viruses that attack cucumbers, muskmelons, and summer squash. The viruses overwinter in perennial plants and in weeds, including catnip, pokeweed, wild cucumber, motherwort, and milkweed. The viruses are spread from plant to plant by aphids and cucumber beetles and can infect plants at any time from the seedling stage to maturity. Infection early in the season is more damaging. Affected fruit tastes bitter. Fruit that is more than half grown at the time of infection is immune to attack.

Solution: No chemical controls virus diseases. Remove and destroy all infected plants immediately. Control aphids and cucumber beetles with an insecticide containing *diazinon*. Repeat at intervals of 7 to 10 days if the plants become reinfested. Remove weeds in and near the garden. Grow resistant cucumber varieties. No summer squash or muskmelon varieties are resistant.

Powdery mildew

Powdery mildew on cucumber.

Problem: A white powdery growth covers the upper surfaces of the leaves. Areas of the leaves and stems turn brown, wither, and die. Fruit may be covered with a white powdery growth.

Analysis: Powdery mildew is caused by either of two fungi (*Erysiphe cichoracearum* or *Sphaerotheca fuliginea*) that thrive in both humid and dry weather. The powdery patches consist of fungal strands and spores. The spores are spread by the wind to healthy plants. The fungus saps plant nutrients, causing yellowing and sometimes death of the leaf. A severe infection may kill the plant. Since these powdery mildews attack several different kinds of plants, the fungus from a diseased plant may infect other types of plants in the garden. (For a list of powdery mildews and the plants they attack, see page 343.) Under favorable conditions, powdery mildew can spread rapidly through a closely spaced planting.

Solution: Treat the plants with ORTHO Multi-Purpose Fungicide Daconil 2787® Plant Disease Control at the first sign of the disease. Continue treatment at intervals of 7 days as long as the disease is a problem. Grow varieties resistant to powdery mildew.

EGGPLANT

Flea beetles

Flea beetle (2× life size).

Problem: Leaves are riddled with shot holes about ⅛ inch in diameter. Tiny (1/16-inch) black beetles jump like fleas when disturbed. Leaves of seedlings and eventually whole plants may wilt and die.

Analysis: Flea beetles jump like fleas but are not related to fleas. Both adult and immature flea beetles feed on a wide variety of garden vegetables, including eggplants. The immature beetle, a legless gray grub, injures plants by feeding on the roots and the lower surfaces of leaves. Adults chew holes in leaves. Flea beetles damage seedlings and young plants most. Leaves of seedlings riddled with holes dry out quickly and die. Adult beetles survive the winter in soil and garden debris. They emerge in early spring to feed on weeds until vegetable seeds sprout or plants are set in the garden. Grubs hatch from eggs laid in the soil and feed for 2 to 3 weeks. After pupating in the soil, they emerge as adults to repeat the cycle. One to four generations occur per year. Adults may feed for up to 2 months.

Solution: Control flea beetles on eggplants with ORTHO Tomato & Vegetable Insect Killer or an insecticidal soap when the leaves first show damage. Watch new growth for evidence of further damage, and repeat the treatment at weekly intervals as needed. Clean all plant debris from the garden after harvest to eliminate overwintering spots for adult beetles.

Tomato hornworm and tobacco hornworm

Tomato hornworm (½ life size).

Problem: Fat green or brown worms, up to 5 inches long with white diagonal side stripes, chew on the leaves. A red or black "horn" projects from the rear end. Black droppings soil the leaves.

Analysis: Tomato hornworms (*Manduca quinquemaculata*) and tobacco hornworms (*M. sexta*) feed on the fruit and foliage of eggplants, peppers, and tomatoes. Although only a few worms may be present, each worm consumes large quantities of foliage and causes extensive damage. The large gray or brown adult moth with yellow and white markings emerges from hibernation in late spring and drinks nectar from petunias and other garden flowers. The worms hatch from eggs laid on the undersides of leaves and feed for 3 to 4 weeks. Then they crawl into the soil and pupate, later emerging as adults to repeat the cycle. One generation occurs per year in the North and two to four in the South.

Solution: Handpicking works well with hornworms. If handpicking is not practical, treat the plants with ORTHO Tomato & Vegetable Insect Killer or the bacterial insecticide *Bacillus thuringiensis*.

GROWING GUIDE

Adaptation: Throughout North America. For regional recommendations, see page 364.

Planting Time: Spring.

Planting Method: Purchase 1-year-old rooted vines, either bare-root or in containers. Plant them 6 to 10 feet apart and with the top of the rootball well below the level of the surrounding soil. Cut the tops back, leaving only 2 or 3 buds.

Soil: Any good garden soil that is high in organic matter. pH 6.0 to 8.0.

Fertilizer: At planting time, use ½ pound of Scotts Vegetable Food per 100 square feet, or ½ tablespoon per plant. For the first year, side-dress with the same amount in midspring and midsummer. In subsequent years, fertilize each spring before new growth begins with ½ pound of fertilizer per plant.

Water: How much: Apply enough water at each irrigation to wet the soil 1 to 1½ to 3 feet deep.
How often: Water when the soil 6 to 12 inches deep is almost dry.

Harvest: Table grapes should taste sweet and be plump with even color. Wine grapes may be slightly tart, but plump and slightly soft with even color.

Black rot

Black rot causes fruit to turn black and shrivel.

Problem: Light brown spots surrounded by a dark brown line appear on the grapes. The grapes turn black, shrivel, and dry up like raisins. They remain attached to the stems. Reddish brown circular spots appear on the leaves. Leaves may wilt. Sunken purple to black elongated lesions spot the canes, leaf stems, and tendrils.

Analysis: Black rot is caused by a fungus (*Guignardia bidwellii*). It is the most destructive disease that attacks grapes, often destroying all the fruit. The fungus spends the winter in infected dormant canes, tendrils on support wires, and mummified (dried on the vine) fruit. In warm, moist spring weather, spores infect new shoots, leaves, tendrils, and eventually the fruit. Fruit is affected in all stages of development but most severely when it is one-half to two-thirds grown. Spores for future infections are produced on infected leaves, canes, and fruit. The severity of the disease depends on the amount of diseased material that survives the winter, and on spring and early summer weather.

Solution: Destroy all infected fruit and prune out infected canes and tendrils. Once the fruit has begun to shrivel, fungicide sprays are ineffective. The following year, spray plants with ORTHO Home Orchard Spray or a fungicide containing *captan*. Spray early in the growing season to keep spread of the disease to a minimum. Treat when new shoots are 6 to 10 inches long. Spray just before and immediately after bloom, and continue at intervals of 10 to 14 days until grapes are full-size. Treat more frequently if leaf symptoms develop or if the weather is wet.

GRAPES (continued)

Grape berry moth

Damage caused by grape berry moth larvae.

Problem: Grapes are webbed together and to leaves. Dark green to purple worms, up to ⅜ inch long with dark brown heads, are found inside the grapes. White cocoons cling to the leaves between flaps of leaf tissue.

Analysis: The grape berry moth (*Endopiza viteana*) is the most serious insect pest of grapes in the East. The worms damage both green and ripening fruit by feeding on the inner pulp and seeds. They web the grapes together and to leaves with silken threads as they feed. The worms spend the winter as pupae on leaves and on the ground. The moths emerge in late spring and lay eggs on blossom stems and small fruit. The larvae that hatch from these eggs feed on the buds, blossoms, and fruit. After 3 to 4 weeks of feeding, they cut a small bit of leaf, fold it over, and make a cocoon inside, where they pupate. Within a few weeks, adult moths emerge to repeat the cycle, this time laying eggs on ripening fruit. This second generation feeds for 3 to 4 weeks, then pupates for the winter.

Solution: Destroy infested grapes. Discontinue cultivation in late summer so that the cocoons remain on top of the ground to be cleaned up with all fallen grape leaves at the end of the season. This will reduce the number of overwintering pupae. The following year, treat the plants immediately after bloom with ORTHO Home Orchard Spray. Repeat the application 7 to 10 days later. To control the second generation, spray again in midsummer. For more precise spray timing, use pheromone traps. Spray when the first moth is captured. Hang traps again in July to monitor the second generation.

Grapeleaf skeletonizers

Grapeleaf skeletonizers (life size).

Problem: Yellow caterpillars with purple or black stripes feed in rows on the leaves. The caterpillars may be covered with black spines. They chew on the upper and lower surfaces of leaves, eventually eating everything but the leaf veins.

Analysis: Grapeleaf skeletonizers (*Harrisina* species) frequently attack grapes in home gardens and abandoned vineyards. The young caterpillars characteristically feed side by side in a row on the leaves. They feed heartily and may defoliate a vine in several days. The loss of leaf tissue slows the growth of the vine and fruit and reduces production. These pests survive the winter as pupae in cocoons on leaves and in debris on the ground. In late spring the metallic green or smoky black adult moths emerge and lay their eggs on the lower surfaces of leaves. The yellow caterpillars that hatch from those eggs feed on the leaves, usually chewing on the upper or lower surfaces; but sometimes, as they mature, they eat all the tissue between the veins. Two to three generations occur per year, so damage continues from mid-May to August.

Solution: Treat infested plants with an insecticide containing *carbaryl* or with the bacterial insecticide *Bacillus thuringiensis* (Bt) as soon as the caterpillars appear. Spray both the upper and lower surfaces of leaves. Repeat the treatment if the plants become reinfested.

Grape leafhoppers

Grape leafhopper (15× life size).

Problem: Areas on the leaves are stippled and turn pale yellow, white, and then brown. Leaves may cup or roll under at the margins. Some of the leaves may fall. Black spots may be evident on berries. On the undersides of leaves are found pale yellow or white ⅛-inch flying or jumping insects with red or yellow marks.

Analysis: Both the young and adult grape leafhoppers (*Erythroneura* species) suck the juices from grape leaves, causing white spots that later turn brown. This damage reduces normal vine growth, resulting in delayed maturity of fruit and poor vine growth the following year. Their black droppings may also mar the fruit, making it unappetizing. Leafhoppers survive the winter as adults in protected places. When new growth begins in the spring, these adults emerge and begin feeding. The adults lay their eggs in the leaves, causing blisterlike swellings. Two or three overlapping generations occur each season, so leafhoppers of all stages of maturity can be found feeding from the time of new growth in the spring until the leaves drop in the fall.

Solution: Treat infested grape vines with ORTHO Bug-B-Gon Insect Killer, ORTHO Home Orchard Spray, or ORTHO Malathion 50 Plus Insect Spray as soon as leafhoppers are noticed. For more precise spray timing, set out yellow sticky traps and spray when catches exceed five per day. Spray both the upper and lower surfaces of leaves. Repeat the treatment at intervals of 10 to 14 days if the plants become reinfested. Clean up and destroy plant debris after harvest to reduce the number of overwintering leafhoppers.

Eutypa dieback

Eutypa dieback.

Problem: Dark, irregular spots develop on the young leaves. These spots may drop out, leaving holes in the leaves. Elongated, sunken spots develop on the current season's canes. Shoot growth may be weak and stunted, and the leaves are small, yellowish, and cupped, with crinkled margins. Later in the season, the leaves may become scorched and tattered. Sunken lesions (cankers) may develop on the woody canes and trunk. Entire branches may die back.

Analysis: Eutypa dieback is caused by a fungus (*Eutypa armeniacae*). The fungus survives the winter in trunk and branch cankers. Fungal spores that form in the cankers are carried by splashing water to pruning wounds, where they infect the plant. Infection causes the formation of cankers that reduce the flow of water and nutrients through the trunk and branches. The portion of the plant above the canker weakens and may eventually die.

Solution: Prune out and destroy infected branches and canes. Late-winter pruning reduces the probability of infection. Make the pruning cut at least 6 inches below the canker and any discolored wood. If cankers are present on the trunk, remove and destroy nearly the entire aboveground plant, cutting the trunk below the lowest canker but above the bud union. Maintain two to four suckers on the trunk. The plant will not produce grapes the current year but will yield a normal crop the following year.

Anthracnose

Anthracnose is caused by a fungus.

Problem: Circular, sunken spots with light gray centers and dark borders appear on shoots, fruit, tendrils, and leafstalks. The fruit remains firm. The leaves may curl downward, and the brown areas drop out.

Analysis: Anthracnose is caused by a fungus (*Elsinoe ampelina*) that may do considerable damage a few years in a row and then disappear. It is often called *birds-eye rot*. It is seldom severe on Concord or muscadine grape vines. The disease first attacks the new growth. The spots on the stems often merge, girdling the stem and killing the vine tips. Anthracnose is prevalent during wet periods in the spring and in poorly maintained vineyards. The fungus survives the winter on old lesions on the canes. Although this disease doesn't kill the grape vines, the infected fruit is often misshapen and unappetizing. Several years of attack from anthracnose sufficiently weaken the vines to make them more susceptible to other problems.

Solution: Discard infected fruit, and prune out diseased canes. Sprays applied after spotting occurs are ineffective. The following spring, before the buds open, spray the vines with a fungicide containing lime sulfur. Treat the plants with a fungicide containing *ferbam* when the shoots are 1 to 2 inches long, when they are 6 to 10 inches long, just before bloom, just after the blossoms fall, and 2 more times at intervals of 2 weeks. Spray all canes and leaves thoroughly.

Downy mildew

Downy mildew reduces vine vigor.

Problem: Small yellow spots appear on the upper surfaces of leaves. The lower surfaces are covered with a white, cottony growth. Older leaves are affected first. Leaves, shoots, and tendrils turn brown and brittle and are often distorted. Grapes may be covered with the white growth, or they may be shriveled and brown, yellow, or red.

Analysis: Downy mildew is caused by a fungus (*Plasmopara viticola*) that attacks grape foliage and fruit from before bloom to the end of harvest. Downy mildew causes leaf defoliation and prevents proper ripening. When the disease is severe, entire grape clusters may be killed. Reduced vine vigor results in poor growth the following season. This disease is more prevalent in cool, moist weather and is always more serious in rainy growing seasons. The fungus survives the winter in diseased leaves on the ground.

Solution: Destroy severely infected leaves and fruit. At the first sign of the disease, spray the vines with a fungicide containing *captan*. Repeat every 10 days until 7 days before harvest. Remove and destroy plant debris at the end of the season to reduce the number of overwintering spores. The following year, spray immediately before bloom, just after the petals fall, when the grapes are about the size of peas, and every 10 days thereafter until 7 days before harvest.

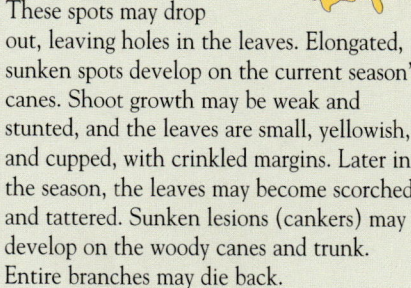

LETTUCE

Bolting

Bibb lettuce bolting.

Problem: A seed stalk emerges from the center of the lettuce plant. The lettuce tastes bitter.

Analysis: Lettuce is a cool-weather crop and grows best between 55° and 60°F. When temperatures rise above 60°F for several days in a row, the plants form a flower stalk if they are mature enough. As the stalk grows, sugars and nutrients are withdrawn from the leaves for the growth of the stalk, making the leaves bitter and tough. The formation of a flower stalk in vegetables grown for their leaves is called *bolting*. Once bolting begins, it cannot be stopped; cutting off the stalk does not help.

Solution: If harvested as soon as bolting begins, the lettuce may still be edible, but quality deteriorates rapidly as the stalk forms. In the future, plant lettuce so that it matures during cool weather, or grow varieties that are slow to bolt.

Cabbage looper

Cabbage looper damage and droppings.

Problem: Leaves have round or irregular holes. Green worms up to 1½ inches long, with light stripes down their backs, feed on the leaves or heads. Masses of green or brown pellets may be found between the leaves.

Analysis: Several worms attack lettuce; the most damaging is the cabbage looper (*Trichoplusia ni*). The looper attacks all varieties of lettuce, as well as members of the cabbage family. Adults lay eggs throughout the growing season. The brownish cabbage looper moth lays pale green eggs on the upper sides of leaves in the evening. The worms eat lettuce leaves and heads. Their greenish brown excrement makes the plants unappetizing. Worms may be present from early spring until late fall. In the South, they may be present year-round. Worms spend the winter as pupae attached to a plant or nearby object.

Solution: Control cabbage loopers with ORTHO Tomato & Vegetable Insect Killer, ORTHO Bug-B-Gon Insect Killer, or the bacterial insecticide *Bacillus thuringiensis* (Bt). Bt is effective only while the caterpillars are small. Repeat treatments at weekly intervals if the plants become reinfested. Use row covers to keep adults from laying eggs. Clean all plant debris from the garden to reduce the number of overwintering pupae.

Cutworms

Cutworm (⅓ life size).

Problem: Young plants are chewed or cut off near the ground. Gray, brown, or black worms, 1½ to 2 inches long, may be found about 2 inches deep in the soil near the base of the damaged plants. The worms coil when disturbed.

Analysis: Several species of cutworms attack plants in the vegetable garden. The most likely pests of lettuce plants in the spring are surface-feeding cutworms. A single surface-feeding cutworm can sever the stems of many young plants in one night. Cutworms hide in the soil during the day and feed only at night. Adult cutworms are dark, night-flying moths with bands or stripes on their forewings. In southern parts of the United States, cutworms may also attack fall-planted lettuce. For more information on cutworms, see page 252.

Solution: Apply ORTHO Diazinon Soil & Turf Insect Control; ORTHO Dursban® Lawn & Garden Insect Control; or ORTHO Bug-Geta Plus Snail, Slug & Insect Killer around the base of undamaged plants when stem cutting is observed. Because cutworms are difficult to control, applications may need to be repeated at weekly intervals. Before planting in the area, apply a preventive treatment of ORTHO Dursban® Lawn & Garden Insect Control, the bacterial insecticide *Bacillus thuringiensis* (Bt), or ORTHO Diazinon Soil & Turf Insect Control, and work it into the soil. Cultivate the soil thoroughly in late summer and fall to expose and destroy eggs, larvae, and pupae.

ONION

GROWING GUIDE

Adaptation: Throughout North America.

Planting Time: See page 361.

Planting Method: Onions: Sow onion seeds ½ inch deep and thin to stand ¼ to 3 inches apart. Plant sets 1 to 2 inches deep, side by side or up to 3 inches apart. Set transplants 2 to 3 inches deep and 3 to 5 inches apart. When growing for green onions, plant bulbs closer than when growing for dry onions. For a continuous harvest, plant every couple of weeks.
Garlic: Plant individual cloves 1 to 2 inches deep and 5 to 6 inches apart.
Shallots: Plant individual sections 1 to 1½ inches apart and 3 to 5 inches apart.
Chives: Sow seeds ½ inch deep. Thin seedlings or set transplants to grow 8 inches apart. Plant transplants so that the top of their rootball is even with the surrounding garden soil.
Leeks: Plant as for onion transplants or seeds.

Soil: Loose, crumbly soil that is rich in organic matter. pH 6.0 to 7.0.

Fertilizer: At planting time, use 1 pound of Scotts Vegetable Food per 100 square feet, or ½ pound per 25-foot row. Side-dress every 3 to 4 weeks with ½ pound per 25-foot row.

Water: How much: Apply enough water at each watering to wet the soil 8 to 10 inches deep.
How often: Water when the soil 2 inches deep is barely moist.

Harvest: Green onions: Pull the largest as needed, when about pencil size.
Dry onions: Harvest when half or more of the tops have turned yellow and fallen over. Dry in a warm, airy spot for 4 to 5 days. Shake off loose soil and skins. Place bulbs in a slatted crate or mesh bag, and continue drying in a well-ventilated, dry area for 2 to 4 weeks. After drying, store in a dry area between 35° and 40°F.
Garlic: Harvest when the tops dry; cure as for dry onions.
Shallots: Treat like dry onions.
Chives: Snip with sharp shears any time there are young, fresh leaves. Use fresh, or dry or freeze for later use.
Leeks: For white stems, gradually hill soil around the plants through the growing season. Pick when the stems are ¾ to 1 inch in diameter. Store in a root cellar or in the refrigerator until use.

Onion maggot

Onion maggot damage.

Problem: Plants grow slowly, turn yellow, wilt, and die. Bulbs may rot in storage. White legless maggots, up to ⅓ inch long, burrow inside the bulb.

Analysis: The onion maggot (*Hylemya antiqua*), a fly larva, is the most serious pest of onions. The larvae burrow into the onion bulb, causing the plant to wilt and die. Once the bulb is damaged by the maggot's feeding, it is susceptible to attack by bacterial soft rot. Early plantings are the most severely injured. When the onions are young and growing close together, the maggots move easily from one bulb to another, destroying several plants. Cool, wet weather favors serious infestations. Maggots spend the winter as pupae in plant debris or in the soil. The brownish gray adult fly emerges in the spring to lay clusters of white eggs at the base of plants. The maggots that hatch from these eggs burrow into the soil and bulbs. After feeding, they pupate in the soil and later emerge as adults to repeat the cycle. Two or three generations occur per year, the last one attacking onions shortly before they are harvested. When maggot-infested bulbs are placed in storage, the maggots continue to feed and damage the bulbs.

Solution: Discard and destroy maggot-infested onions. Clean all debris from the garden at the end of the season to reduce the number of overwintering pupae. At planting time, treat the soil in the row with ORTHO Diazinon Soil & Turf Insect Control or with an insecticide containing *chlorpyrifos*. Repeat the treatment 7 to 10 days after plants emerge through the soil. Adult flies can be kept from laying eggs with row covers.

ONION (continued)

Fusarium basal rot

Fusarium basal rot, a fungal disease.

Problem: Leaf tips wilt and die back. The neck of the bulb is soft. A white fungal growth may appear on the outside of the base of the bulb. The bulb is soft and brown inside. Bulbs may also be affected in storage.

Analysis: Fusarium basal rot is caused by a soil-inhabiting fungus (*Fusarium* species) that attacks onions, shallots, garlic, leeks, and chives. It persists indefinitely in the soil. The disease is spread by contaminated bulbs, soil, and equipment. The fungus enters the bulb through wounds from maggots and old root scars and spreads up into the leaves, resulting in leaf yellowing and dieback. Bulbs approaching maturity are the most susceptible to attack. If infection occurs during or after harvest, the rot may not show until the bulbs are in storage. Fusarium basal rot is most serious when bulbs are stored in a moist area at temperatures above 70°F.

Solution: No chemical controls are available. Destroy infected plants and bulbs. Harvest healthy bulbs promptly at maturity. Dry according to the instructions on page 287. *Fusarium* can be removed from the soil only by fumigation or solarization techniques. To solarize soil, cultivate the area and cover it with a layer of clear polyethylene film during the hottest time of the year. Leave the film in place for 4 to 6 weeks. Control onion maggots to reduce chances of infection. Store bulbs in a cool (35° to 40°F), dry area. Rotate the planting site if possible.

Pink root

Pink root.

Problem: Plants grow slowly, and their tops may be stunted. Roots turn light pink and shrivel, then a darker pink and die. Leaves may turn yellow or white and die.

Analysis: Pink root is caused by a fungus (*Pyrenochaeta terrestris*) that attacks onions, garlic, shallots, leeks, and chives. It persists indefinitely in the soil and infects plants at all stages of growth. Mature and weakened bulbs are the most susceptible to attack. Plants infected early in their life seldom produce large bulbs. Pink root thrives in warm weather (60° to 85°F).

Solution: No chemical control is available. Discard all infected plants. Pink root fungus can be removed from the soil only by fumigation and solarization techniques. To solarize soil, cultivate the area and cover it with a layer of clear polyethylene film during the hottest time of the year. Leave the film in place for 4 to 6 weeks. Weak plants are the most susceptible to attack, so keep the plants healthy with adequate water and nutrition, as discussed on page 287. Grow varieties that are resistant to pink root.

Onion thrips

Onion thrips damage.

Problem: White streaks or blotches appear on onion leaves. Tips may be distorted. Plants may wilt, wither, turn brown, and die. Bulbs may be distorted and small. Small pale green to white insects are observed at the base of the leaves.

Analysis: Onion thrips (*Thrips tabaci*) attack many vegetables, including onions, peas, and cabbage. Thrips are barely visible insects, less than 1/25 inch long and dark brown to black. They reduce the quality and yield of onion bulbs by rasping holes in the leaves and sucking out the plant sap. This rasping causes the white streaks. Plants often die when thrips populations are high. Damage is most severe in the leaf sheath at the base of the plant. Thrips favor this protected area where the elements and pesticides have difficulty reaching them. Onion thrips survive the winter in grass stems, plant debris, and bulbs in storage. Thrips are active throughout the growing season; in warm climates they are active all year.

Solution: Treat infested onion plants with ORTHO Bug-B-Gon Insect Killer or ORTHO Diazinon Ultra Insect Spray at the first sign of thrips damage. Repeat at weekly intervals until the new growth is no longer damaged.

PEAS

Powdery mildew

Powdery mildew can spread rapidly.

Problem: A white powdery coating develops first on the upper surfaces of the lower leaves. Stems, pods, and other leaves may then become infected. Leaves may turn yellow and be malformed. Pods may be distorted, with dark streaks or spots.

Analysis: Powdery mildew is caused by a fungus (*Erysiphe polygoni*) that thrives in both humid and dry weather. The powdery coating consists of fungal strands and spores. The spores are spread by the wind to healthy plants. The fungus saps plant nutrients, causing yellowing and sometimes death of the leaf. A severe infection reduces pea yield considerably and may kill the plant. Fall crops are most susceptible to serious damage; spring crops are attacked late in the season. Since this powdery mildew attacks many vegetables, the fungus from a diseased plant may infect other plants in the garden. Under favorable conditions, powdery mildew can spread rapidly through a planting.

Solution: Treat plants with garden sulfur. Repeat at intervals of 7 to 10 days as needed. This fungicide does not cure infected leaves but does protect healthy leaves from infection.

Pea aphid

Pea aphids (2× life size).

Problem: Plants turn yellow and wilt. Pods may not be completely filled with peas. Clustered on the leaves—especially the young, tender ones—are tiny, soft-bodied, light to deep green insects with red eyes. Leaves may be covered with a sticky material or black mold.

Analysis: Several species of aphids attack peas, but the most common and damaging is the pea aphid (*Acyrthosiphon pisum*). Aphids generally do little damage in small numbers. They are extremely prolific, however, and populations can build up to damaging numbers during the growing season. Damage occurs when the aphid sucks the juices from the pea leaves, stems, blossoms, and pods. This feeding results in stunted plants and fewer and smaller pods that may be only partially filled with peas. The aphid is unable to digest fully all the sugar in the plant sap, and it excretes the excess in a fluid called *honeydew*. The honeydew often drops onto the leaves below. Ants feed on this sticky substance and are often present where there is an aphid infestation. Some aphids also transmit viruses, which can be quite damaging. Infected pea plants and pods are distorted, and plants sometimes die.

Solution: Spray infested plants with an insecticidal soap, ORTHO Diazinon Ultra Insect Spray, or ORTHO Malathion 50 Plus Insect Spray as soon as the aphids appear. Repeat the spray if the plants become reinfested. Use yellow sticky traps to monitor aphid arrival.

Pea weevil

Pea weevil (5× life size).

Problem: Peas have small round holes. The peas are partially or completely hollow, and fat white legless grubs with brown heads may be found inside. When the pea plants are blooming, ⅓-inch-long dark brown beetles with light markings may be seen crawling or flying about the plants.

Analysis: The pea weevil (*Bruchus pisorum*) attacks all varieties of edible peas. Weevils emerge from hibernation as peas are beginning to bloom. The adults feed on pea nectar and pollen. This feeding does not harm the plant. Adults can migrate up to 3 miles in search of food. They lay orange to white eggs on the developing pods. The white grubs that hatch from these eggs eat through the pod and into the pea. They continue feeding for 6 to 8 weeks, then pupate inside the hollow pea. The adult weevil emerges in 1 to 3 weeks and the following year hibernates to repeat the cycle. Only one generation occurs per year. Infested peas are inedible.

Solution: Insecticides must be applied to kill the adults before they lay eggs. Once the eggs are laid on the pods, it is too late to prevent injury. Treat the plants with an insecticide containing *diazinon* or *rotenone* soon after the first blooms appear and before pods start to form. Additional sprays may be needed to control migrating weevils. Destroy infested peas.

PEPPERS

GROWING GUIDE

Adaptation: Throughout North America.

Planting Time: See page 361.

Planting Method: Sow seeds ¼ inch deep. Thin seedlings or set transplants to grow 18 to 24 inches apart.

Soil: Any good garden soil that is rich in organic matter. pH 5.5 to 7.5.

Fertilizer: At planting time, use 3 pounds of Scotts Vegetable Food per 100 square feet, or ½ pound per 25-foot row. Side-dress every 3 to 4 weeks with ½ pound per 25-foot row, or 1 tablespoon per plant.

Water: How much: Apply enough water at each irrigation to wet the soil 10 to 12 inches deep.
Containers: 10 percent of the applied water should drain out the bottom.
How often: Water when the soil 2 inches deep is barely moist.

Harvest: Green or yellow peppers: Pick when of usable size with a rich color. Fruit should be slightly soft.
Red peppers: Red peppers are green peppers that have remained on the plants and matured. Pick when they have a rich red color. Cut peppers from the plant with pruning shears or a sharp knife, leaving ½ to 1 inch of the stem attached. Cool in the refrigerator.

Blossom drop

Blossom drop delays fruit production.

Problem: Little or no fruit develops. Fruit that does develop may have rough skin or be misshapen. Plants remain vigorous, with lush foliage.

Analysis: Pepper blossoms are sensitive to temperature fluctuations during pollination. Normal pollination and fruit set don't occur when night temperatures fall below 58°F and daytime temperatures rise above 85°F. Under these temperature conditions, the blossoms fall off, often before pollination. If pollination has occurred and the fruit has begun to set but isn't completely fertilized at the time the blossoms drop, rough and misshapen fruit results.

Solution: Blossom drop causes only a delay in fruit production. When the temperatures are less extreme, a full crop of fruit will set, and the plants will be productive the rest of the season. Discard rough or misshapen fruit; it will never develop fully. Irrigation for cooling during hot periods can help to reduce losses.

Sunscald

Control leaf diseases to reduce sunscald.

Problem: An area on the pepper fruit becomes soft, wrinkled, and light in color. Later, this area dries and becomes slightly sunken, with a white, paperlike appearance. An entire side of the fruit may be affected. Black mold may grow in the affected areas.

Analysis: Pepper fruit exposed directly to sunlight may be burned by the heat of the sun. The fruit may be exposed to the sun as a result of leaf diseases that cause leaf drop. Early fruit on small plants without enough protective foliage may be burned. Also, some varieties do not produce enough foliage to shade the fruit. Rot organisms sometimes enter the fruit through the damaged area, making the fruit unappetizing or inedible. Sunscalded fruit without these molds is still edible if the discolored tissue is removed.

Solution: Control leaf diseases that may defoliate the plants. Fertilize according to the guidelines at left to keep plants healthy with lush foliage. Select pepper varieties that form a protective canopy of leaves.

Blossom-end rot

Blossom-end rot.

Problem: A round, sunken, water-soaked spot develops on the bottom of the fruit. The spot enlarges, turns brown to black, and feels leathery. Mold may grow on the rotted surface.

Analysis: Blossom-end rot occurs on peppers, tomatoes, squash, and watermelons from a lack of calcium in the developing fruit. This lack results from slowed growth and damaged roots caused by the following factors:
1. Extreme fluctuations in soil moisture, either very wet or very dry.
2. Rapid plant growth early in the season, followed by extended dry periods.
3. Excessive rains that smother root hairs.
4. Excess soil salts.
5. Cultivation too close to the plant.
The first fruits are the most severely affected. The disorder always starts at the blossom end (the end farthest from the stem), and it may enlarge to affect up to half the fruit. Moldy growths on the rotted area are from fungi or bacteria that frequently invade the damaged tissue. The unrotted part of the fruit is edible.

Solution: Blossom-end rot is difficult to eliminate, but it can be controlled by following these guidelines.
1. Water regularly.
2. Avoid overuse of high-nitrogen fertilizers and large quantities of fresh manure.
3. Plant in well-drained soil.
4. If your soil or water is salty, provide more water at each watering to help leach salts through the soil.
5. Do not cultivate deeper than 1 inch within 1 foot of the plant.

Corn earworm

Corn earworm larva (life size).

Problem: Holes are chewed in the fruit. Inside the fruit are striped yellow, green, or brown worms from ½ to 2 inches long. These worms may also be feeding on the leaves.

Analysis: The corn earworm (*Helicoverpa zea,* formerly *Heliothis zea,* also known as the *tomato fruitworm* or *cotton bollworm*) attacks many vegetables and flowers. The worms feed on the foliage and also chew holes in the fruit, making it worthless. The worm is the larva of a light gray-brown moth with dark lines on its wings. In the spring, the moth lays yellow eggs on the leaves and stems. The worms that hatch from these eggs feed on the new leaves. When they are about ½ inch long, the worms move to the fruit and bore inside. After feeding for 2 to 4 weeks, they drop to the ground and pupate 2 to 6 inches deep in the soil. After several weeks they emerge as adults to repeat the cycle. Several generations occur per year. In the South, where earworms survive the winter, early and late plantings suffer the most damage. Adult moths migrate into northern areas, where they do not survive the winter.

Solution: Once the worms are inside the fruit, sprays are ineffective. Pick and destroy infested fruit. Clean all plant debris from the garden after harvest to reduce the number of overwintering adults. The following year, treat the plants with an insecticide containing *carbaryl* when the worms are feeding on the foliage or when fruit is about 1 to 2 inches in diameter. Repeat treatment at intervals of 10 to 14 days if the plants become reinfested. Use row covers to keep adults from laying eggs.

European corn borer

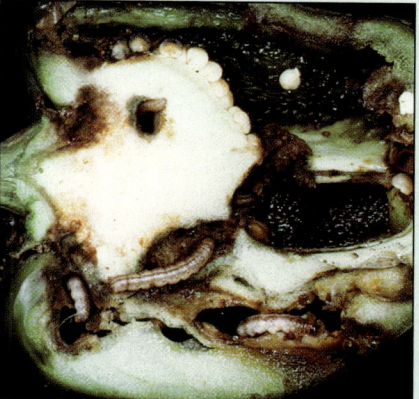

European corn borers (½ life size).

Problem: Pink or tan worms up to 1 inch long with dark brown heads and two rows of brown dots feed in the seed cavity of the pepper. The fruit is decayed inside.

Analysis: The European corn borer (*Ostrinia nubilalis*), a moth larva, is a destructive pest of peppers, corn, tomatoes, beans, and eggplant. The worms feed and promote decay inside the fruit. The borers spend the winter in plant debris, pupate in the spring, and emerge as adults in early summer. The adults are tan moths with dark wavy lines on their wings. The moths lay clusters of fifteen to thirty white eggs on the undersides of leaves. Overlapped eggs resemble fish scales. In midsummer, the eggs hatch, and young borers enter the fruit at the cap. After feeding for 1 month inside the fruit, they pupate; later they emerge as adults to repeat the cycle. Two to three generations occur per year. Cool, rainy weather in early summer inhibits egg laying and washes the hatching larvae from the plants, reducing borer populations. Very dry summers and cold winters also reduce borer populations. Peppers are edible if the damaged part is cut away.

Solution: Remove infested fruit. Treat the plants with an insecticide containing *carbaryl* when the fruit is 1 to 1½ inches in diameter. Repeat the treatment two more times, 7 days apart. For more precise timing, monitor adult moths with pheromone traps. Spray when numbers trapped exceed 25 moths in 5 days. Keep adults from laying eggs by using row covers. Clean all debris from the garden after harvest to reduce overwintering sites for larvae.

PEPPERS *(continued)*

Tomato hornworm or tobacco hornworm

Tomato hornworm (⅓ life size).

Problem: Fat green or brown worms, up to 5 inches long with white diagonal side stripes, chew on the leaves. A red or black "horn" projects from the rear end. Black droppings are found on the leaves and the soil surface beneath the damaged foliage.

Analysis: Tomato hornworms (*Manduca quinquemaculata*) and tobacco hornworms (*M. sexta*) feed on the fruit and foliage of peppers, eggplant, and tomatoes. Although only a few worms may be present, each worm consumes large quantities of foliage and causes extensive damage. The large gray or brown adult moth with yellow-and-white markings emerges from hibernation in late spring and drinks nectar from petunias and other garden flowers. The worms hatch from eggs laid on the undersides of leaves and feed for 3 to 4 weeks. Then they crawl into the soil and pupate; later they emerge as adults to repeat the cycle. One generation occurs per year in northern parts of the United States, two to four in southern areas.

Solution: Hornworms can be handpicked effectively. If that is not practical, treat the plants with ORTHO Tomato & Vegetable Insect Killer or the bacterial insecticide *Bacillus thuringiensis* (Bt).

POTATOES

GROWING GUIDE

Adaptation: Throughout North America.

Planting Time: See page 361.

Planting Method: Plant 1½- to 2-ounce seed pieces (about the size of a medium egg) in trenches 3 to 4 inches deep. Allow 12 to 18 inches between pieces. Plant pieces whole or cut, with at least one eye on each piece. Purchase new seed pieces each year.

Soil: Any good garden soil that is rich in organic matter. pH 4.8 to 6.8.

Fertilizer: At planting time, use 1 pound of Scotts Vegetable Food per 100 square feet, or ¼ pound per 25-foot row. Side-dress when the plants are 4 to 6 inches tall.

Water: How much: Apply enough water at each irrigation to wet the soil 8 to 12 inches deep.
How often: Water when the soil 2 inches deep is barely moist.

Harvest: Dig potatoes as needed about 2 to 3 weeks after flowering. Dig those to be stored after the tops naturally turn yellow and die. Carefully dig with a spade or pitchfork, being careful not to wound the tubers.

Storage: Cure in the dark for 1 week at 70°F and high humidity. Then store in a humid area at between 40° and 45°F until used.

Common scab

Scab, a bacterial disease.

Problem: Brown corky scabs or pits occur on potato tubers. Spots enlarge and merge together, sometimes covering most of the tuber. Leaves and stems are not affected.

Analysis: Common scab is caused by a bacterium (*Streptomyces scabies*) that persists in the soil for long periods of time. Besides potatoes, scab also infects beets, carrots, and parsnips. Scab affects only the tubers, not the leaves or stems. The bacteria spend the winter in the soil and in infected tubers left in the garden. Infection occurs through wounds and through the breathing pores in the tuber skins when the tubers are young and growing rapidly. Scab is most severe in warm (75° to 85°F), dry soil with a pH of 5.7 to 8.0. The severity of scab often increases when the pH is raised with lime or wood ashes. Scab is not a problem in acid soils with a pH of 5.5 or less. Poorly fertilized soil also encourages scab. The bacteria withstand temperature and moisture extremes. Because they pass intact through the digestive tracts of animals, manure can spread the disease. Tubers infected with scab are edible, but much may be wasted as the blemishes are removed.

Solution: No chemical control is available. Test your soil pH, and if necessary correct it to 5.0 to 5.5 with aluminum sulfate. Keep the soil moist for 1 to 2 months after tuber set. Avoid alkaline materials such as wood ashes and lime. Do not use manure on potatoes. Plant potatoes in the same area only once every 3 to 4 years. Use certified seed pieces that are resistant to scab.

Wireworms

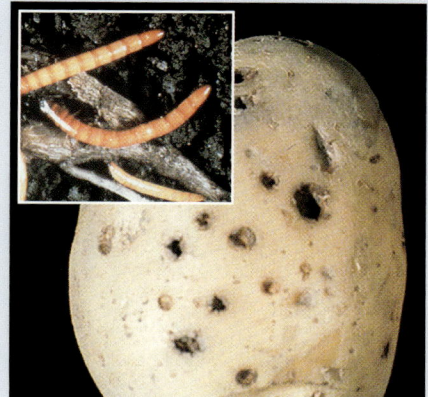

Damaged potato. Inset: Wireworms (life size).

Problem: Plants are stunted and grow slowly. Tunnels wind through stems, through roots, and in and on the surface of tubers. Shiny, hard, jointed, yellow to reddish brown worms up to 2 inches long are found in the tuber and in the soil.

Analysis: Wireworms feed on potatoes, corn, carrots, beets, peas, beans, lettuce, and many other plants. Wireworms feed entirely on underground plant parts, devouring seed potatoes, underground stems, tubers, and roots. Infestations are most extensive in soil where grass or grassy weeds have grown during the preceding 2 years. Adults, about ½ inch long, are brown, gray, or blackish beetles with a body that tapers at both ends. They are known as *click beetles* because they make a clicking sound when turning from their back to their feet. Wireworm larvae feed for 2 to 6 years before maturing into adult beetles, so all sizes and ages of wireworms may be found in the soil at the same time. Undamaged portions of the tubers are edible.

Solution: It is difficult to control wireworms after the crop is planted. Use ORTHO Diazinon Soil & Turf Insect Control, incorporating it into the top 4 to 8 inches of soil. Do not store damaged tubers. The following year, apply the control before planting.

Ring rot

Bacterial ring rot.

Problem: Shoot tips are stunted, forming rosettes. Leaves turn yellow, then brown between the veins. Leaf margins curl upward, and stems may wilt. Stems cut at the ground level exude a creamy white, odorless ooze. Only a few stems on a plant may show symptoms. Tubers may be cracked and, when cut near the stem end, reveal a yellow to light brown ring of crumbly decay.

Analysis: Ring rot is caused by a bacterium (*Clavibacter michiganense* subsp. *sepedonicus*) that attacks both tubers and stems. Infected tubers are inedible. Tuber decay may be evident at harvest, or it may not develop until after several months in storage. Other rot organisms frequently invade and completely rot the tubers. Ring rot bacteria enter plants through wounds, especially those caused by cutting seed pieces before planting. They do not spread from plant to plant in the field. The bacteria survive between seasons in infected tubers and storage containers.

Solution: Discard all infected tubers and plants at the first sign of the disease. Do not plant potatoes in the location where ring rot occurred for the following 2 years. Use only certified potato seed pieces for plants. Plant whole, or if you cut seed pieces, disinfect the knife between cuts. Disinfect storage containers with chlorine bleach. Wash storage bags in hot, soapy water.

Colorado potato beetle

Adult Colorado potato beetle (½ life size). Inset: Larva (life size).

Problem: Yellow-orange beetles with black stripes, about ⅜ inch long, are eating the leaves. Fat red humpbacked larvae with two rows of black dots may also be present.

Analysis: The Colorado potato beetle (*Leptinotarsa decemlineata*), also known as the *potato bug*, often devastates potato, tomato, eggplant, and pepper plantings. Both adults and larvae damage plants by devouring leaves and stems. Small plants are most severely damaged. The beetle was originally native to the Rocky Mountains and spread eastward in the late 1800s as potato plantings increased. Now it is found in all states except California and Nevada. In some areas of the country, the beetle population may reach epidemic proportions. The beetles lay their yellow-orange eggs on the undersides of leaves as the first potatoes emerge from the ground in the spring. The larvae that hatch from these eggs feed for 2 to 3 weeks, pupate in the soil, and emerge 1 to 2 weeks later as adults, which lay more eggs. One generation is completed in a month. One to three generations occur per year, depending on the area of the country.

Solution: Potato beetles are increasingly developing resistance to insecticides, so control may be difficult. Spray with ORTHO Bug-B-Gon Insect Killer, ORTHO Diazinon Ultra Insect Spray, or ORTHO Tomato & Vegetable Insect Killer. Treat when the insects are first noticed, and repeat every 7 days for as long as infestation continues. Floating row covers will also keep beetles away from potatoes. Varieties of potato resistant to potato beetles are available.

POTATOES (continued)

Flea beetle

Flea beetle damage.

Problem: Leaves are riddled with shot holes about ⅛ inch in diameter. Tiny (¹⁄₁₆-inch) black beetles jump like fleas when disturbed. Leaves of seedlings and eventually whole plants may wilt and die.

Analysis: Flea beetles jump like fleas but are not related to fleas. Both adult and immature flea beetles feed on a wide variety of garden vegetables, including potatoes. The immature beetle, a legless gray grub, injures plants by feeding on the roots and the lower surfaces of leaves. Adults chew holes in leaves. Flea beetles damage young plants most. Adult beetles survive the winter in soil and garden debris. They emerge in early spring to feed on weeds until potatoes sprout. Grubs hatch from eggs laid in the soil and feed for 2 to 3 weeks, damaging the tubers. After pupating in the soil, they emerge as adults to repeat the cycle. One to four generations occur per year. Adults may feed for up to 2 months.

Solution: Control flea beetles on potatoes with ORTHO Bug-B-Gon Insect Killer or ORTHO Diazinon Ultra Insect Spray when the leaves first show damage. Spray carefully at the base of stems. Watch new growth for evidence of further damage, and repeat the treatment at weekly intervals as needed. Clean all plant debris from the garden after harvesting to eliminate overwintering sites for adult beetles.

Potato leafhopper

Potato leafhopper damage. Inset: Potato leafhopper (4× life size).

Problem: Spotted pale green insects up to ⅛ inch long hop, run sideways, or fly away quickly when a plant is touched. The leaves are stippled or appear scorched, with a green midrib and brown edges curled under.

Analysis: The potato leafhopper (*Empoasca fabae*) feeds on potatoes and beans and on some fruit and ornamental trees. It sucks plant sap from the undersides of leaves, causing leaf stippling. This leafhopper is responsible for hopperburn and the browning and curling of the edges of potato leaves. The leafhopper injects toxic saliva into the nutrient-conducting tissue, interrupting the flow of food within the plant. Potato yields may be reduced drastically by hopperburn. Leafhoppers at all stages of maturity are active during the growing season. Both early and late varieties of potatoes are infested. Leafhoppers live year-round in the Gulf states and migrate northward on warm spring winds, so even areas that have winters so cold that the eggs cannot survive may be infested.

Solution: Spray infested plants with ORTHO Tomato & Vegetable Insect Killer, ORTHO Malathion 50 Plus Insect Spray, or ORTHO Bug-B-Gon Insect Killer. Be sure to cover the lower surfaces of leaves. Repeat the spray as often as necessary to keep the insects under control. Allow at least 10 days between applications. Eradicate nearby weeds that may harbor leafhopper eggs.

Early blight

Early blight.

Problem: Irregular dark brown to black spots, ⅛ to ½ inch in diameter, appear on the lower leaves. Concentric rings develop in the spots. Spots may enlarge, causing the leaf to die and fall off. Tubers may develop dark, sunken spots, often with a purplish raised border.

Analysis: Early blight is caused by a fungus (*Alternaria solani*) that attacks both vines and tubers. The same fungus causes early blight of tomatoes. It is most severe toward the end of the growing season when the vines approach maturity and after tubers are formed. Many leaves may be killed. The potato yield is reduced, but the plant seldom dies. Tubers are frequently infected through wounds inflicted during harvest. Early blight thrives in alternating periods of wet and dry weather and by temperatures from 65° to 85°F. The fungal spores spend the winter in plant debris left in the garden. Infected tubers are inedible.

Solution: Spray plants with ORTHO Multi-Purpose Fungicide Daconil 2787® Plant Disease Control as soon as leaf spotting occurs. Repeat the treatment every 7 to 10 days until the leaves die back naturally. Clean up and destroy plant debris after harvest. Do not store infected tubers. Avoid overhead watering by using drip or furrow irrigation. Maintain adequate fertility as outlined on page 292. Use seed potatoes certified by state departments of agriculture to be free of diseases.

Late blight

Late blight, a serious fungal disease.

Problem: Brownish water-soaked spots appear on the leaves. Spots enlarge rapidly, turn black, and kill first leaves and then leafstalks and main stems. In moist weather, a gray mildew grows on the lower surfaces of leaves. Tuber skins are infected in the ground or in storage with brownish purple spots that become a wet or dry rot.

Analysis: Late blight is caused by a fungus (*Phytophthora infestans*) that injures potatoes and tomatoes. The disease was responsible for the great potato famine in Ireland from 1845 to 1850 and is the most devastating disease of potatoes worldwide. The fungus spreads very rapidly, killing an entire planting in a few days. Infected tubers are inedible. The tubers are infected when the spores wash off the leaves and into the soil. Tubers may also be attacked during harvest and rot in storage. A soft rot often invades the damaged tubers. Foggy, misty weather and heavy dew provide enough moisture for infection. The spores survive the winter in infected tubers in the garden or compost pile.

Solution: If late blight is an annual problem in your area, spray plants with ORTHO Multi-Purpose Fungicide Daconil 2787® Plant Disease Control when the plants are 6 inches tall. Continue at intervals of 7 to 10 days until the plants naturally turn yellow and die. Avoid overhead watering; use drip or furrow irrigation. Wait at least a week after plants die naturally before digging the tubers. This allows time for the spores to die. Handle the tubers gently to avoid wounding them. Clean up and destroy plant debris. Plant certified disease-free seed pieces.

Flea beetle

Flea beetle damage.

Problem: Leaves are riddled with shot holes about ⅛ inch in diameter. Tiny (1⁄16-inch) black beetles jump like fleas when disturbed.

Analysis: Flea beetles jump like fleas but are not related to fleas. Both adult and immature flea beetles feed on a wide variety of garden vegetables. The immature beetle, a legless gray grub, injures plants by feeding on the roots and the lower surfaces of leaves. Adults chew holes in leaves. Adult beetles survive the winter in soil and garden debris. They emerge in early spring to feed on weeds until vegetable seeds sprout. Grubs hatch from eggs laid in the soil and feed for 2 to 3 weeks. After pupating in the soil, they emerge as adults to repeat the cycle. One to four generations occur per year. Adults may feed for up to 2 months.

Solution: Control flea beetles on radishes with ORTHO Tomato & Vegetable Insect Killer, ORTHO Bug-B-Gon Insect Killer, or ORTHO Diazinon Ultra Insect Spray when the plants first emerge through the soil or at the first sign of damage. Watch new growth for evidence of further damage, and repeat the treatment at weekly intervals as needed. Clean up and destroy plant debris after harvest to eliminate overwintering sites for adult beetles.

Root maggots

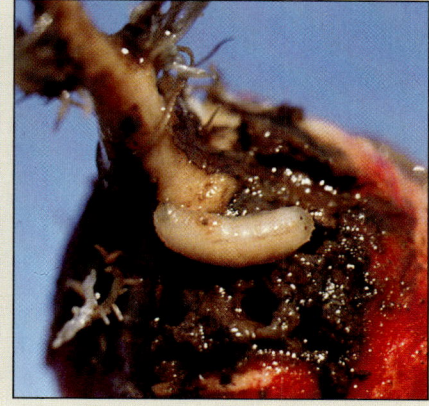

Root maggot (life size).

Problem: Young plants wilt in the heat of the day. They may later turn yellow and die. Soft-bodied, yellow-white maggots, about ¼ inch long, are feeding in the roots. The roots are honeycombed with slimy channels and scarred by brown grooves. Decay often accompanies feeding damage. Infested roots may break off when plants are pulled up. Younger plants are more severely affected than older ones.

Analysis: Root maggots (*Delia* species) are most numerous during cool, wet weather in spring, early summer, and fall. Early maggots attack the roots, stems, and seeds of radishes, cabbage, broccoli, and turnips in the spring and early summer. Later insects damage fall crops. The adult is a gray fly slightly smaller than a housefly, with black stripes and bristles down its back. It lays eggs on stems and nearby soil. The maggots hatch in 2 to 5 days and tunnel into radish roots, making them inedible.

Solution: Once the growing plant wilts and turns yellow, nothing can be done. To control maggots in the next planting of radishes, mix ORTHO Dursban® Lawn & Garden Insect Control or ORTHO Diazinon Ultra Insect Spray 4 to 6 inches into the soil before seeding. Control lasts about 1 month. Adult flies can be kept from laying eggs with row covers.

RHUBARB

Crown rot

Crown rot thrives in heavy, waterlogged soils.

Problem: Leaves wilt. Brown, sunken, water-soaked spots appear on the base of the leafstalks. Leaves yellow, and stalks collapse and die. The whole plant eventually dies.

Analysis: Crown rot, also called *stem rot*, *foot rot*, or *root rot*, is caused by a fungus (*Phytophthora* species) that lives in the soil. It thrives in waterlogged, heavy soils and attacks the crown and base of the stems. The stems and eventually the roots rot, resulting in wilting and finally the death of the plant. The fungus is most active in warm (60° to 75°F), moist soils in the late spring and early summer. The spores are spread to healthy plants in running or splashing water or by contaminated soil or infected plants brought into the garden.

Solution: Remove and destroy dying leaves and plants. Apply a drench of fungicide containing *captan* or basic copper sulfate to the crown or base of the plant and to the surrounding soil. When replanting, purchase disease-free plants from a reputable company, and plant in well-drained soil. Avoid overwatering.

Small stalks

Small stalks can have several causes.

Problem: Rhubarb stalks are thin and small.

Analysis: Rhubarb stalks may be small for any of several reasons.
1. End of the harvest season: Rhubarb stalks are produced from food stored in the roots. Toward the end of the harvest season, the food is used up and the stalks become smaller and smaller.
2. Lack of fertilization: Rhubarb plants are heavy feeders, requiring large amounts of fertilizer to encourage healthy growth so that an adequate food supply will be stored for the following year's harvest.
3. Young plants: Rhubarb stalks are small until the plants establish their roots and are able to store an adequate amount of food. This may take 2 years after planting.
4. Overcrowding: Rhubarb plants are vigorous growers with deep roots and may compete with each other if not divided every 5 to 7 years.
5. Crown rot: This disease reduces plant vigor and kills the roots. For more information, see at left.
6. Poor soil drainage: Rhubarb does not tolerate wet soil. The roots rot and the plants eventually die.

Solution: To improve your harvest, fertilize at planting time using 2 pounds of Scotts Vegetable Food per 100 square feet, or 1 pound per 50-foot row. In the following years, fertilize with 1 pound each spring before the new leaves begin to grow. After harvest, side-dress with blood meal using 1 ½ pounds per 50-foot row. Control crown rot and improve the soil drainage. Water when the soil 2 inches deep is barely moist.

SPINACH

Leafminers

Leafminers (life size).

Problem: Irregular tan blotches, blisters, or tunnels appear in the leaves. The tan areas peel apart like facial tissue. Tiny black specks and white or yellow maggots are found inside the tunnels.

Analysis: Leafminers belong to a family of leafmining flies. The tiny black or yellow adult fly lays her white eggs on the undersides of leaves. The maggots that hatch from these eggs bore into the leaf and tunnel between the upper and lower surfaces, feeding on the inner tissue. The tunnels and blotches are called *mines*. The black specks inside are the maggots' droppings. Several overlapping generations occur during the growing season, so larvae are present continuously from spring until fall. Infested leaves are not edible.

Solution: Control leafminers on spinach with ORTHO Bug-B-Gon Insect Killer or ORTHO Diazinon Ultra Insect Spray when the white egg clusters are first seen under the leaves. Repeat two times at weekly intervals to control succeeding generations. Once leafminers enter the leaves, sprays are ineffective. Spraying after the mines first appear will control only those leafminers that attack after the application. Clean all plant debris from the garden after harvest to reduce overwintering spots for the pupae. Adult flies can be kept from laying eggs with row covers.

STRAWBERRIES

GROWING GUIDE

Adaptation: Throughout North America.

Planting Time: Early spring.

Planting Method: Select dormant or growing plants. Pick off all but 2 or 3 of the healthiest leaves. Prune away a third of the roots. Set the plants 18 inches apart with the roots fanning outward. The crown should be just above the soil level. As runners develop, maintain the rows no wider than 18 to 24 inches. Runners can also be removed throughout the season and established as separate plants.

Soil: Any good, well-drained garden soil that is high in organic matter. pH 5.5 to 7.5.

Fertilizer: At planting time, add 1 pound of Scotts Vegetable Food per 100 square feet, or ¼ pound per 25-foot row. Side-dress after harvest each year with ½ pound per 25-foot row.

Water: How much: Apply enough water at each irrigation to wet the soil 8 to 10 inches deep.
How often: Water when the soil 1 inch deep is barely moist.

Harvest: For maximum production the second year, remove blossoms and runners the first planting year. Plants will usually be productive for 3 to 4 years.

Snails and slugs

Slug (3× life size).

Problem: Stems and leaves may be sheared off and eaten. Holes are often found in ripening berries, especially under the berry cap. Silvery trails wind around on the plants and soil nearby. Snails or slugs may be seen moving around or feeding on the plants, especially at night. Inspect the garden for them at night by flashlight.

Analysis: Snails and slugs are mollusks and are related to clams, oysters, and other shellfish. They feed on a wide variety of garden plants. Like other mollusks, snails and slugs need to be moist all the time. For this reason, they avoid direct sun and dry places and hide during the day in damp places, such as under flowerpots or in thick ground covers. They emerge at night or on cloudy days to feed. Snails and slugs are similar except that the snail has a hard shell into which it withdraws when disturbed. Slugs lay white eggs encased in a slimy mass in protected places. Snails bury their eggs in the soil, also in a slimy mass. The young look like miniature versions of their parents.

Solution: Apply ORTHO Bug-Geta Snail & Slug Killer in the areas you wish to protect. Also apply in areas where snails or slugs might be hiding, such as in dense ground covers, weedy areas, compost piles, or pot storage areas. Before applying, wet down the areas to be treated to encourage snail and slug activity that night. Repeat the application every 2 weeks for as long as snails and slugs are active. Do not apply to leaves.

Gray mold

Gray mold infects strawberries at all stages.

Problem: Light tan spots appear on berries. Some berries are soft, mushy, and rotting. A fluffy gray mold may cover rotting berries.

Analysis: Gray mold is caused by a fungus (*Botrytis cinerea*). It is the most damaging rot of strawberries, attacking both flowers and berries, and greatly reduces the amount of edible fruit. The flowers are infected when in bloom and may not produce berries. Berries are attacked at all stages of development. They are infected directly when a healthy berry touches a decaying one, the ground, or a dead leaf. Infected berries are inedible. The fuzzy gray mold on the berries is composed of fungal strands and millions of microscopic spores. The fungus is most active in cool, humid weather and is spread by splashing water, people, or infected fruit. Crowded plantings, rain, and overhead watering enhance its spread.

Solution: Destroy infected fruit. To reduce the spread of the fungus to uncontaminated fruit, treat with ORTHO Home Orchard Spray or with a fungicide containing *captan* at the first sign of disease. Continue treating every 8 to 10 days. Pick berries as they ripen. Avoid overhead watering by using soaker or drip hoses. Mulch with straw, pine needles, or other material to keep fruit off the ground. To help prevent infection the following year, remove and destroy plant debris in the fall and treat the plants when in bloom. Repeat treatment every 8 to 10 days. When setting out new plants, provide enough space between plants to allow for good air circulation and rapid drying after rain and irrigation.

STRAWBERRIES (continued)

Strawberry bud weevil

Weevil-damaged buds. Inset: Adult (5× life size).

Problem: Flower buds droop, turn brown and dry, and hang from the plant or fall to the ground. Small holes appear in the sides of the buds. Dark reddish brown, ⅛-inch weevils with curved snouts crawl on the plants.

Analysis: The strawberry bud weevil (*Anthonomus signatus*), an insect pest of strawberries, dewberries, blueberries, and wild blackberries, is also known as the clipper because it clips the flower bud stems, causing buds to droop and fall to the ground. By destroying the flower buds, they reduce the berry crop. Adult weevils survive the winter in debris in and near the garden. In the spring, they puncture holes in the sides of unopened flower buds and lay an egg in each hole. Then the adult cuts a notch in the flower stem ⅛ to ¼ inch below the bud. The buds droop for a few days and then fall to the ground. Within a week the eggs hatch and fat, white grubs feed on the pollen inside the bud. After feeding for about 4 weeks, the grubs pupate and emerge as adults in early to midsummer. These adults feed on blackberry and dewberry pollen, hibernate, and emerge in the spring to repeat the cycle. Only one generation occurs per year.

Solution: Treat plants with ORTHO Home Orchard Spray when cut buds first appear. Repeat through the closed-bud stage as long as damage occurs. Clean up and destroy plant debris in and near the garden at the end of the season to reduce overwintering locations for the adults.

Spider mites

Spider mite webbing.

Problem: Leaves are stippled, yellowing, and dirty. Leaves may dry out and drop. There may be webbing between leaves or on the lower surfaces of leaves. Few berries are produced. To determine if a plant is infested with mites, examine the bottoms of the leaves with a hand lens. Or hold a sheet of white paper underneath an affected leaf and tap the leaf sharply. Minute specks the size of pepper grains will drop to the paper and begin to crawl around. These pests are easily seen against the white background.

Analysis: Spider mites, related to spiders, are major pests of many garden and greenhouse plants. Spider mites are larger than cyclamen mites, which also attack strawberries. Spider mites attack the older leaves, while cyclamen mites attack the younger leaves. Spider mites cause damage by sucking sap from the undersides of leaves. As a result of their feeding, green leaf pigment disappears, producing the stippled appearance. Spider mite webbing traps cast-off skins and debris, making the plant messy. Mites are active throughout the growing season but thrive in hot, dry weather (70°F and up). By midsummer, they may have built up to tremendous numbers.

Solution: Treat infested strawberry plants with ORTHO Diazinon Ultra Insect Spray, ORTHO Bug-B-Gon Insect Killer, or ORTHO Malathion 50 Plus Insect Spray when damage first appears. Repeat the treatment at intervals of 7 to 10 days until harvest or until damage no longer occurs. Be sure the pesticide contacts the lower surfaces of the leaves.

Viruses

Virus disease.

Problem: Strawberry leaves are crinkled, yellowed, or distorted. Plants may be stunted with cupped leaves or may grow in a tight rosette. Small, dull fruit or no fruit is produced. Plants produce few runners.

Analysis: Two types of viruses attack strawberries: killer viruses and latent viruses. Killer viruses have obvious symptoms and kill individual mother plants and daughter plants attached to the ends of runners. Latent viruses show few or no symptoms. A single latent virus may merely reduce the number of runners and fruit produced. Plants infected with more than one kind of latent virus, however, are so weakened that they are frequently attacked and killed by other diseases. Viruses are transmitted by aphids and nematodes through runners from mother to daughter plants. Aphids are most numerous in cool spring and fall weather.

Solution: Remove and destroy all infected plants. If aphids are present, control them with ORTHO Diazinon Ultra Insect Spray, ORTHO Home Orchard Spray, or ORTHO Malathion 50 Plus Insect Spray. When replanting, buy certified plants from a reputable company. Certified plants are grown under special isolated conditions and are essentially virus free. Or choose virus-tolerant varieties.

SWEET POTATOES

Black rot

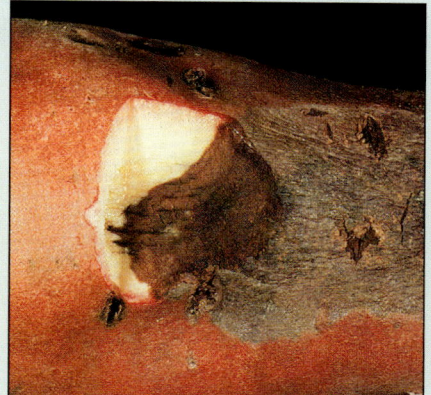

Black rot on sweet potato.

Problem: Black spots occur on roots in the ground and in storage. Tiny black specks appear in the centers of blackened areas. Black lesions occur on stems at the soil line. Leaves are yellow and dwarfed.

Analysis: Black rot is caused by a fungus (*Ceratocystis fimbriata*) that attacks all underground parts of sweet potato plants. The fungus usually enters plants through injuries, but healthy tissue may also be attacked under favorable conditions. Infections can occur between 50° and 93°F during periods of high soil moisture. Infected roots may not show symptoms when dug up, but once in storage they develop sunken, blackened areas. Both discolored tissue and the surrounding healthy tissue taste bitter. The fungus survives in sweet potato debris, manure, and weeds for at least 2 years. Spores are spread by wind, water, and insects, including the sweet potato weevil.

Solution: This disease can be prevented but not cured. Remove and discard all infected plants and roots. Disinfect the storage area thoroughly. Purchase certified slips or seed potatoes. Do not plant sweet potatoes in the same soil more than once in 3 years.

Scurf

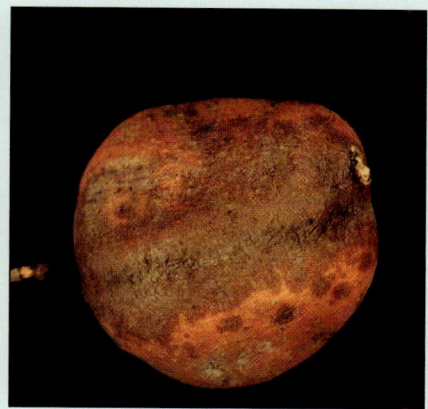

Scurf is most severe in poorly drained soils.

Problem: Dark brown to black spots or irregular patches stain sweet potato skins. Discoloration affects only the skin, not the inside of the root.

Analysis: Scurf, also called *soil stain*, is caused by a fungus (*Monilochaetes infuscans*) that attacks only the roots. Although the appearance of scurf is unappetizing, the roots are still edible and the flavor is not affected. Scurf is most severe in wet, poorly drained soils and in soils high in organic matter. The disease develops most rapidly at temperatures around 75°F. Spots may enlarge in storage. Injury to the root surface allows rapid water loss, causing roots to shrivel. The fungus is introduced into the garden on infected roots and slips. The spores survive for 1 or 2 years in infected vines rotting in the garden and in humus or partially decomposed organic matter.

Solution: Eat scurf-infected roots promptly, without storing them. Use sweet potatoes free of the disease as seed. Cut transplants at least ½ inch above the soil line. Avoid planting in heavy soil. Improve soil drainage. If you add organic matter, till it into the soil a year before planting so it will decompose. Plant sweet potatoes in the same place only once in every 3 or 4 years.

Sweet potato flea beetle

Flea beetle larva (½ life size).

Problem: Yellow irregular channels are chewed in the surface of the leaves. Tiny (¹⁄₁₆ - inch) black beetles jump like fleas when disturbed. Shallow, dark tunnels scar the root.

Analysis: Sweet potato flea beetles (*Chaetocnema confinis*) jump like fleas but are not related to them. Both adults and immature flea beetles feed on sweet potatoes. The immature beetle, a legless white grub, feeds on the surface tissue of the roots, making them unattractive but still edible. Adults chew channels in leaves. Flea beetles are present nearly everywhere sweet potatoes are grown. They are most damaging to young plants. Adult beetles survive the winter in soil and garden debris. They emerge in early spring to feed on weeds until slips (young sweet potato plants) are set in the garden. Grubs hatch from eggs laid in the soil and feed for 2 to 3 weeks. After pupating in the soil, they emerge as adults to repeat the cycle. Several generations occur per year.

Solution: Control flea beetles on sweet potatoes with an insecticidal spray containing *methoxychlor* when the leaves first show damage. Watch new growth for evidence of further damage, and repeat the treatment at weekly intervals as needed. Clean all plant debris from the garden after harvest to eliminate overwintering spots for the adult beetles. Treat soil with ORTHO Dursban® Lawn & Garden Insect Control.

TOMATOES

GROWING GUIDE

Adaptation: Throughout North America.

Planting Time: See page 361.

Planting Method: Sow seeds ½ inch deep. Thin seedlings when they have 2 or 3 leaves, or set transplants 18 to 36 inches apart. Set transplants horizontally with only the top third of the plant above the soil. Remove any fruit and open flowers from transplants before planting.

Soil: Any good garden soil that is high in organic matter. pH 5.5 to 7.5.

Fertilizer: At planting time, use 2 pounds of Scotts Vegetable Food per 100 square feet, or ½ pound per 25-foot row. In hot-summer areas, side-dress every 3 to 4 weeks with ½ pound per 25-foot row, or 1 tablespoon per plant. In cool-summer areas, side-dressing may produce excess foliage and green fruit that doesn't ripen.
Containers: Water thoroughly every 2 to 4 weeks with a liquid or soluble plant food.

Water: How much: Apply enough water at each irrigation to wet the soil 18 to 20 inches deep.
How often: Water when the soil 2 inches deep is barely moist.

Harvest: Harvest tomatoes when they have a deep, rich color. Store at room temperature until use.

Poor fruit set

Poor fruit set can have several causes.

Problem: Little or no fruit develops. The plants are healthy and may even be extremely vigorous.

Analysis: Poor fruit set occurs on tomatoes for any of several reasons.
1. Extreme temperatures: The blossoms drop off without setting fruit when night temperatures fall below 55°F or day temperatures rise above 90°F for extended periods.
2. Dry soil: Blossoms dry and fall when the plants don't receive enough water.
3. Shading: Few blossoms are produced when the plants receive less than 6 hours of sunlight per day.
4. Excessive nitrogen: High levels of nitrogen in the soil promote leaf growth at the expense of blossom and fruit formation.

Solution: The numbered solutions below correspond to the numbered items in the analysis.
1. Plant early-, mid-, and late-season varieties at the appropriate time of year. For a list of these varieties, see page 365.
2. Water tomatoes regularly, never allowing the soil to dry out. Mulch with straw, black plastic, or other material to reduce the need for watering.
3. Plant tomatoes in an area that receives at least 6 hours of sunlight per day. If your yard is shady, grow tomatoes in containers on your porch or patio.
4. Correct the nitrogen imbalance with superphosphate or 0–10–10 fertilizer. To prevent future problems, follow the fertilizing recommendations at left.

Sunscald

Sunscald on tomatoes.

Problem: On green and ripening fruit, a light patch develops on the side facing the sun. This area blisters and finally becomes slightly sunken and grayish white, with a paperlike surface. A black mold may grow on the affected area, causing the fruit to rot.

Analysis: Sunscald occurs on tomatoes when they are exposed to the direct rays of the sun during hot weather. It is most common on green fruit, but ripening fruit is also susceptible. It is most prevalent on varieties with sparse foliage and on staked plants that have lost their foliage because of leaf diseases such as early blight, late blight, septoria leaf spot, fusarium wilt (see page 303), verticillium wilt, or leaf roll. Fruit on plants that have been pruned to hasten ripening is also subject to sunscald. Tomatoes are still edible if the sunscalded area is removed. Rot fungi frequently invade the damaged tissue, resulting in moldy, inedible fruit.

Solution: Cover exposed fruit with straw or other light material to protect it from the sun's rays. Do not prune leaves to hasten ripening. Control leaf diseases. Grow verticillium- and fusarium-wilt-resistant varieties.

Growth cracks

Growth cracks.

Problem: Circular or radial cracks mar the stem end (top) of ripening fruit. Cracks may extend deep into the fruit, causing it to rot.

Analysis: Tomatoes crack when certain environmental conditions encourage rapid growth during ripening such as a drought followed by heavy rain or watering. Tomatoes are most susceptible to cracking after they have reached full size and begin to change color. Some varieties crack more easily than others. Cracking is more severe in hot weather. Some cracks may be deep, allowing decay organisms to enter the fruit and rot it. Shallow cracks frequently heal over but may rupture if the fruit is roughly handled when picked. Cracked tomatoes are edible.

Solution: Maintain even soil moisture with regular watering, according to the instructions on page 300. Grow varieties less subject to cracking.

Blossom-end rot

Blossom-end rot can have several causes.

Problem: A round, sunken, water-soaked spot develops on the bottom of the fruit. The spot enlarges, turns brown to black, and feels leathery. Mold may grow on the rotted surface.

Analysis: Blossom-end rot occurs on tomatoes, peppers, squash, and watermelons from a lack of calcium in the developing fruit. This results from slowed growth and damaged roots caused by any of several factors.
1. Extreme fluctuations in soil moisture, from very wet to very dry.
2. Rapid plant growth early in the season, followed by extended dry weather.
3. Excessive rains that smother root hairs.
4. Excess soil salts.
5. Cultivation too close to the plant.
The first fruits are the most severely affected. The disorder always starts at the blossom end, and it may enlarge to affect up to half of the fruit. Moldy growths on the rotted area are from fungi or bacteria that invade the damaged tissue. The unrotted part of the fruit is edible.

Solution: To prevent future blossom-end rot, follow these guidelines.
1. Maintain uniform soil moisture by following the watering guidelines on page 300.
2. Avoid using high-nitrogen fertilizers or large quantities of fresh manure.
3. Plant in well-drained soil.
4. If your soil or water is salty, provide more water at each watering to help leach salts through the soil.
5. Do not cultivate deeper than 1 inch within 1 foot of the plant.

Anthracnose

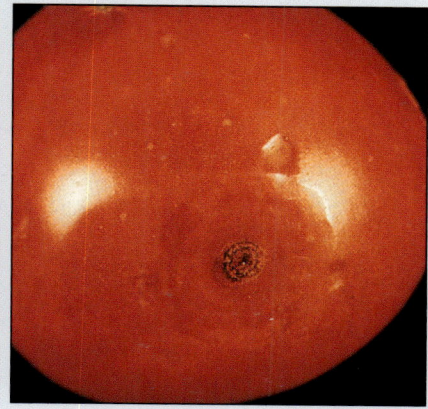

Anthracnose is most active in wet weather.

Problem: Sunken spots up to ½ inch in diameter occur on ripe tomatoes. The centers of the spots darken and form concentric rings. Spots may merge, covering a large part of the tomato.

Analysis: Anthracnose is caused by a fungus (*Colletotrichum coccodes*) that rots ripe tomatoes. Green tomato fruit is also attacked, but the spots don't appear until the fruit ripens. Infected fruit is inedible. Anthracnose is most common on overripe fruit and fruit close to the ground. Fruit on plants partially defoliated by leaf spot disease is also prone to infection. The leaves may be infected but are usually not severely damaged. The fungus is most active in wet weather with a temperature range of 60° to 90°F. When infection is severe, fruit is damaged in a short period of time. Infections frequently become epidemic in hot, rainy weather. Water from heavy dew, overhead watering, and frequent abundant rain provides the moisture necessary for infection.

Solution: Spray tomato plants at the first sign of the disease with ORTHO Multi-Purpose Fungicide Daconil 2787® Plant Disease Control. Repeat at intervals of 7 to 10 days until harvest. Destroy all infected fruit. Pick tomatoes as they mature, and use promptly. To reduce the spread of the disease, do not work around wet plants. Avoid overhead irrigation by using ditch or drip irrigation. Clean up and destroy plant debris after harvest.

TOMATOES *(continued)*

Bacterial spot

Bacterial spot attacks green tomatoes.

Problem: Dark, raised, scablike spots, ⅛ to ¼ inch in diameter, appear on green tomatoes. The centers of the spots are slightly sunken. Dark, greasy, ⅛-inch spots occur on the older leaves, causing them to drop.

Analysis: Bacterial spot is caused by a bacterium (*Xanthomonas vesicatoria*) that attacks green but not red tomatoes. Peppers are also attacked. The bacteria infect all aboveground parts of the plant at any stage of growth. Infected blossoms drop, reducing the fruit yield. Fruit is infected through skin wounds caused by insects, blowing sand, and mechanical injuries. Infected fruit does not ripen properly and is frequently invaded by rot organisms. The bacteria are most active after heavy rains and in temperatures of 75° to 85°F. They spread rapidly in the rain, often resulting in severe defoliation that weakens the plant and exposes the fruit to sunscald. The bacteria spend the winter in the soil and are carried on tomato seeds.

Solution: Pick and destroy all infected green fruit. If infection is severe, spray with a fungicide containing basic copper sulfate at the first sign of the disease. Repeat at intervals of 7 to 10 days as long as weather conditions favorable to the spread of bacterial spot continue. To reduce the spread of the disease, do not work around wet plants. Avoid overhead watering by using drip or furrow irrigation. Purchase disease-free plants or seeds from a reputable dealer. Do not save seeds from infected fruit.

Tomato fruitworm

Tomato fruitworm damage.

Problem: Deep holes are chewed in the fruit. Striped yellow, green, or brown worms from ¼ to 2 inches long are feeding in them. Worms may also be feeding on the leaves.

Analysis: Tomato fruitworms (*Helicoverpa zea*), also known as *corn earworms* or *cotton bollworms*, attack many plants, including tomatoes, corn, and cotton. The worms feed on the foliage and also chew deep holes in the fruit and feed inside. Damaged fruit is inedible. The adult tomato fruitworm is a light grayish brown moth with dark lines on the wings. In the spring, the adults lay white eggs on the leaves and stems. The worms that hatch from these eggs feed on the leaves until they are about ½ inch long. Then they move to the fruit and bore inside. After feeding for 2 to 4 weeks, they drop to the ground, burrow 2 to 6 inches deep, and pupate; they emerge in several weeks as adults. Several generations of worms occur per year, so damage can continue until fall.

Solution: Once fruitworms are inside the tomatoes, you cannot do anything about them. Destroy infested fruit. Clean all plant debris from the garden after harvest to reduce the number of overwintering adults. If fruitworms were numerous in the current year, treat your plants the following year with an insecticide containing *methoxychlor* or *carbaryl* or with the bacterial insecticide *Bacillus thuringiensis* (Bt) when the worms are feeding on the foliage and fruit is about ½ inch in diameter. Repeat in 2 and 4 weeks if the plants become reinfested.

Tomato hornworm and tobacco hornworm

Tomato (top) and tobacco hornworms (life size).

Problem: Fat green or brown worms, up to 5 inches long with white diagonal stripes, chew on the leaves. A red or black "horn" projects from the rear end. Black droppings soil the leaves.

Analysis: Tomato hornworms (*Manduca quinquemaculata*) and tobacco hornworms (*M. sexta*) feed on the fruit and foliage of tomatoes, peppers, and eggplants. Although only a few worms may be present, each worm consumes large quantities of foliage and causes extensive damage. The large gray or brown adult moth with yellow-and-white markings emerges from hibernation in late spring and drinks nectar from petunias and other garden flowers. The worms that hatch from eggs laid on the undersides of leaves feed for 3 to 4 weeks. Then they crawl into the soil and pupate, later emerging as adults to repeat the cycle. One generation occurs per year in the North and two to four in the South.

Solution: Handpicking is usually effective against hornworms. If not, spray the plants with an insecticide containing *methoxychlor* or with the bacterial insecticide *Bacillus thuringiensis* (Bt).

Flea beetle

Flea beetle damage.

Problem: Leaves are riddled with shot holes about ⅛ inch in diameter. Tiny (1/16-inch) black beetles jump like fleas when disturbed. Leaves of seedlings may wilt and die.

Analysis: Flea beetles jump like fleas but are not related to fleas. Both adult and immature flea beetles feed on a wide variety of garden vegetables, including tomatoes. The immature beetle, a legless gray grub, injures plants by feeding on the roots and the lower surfaces of leaves. Adults chew holes in leaves. Flea beetles damage seedlings and young plants most. Leaves of seedlings riddled with holes dry out quickly and die. Adult beetles survive the winter in soil and garden debris. They emerge in early spring to feed on weeds until vegetables sprout or plants are set in the garden. Grubs hatch from eggs laid in the soil and feed for 2 to 3 weeks. After pupating in the soil, they emerge as adults to repeat the cycle. One to four generations occur per year. Adults may feed for up to 2 months.

Solution: Control flea beetles on tomatoes with ORTHO Bug-B-Gon Insect Killer, ORTHO Diazinon Ultra Insect Spray, or an insecticidal soap. Treat when the leaves first show damage. Watch new growth for evidence of further damage, and repeat the treatment at weekly intervals as needed. Clean all plant debris from the garden after harvest to eliminate overwintering spots for adult beetles.

Aphids

Aphids (life size).

Problem: Leaves turn yellow and may be curled, distorted, and puckered. Pale green, yellow, or purple soft-bodied insects cluster on stems and on the undersides of leaves.

Analysis: Aphids do little damage in small numbers. They are extremely prolific, however, and populations can rapidly build up to damaging numbers during the growing season. Damage occurs when the aphid sucks the juices from tomato leaves. Aphids feed on nearly every plant in the garden and are spread from plant to plant by wind, water, and people. Aphids can also spread diseases.

Solution: Control aphids on tomatoes with ORTHO Bug-B-Gon Insect Killer, ORTHO Malathion 50 Plus Insect Spray, ORTHO Tomato & Vegetable Insect Killer, or an insecticidal soap. Treat at the first sign of infestation, and repeat at intervals of 7 to 10 days if plants become reinfested.

Fusarium wilt

Fusarium wilt; control with resistant varieties.

Problem: Lower leaves turn yellow, wilt, and die. Then upper shoots wilt, and eventually the whole plant dies. Wilting usually occurs first on one side of the leaf or plant, then on the other. When the stem is sliced lengthwise near the soil line, the tissue ⅛ inch under the surface is found to be dark brown.

Analysis: Fusarium wilt is caused by a fungus (*Fusarium oxysporum* ssp. *lycopersici*) that infects only tomatoes. The fungus persists indefinitely on plant debris or in the soil. *Fusarium* is most prevalent in warm-weather areas. The disease is spread by contaminated soil, seeds, plants, and equipment. The fungus enters the plant through the roots and spreads up into the stems and leaves through water-conducting vessels in the stems. These vessels become discolored and plugged. This plugging cuts off the flow of water and nutrients to the leaves, causing leaf yellowing and wilting. Affected plants may or may not produce fruit. Fruit that is produced is usually deformed and tasteless. Many plants die.

Solution: No chemical control is available. Destroy infected plants promptly. *Fusarium* can be removed from the soil only by fumigation and solarization techniques. To solarize soil, cultivate the area and cover it with a layer of clear polyethylene film during the hottest time of the year. Leave the film in place for 4 to 6 weeks. The best solution is to use plants that are resistant to fusarium wilt. This is denoted by the letter *F* after the tomato variety name. For a list of wilt-resistant varieties, see page 362.

TOMATOES *(continued)*

Nematodes

Nemetode damage to roots.

Problem: Plants are stunted, are yellow, and wilt in hot, dry weather. Round and elongated nodules occur on roots.

Analysis: Nematodes are microscopic worms that live in the soil. Some are highly beneficial, some highly destructive. They are not related to earthworms. Destructive nematodes feed on plant roots. The damaged roots can't supply sufficient water and nutrients to the aboveground plant parts, and the plant is stunted or slowly dies. Nematodes are found throughout the country but are most severe in the Southeast. They prefer moist, sandy loam soils. They can move only a few inches each year on their own, but may be carried long distances by soil, water, tools, or infested plants. Testing roots and soil is the only method for confirming their presence. Contact your local county extension service for sampling instructions and addresses of testing laboratories. Soil and root problems such as poor soil structure, drought stress, nutrient deficiency, and root rots, can produce symptoms of decline similar to those caused by nematodes. Eliminate these problems as causes before sending soil and root samples for testing.

Solution: No chemicals available to homeowners kill nematodes in planted soil. They can be controlled before planting, however, by soil fumigation or solarization. To solarize soil, cultivate the area and cover it with a layer of clear polyethylene film during the hottest time of the year. Leave the film in place for 4 to 6 weeks. Some varieties are resistant to nematodes. An *N* after the variety name indicates resistance.

Cutworms

Cutworm damage.

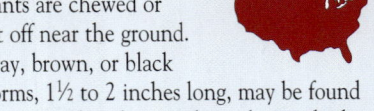

Problem: Young plants are chewed or cut off near the ground. Gray, brown, or black worms, 1½ to 2 inches long, may be found about 2 inches deep in the soil near the base of the damaged plants. The worms coil when disturbed.

Analysis: Several species of cutworms attack plants in the vegetable garden. The most likely pests of young tomato plants set out early in the season are surface-feeding cutworms. A single surface-feeding cutworm can sever the stems of many young plants in one night. Cutworms hide in the soil during the day and feed only at night. Adult cutworms are dark, night-flying moths with bands or stripes on their forewings.

Solution: Apply ORTHO Diazinon Soil & Turf Insect Control or ORTHO Bug-Geta Plus Snail, Slug & Insect Killer around the base of undamaged plants when stem cutting is observed. Because cutworms are difficult to control, applications may need to be repeated at weekly intervals. Before transplanting into the area, apply a preventive treatment of ORTHO Diazinon Soil & Turf Insect Control, ORTHO Dursban® Lawn & Garden Insect Control, or the bacterial insecticide *Bacillus thuringiensis* (Bt). Cultivate the soil thoroughly in late summer and fall to expose and destroy eggs, larvae, and pupae. Further reduce damage with cutworm collars made of stiff paper or aluminum foil bent into a cylinder or out of tin cans or paper cups with the bottoms removed. These collars should be at least 2 inches high, surround the plant stem fairly closely, and be pressed into the soil.

Damping-off

Damping-off, a common problem in wet soil.

Problem: Seeds don't sprout, or seedlings fall over soon after they emerge. The stem at the soil line is water-soaked and discolored. The base of the stem is soft and thin.

Analysis: Damping-off is a common problem in wet soil with a high nitrogen level. Wet, rich soil promotes damping-off in two ways: the fungi are more active under these conditions, and the seedlings are more succulent and susceptible to attack. Damping-off is often a problem with crops that are planted too early in the spring, before the soil has had a chance to dry and warm sufficiently for quick seed germination. Damping-off can also be a problem when the weather remains cloudy and wet while seeds are germinating or if seedlings are too heavily shaded.

Solution: To prevent damping-off, take the following precautions.
1. Allow the surface of the soil to dry slightly between waterings.
2. Do not start seeds in soil that has a high nitrogen level. Add nitrogen fertilizers after the seedlings have produced their first true leaves.
3. Plant seeds after the soil has reached at least 70°F, or start seeds indoors in sterilized potting mix.
4. Protect seeds during germination by coating them with a fungicide containing *captan* or *thiram*. Add a pinch of fungicide to a packet of seeds and shake well to coat the seeds with the fungicide.

TURNIPS

Root maggots

Root maggots (½ life size).

Problem: Young plants wilt in the heat of the day. They may later turn yellow and die. Soft-bodied, yellow-white maggots, ¼ to ⅓ inch long, are feeding in the roots and on seeds. The roots are honeycombed with slimy channels and scarred by brown grooves.

Analysis: Root maggots (*Hylemya* species) are a damaging pest of turnips in the northern United States. They are most numerous during cool, wet weather in the spring, early summer, and fall. Early maggots attack the roots, stems, and seeds of turnips, cabbage, broccoli, and radishes in the spring and early summer. Later insects damage crops in the fall. The adult is a gray fly somewhat smaller than a housefly, with black stripes and bristles down its back, that can often be seen clinging to plants. It lays eggs on stems and in nearby soil. The maggots hatch in 2 to 5 days to feed and tunnel into the roots, rendering them inedible.

Solution: Once the maggots are in the roots, nothing can be done. To control maggots in the planting of turnips, mix ORTHO Dursban® Lawn & Garden Insect Control or ORTHO Diazinon Soil & Turf Insect Control into the top 4 to 6 inches of soil before seeding. Control lasts about 1 month. To prevent egg laying, screen adult flies from the seedbed with a row cover.

Wireworm

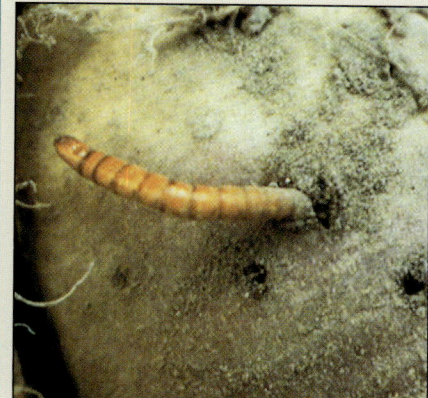

Wireworm (2× life size).

Problem: Plants are stunted and grow slowly. Holes are drilled into the base of the plants. In the soil and throughout the roots are hard, jointed, shiny, cream-colored to yellow worms up to ⅝ inch long.

Analysis: Wireworms feed on turnips, corn, carrots, beets, peas, and many other plants. They feed on the roots and seeds of turnips. Infestations are most extensive in soil where turfgrass has previously grown, in poorly drained soil, and in soil that is high in organic matter. The adult is known as the *click beetle* because it makes a clicking sound when turning from its back to its feet. Adults lay their eggs in the spring. Wireworms feed for 2 to 6 years before maturing into adult beetles, so all sizes and ages may be present in the soil at the same time. Infestations are often spotty.

Solution: Control wireworms with an insecticide containing *diazinon* or *chlorpyrifos*. Both granular and liquid forms are effective. Treat the soil 1 week before planting, working the insecticide into the top 6 inches of soil.

Cabbage caterpillars

Cabbage looper (life size).

Problem: Leaves have ragged, irregular, or round holes. Green caterpillars, up to 1½ inches long, are feeding on the leaves. Masses of greenish brown pellets may be found when the outer leaves are parted.

Analysis: These destructive caterpillars are either the cabbage looper (*Trichoplusia ni*), native to the United States, or the imported cabbageworm (*Pieris rapae*), introduced into North America from Europe around 1860. Both species attack all members of the cabbage family, including turnips. Adults lay their eggs singly on leaves in the spring. The white cabbage butterfly attaches its tiny yellow, bullet-shaped eggs to the undersides of leaves; the brownish cabbage looper moth lays its pale green eggs at night on the upper surfaces of leaves. The larvae that hatch from these eggs feed on leaves. The cabbageworm also causes damage by contaminating plants with its greenish brown excrement. As many as five generations occur per season, so caterpillar damage can occur from early spring through late fall. The caterpillars spend the winter as pupae attached to a plant or nearby object. Cabbage loopers migrate from the South in early summer.

Solution: When the caterpillars first appear, treat the turnip leaves with ORTHO Tomato & Vegetable Insect Killer, ORTHO Bug-B-Gon Insect Killer, or the bacterial insecticide *Bacillus thuringiensis* (Bt). Or, if practical, handpick caterpillars from the leaves and destroy them. Keep adults from laying eggs with row covers. Clean all plant debris from the garden to reduce the number of overwintering pupae.

Clockwise from top left: booklouse, scorpion, house spider, and drain fly.

Improved living standards have markedly diminished parasite and fly problems. Yet the same buildings that shelter people also shelter household pests, which are often attracted to homes by conditions in the yard and garden.

INSPECTING THE OUTDOORS

Piles of grass clippings or other damp trash, along with weather conditions, can encourage sowbugs and pillbugs; windowsill dampness can lead them into your home. Caterpillars of all sorts may wander indoors, looking for warmth and shelter. Larvae may even be blown in by the wind. Spiders and termites can inhabit your woodpile or wainscoting. Because many indoor pest problems begin outdoors, begin your pest-control efforts outside your home. Is landscaping overgrown with weeds or plants that badly need thinning or trimming? Insects find such areas prime nesting habitat. If there is outdoor pet food or an open garbage can, you may be providing food and water, too. If you feed pets outdoors, remove food at night. Cover garbage cans tightly, so raccoons and stray dogs cannot invade and scatter contents. Sometimes the food you leave out for the birds becomes a pest attractant; remove excess if it is on the ground.

Wear protective gloves at all times when cleaning yard trash. If you live in an area possibly populated by brown recluse spiders, black widow spiders, poisonous snakes, biting ants, ground wasps, or scorpions, work with long sleeves and a long stick when turning over logs or construction remnants.

Put leaves and lawn clippings in a secure closable container or in a properly constructed compost pile. Be aware that a compost pile must be functioning properly to reach temperatures high enough to destroy pests among the clippings and debris. If your compost pile is not degrading adequately, destroy debris or haul it away promptly. Moving trash from one section of the yard to another merely moves the problem.

Standing water in fishponds, decorative ponds, or unused above- or below-ground pools attracts all kinds of water-loving insects, from water beetles to mosquitoes. Bees and wasps like standing water, too— they are attracted by both reflected light and moisture.

When inspecting the building exterior, pay particular attention to the area where building meets earth. This is a favorite pest entrance. Outdoor wood invites termites, carpenter beetles, carpenter ants, and other cellulose chewers.

Discard excess construction materials or move them away from the house. If you are saving wood for fireplace use, stack it well away from the house, on an elevated platform. If firewood will not be used for some time, spray it with a pesticide such as diazinon to hold down pest populations. Follow label directions regarding time to allow between spraying and burning.

Keep shrubs that are close to the house trimmed back so they do not touch the walls. Use mulch sparingly on any plants near the foundation of the home. Consider treating areas around the foundation with long-acting pesticides such as chlorpyrifos or diazinon to prevent pest entry; heed warnings regarding pets and beneficial wildlife such as birds.

Are there cracks or holes around doors or windows where insect pests can gain entry? Check all exterior doors, including garage doors and side entry doors. Check any areas where utility lines, such as power or phone lines, make entry. Use a good-quality caulk and a caulking gun to seal every crack you find. Weather-strip around doors and windows. This will not only help keep pests out, it will help keep heat in and thereby reduce your winter heating bills.

Inspect screens on doors, windows, and vents. Patch or replace them as necessary with screening small enough to keep tiny insects out. Buy or build a spark arrester screen for your chimney, preferably of 1-inch mesh hardware cloth. Be sure your clothes dryer vent has a flap closure in proper working order.

How you effectively eliminate indoor pests depends on many factors. These include the pest type, its life cycle, how it entered the home, how it moves about, and whether pets or children limit treatment options. A thorough periodic inspection indoors will help keep potential entries closed to pests and ensure that you spot problems before they become severe.

Once pests do enter your home, they require food, water, and appropriate hiding places.

Eliminating these necessities will go a long way toward eliminating the pests.

Caulk cracks along baseboards and cabinets to destroy potential insect homes. Be aware that items stored for recycling can be attractive havens for pests. Piles of newspapers and bins of empty soft-drink cans offer both hiding places and food. If your community offers weekly curbside recycling, put out whatever materials you have collected each week. If you must take your efforts to a recycling center, add a trip there to your list of weekly chores—perhaps you can turn in recyclables on your way to do the food shopping. Do not allow recyclables to sit in your home for long periods.

Leaky pipes anywhere in the home provide the damp situation many insects seek. These pipes can become highly populated breeding grounds. Repair or replace faulty plumbing as soon as possible.

Drop ceilings, often used in converted basements or garages, seem to be particularly inviting homes for rodents. Push one of the sections aside and flash a light around the space above the ceiling; you may find it littered with mouse or rat droppings and chewed insulation. Since neither children nor pets can reach this area, it is a fairly safe place for setting out traps or bait. Handle either with care.

Specific rooms in the house present particular pestproofing problems. The sections that follow constitute a room-by-room tour and will alert you to potential pest attractants.

Firewood can be a source of indoor pests. It is best to stack it well away from the house.

Top: Keep pantry food sealed in lock-tight plastic containers. Right: Before using any control spray in cabinets, remove all food, dishes, and other articles.

The kitchen: A seemingly immaculate kitchen may still be host to a daunting number of ants, cockroaches, grain moths, and beetles. Pests can gain entrance in packages of food or even in the grocery bags themselves. Cockroaches and silverfish hide within the seams of cardboard cartons.

If bringing cartons into the house is unnecessary or if you suspect the cleanliness, leave them outdoors.

Grains are particularly prone to insect pests. Check all flours, cake mixes, and cereals for infestation when you bring them into the home. Destroy contaminated foods immediately. Or, if you intend to return them to the store, keep them in an animalproof container outdoors. If foods seem pest-free, seal each product in a separate lock-tight plastic container. Sealing these foods keeps insects out or, if the products are infested, keeps insects from spreading to other packages. Do not store grain products for long periods. Check them often so that any infestation that does occur can be stopped as quickly as possible. Store large bags of pet food in a metal or plastic garbage can with a tight-fitting lid. Plastic is usually effective, but desperate rodents may actually chew through it and force you to turn to metal.

Of course, crumbs littering the counter, fruit ripening in a bowl, vegetable trimmings in the sink, and cake or bread protected only by plastic wrap offer a free banquet to pests. Vigorous cleaning and a constant eye toward food protection will help starve invaders out of your home.

Pests need water as well as food.

Use caulk and weather stripping to seal cracks around exterior doors.

Presumably, you have already repaired leaky pipes. The kitchen offers other water sources, however. Water in the dish rack or the soggy sponge at the sink can provide plenty of liquid refreshment. Water may be collecting in the drain pan of your frost-free refrigerator. If you have pets, the water you put out for them can be used by less desirable animals.

Make a thorough stove check, both back and sides. Foods tend to drop or spill here and act as household pest attractants. If possible without disturbing gas or other connections, pull the stove away from the wall. Clean walls and stove thoroughly and regularly.

Undersink areas often contain sweaty or leaky pipes, paper bags, and other items among which pests hide. Use a flashlight to inspect pipe and utility line entryways. Home centers sell caulk and other products with which to fill or cover the area between pipe and wall so pest insects cannot enter from outdoors. Clean trash containers daily. Clean undersink areas thoroughly on a regular basis. Also clean the trash container often kept under the sink.

Every so often, clean behind kitchen drawers. Caulk around cupboards where they meet the wall. Purchase caulk that can be painted to match the surroundings.

As you can see, keeping a clean kitchen is not enough to prevent pests. You must also eliminate pest access, make regular sweeps through their hiding places, remove sources of food and water, and react quickly to pest incursions.

The bathroom and laundry room: Although these rooms do not usually contain food for pests, they do provide an array of possible water sources. Repair leaky faucets as well as cracked toilet backs or bowls. A leak provides continual dampness for destructive insects and fungi attracted to water, and even a tiny long-term drip can cause severe damage to your home. Caulk cracks where pipes enter walls, and seal any open areas between toilet bowls and floor. Empty and clean bathtub and sink drain traps every week. Hair and other organic matter trapped here can provide a breeding ground for small pest flies of all kinds.

Check water connections to washing machines. Drips can occur anywhere along the lines. Clean filters in both washer and dryer regularly.

An often-neglected infestation source is laundry brought home from a vacation or by youngsters returning from camp or college. Wash it immediately. Moths are the most common pest cargo in dirty clothing, though it may harbor other insects too.

The living room and bedrooms: Unfortunately, garage sale finds, thrift store treasures, antiques, and donations from friends or relatives can be alive with pests. Inspect all used furniture carefully. Consider spraying newly acquired pieces with insecticide before bringing them indoors.

Vacuuming regularly is a good front-line pest preventive. Use a corner attachment to get into hard-to-reach areas, where carpet insects and fleas tend to hide. Get some assistance, and move heavy furniture occasionally so you can vacuum underneath it, including where furniture legs meet carpet. Vacuum under furniture cushions and into furniture folds. After each vacuuming, empty the vacuum-cleaner bag outdoors and place the contents in a sealed bag for disposal. Many insects find the vacuum-cleaner bag a lovely place to breed.

Regular dusting is also important. Scientists are giving the dust mite increasing attention as a source of human allergy.

In sensitive individuals, dust mites may cause asthmatic symptoms. Wash bedclothes frequently to keep dust mites from accumulating in mattresses and pillows.

The attic: Even if it is only a storage place, clean your attic regularly. Pest bugs frequently move into attics because they are dark, quiet, and often somewhat damp. Discard attic debris, and use a broom to sweep spiderwebs out of the corners. Consider using a high-powered indoor insect fogger on a regular basis to discourage problems. Seal all cracks, and put small-mesh screen on vents.

In addition to damaging items in your home and encouraging rot and other structural damage, roof and plumbing leaks can provide water for many unwanted attic residents. Trace a leak to its source—water tends to run along rafters or framing before it seeks an exit—and correct the problem as soon as possible.

One potential inhabitant of the attic should be encouraged to stay outside, but not driven away. Bats devour an immense number of insect pests, and can be highly beneficial. If one or more has taken up residence in your attic, wait until they have left on their nightly feeding flight and install screening to prevent re-entry. Purchase a bat roost and hang it in an appropriate site. Your insect-control program will get a welcome boost.

The basement: Like attics, basements are dark, damp, quiet, and often full of stored materials. These provide ideal hiding and breeding places for pests. Use an indoor fogger in your basement if it appears to be harboring large numbers of insects.

Regularly inspect basement floors for damp areas and cracks. These can indicate overhead leakage from plumbing. Correct leaks and patch cracked areas so that insects do not crawl up from under the soil. Check all exposed wood, particularly in areas close to the ground. Decay organisms often enter a home at low points and move to the rest of the house.

CONTROLLING PESTS SAFELY

The improper use and disposal of insect and rodent control materials may cause more problems than the pest.

Control preparation: Before using any control, read all the instructions thoroughly. If the instructions are unclear, ask a knowledgeable salesclerk or an agent from your county extension service to explain them to you. Follow instructions exactly. If you are physically unable to use the material as stated, call a licensed pest-control contractor for assistance.

To avoid having to store mixed sprays, try to use up all sprays on the day you prepare them. Store mixed and unmixed controls where children cannot reach them by any means, including by climbing on something. Do not transfer control materials from one container to another—children and pets can easily spill pesticides or rodenticides in cups, glasses, or open bottles. Immediately return any undiluted excess to the original labeled container. Never leave control materials in an unlabeled container.

Safe use: To avoid inhaling chemical vapors, have an ample fresh-air supply available when using controls. Open windows and doors for cross-ventilation. Make certain, before using an aerosol, that the spray opening is pointed away from your face. Do not use a spray near an open flame, and do not smoke while spraying—some controls are extremely flammable. Do not

throw aerosol cans into fire, where heat may make them explode. If any chemical control gets on your body, wash it off immediately. After handling chemicals, wash your hands before handling food.

Safe disposal: Pest control presents two disposal problems: disposal of chemicals and their containers and disposal of dead pests.

■ **Disposal of chemicals:** In some communities throwing chemicals or their containers in the trash or pouring them down the drain is illegal. Most of these communities have set aside special days for the collection of hazardous wastes; on these days you can dispose of your pest controls and pest-control containers along with used motor oil, empty paint cans, and the like. If such a service is unavailable in your area, take your controls and containers to a local collection site as soon as possible.

If pest-control containers can go into the regular trash, rinse them out thoroughly. Then wrap them in newspaper before putting them out for pickup with the rest of the rubbish.

■ **Disposal of dead pests:** Don't forget that the end product of your efforts is a chemically destroyed insect or rodent. Sweep or vacuum up dead insects and seal them in a plastic bag. Keep dead rodents away from children and pets by sealing carcasses in bags and discarding them in a rubbish container with a tight-fitting lid. If this is not possible, or if rubbish pickup is infrequent, bury rodents deep in the soil so animals cannot dig them up.

When storing clothing in plastic bags, wrap mothballs in paper before inserting them into clothing, making sure that they don't touch the fabric or plastic.

PESTS AROUND THE HOME

Housefly

Housefly (7× life size).

Problem: Flies are present in the home and other living areas.

Analysis: The housefly (*Musca domestica*) is common throughout the world. In addition to being annoying, it can spread several serious human diseases and parasites, such as diarrhea, dysentery, typhoid, cholera, intestinal worms, and salmonella bacteria. Several other closely related fly species, including face flies (*M. autumnalis*) and little houseflies (*Fannia canicularis*), may also infest the home. Flies feed on and lay their eggs in decaying organic materials. The eggs hatch within several days, even within 12 hours if conditions are ideal. The creamy white maggots (up to ⅓ inch long) burrow into and feed on the decaying material for several days, pupate, and then emerge as adult flies. Under warm conditions, the entire life cycle may be completed within 14 days. Cooler conditions will greatly extend this period, however. The adults usually live for 15 to 25 days.

Solution: To reduce the fly population, maintain sanitary conditions in the home and garden. Keep garbage tightly covered, and dispose of it regularly. Keep the undersides of lawn mower decks clean of grass clippings. Maintain mulch piles well away from the living area. Keep door and window screens in good condition. Kill flies indoors with ORTHO HomeDefense Home & Garden Insect Killer, ORTHO HomeDefense Hi-Power Brand Indoor Insect Fogger, ORTHO HomeDefense Flying & Crawling Insect Killer, or ORTHO HomeDefense Indoor & Outdoor Insect Killer. Kill flies outdoors with ORTHO Malathion 50 Plus Insect Spray or ORTHO Outdoor Insect Fogger.

Vinegar flies

Vinegar fly (20× life size).

Problem: Tiny (up to ⅙-inch), yellowish brown, clear-winged insects fly around rotting fruit and vegetables, garbage cans, and other wet, fermenting, or rotting materials. These insects fly in a slow, hovering manner.

Analysis: Vinegar flies (*Drosophila* species), also known as fruit flies, do not constitute a serious health menace but can be annoying in locations where garbage, fruit, or vegetables are allowed to rot and ferment. The adult female flies lay their eggs in the decaying fruit or vegetables. The eggs hatch in a few days, and the tiny maggots feed on yeasts growing in the decaying food. The maggots pupate and become adults; the entire life cycle takes only 10 to 20 days.

Solution: To reduce the fly population, maintain sanitary conditions. Keep garbage tightly covered, and dispose of it regularly. Wash garbage cans on a regular basis. Kill flies with a spray or fogger containing *allethrin*, *chlorpyrifos*, or *pyrethrins*.

Household spiders

House spider with web case.

Problem: Spiderwebs and spiders are found in secluded, rarely disturbed areas in and around the home.

Analysis: Many kinds of spiders wander into the home. With only a few exceptions, these familiar creatures are harmless and cannot reproduce in the house. Spiders are often beneficial, feeding on other spiders and insects, including such household pests as flies and moths. The more insects there are inside the home, the more likely spiders will live there. Most spiders spin silken webs. Some, such as tarantulas, are active, hunting spiders that do not spin webs. When spiders bite humans, it is usually because they have been squeezed, lain on, or somehow provoked. Only a few spiders, particularly the black widow and brown recluse, are dangerous to people, but their bites are rarely fatal. (For more information on black widow and brown recluse spiders, see pages 327 and 328.)

Solution: Knock down webs with a broom or duster. Kill spiders and the insects they feed on by spraying infested areas with ORTHO HomeDefense Hi-Power Indoor Insect Fogger or ORTHO HomeDefense Indoor & Outdoor Insect Killer. To reduce the number of spiders entering the home, seal cracks in the home, inspect and repair window screens, and clean up accumulations of debris outdoors that may harbor spiders or their prey. Spray with ORTHO HomeDefense Indoor & Outdoor Insect Killer around doors, windows, and foundations where spiders may enter.

Daddy longlegs

Daddy longlegs (2× life size).

Problem: Spiderlike creatures with small bodies and long, delicate legs are found in or around the home and garden. They are sometimes seen in large gatherings, standing with their legs interlaced.

Analysis: Daddy longlegs, also called *harvestmen*, are closely related to spiders but are not true spiders. They cannot spin silken webbing. They are most common in areas near a source of water. In northern states, most daddy longlegs die in autumn after the female lays eggs. In southern states, females usually spend the winter under ground litter and lay eggs in the spring. They lay eggs in the soil under stones, wood, and other debris. Daddy longlegs feed mainly on small insects. They are most active at night and do not bite humans. Occasionally they wander indoors, but they do not cause any damage.

Solution: Reduce the number of daddy longlegs coming indoors by cleaning up wood, trash, and other debris that may harbor them outside the home. Trim plant growth away from the house. Seal cracks and crevices around windows and doors, and repair broken screens.

Crickets

House cricket (life size).

Problem: Crickets are chirping in the home. These insects, which look like grasshoppers, are light to dark brown, are ½ to ¾ inch long, and have long antennae that curve back along the sides of the body.

Analysis: The two types of crickets that may invade the home are field crickets (*Gryllus*) and house crickets (*Acheta domesticus*). Field crickets usually live outdoors, feeding on vegetation and plant debris. In the fall, when their natural food supply fails, or during periods of heavy rainfall, they may invade buildings in search of food. Field crickets cannot reproduce in the house and usually die by winter. House crickets, however, survive and reproduce indoors. During the day, both types of crickets hide in dark, warm locations, such as behind baseboards, in closets, and in attics. Male crickets make a chirping sound by rubbing the file and scraper on their forewings together. Crickets may chew on fabrics and paper items, and large numbers of them may cause serious problems.

Solution: To control crickets outdoors, apply ORTHO Ant-Stop Ant Killer Dust, ORTHO Flea-B-Gon Outdoor Flea & Tick Killer, or ORTHO Diazinon Granules along the foundation of the house. To control crickets indoors, spray with ORTHO HomeDefense Hi-Power Brand Roach, Ant & Spider Killer or ORTHO HomeDefense Indoor & Outdoor Insect Killer. Remove dense vegetation and debris from around the building foundation where crickets may hide. Seal openings around doors and windows. Apply a barrier spray around the house using an insecticide containing *diazinon*.

Night-flying insects

June beetle

Problem: Numerous insects are flying around indoor or outdoor lights at night. Their physical presence as well as their buzzing or droning is bothersome. Dead insects may accumulate below lights, attracting ants and other insects to the site.

Analysis: Many night-flying insects use the moon and stars to help them discern direction. When they see a brighter source of light, such as a lightbulb, they mistake it for one of these objects and orient to it, eventually striking the light. Lights attract a wide variety of night-flying insects, including most moths and certain beetles, mosquitoes, flies, gnats, and leafhoppers. When many insects become adults during a short period, large numbers may be attracted to lights.

Solution: Spray ORTHO Outdoor Insect Fogger or burn citronella candles to temporarily eliminate outdoor flying insects. Spray ORTHO HomeDefense Home & Garden Insect Killer to control flying insects indoors. Replace white lightbulbs with yellow ones; yellow light is less visible to insects and therefore less attractive to them. Use lightbulbs of lower wattage, and turn off lights when not needed. Locate outdoor lights at least 25 feet from doors and windows. Install or repair screens to prevent insects from moving indoors. If entertaining outdoors, consider using candles for light. Their lower light intensity is less attractive to insects.

PESTS AROUND THE HOME *(continued)*

Swimming pool pests

Water boatman (8× life size).

Problem: Insects are in the swimming pool. The insects may be actively swimming in the pool, or they may be floundering or dead. Even well-maintained pools can have this problem. Some of these insects may inflict painful bites.

Analysis: Many insects and related organisms become pests in swimming pools. They may either fall in the pool and drown or live in the pool. Sowbugs, millipedes, springtails, and other insects living in nearby vegetation may crawl into the pool and drown. They are particularly common if there is an abundance of organic matter under shrubbery near the pool. Bees and wasps may fall into the pool as they search for water. Insects that can live even in chlorinated and clean pools include some beetles and several bugs: back swimmers, giant water bugs, water boatmen, and water striders. Back swimmers and giant water bugs can inflict painful bites similar to bee stings. These insects, as well as many moths, are attracted to pool lights. If a swimming pool is not kept chlorinated and clean, mosquitoes and midges may breed in the water.

Solution: Skim insects off the surface of the water with a dip net. Use lights sparingly near pools, and switch to yellow lights (which are less attractive to insects) if you have a continual problem. Or place a very bright light source a few hundred feet away to attract night fliers away from the pool. Keep the pool chlorinated and reasonably clean. Keep grass and shrubbery trimmed near the pool. Control insects in nearby shrubbery. Do not spray the pool directly.

Earwigs

Earwig (2× life size).

Problem: Reddish brown, flat, elongated insects up to 1 inch long with straight or curved pincers projecting from the rear of the body are present in the home. They are often found in dark, secluded places such as in pantries, closets, and drawers and even in bedding. They may be seen scurrying along baseboards or moving from room to room.

Analysis: Several species of earwigs, a nocturnal insect, may infest the home, including the European earwig (*Forficula auricularia*), the ringlegged earwig (*Euborellia annulipes*), and the striped earwig (*Labidura riparia*). Earwigs are usually found in the garden, where they feed on mosses, decaying organic matter, vegetation, and other insects. Large numbers of earwigs may invade homes, however, through cracks or openings in the foundation, doors, and window screens, especially during hot, dry spells. Although they do not damage household furnishings, their presence is annoying, and they may feed on stored food items or hide in areas where food is kept. They may inflict painful pinches when provoked.

Solution: Store food in sealed containers. Repair cracks or openings in window screens, doors, and the building foundation. Use a barrier spray to prevent entrance to the house. Control earwigs indoors by spraying with ORTHO HomeDefense Indoor & Outdoor Insect Killer or dusting with ORTHO Ant-Stop Ant Killer Dust or ORTHO HomeDefense Hi-Power Brand Roach, Ant & Spider Killer. Use according to label directions. To prevent reinfestations, control earwigs outdoors.

Outdoor-originating cockroaches

American cockroach (life size).

Problem: Cockroaches are found outdoors in woodpiles, ground covers, leaf litter, and other protected areas. Occasionally they wander indoors.

Analysis: Outdoor-originating cockroaches, often called *wood roaches*, live mainly outdoors; they occasionally wander inside but cannot reproduce there. The American cockroach (*Periplaneta americana*) and the smokybrown cockroach (*P. fuliginosa*) are two species that can live equally well indoors and outdoors in warm climates. The smokybrown cockroach, in particular, is a pest in many southern states. It moves indoors when weather conditions outside become adverse. Outdoor-living cockroaches are more likely to wander indoors if there are suitable places for them to live and breed next to the house. Favorite habitats include ground covers and piles of wood, compost, and other debris. These cockroaches are general scavengers, eating decaying plant and animal material.

Solution: To keep cockroaches from coming indoors, move compost and woodpiles away from the house. Clean up litter and debris near the home. Apply ORTHO Diazinon Ultra Insect Spray or ORTHO Ant-Stop Ant Killer Dust as a 2- to 5-foot barrier around the foundation and in nearby ground covers. Make sure ground cover is not sensitive to the chemical. Appropriate-use plants are listed on the product label. Repair windows, door screens, cracks, and crevices in the walls and foundation. Inspect firewood before bringing it indoors (see page 330).

House centipede

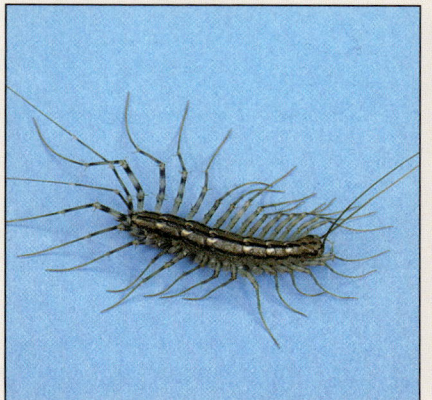

House centipede (life size).

Problem: A thin centipedelike creature with long legs is crawling on the floor or wall. It is up to 1½ inches long, with very long antennae and legs in comparison with its body size. The antennae and rear set of legs can be more than twice the length of the body. The creature runs quickly, with sudden stops.

Analysis: The house centipede (*Scutigera coleoptrata*) is found both indoors and outdoors in warm regions of the United States but only indoors where winters are colder. Unlike other centipedes, which wander indoors but cannot reproduce there, house centipedes live and reproduce indoors. House centipedes prey on other insects and become most numerous where there is an abundance of insects to feed on. They prefer dampness and thrive in typically moist areas such as cellars, closets, or bathrooms. Outdoors they are common in moist piles of compost and other debris. House centipedes are most active at night. They seldom bite humans, but when they do, the bite is no more severe than a bee sting.

Solution: To control house centipedes and the insects they feed on, spray indoor areas with an insecticide containing *diazinon* or *pyrethrins*. Eliminate moist areas in and around the home. Air out damp places. Outdoors, remove piles of compost and other materials that provide hiding places near the house.

Silverfish and firebrats

Silverfish (3× life size).

Problem: Paper and fabric products, especially those made with glue, paste, or sizing, are  stained yellow, chewed, or notched and may be covered with excrement and silver or gray scales. Flattened, slender, wingless insects up to ½ inch long may scurry away when infested products are moved. These insects have long, thin antennae and are either silvery and shiny or dull and mottled gray.

Analysis: Silverfish (*Lepisma* species) and firebrats (*Thermobia* species) are similar in size, shape, and feeding habits except for their coloration and hiding places. Silverfish are silvery and prefer damp, cool to warm locations (70° to 80°F) such as basements and wall voids. Firebrats are mottled gray and prefer damp, hot locations (90° to 105°F) such as hot-water pipes and areas near the oven or furnace. These pests are active at night and hide during the day. They feed on a wide range of foods, especially products high in starches, including human food, paper, paste, and linen and other fabrics. Silverfish damage books by feeding on the bindings. These pests crawl throughout the house along pipes and through holes or crevices in the walls or floor. The adult females lay eggs in cracks or openings behind baseboards and in other protected areas.

Solution: Treat with ORTHO HomeDefense Indoor & Outdoor Insect Killer or ORTHO Ant-Stop Ant Killer Dust according to label directions. Where practical, seal all cracks and crevices in the infested areas. Store valued papers and clothes in tightly sealed plastic bags.

Booklice

Psocid (15× life size).

Problem: Numerous tiny insects the size of a pinhead are crawling in stored food products or  around books. These insects may emerge from behind walls for several months after construction of a building. They run along surfaces in a hesitating, jerky manner.

Analysis: Booklice, also known as *psocids*, thrive in warm, damp, undisturbed places. They feed mainly on microscopic molds that may develop on certain kinds of adhesives used in book binding and wallpaper. Booklice sometimes infest damp Spanish moss, straw, or other vegetable matter used in making upholstered furniture. They feed directly on cereals and other starchy materials (particularly if these products are stored for a long period in damp conditions) and on dead insects. Although booklice contaminate stored food with their body parts, they do not cause other damage. Insect damage seen on books is caused by other insects, such as silverfish (see column at left) or cockroaches (see page 330). Booklice do not bite or carry disease organisms.

Solution: Dry out infested areas of the home. If booklice are in the food pantry, search for and throw out any infested food. Ventilation and artificial heat can aid in drying out cupboards. Booklice will disappear from new homes as the structure dries. Control infested furniture by thoroughly drying the item in sunlight for several days, having it fumigated by a pest control operator, or, if practical, discarding the infested stuffing. Infested areas also can be treated with an insecticide containing *pyrethrins*. Make sure the site you spray is listed on the product label.

PESTS AROUND THE HOME (continued)

Carpet beetles

Carpet beetles and larvae (4× life size).

Problem: Irregular holes are chewed in carpets, blankets, clothing, and articles made of animal fur, hair, feathers, or hides. Light brown to black grubs up to ¼ inch long may be seen crawling on both damaged and undamaged items. The grubs are distinctly segmented and covered with circular rows of stiff, dark hairs.

Analysis: The larvae of carpet beetles (*Attagenus megatoma* and *Anthrenus* species) damage carpets, clothes, upholstery, and other products of animal origin. Some species also feed on stored foods. The adult beetles are about ⅛ inch long and may be black or mottled gray, brown, and white. They usually live outdoors, feeding on pollen and nectar. The beetles fly into homes during the late spring or early summer and lay their eggs in cracks or crevices or on clothes, carpets, or other materials. The emerging larvae seek out dark, undisturbed locations in which to feed. They shed their skins several times during their development. Most of the larvae hibernate during the winter and pupate in the spring.

Solution: Shake out, brush, and air infested clothes and blankets. To kill remaining grubs, dry-clean infested items. Vacuum infested rooms, particularly under furniture and in corners. Destroy the sweepings immediately. Kill remaining insects with ORTHO HomeDefense Indoor & Outdoor Insect Killer.

Ants

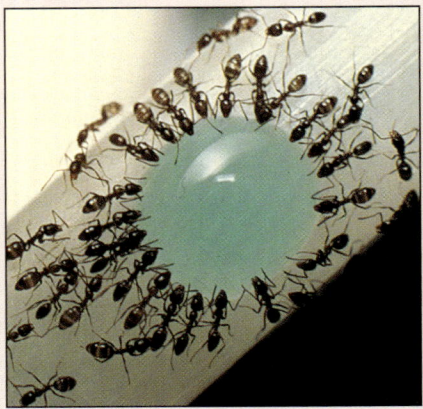

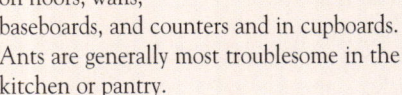
Black ants (2× life size).

Problem: Ants are present in the house. Trails of ants may crawl on floors, walls, baseboards, and counters and in cupboards. Ants are generally most troublesome in the kitchen or pantry.

Analysis: Several species of ants invade households. Most ants are strongly attracted to sweets, starches, fats, and grains and invade households in order to carry these foods back to their nests. Adverse outdoor conditions, such as flooding or drought, may cause ants to move their nests or colonies into buildings. Ant colonies are often built underground in the garden but may also be found under flooring and in building foundations, wall partitions, attics, and other protected locations. Ant colonies may contain from several hundred to several thousand individuals. Colonies may be located by following the established ant trails to their source.

Solution: Destroy ant colonies by dusting with ORTHO Ant-Stop Ant Killer Dust or by spraying the nests and ant trails with ORTHO HomeDefense Home & Garden Insect Killer, ORTHO HomeDefense Indoor & Outdoor Insect Killer, or ORTHO Ant-Stop Ant Killer Spray. Eliminate ant colonies from the garden by treating the anthills with ORTHO Diazinon Granules or by spraying the nests and surrounding soil with ORTHO Diazinon Ultra Insect Spray. Store food in sealed containers, and keep kitchens and pantries free of exposed foods. Place ORTHO Ant-Stop Ant Killer Bait in areas where ants have been observed.

Clothes moths

Clothes moths (life size).

Problem: Holes are chewed in clothing, blankets, carpets, pillows, upholstery, and other items. Infested articles may be covered with a webbing of silken tubes, cases, or strands. Shiny white caterpillars up to ½ inch long may be seen crawling on damaged items.

Analysis: The larvae of the small, yellowish to tan clothes moths (*Tineola bisselliella* and *Tinea pellionella*) damage clothes and other items made of fur, wool, feathers, and leather. The female moths attach their eggs to the fabric. Soon after the larvae emerge they spin silken tubes, strands, or cases. The larvae usually feed from within these protective casings but may crawl out to feed unprotected. The larvae pupate in cocoons attached by silken threads to the infested item and emerge as adult moths. New infestations occur when moths lay eggs on clothing, carpets, and other articles and when moth-, larva-, and egg-ridden items are stored with uninfested articles. Clothes moths are not attracted to light.

Solution: Shake out, brush, and air infested clothes and blankets in a sunny location. To kill remaining moths, dry-clean infested items. Place in airtight closets or containers with mothballs or flakes. Vacuum or sweep infested rooms, and destroy the sweepings immediately. Seek professional help to protect carpets and furs.

Clover mite

Clover mite (30× life size).

Problem: Reddish brown mites smaller than a pinhead, with long front legs, are present in the home in the fall. They may be found on walls, windowsills, floors, and furniture and even in bedding and clothes. When crushed, these mites leave a blood red stain. They are often so numerous that they give infested surfaces a reddish appearance.

Analysis: The clover mite (*Bryobia praetiosa*), related to spiders, is found throughout the United States. Clover mites feed and reproduce on clover, grasses, and other plants. They are most active in the spring and fall and on warm winter days. The mites lay eggs during the summer and fall. The young that hatch from these eggs feed on vegetation and then migrate into homes and other protected areas during the early fall. Mites enter homes through cracks or openings in the foundation and around doors and windows. Mite activity usually decreases when temperatures rise above 85°F or fall below 40°F.

Solution: Remove mites from household furnishings by vacuuming them from infested surfaces. Do not crush the mites, because they will leave a red stain. Kill mites indoors by spraying them directly with ORTHO HomeDefense Indoor & Outdoor Insect Killer. Treat outdoor areas with ORTHO HomeDefense Indoor & Outdoor Insect Killer, ORTHO Ant-Stop Ant Killer Dust, or ORTHO Malathion 50 Plus Insect Spray as a barrier treatment. Keep a strip of soil 18 to 24 inches wide around the foundation of the building free of vegetation and debris to reduce mite movement into homes.

Boxelder bugs

Boxelder bug (4× life size).

Problem: During the fall, hordes of brownish black bugs ½ inch long with red stripes on their wings swarm into the home and outdoor living areas. They congregate on walls, walks, furniture, drapes, and other objects. When crushed, they emit a strong, unpleasant odor.

Analysis: Boxelder bugs (*Leptocoris* species) are common in all parts of the country. They are most numerous in areas where boxelder trees (*Acer negundo*) grow. In the spring, the female bugs lay their eggs in the bark of boxelders or sometimes of maples, ash, and fruit trees. The young feed on tender twigs, foliage, and seeds through the spring and summer. During the fall, especially on bright, sunny days, the bugs migrate in large numbers into tree trunks, homes, buildings, or other dry, protected locations to hibernate for the winter. Boxelder bugs do not feed on fabric or furniture, but they may stain household items with their excrement. Boxelder bugs occasionally bite, and they may feed on houseplants.

Solution: Vacuum boxelder bugs with a tank-type vacuum cleaner and then destroy the bag. Or spray the bugs with an insecticide containing *diazinon* or *pyrethrins*. Spray outdoor areas with ORTHO Malathion 50 Plus Insect Spray or ORTHO Diazinon Ultra Insect Spray. Keep the doors and windows screened and the cracks around them well sealed.

Cliff swallow

Swallow nests.

Problem: Mud nests are found beneath eaves of the building. The nests are about 6 inches in diameter, are shaped like gourds, and have necklike entranceways with round holes in them. Nests are usually grouped together. Bird droppings and mud are scattered beneath the mud structures. In spring, birds fly in and out of the nests.

Analysis: The cliff swallow (*Hirundo pyrrhonota*), also known as *mud swallow*, spends its winters in South America and annually migrates northward to the United States. From March through June the birds build their mud nests, usually against a vertical wall just beneath an overhang such as an eave. The same nesting sites are used year after year, and many birds return to the same area they nested in the previous year. The birds abandon the nests by the end of June.

Solution: Swallows are protected by state and federal regulations, so obtain a depredation permit from the U.S. Fish and Wildlife Service before removing swallow nests. Wash the nests from under the eaves with a strong stream of water. This must be done consistently over an extended period, or the birds will rebuild the mud colony. Or, after washing the nests away, string a wire across the nest area (usually that means stringing a wire along the junction of the wall and the roof overhang) and drape a 12-inch curtain of aluminum foil or polyethylene sheeting over it. This prevents cliff swallows from attaching their nests to the wall—the new surface is too smooth. If the problem continues, contact a licensed pest control operator.

PESTS AROUND THE HOME *(continued)*

Birds

Pigeon.

Problem: Birds are roosting or nesting in wall voids, under eaves, and in other areas around the home.

Analysis: Most birds are harmless and pleasant, but a few species, especially pigeons, starlings, and sparrows, may become a nuisance around the home. These birds are adapted to urban and suburban environments. They roost and build nests on chimneys, ledges, rafters, eaves, drainpipes, and similar locations and often return to the same nesting site year after year. In addition to their messy droppings and irritating chirping, birds may transmit to humans such diseases as pigeon psittacosis, aspergillosis, encephalitis, and histoplasmosis.

Solution: Where possible, exclude birds by installing screens. Apply a bird repellent adhesive or jelly to roosting and nesting areas. Clean up possible food or nest-building materials (such as dried weeds or vegetation) to discourage bird activity in the vicinity. Place flashings of hard slippery plastic or metal at a 45-degree angle on nesting or roosting areas. Hang protective netting from roof eaves. Birds may also be trapped or poisoned. If you are considering poison, contact a licensed pest control operator or your county agricultural commissioner's office for regulations pertaining to your area.

Raccoons

Raccoon.

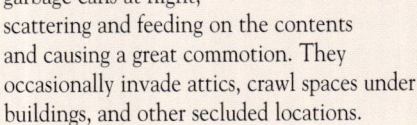

Problem: Raccoons are present around the home. They overturn garbage cans at night, scattering and feeding on the contents and causing a great commotion. They occasionally invade attics, crawl spaces under buildings, and other secluded locations.

Analysis: Raccoons (*Procyon lotor*) are a nuisance mainly in rural and suburban areas. They generally live near a source of natural water, such as a stream, marsh, or pond. Raccoons are dexterous and inquisitive animals. They sometimes take up residence in attics, basements, barns, or similar locations. Although these animals carry fleas and ticks, they are not considered a serious health threat. Raccoons can be dangerous if cornered. They can be carriers of rabies, so avoid all animals exhibiting odd behavior.

Solution: Keep garbage cans securely anchored in racks or immovable frames. Lids should be tightly secured to the can. Screen or seal openings into buildings. Consult your state department of fish and game to find out about local raccoon control restrictions and regulations. Live-catch traps baited with pieces of melon, prunes, honey-coated bread, or smoked fish are usually effective in controlling raccoons. Attach traps to a tree, stake, or fence post. If possible, push the trap back and forth in the ground until the soil covers the wire mesh on the bottom of the trap. Wait a few days before setting the trap; this allows the animal to become accustomed to it. Transport the trapped animal to a wooded area at least several miles away.

Tree squirrels

Gray squirrel.

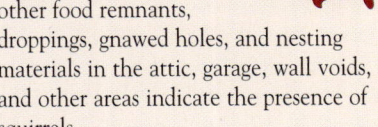

Problem: Squirrels are seen or heard in the building. Or nuts or other food remnants, droppings, gnawed holes, and nesting materials in the attic, garage, wall voids, and other areas indicate the presence of squirrels.

Analysis: Several species of tree squirrels invade houses, including the fox squirrel (*Sciurus niger*), the eastern gray squirrel (*S. carolinensis*), and flying squirrels (*Glaucomys* species). Squirrels enter buildings through vents, broken windows, construction gaps under eaves and gables, and occasionally chimneys and fireplaces. They may build nests or store food in attics, wall voids, garages, and similar locations, and they may damage items stored in attics or garages.

Solution: Contact your state department of fish and game for regulations governing the control of tree squirrels in your area. Eliminate animals inside the building by placing traps in the areas they are inhabiting. Bait the traps with nutmeats, chunk-style peanut butter, sunflower seeds, or raisins. If tree squirrels are entering the building via trees or power lines, secure traps to tree limbs or the rooftop to intercept them. Once squirrels have been eliminated from the building, seal entry routes into the home with sheet metal or hardware cloth. Prune tree limbs at least 6 feet away from the roof or any other part of the building.

House mice

House mouse.

House mouse.

Skunks

Skunk.

Problem: Mice are seen in the garage or home, or signs of mouse infestation—including droppings, tracks, or gnawed doors, baseboards, or kitchen cabinets—are found. Books, fabrics, furniture, and other objects may be chewed or shredded, and packages of food may be gnawed open and the contents eaten.

Analysis: House mice (*Mus musculus*) often go unnoticed if only a few are present but may cause significant damage when their numbers are large. In addition to gnawing on clothing, furniture, and other items, mice contaminate food with their urine and droppings and may spread parasites and diseases. Mice are generally active at night. Under ideal conditions, the females produce up to 50 young in a year. Mice are agile and can jump as high as 12 inches off the ground, run up almost any rough vertical surface, swim, and squeeze through openings slightly larger than ¼ inch. House mice feed primarily on cereal grains but will eat many other kinds of foods, including butter, fat, meat, sweets, and nuts.

Solution: Apply anticoagulant bait or a bait consisting of cereal grains treated with *cholecalciferol* (vitamin D-3), a new-generation rodenticide. This cereal bait is safer to use around pets and other domestic animals than anticoagulant baits. The bait is contained in packets that should be placed in the same areas in which traps would be placed but out of the reach of children and pets. Another way to eliminate mice in the home is to trap them. Mice are more likely to seek bait in traps if their normal source of food is scarce. Remove food from areas where mice can get to it, and store grains in sealed metal, glass, or heavy plastic containers. Place traps where mouse droppings, gnawings, and damage indicate the presence of mice. These include such areas as behind refrigerators and other protective objects, in dark corners, along baseboards, and in cupboards. Bait the traps with pieces of bacon, nutmeats, raisins, or peanut butter. Tie the bait to the trigger so the mouse won't be able to remove the bait without springing the trap. Check the traps daily. Wear gloves when handling dead mice, or use tongs to pick them up to avoid bites from mouse parasites. If you are unable to eliminate all the mice, contact a professional pest control operator. After the mice have been eliminated, prevent them from returning by sealing holes or cracks larger than ¼ inch in walls, floors, windows, doors, and areas of the foundation that open to the outside. For details on mouse-proofing your home, contact your local county extension service.

Problem: Skunks are observed living beneath the building, or their tracks and strong scent are present around the home.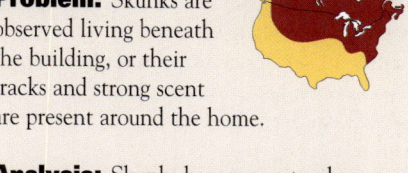

Analysis: Skunks become pests when they take up residence under a house. They are most likely to make a den under a house when natural burrows or dens are not readily available. The strong scent they spray when threatened may cause nausea and even temporary blindness. Skunks can eject this potent fluid as far as 10 feet. Skunks carry a variety of diseases, including rabies. They may transmit rabies to humans and pets. Rabid skunks often show abnormal behavior such as listlessness, unprovoked aggressiveness, or a tendency to wander around during the day. Such animals will bite if handled.

Solution: Contact your state department of fish and game office for skunk control regulations. Eliminate skunks by placing mothballs, open pans of household ammonia, or several floodlights under the building to drive them out. Live-catch, box-type traps may also be used. When handling skunks, wear old clothing and goggles. After the animals have been eliminated, screen off or seal openings into the building. Have skunk bites treated immediately by a physician or veterinarian. Skunk scent may be neutralized with tomato juice or *neutroleum alpha*, a compound that may be obtained through a hospital supply outlet, or specially formulated products available at pet stores.

PESTS AROUND THE HOME (continued)

Rats

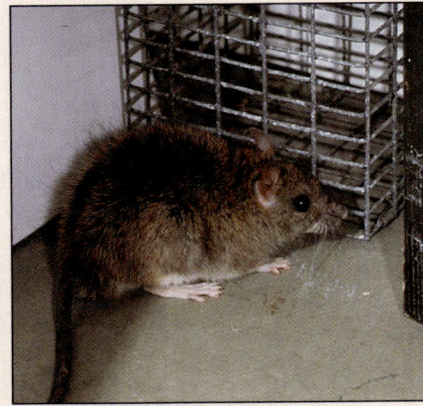

Roof rat.

Norway rat.

Problem: Rats are seen or heard in the attic, garage, basement, wall voids, or other areas of the home. Signs of rat infestation include droppings, tracks, and loosely constructed nests made of rags, paper, and other scraps. Pipes, beams, and wiring may be gnawed. Books, fabrics, furniture, and other objects may be chewed or shredded, and packages of food may be gnawed open and the contents eaten.

Analysis: Rats (*Rattus* species) are distributed worldwide and infest well-maintained residences as well as run-down houses and apartments. The species that most frequently infest houses are the Norway rat (also known as the brown, house, wharf, or sewer rat) and the roof rat. Rats enter buildings through any opening, including toilets, pipes, chimneys, and garbage chutes. They are excellent climbers and can gain access to homes from nearby trees. Young rats can squeeze through openings as small as ½ inch wide. These animals make their nests and breed in wall voids, attics, crawl spaces, basements, and other secluded locations. They also breed in heavy vegetation, such as ivy or juniper ground covers, near the home. Their long front teeth grow constantly. To keep them worn down, rats gnaw on almost anything, including clothing, furniture, and electrical wires. They can also gnaw through gas lines, causing gas leaks. Rats are notorious for contaminating food with their urine, droppings, and hair, spreading parasites and diseases. They occasionally bite people, especially sleeping infants. The bites are dangerous and must be treated by a doctor.

Solution: Apply anticoagulant bait or a bait consisting of cereal grains treated with *cholecalciferol* (vitamin D-3), a new-generation rodenticide. This cereal bait is safer to use around pets and other domestic animals than anticoagulant baits. The bait is contained in packets that should be placed in the same areas as traps would be placed but out of the reach of children and pets. Another way to control rats in the home is by trapping them. Use rat traps; the smaller mouse traps will not be effective. Rats are more likely to seek bait in traps if their normal source of food is scarce. Remove food from areas where rats can get to it easily. Store food in glass or tin containers with tight lids. Place traps along rat runways, anchoring the trap securely to a nearby object so the animal won't drag it away. Bait traps with pieces of beef, bacon, fish, nutmeats, or carrots. Tie the bait to the trigger so the rat can't remove the bait without springing the trap. Check the traps daily. Rats are wary creatures. Unlike mice, they will not accept strange objects in their territory, even if baited. Be patient: it may take several days to a few weeks before these animals move into a live trap or take bait from a snap trap. Wear gloves or use tongs to pick up rats to avoid bites from rat parasites such as fleas and mites. If you are unable to eliminate the rats, contact a professional pest control operator. Rat-proof the building. Clear landscape plantings to at least 18 inches from the structure. Identify ground burrows, place bait inside, and cover the hole. Rat-proofing may involve much expense and work because it involves sealing all openings larger than ¼ inch leading into the building from the outside. Sealing such small openings also keeps out mice. For details, contact your county extension service.

PET AND BODY PESTS

Chiggers

Chigger-infested field.

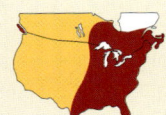

Problem: Welts and hard raised bumps (papules) appear on the skin, particularly on parts of the body where clothing is binding or where body parts come in contact, such as at the belt line, the armpits, or the backs of the knees and under cuffs or collars. Itching is severe and may last as long as 2 weeks. Welts and itching often develop within several hours to a day after the affected person has been in a scrubby, thicket-covered, or otherwise heavily vegetated area.

Analysis: Chiggers (*Trombicula* species), also known as red bugs, are the larval forms of several closely related microscopic mites. Only the larvae are harmful. They hatch from eggs laid in the soil of uncultivated, scrubby woodland or marshy areas and attach themselves to people and other hosts as they pass by. Chiggers insert their mouthparts into the skin and feed on blood for several days until they become engorged and drop off.

Solution: To remove chiggers from your skin, bathe thoroughly in hot, soapy water. Contact your druggist for compounds to relieve the itching. When walking through chigger-infested areas, wear protective clothing, and tightly button or tape sleeves, pant cuffs, and collars. Apply a repellent containing *diethyl toluamide* or *ethyl hexanediol* to the skin and clothing, especially around the ankles, underarms, waist, sleeves, and cuffs. Treat infested areas around your home with ORTHO Diazinon Soil & Turf Insect Control or ORTHO Diazinon Ultra Insect Spray according to label directions.

Fleas

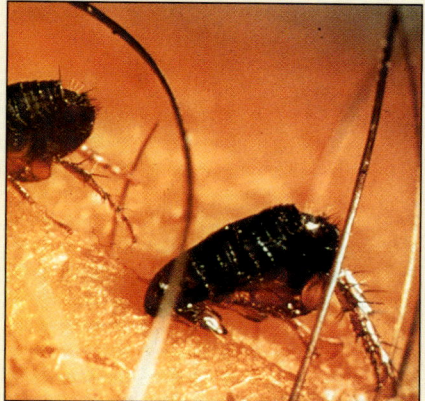

Fleas.

Flea larvae (6× life size).

Bedbugs

Bedbugs (6× life size).

Problem: Fleas infest pets, pet quarters, rooms, carpets, upholstered furniture, or the garden.

Analysis: Fleas, which are pests of humans, dogs, cats, and many other warm-blooded animals, are found throughout the world. In addition to causing annoying bites, they can transmit sserious diseases and parasites such as bubonic plague, murine typhus, and tapeworms. The cat flea (*Ctenocephalides felis*), the dog flea (*C. canis*), and the human flea (*Pulex irritans*) are the most common species found around the home. These fleas have a wide host range, attacking humans, dogs, cats, and several other animals. The female fleas lay eggs shortly after feeding upon animal blood. The eggs are usually laid on the host's body or in the host's bedding. The eggs often fall off the host's body into floor crevices, dog and cat boxes, carpets, and other areas where the infested animals spend time. Within 10 days, the eggs hatch into tiny, wormlike larvae that feed on dried blood and excrement. Pupation occurs after 1 week to several months. The adult fleas may emerge after only 1 week if conditions are favorable, or emergence may be delayed up to 1 year. The adults often remain in their pupal cocoons until a host is present. A flea's life cycle may vary from 2 weeks up to 2 years. Because fleas have the ability to survive for many months in their cocoons, they can remain in vacated residences for long periods of time, waiting to emerge and bite returning pets and humans. Fleas are mainly spread by infested animals. Uninfested animals can easily pick up fleas when visiting flea-ridden areas. Fleas may also be spread by infested articles of clothing or furniture.

Solution: Treat infested pets. Spray them with ORTHO Flea-B-Gon Total Flea Killer Indoor Spray or ORTHO Flea-B-Gon Pet Flea & Tick Killer. Spray infested animal quarters with ORTHO Flea-B-Gon Flea & Tick Killer Indoor & Outdoor Spray, or dust with ORTHO Ant-Stop Ant Killer Dust. Read and follow label directions carefully. Flea-repellent collars may help to control fleas on animals. Destroy infested pet bedding or wash it thoroughly in hot, soapy water. Vacuum carpeting, chairs, sofas, and other areas or objects that may contain eggs and larvae-ridden lint or debris, and then dispose of the vacuum bag immediately. Kill remaining fleas with ORTHO Flea-B-Gon Total Flea Killer Indoor Spray, ORTHO HomeDefense Hi-Power Brand Indoor Insect Fogger, or ORTHO Flea-B-Gon Total Flea Killer Indoor Fogger. Kill fleas in the yard by spraying with ORTHO Flea-B-Gon Outdoor Flea & Tick Killer, ORTHO Malathion 50 Plus Insect Spray, or ORTHO Diazinon Ultra Insect Spray. To prevent reinfestations, do not allow infested animals to enter the house and yard, and keep pets away from infested areas.

Problem: Painful swellings develop on the body. Dark brown to black spotted stains appear on pillows, sheets, and other bedding. When the lights are turned on at night, reddish brown, oval-shaped bugs about ¼ inch long may be seen crawling on the skin and bedding.

Analysis: Bedbugs (*Cimex lectularius*) have become an infrequent problem in the United States. They still infest homes where living conditions are unsanitary, however, and may be transported to well-maintained residences on infested clothing and furniture. Bedbugs hide during the day behind baseboards and in mattresses and upholstered furniture, cracks and crevices in the floor, bed frames, ceilings, and similar locations. These pests are nocturnal and become active at night. They crawl onto a sleeping person, pierce the skin, and suck the blood for several minutes. Bedbugs usually deposit telltale masses of dark excrement on bedding or resting sites after they feed. They may crawl from room to room during the night.

Solution: Spray baseboards, wall crevices, and floor cracks with an insecticide containing 2 percent *malathion* or *ronnel*. Spray bed frames and springs with an insecticide containing 1 percent *malathion* or *ronnel*. Mist mattresses with an insecticide containing 1 percent *malathion*, but do not soak them. Be sure to spray mattress seams and tufts. Let mattresses dry completely before reusing. Do not spray infant cribs and bedding. If bedbugs reinfest the home, treat again after an interval of 2 weeks. Launder all bedding thoroughly. Keep the house clean.

STRUCTURAL PESTS

Ticks

Pajaroello tick (2× life size).

Deer tick larva, nymph, and adult.

Problem: Leathery, oval-shaped reddish brown to dark brown pests are walking over or burrowed into the skin of a person or pet. Ticks range in size from 1/16 to 1/8 inch before feeding. On humans, they are usually found where skin is exposed; on pets, they are most often on the ears or neck. Ticks are most likely to be discovered after the host has walked through tall grass or vegetation, especially at the end of summer or in fall. In some instances a rash may break out at the site of the bite. The bite itself forms a strawberry-colored bull's-eye in the midst of the rash. The rash may also appear with no sign of a tick bite. Typical cold symptoms (fever, aches, pains, runny nose, cough) may follow and then disappear after a few days. Months to years later, pains in joints or periodic muscle pains can develop. For susceptible individuals, effects can be severely debilitating.

Analysis: Blood-sucking ticks (members of the *Ixodidae* family) are related to spiders. Ticks feed on humans, dogs, cats, and other animals by sinking their mouthparts and heads into the flesh of their hosts. If left undisturbed, they may continue to suck blood for as many as 15 days before dropping off. Severely infested dogs may become weak and even die if ticks are not removed. Ticks can live for up to 18 months without food or water. Lyme disease is spread mainly by the deer tick (*Ixodes dammini*), although other tick species have also been found to be vectors for the disease. The bacterium that causes the disease, *Borrelia burgdorferi*, is a spirochete that affects large warm-blooded animals, causing joint stiffness, neurological problems, or both. In many cases, tick bites and the rash that accompanies them go unnoticed. Only when serious health problems appear, months to years later, is Lyme disease diagnosed. Even then, it is often mistaken for arthritis or a cold.

Solution: Reduce the danger of tick infestation by clearing yards of tall grass and weeds. Avoid setting out food for animal hosts, especially deer. Treat pets with tick repellents during tick season. If walking in fields, wear protective clothing and apply tick repellent. Tuck trousers into socks and shirts into trousers to avoid exposing skin. After spending time in a tick habitat, inspect your body and clothing for ticks. Remove ticks by pulling them straight out with tweezers. Do not attempt to remove ticks by using a lighted cigarette, flame, or alcohol; you could induce the tick to regurgitate infection into the host. Save the removed tick in a jar in the refrigerator to show to a doctor if problems develop. If a rash appears within a month after a tick bite, see a physician. Early treatment with an antibiotic can destroy the spirochete. If animals are severely infested, consult a veterinarian. Control ticks in lawns and gardens with ORTHO Dursban® Lawn Insect Spray. A vaccine is now available for Lyme disease.

Carpenter ants

Slitlike holes. Inset: Carpenter ant (2× life size).

Problem: Black or reddish black, winged or wingless ants up to 1/2 inch long are seen around the home. Piles of sawdust may be found in the basement or attic, under porches, or near supporting girders or joists. Slitlike holes are often present in woodwork. On warm spring days, swarms of winged ants may cluster around windows. Unlike termites, these pests have constricted waists.

Analysis: Many closely related species of the wood-damaging carpenter ant (*Camponotus* species) are found throughout the United States. Carpenter ants bore into moist, decaying wood, forming extensive galleries in which they make nests. They do not eat their sawdustlike wood borings but feed on other insects, plant sap, pollen, and seeds. When ant colonies grow too large, part of the colony migrates, often invading nearby homes through windows and similar entry points. They either colonize undisturbed hollow spaces such as walls or bore into structural timbers, ceilings, and floor areas. They require damp and rotted wood. In addition to weakening wood, carpenter ants may infest pantries, and they can inflict painful bites.

Solution: Dust baseboards, windowsills, door frames, and other places where ants crawl with ORTHO Ant-Stop Ant Killer Dust. Dust into nests if possible. Remove nearby logs, stumps, and woodpiles. Seal openings in the foundation, windows, and other access areas into the home. Spray along the foundation and around the outside of door and window frames and sills with ORTHO Ant-Stop Ant Killer Spray.

Subterranean termites

Termites.

Termite-damaged book.

Carpenter bees

Galleries. Inset: Adult carpenter bee (life size).

Problem: On warm, sunny spring or fall days, brown to brownish black winged insects, about ⅜ inch long, swarm in and around the house. These insects resemble flying ants but have thick rather than constricted waists and beaded rather than elbowed antennae. Their discarded wings may be found around the building. Earthen tubes extend from the soil up along the foundation and any other termite-proof surface to the infested wooden structures. These tubes are commonly found in basements and crawl spaces under buildings. When broken off, they are rebuilt within several days. Dark or blistered areas may develop in the flooring.

Analysis: The wood-feeding subterranean termite causes more structural damage to buildings than any other insect. Termites live in colonies as deep as 5 feet in the ground and move up to infest wooden structures through tubes of soil they build over masonry or metal to bridge the gap from soil to wood. Except for the dark, winged swarmers, these termites are white, wingless, and sensitive to moisture loss. They always remain within the nest, soil tubes, or infested wood, protected from desiccation and insect predators. Within their colony, termites maintain a complicated caste system that includes sterile workers and soldiers, winged reproducers, and an egg-laying queen. Colonies are formed when a pair of winged reproducers leaves the parent colony and excavates a nest in a piece of wood on top of or buried in the ground. As the new colony develops, galleries are formed deep in the soil. Termite colonies develop slowly—3 or 4 years usually pass before the reproductive swarmers develop, and structural damage may not be noticed for several more years. When buildings are erected over established termite colonies, however, serious damage may occur within a year. Termites hollow out the inside of a wooden structure, leaving only an outer shell. Damage is most severe when they infest main supporting wooden beams and girders. One species of subterranean termite, the Formosan termite (*Coptotermes formosanus*), is not native to the United States but is present in areas of the Southeast and Southwest. This termite is more vigorous and aggressive than native North American species and is more difficult to control.

Solution: Termite infestations can be treated most effectively only after a thorough and accurate diagnosis of the damage is made. Consult a professional termite or pest control operator. Once the termite colony has been located and the damage revealed, a physical or chemical barrier is placed between the soil and the building to prevent the termites from reaching the building. An insecticide containing *chlorpyrifos* is applied to the soil around and underneath the building. Infestations should be treated by professional pest control operators. Prevent termite damage with ORTHO HomeDefense Ortho-Klor Insect & Termite Killer used according to label directions. Discourage infestations by keeping the area under and around the house free of wood debris above and below the ground. If the soil around the foundation remains moist because of faulty plumbing or improper grade, repair the plumbing and alter the grade; termites prefer moist soil. For details on termite-resistant construction methods, contact a reliable building contractor or your local county extension service.

Problem: Metallic blue or black buzzing bees fly around the home and yard. They may be seen entering and leaving holes about 1 inch wide in decks, posts, beams, rafters, and other wooden structures. When damaged wood is sliced open, partitioned galleries may be seen. The partitions may contain immature bees.

Analysis: Carpenter bees (*Xylocopa* species) do not usually cause serious damage; continued burrowing and gallery formation year after year will eventually weaken wooden structures, however. These insects burrow into wood to make their nests. The female bees partition the galleries into small cells in which the carpenter bee larvae mature. When bee nests are approached, the males hover around the head of the intruder. Although they are frightening because of their loud buzzing and large size, the male bees do not sting, and the females sting only when handled.

Solution: Paint wood surfaces once a year to discourage bee tunneling. Flood galleries in exposed wood with ORTHO HomeDefense Ortho-Klor Insect & Termite Killer. Close the holes with putty, caulking compound, dowel pins, or plastic wood to prevent bees from returning to the nest.

STRUCTURAL PESTS *(continued)*

Powderpost beetles

Powderpost beetle and larva (3× life size).   *Emergence holes.*

Woodwasps

Woodwasp (life size).

Problem: Wood flooring, structural timbers, cabinets, furniture, and other items are riddled with round holes that range in size from 1/16 to 3/8 inch long. Wood powder or tiny pellets may be piled around the holes or on the floor below. When the infested item is tapped, wood powder or additional pellets are expelled from the holes. Tiny red, brown, or black beetles ranging in size from 1/12 to 1/3 inch long may be seen crawling around the infested wood or, in the evening, flying around windows and electric lights. When the damaged wood is cut open, the inside is found to be riddled with sawdust-filled tunnels; or it may be pulverized into a mass of wood powder or pellets.

Analysis: Wood-feeding beetles, including the powderpost, false powderpost, and deathwatch beetles, damage wood houses and household furnishings throughout the country. Powderpost beetles feed only on deadwood. They are brought into the home in infested timber or furnishings, or they may fly from infested lumber or woodpiles in the yard. The female beetles deposit their eggs in unfinished wood. The grubs that hatch from the eggs tunnel through the wood, leaving masses of wood powder or pellets behind them. They pupate just under the surface of the wood and emerge as adult beetles through the round holes they chew in the wood. Beetle eggs or larvae present in wood before it has been coated with paint, varnish, shellac, or other finishings can chew through the finished surface when they have matured, leaving round emergence holes. They do not lay eggs in coated wood surfaces, however, and reinfestation cannot occur.

Solution: If the infestation is localized, remove and destroy badly infested timbers. Replace them with kiln-dried or insecticide-treated wood. Or treat unfinished exterior wood yourself by painting or spraying it with ORTHO HomeDefense Ortho-Klor Insect & Termite Killer. Wherever possible, apply paint, shellac, varnish, paraffin wax, or other wood coatings to unfinished wood around the home to prevent further infestation. Inspect woodpiles periodically for signs of powderpost beetle infestation. Infested wood may also be treated with ORTHO HomeDefense Ortho-Klor Insect & Termite Killer. If infestation is widespread, contact a professional pest control operator to fumigate the building. Individual pieces of furniture may also be fumigated to kill beetle eggs and larvae. Many pest control operators maintain fumigation chambers for movable items. Eggs and larvae in small wooden items may be killed by placing the items in the freezer for 4 days. When purchasing furniture, get assurances from the dealer that all woodwork is made from kiln-dried stock.

Problem: Round holes about 1/4 inch in diameter appear in wood floors, walls, doors, and other surfaces. Or holes appear in wallpaper, linoleum, carpeting, and other types of coverings over wood. Metallic blue, black, or multicolored wasplike insects may be seen flying around the home. These buzzing insects are 1 to 2 inches long and may have hornlike "tails."

Analysis: Woodwasps, also known as *horntails*, do not cause structural damage; the holes they make in wood or covered wood surfaces are of cosmetic concern only. Woodwasps lay their eggs in weak and dying forest trees. The adult insects emerge from the wood 2 to 5 years later, often long after the tree has been used for construction. Most woodwasp holes occur within the first 2 years after the cut wood has been used, however. These insects lay their eggs only in forest trees; they do not reinfest buildings.

Solution: Seal or fill emergence holes. You cannot do anything to prevent the woodwasps from emerging. For future construction, purchase lumber that has been kiln dried or vacuum fumigated. These processes kill woodwasp larvae embedded in the wood.

Roundheaded borers

Old house borer (4× life size).

Old house borer damage.

Dryrot

Dry rot.

Problem: Oval holes ¼ to ⅓ inch wide appear in walls and flooring. Or holes appear in wallpaper, plaster, linoleum, or other types of wood coverings. Sawdustlike borings may be piled around the holes. In some cases, rasping or ticking sounds may be heard before the holes appear, and the wood may be blistered or rippled. Grayish brown to black beetles, 1 inch long, with antennae, may be seen around the house.

Analysis: The larvae of these beetles, including the new house borer (*Arhopalus productus*) and the old house borer (*Hylotrupes bajulus*), cause damage to fir, pine, and other softwood structural timbers. The adults lay their eggs in the bark of weak and dying forest trees and, in the case of old house borers, in seasoned lumber. The yellow grubs tunnel into wood that is later incorporated into a building before the adult beetles emerge. Sometimes the grubs make rasping or clicking noises while they feed. If they are tunneling close to the surface, wood blistering or rippling may result. New house borers continue to emerge through holes in wood and wood coverings for up to a year after construction. New house borers generally do not cause structural damage, however. The holes they make in wood or covered wood surfaces are of cosmetic concern only, and the borers cannot reinfest the building. Old house borer beetles generally do not emerge from timbers until 3 to 5 years after the building has been constructed. They are the only species of roundheaded borer that reinfests wood, and they may cause serious structural damage.

Solution: If damage occurs within a year after construction, new house borers are the problem. To repair new house borer damage, seal or fill emergence holes. Localized areas may be painted with ORTHO HomeDefense Ortho-Klor Insect & Termite Killer, but because damage is only cosmetic and will stop within a year, this procedure is seldom justified. If damage occurs 3 or more years after construction, old house borers are the problem. Buildings infested with old house borers must be fumigated. Contact a professional pest control operator to fumigate the building. To prevent future infestations of old house borers when building new structures, use pressure-treated wood.

Problem: Foundation timbers, paneling, flooring, and other wooden structures are damp and soft, or they are dry, cracked, brown, and crumbling. Often the wood is broken into small, cubical pieces. Thin mats of white fungal strands may be seen on the rotted wood. Thick white, brown, or black fungal cords up to 2 inches wide may extend across the rotted area. These cords often extend over impenetrable surfaces such as brick and concrete to reach wood surfaces beyond.

Analysis: Dry rot is caused by fungi that live in the soil and grow into wood that is in direct contact with damp soil. The white fungal strands penetrate and decay the wood fibers, causing a soft rot. In some cases, the fungus draws water from the soil up through thick fungal cords that extend across the rotted area. The water is used to moisten dry wood, providing the damp condition in which the fungus thrives. After the fungus dies, badly rotted wood cracks and crumbles into chunks when handled.

Solution: Remove any water-conducting fungal cords from the wood. Eliminate moist soil conditions around wood structures as much as possible by improving ventilation, changing soil grade and drainage, or fixing leaky plumbing. As soon as the soil and wood dry out, the fungus will become dormant. Replace badly rotted wood with wood that has been pressure-treated with preservatives. Remove all wood scraps around the building foundation. When building new structures, use pressure-treated wood in all areas where wood and soil make contact.

STRUCTURAL PESTS
(continued)

BITING AND STINGING PESTS

Firewood insects

Cucujid beetle grub on firewood.

Problem: Firewood is riddled with holes. Small piles of sawdust accumulate around the holes or on the ground around the firewood. If the wood has been stored indoors, insects may be crawling around the firewood pile or flying around lights or windows.

Analysis: Many insects develop in and emerge from cut firewood. If infested wood is stored either indoors or outdoors so that it rests against the house, insects may invade the wooden structure of the house. Insects capable of moving from firewood into the structure include carpenter ants (see page 320), termites (see page 321), and powderpost beetles (see page 322). Some insects, such as bark beetles, most flatheaded and roundheaded borers (see page 323), and woodwasps (see page 322), emerge from firewood but attack only living or recently killed trees; these insects will not damage structural wood or household articles.

Solution: Spray infested firewood outdoors with ORTHO HomeDefense Ortho-Klor Insect & Termite Killer or ORTHO Diazinon Ultra Insect Spray. Sprayed firewood may be burned 2 weeks following treatment. Store all firewood outdoors unless you plan to burn it within a couple of days. Do not lean an outdoor woodpile against the home; stack it so that there is at least an inch between the wood and the structure. If practical, choose a location for the woodpile at least 10 feet away from the house.

Yellowjackets

Yellowjackets (2× life size).

Yellowjacket (2× life size).

Problem: Yellowjackets are present around the home. They hover around patios, picnic areas, garbage cans, and other areas where food or garbage is exposed. Yellowjackets may be seen flying into underground nests. They will also nest in wall voids or heating ducts if access is available. Yellowjackets inflict painful stings when threatened or harmed or when their nests are approached.

Analysis: Unlike most other species of wasps, yellowjackets (*Vespa* and *Vespula* species) live in large colonies, often numbering in the thousands. Some species of yellowjackets feed their offspring insects and spiders, while others scavenge scraps of meat from recreational areas or dump sites. These pests may also feed on nectar, sap, and other sugary fluids and may be seen hovering around soft drinks and cut fruit. Yellowjackets can inflict painful stings and are capable of repeated stings. The venom injected along with the sting causes reddening, swelling, and itching of the affected area. Some people who are very sensitive to the stings experience extreme swelling, dizziness, difficulty in breathing, and even death. Yellowjackets are protective of their nests, and large numbers may emerge to sting intruders. Some species of yellowjackets build their nests underground; the only evidence of the nest is a raised mound of dirt surrounding a depression several inches deep. Other species build football-shaped nests in trees or shrubs or under eaves. Most yellowjackets die in the late fall, and overwintering queens start new nests the following spring in a different location.

Solution: Keep food and garbage covered, and empty garbage frequently. To kill yellowjackets before picnicking, spray with ORTHO Outdoor Insect Fogger, or place a properly baited yellowjacket trap downwind from the picnic or barbecue area (yellowjackets will be attracted to the bait and die within minutes after entering the trap). To remove yellowjackets from the vicinity, you must eliminate the nests. After locating the nests, spray them at dusk or during the night with ORTHO Hornet & Wasp Killer. Stay 8 feet away from the nest, and spray directly into the entrance hole. If you need to illuminate the area, use a flashlight covered with red cellophane. Use it for only short periods of time. Stop spraying when the yellowjackets begin to emerge; leave the nest area quickly by walking, not running away. Repeat the spraying every evening until the insects fail to emerge, then quickly cover the hole with moistened soil. Contact a professional pest control operator to remove yellowjacket nests from difficult locations or from inside the home. If you are stung, apply a cold compress or ice pack to the affected area. If a severe reaction develops, call a doctor.

Honeybee

Honeybee.

Problem: Bees hover around flowering plants in the garden and may inflict stings when threatened or harmed. Large numbers of bees may cluster on shrubs or trees. Hives may be located in attics, chimneys, or wall voids.

Analysis: Familiar and often feared honeybees (*Apis mellifera*) provide honey and wax and are important pollinators. On warm, sunny days, bees forage for nectar among flowering garden plants, then return to their hives in the evening. When an established hive gets too crowded, thousands of bees leave in a swarm. Swarms fly for a mile or so before settling in a new location. En route, they often rest in a tight cluster on a tree branch or other object. Some people are allergic to bee stings and experience extreme swelling, dizziness, difficulty in breathing, and possibly death. A bee does not sting more than once, because the stinger and venom sac rip out of its body when it flies away. The injured bee subsequently dies.

Solution: When stung, scrape the stinger off the skin with a knife or fingernail. Avoid squeezing it; this forces more venom into the wound. Apply cold compresses or ice packs to the swollen area. If a severe reaction develops, call a doctor immediately. Avoid using plants that are attractive to bees, especially around pools, patios, and other recreational areas. Do not try to remove hives or swarms yourself. Contact a professional beekeeper or pest control operator if you find a swarm or if bees are nesting in your home or garden.

Paper wasps

Paper wasps (life size).

Problem: Single-layered paper nests are suspended from eaves, ceilings, or branches.  These nests are composed of exposed, open cells and have a honeycomb appearance. They are often umbrella-shaped. Black or brown wasps with yellow or red stripes may be seen hovering around or crawling on the nests.

Analysis: Paper wasps (*Polistes* species), also known as *umbrella wasps*, usually live in small colonies consisting of 20 to 30 wasps. They build paper nests in which they raise their young. Paper wasps feed on insects, nectar, and pollen. They are not as aggressive and protective of their nests as are yellowjackets. (For more information on yellowjackets, see page 324.) They will sting if threatened, however, causing swelling, itching, and more generalized symptoms in sensitive individuals.

Solution: To remove paper wasps from the vicinity, you must eliminate their nests. Spray the nests at dusk or during the night with ORTHO Hornet & Wasp Killer or ORTHO Outdoor Insect Fogger. Spray the nest from a distance of 8 feet. Stop spraying when wasps begin to emerge; leave the nest area quickly by walking, not running, away. Repeat the spraying every evening until the insects fail to emerge. Remove the nest and dispose of it. If you are stung, apply a cold compress or ice pack to the swollen area. If a severe reaction develops, call a doctor.

Solitary wasps

Mud dauber nest.

Problem: Wasps are flying around the garden. Some of these insects are black with yellow, red, or white markings. Others are black, brown, blue, red, or yellow. Many wasps have very thin, elongated waists. Their mud nests may be found under eaves or on plants or rocks.

Analysis: Unlike the social yellowjackets, many species of wasps, including the potter and mason wasps, mud daubers, and spider wasps, live alone. Most of them build nests of mud and sand in which they raise their young. Many of these wasps feed on insects; a few feed on pollen and nectar. These wasps are not as aggressive and protective of their nests as are yellowjackets and do not sting as readily. (For more information on yellowjackets, see page 324.) If highly provoked, however, they can sting, causing swelling, itching, and more generalized symptoms in sensitive individuals.

Solution: Eliminate mud nests by hosing or knocking them down. Kill wasps by spraying them with ORTHO Hornet & Wasp Killer or ORTHO Outdoor Insect Fogger according to label directions. Even though many solitary wasps are docile, they may be confused with yellowjackets; it is best to avoid threatening or provoking them. If you are stung, apply a cold compress or ice pack to the swollen area. If a severe reaction develops, call a doctor.

325

BITING AND STINGING PESTS *(continued)*

Stinging caterpillars

Saddleback caterpillar.

Problem: An irritating rash forms where a hairy or spiny caterpillar has touched the skin. Reactions vary depending on the caterpillar and the individual and include mild itching, rash, swelling, local severe pain, local lesions, and fever.

Analysis: There are about 25 species of stinging caterpillars. These insects have hollow hairs that contain a mild poison. The hairs release the irritating substance when people handle the caterpillars or accidentally brush against them. Most of these caterpillars cause only a mild itching or skin rash. The puss caterpillar (*Megalopyge opercularis*), however, causes a severe reaction. It can cause intense itching, swelling, local numbness, nausea, and fever (especially in children). This caterpillar is widely distributed in the southeastern and south-central states and feeds on a wide variety of deciduous trees and shrubs. During some years it increases to unusually large numbers. The saddleback caterpillar (*Sibine stimulea*), the io moth caterpillar (*Automeris io*), and the flannel moth caterpillar (*Norape ovina*) are other caterpillars known for their stinging hairs.

Solution: Avoid handling hairy and spiny caterpillars. Spray shrubs and trees on which these caterpillars are found with an insecticide containing *carbaryl*. Make sure that your plant is listed on the product label. Call a doctor if a severe reaction begins to develop.

Biting flies

Horsefly (3× life size).

Problem: Black, brown, or black-and-white biting flies about ½ to 1 inch long are present around the home. They are especially bothersome in areas where horses and domestic animals are common.

Analysis: Biting flies, including several species of horseflies and deerflies, attack humans and domestic animals in rural and suburban areas. The female flies deposit their eggs in still pools of water, in moist soil, or on vegetation. The larvae feed on other insects or decaying vegetation and pupate in damp plant debris. The adult flies inflict painful bites that often continue to bleed after the fly has left. Some people bitten by horseflies may suffer from fever and general illness.

Solution: Keep doors and windows tightly screened. Spray outdoor living areas with ORTHO Outdoor Insect Fogger or ORTHO HomeDefense Home & Garden Insect Killer. Kill biting flies indoors by spraying with ORTHO HomeDefense Home & Garden Insect Killer or ORTHO HomeDefense Hi-Power Brand Indoor Insect Fogger. Use insect repellent. Eliminate fly breeding areas by cleaning up stagnant pools of water and wet, decaying vegetation around the yard.

Biting midges

Biting midge (15× life size).

Problem: During the spring and summer, tiny black biting midges ½5 to ⅛ inch long infest the yard. They are most common in coastal areas and near lakes, streams, marshlands, and swamps.

Analysis: Biting midges are also known as *no-see-ums*, *sand flies*, or *black gnats*. They feed on warm-blooded animals, including humans and birds. Biting midges breed in wet sand or mud, damp rotting vegetation, shallow stagnant or brackish water, and similar locations. They rarely infest the home but are bothersome outdoors, inflicting bites around the feet, legs, ears, and eyes and under clothes, especially in cuff, collar, and belt areas where clothing binds. The bites are rarely painful but produce tiny swellings or blisters that continue to itch for several days.

Solution: Spray infested areas of the yard with ORTHO Outdoor Insect Fogger according to label directions. Use insect repellent.

Black flies

Black flies (3× life size).

Problem: During the late spring and summer, many black or gray humpbacked flies 1/25 to 1/5 inch long are present around the home. They inflict painful bites.

Analysis: Black flies, also known as *buffalo gnats* or *turkey gnats*, attack humans and domestic animals in rural and suburban areas throughout the United States. The female flies deposit their eggs in swiftly running water, including streams and irrigation ditches. The larvae develop in the water and emerge as adult flies during the spring and summer. Black flies may be blown many miles from their breeding areas. They bite any exposed part of the body and may also bite under clothing, especially where clothes are binding, such as around belts and collars. These irritating bites often swell and itch for several days. The victim may suffer headaches, fever, and nausea.

Solution: Keep windows and doors tightly screened. Spray outdoor living areas with ORTHO Outdoor Insect Fogger or ORTHO HomeDefense Home & Garden Insect Killer. Kill black flies indoors by spraying with ORTHO HomeDefense Home & Garden Insect Killer. Use insect repellent.

Mosquitoes

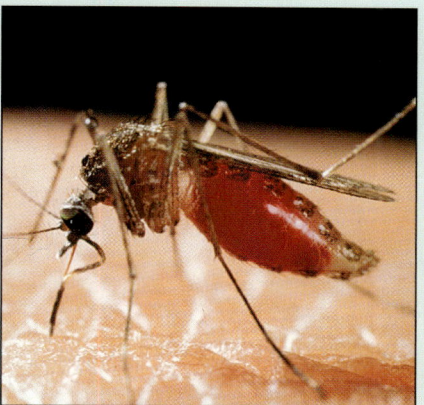

Mosquito (8× life size).

Problem: Mosquitoes are present in the home and yard. They are most bothersome at dusk and during the night.

Analysis: Many different species of mosquitoes occur throughout the world. In addition to having an annoying bite, they transmit encephalitis within the United States and other serious diseases, such as yellow fever and malaria, in other parts of the world. Adults emerge from hibernation with warm spring weather. The males feed on nectar, honeydew, and plant sap; the females require a blood meal in order to produce their eggs. Larval development takes place exclusively in water. Typically, eggs are laid in shallow accumulations of fresh, stagnant, or salty water. The larvae may mature within 5 days or may take months to mature.

Solution: Maintain door and window screens in good repair. Kill mosquitoes indoors by fogging with ORTHO HomeDefense Hi-Power Brand Indoor Insect Fogger or spraying with ORTHO HomeDefense Home & Garden Insect Killer. Kill mosquitoes outdoors by spraying with ORTHO Dursban® Lawn Insect Spray or ORTHO Malathion 50 Plus Insect Spray around the lawn and foundation of the house. Spray resting areas under eaves with ORTHO HomeDefense Home & Garden Insect Killer or ORTHO Outdoor Insect Fogger. Apply insect repellents to the skin. Drain unnecessary accumulations of water. Stock ornamental ponds with mosquitofish (*Gambusia affinis*), which eat mosquito larvae. Goldfish also eat mosquito larvae but are not as effective as mosquitofish.

Black widow spiders

Black widow spider (2× life size).

Problem: Black widow spiders are shiny, black, about the size of a quarter, and have a red hourglass marking on the underside of the abdomen. Outdoors, they live under rocks or clods of dirt and in wood and rubbish piles. Indoors, they are found in garages, attics, cellars, and other dark secluded places, such as under boards or cluttered debris, in old clothing, or in crevices. Black widow webs are coarse and irregular, about 1 foot wide. The strands break with a snap when the web is torn.

Analysis: Several species of the poisonous black widow spider (*Latrodectus* species) are found throughout the United States. Black widows live in secluded locations and feed on insects trapped in their webs. If the spiders are accidentally touched, or if their webs are disturbed, they will bite the intruder. The venom may cause serious illness and, on rare occasions, death. The females produce egg sacs that contain hundreds of eggs. The tiny spiderlings that emerge are also capable of inflicting poisonous bites. They may be carried long distances by the wind.

Solution: Kill spiders by spraying webs and infested areas with ORTHO HomeDefense Hi-Power Brand Roach, Ant & Spider Killer or ORTHO HomeDefense Indoor & Outdoor Insect Killer, or dust areas where spiders may hide with ORTHO Ant-Stop Ant Killer Dust. Remove loose wood, trash, and clutter from areas where spiders might hide. Wear gloves and protective clothing when cleaning up infested areas. Vacuum infested areas to remove egg sacs, and destroy the contents of the vacuum cleaner bag. Put ice on spider bites and call a doctor immediately.

BITING AND STINGING PESTS (continued)

Brown recluse spider

Brown recluse spider (2× life size).

Problem: Brown recluse spiders are light to dark brown, ⅓ to ½ inch in length, with a violin-shaped marking behind the head. Outdoors, they are found under rocks. Indoors, they are found in old boxes, among papers and old clothes, behind baseboards, underneath tables and chairs, and in other secluded places. Grayish, irregular, sticky webs and round white egg sacs ¾ inch wide may be found in infested areas.

Analysis: The poisonous brown recluse spider (*Loxosceles reclusa*) is found in the Midwest and Southeast. Other related but less poisonous spiders (*Loxosceles* species) are found throughout most of the United States. Brown recluse spiders live in secluded places and are very shy, moving away quickly when disturbed. If they are touched or trapped in shoes, clothing, or bedding, they may bite. Their venom is rarely fatal but causes a severe sore that is slow to heal and sometimes causes illness.

Solution: Kill spiders by spraying webs and infested areas with ORTHO HomeDefense Indoor & Outdoor Insect Killer or ORTHO HomeDefense Hi-Power Brand Roach, Ant & Spider Killer. Or spray outdoor areas where spiders may hide with ORTHO Outdoor Insect Fogger or ORTHO Diazinon Ultra Insect Spray. Remove loose wood, trash, or clutter from areas where spiders might hide. Wear gloves and protective clothing when cleaning up infested areas. Spray outdoor living areas and clean up debris around the home that may harbor spiders. Vacuum infested areas to remove egg sacs, and destroy the contents of the vacuum bag. Put ice on spider bites and call a doctor immediately.

Fire ants

Fire ant mound. Inset: Fire ant (4× life size).

Problem: Small (¼-inch) reddish to black ants crawl to and from large mounds of soil in the lawn and garden. These ants inflict painful stings to people or animals disturbing the mounds.

Analysis: Fire ants (*Solenopsis* species) are notorious for their large mounds and painful stings. The mounds, which are their nests, are usually found in lawns and gardens. Occasionally the ants move into or underneath homes during periods of rain or drought. They feed mainly on other insects but will also feed on young succulent plants, seeds, fruit, household foods, and even small, weak animals such as newly hatched birds. Because fire ants feed on other insects, they are of some benefit in the yard. However, their presence can greatly limit use of the garden without the threat of painful stings. The sting results in a pustule that develops within 24 hours. It can take several weeks or longer to heal.

Solution: Control fire ants in the lawn by sprinkling powdered ORTHO Ant-Stop Orthene Fire Ant Killer or by drenching the mounds with ORTHO Dursban® Lawn Insect Spray, ORTHO HomeDefense Ortho-Klor Insect & Termite Killer, or ORTHO Diazinon Ultra Insect Spray. Also treat the area surrounding the mounds out to a distance of 4 feet. Spray ants indoors with ORTHO Ant-Stop Ant Killer Dust, or sprinkle with ORTHO Diazinon Granules. If a severe reaction to fire ant stings develops, call a doctor.

Harvester ants

Harvester ant mound.

Problem: Large ants up to ½ inch long are crawling on cleared areas on the ground. Ants are seen entering holes or large craters in the ground. The holes are surrounded by a large cleared area from 3 to 35 feet in diameter. The cleared area may be strewn with small pebbles and seed husks. Ant trails radiate from the nest in all directions. If there is a mound, it is usually low. If disturbed, the ants will inflict a painful sting.

Analysis: Harvester ants (*Pogonomyrmex* species) do not invade the home but may be a problem in lawns and gardens. These ants eat all tender vegetation surrounding their nests, resulting in a large cleared area where they dump small pebbles removed from their nests as well as husks and other inedible portions of seeds. They are primarily seed eaters. The holes in the center of the cleared areas lead to their underground nests. Harvester ants aggressively sting and bite anyone who disturbs their nests. They have been known to kill very small animals that accidentally wander over their nests.

Solution: Treat the entrances to the ant nests and the cleared area around them with ORTHO Dursban® Lawn Insect Spray, ORTHO Ant-Stop Orthene Fire Ant Killer, ORTHO Ant-Stop Ant Killer, ORTHO Diazinon Granules, ORTHO Diazinon Ultra Insect Spray, or ORTHO Ant-Stop Ant Killer Dust. If you are stung, apply a cold compress or ice pack to the swollen area. If a severe reaction develops, call your doctor.

Scorpions

Scorpion (1.5× life size).

Problem: Scorpions are found in the garden under rocks, boards, and protective debris. Indoors, they may be found in attics or crawl spaces under the home. They may move down into living areas such as kitchens and bathrooms when attic temperatures rise above 100°F.

Analysis: All scorpions are capable of inflicting stings; only a few species, found in the Southwest, are dangerous, however. Scorpions are nocturnal creatures, feeding at night on insects and small animals. During the day, they hide in dark, protected locations. They are shy and sting only when touched, trapped, or otherwise provoked. Except for a few fatally poisonous scorpions (*Centruroides* species), most scorpions deliver stings that are no more serious than a bee sting. The venom varies in potency from season to season, however, and like many insect stings may cause severe illness in a sensitive individual. Although scorpions live mostly in the garden, they may crawl into the home through open or loose doors and windows.

Solution: Remove loose boards, rocks, clutter, and other debris around the yard and in the home to eliminate scorpion hiding places. Wear gloves and protective clothing when cleaning up infested areas. Spray locations where scorpions might hide in the yard with ORTHO Diazinon Ultra Insect Spray and in the home with an insecticide containing *carbaryl*. Maintain window and door screens and weather stripping in good repair. Call a doctor if you are stung by a scorpion.

Tarantulas

Black Mexican tarantula.

Problem: A large, hairy spider, up to 5 inches across, is crawling around on the floor indoors or outside in the garden.

Analysis: Tarantulas are often feared because of their large size and hairy bodies. Although a few South American species can give a very painful bite, all tarantulas occurring in the United States inflict a bite that is like a bee sting. Their hairs, which easily rub off their bodies, can irritate the skin. When cornered, tarantulas may make a purring sound or rear up on their back legs. Tarantulas are sluggish, bite only rarely, and can be handled with ease. Female tarantulas may live 20 years or more in captivity. Males are shorter-lived. Tarantulas are nocturnal, living in dark cavities or burrows during the day and hunting at night.

Solution: Capture the spider in a large jar or box and release it in a secluded area. Chances of being bitten while catching it are minimal, but it is wise to wear protective clothing, such as long rubber gloves, and to avoid sudden, quick movement. If a severe reaction to a tarantula bite develops, call your doctor. To control spiders, spray a 5-foot band around the home with ORTHO Diazinon Ultra Insect Spray, or dust a 2-foot band around the foundation with ORTHO Ant-Stop Ant Killer Dust.

Flour moths

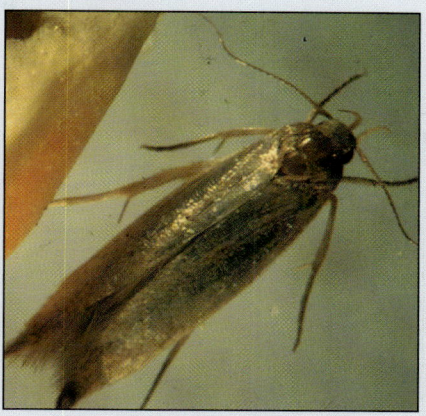

Angoumois grain moth (5× life size).

Problem: Pinkish or greenish caterpillars up to ⅝ inch long are feeding inside silken webbing in stored grain, flour, cereals, and other grain products. Beige-, gray-, and coppery-winged moths ⅓ inch long may be seen flying in the home.

Analysis: The larvae of Indian meal moths (*Plodia interpunctella*) and Mediterranean flour moths (*Anagasta kuehniella*) damage ground or broken grain products, dried cereals, dried fruits, powdered milk, and other pantry items. The larvae of angoumois grain moths (*Sitotroga cerealella*) infest whole wheat and corn kernels. The adult moths lay eggs in stored grain products. The larvae that emerge spin silken webs, under which they feed. When mature, they usually leave the infested food to pupate in a corner or crack in the cupboard.

Solution: Discard all infested food. Clean out cupboards thoroughly before restocking. If infestation is widespread, remove all food and utensils and fog the infested area with ORTHO HomeDefense Hi-Power Brand Indoor Insect Fogger; or treat cracks and crevices along shelves in the pantry with ORTHO HomeDefense Home & Garden Insect Killer. Do not treat countertops or other food work areas. Reline shelves with paper and replace food after the spray has dried. If you suspect that food is infested, kill the eggs, larvae, and pupae by deep-freezing food for 4 days or heating it in a shallow pan in the oven at 150°F for half an hour. Keep foods in airtight glass, plastic, or metal containers. Keep the pantry clean, and avoid buying broken packages; they are more likely to be infested.

PANTRY PESTS *(continued)*

Mealworms

Mealworms (life size).

Problem: Shiny yellow to brown grubs up to 1¼ inches long are feeding in damp or moldy flour, grain, or cereal products. Flat, shiny, brown to black beetles ¼ to ¾ inch long may also be found.

Analysis: Mealworms (*Tenebrio* and *Alphitobius* species) prefer to feed on damp or moldy grain products stored in dark, rarely disturbed, dusty locations such as warehouses. If infested food items are brought into the pantry, however, the mealworms and beetles may migrate to infest and reproduce in poorly sealed bags of flour, bran, crackers, and other grain products.

Solution: Chemical control is not necessary. Discard infested food items. Clean out cupboards thoroughly. Keep grain products in dry, tightly sealed glass, plastic, or metal containers.

Cockroaches

German cockroaches (life size).

Oriental cockroach (2× life size).

Problem: Cockroaches infest the kitchen, bathroom, and other areas of the home. These flat, shiny insects range in size from ½ to 1¾ inches long. They may be light brown, golden tan, reddish brown, or black. In large numbers, they emit a fetid odor.

Analysis: Cockroaches thrive in human habitations throughout the world. The most important household species in the United States are the German cockroach (*Blattella germanica*), the brown-banded cockroach (*Supella longipalpa*), the Oriental cockroach (*Blatta orientalis*), the American cockroach (*Periplaneta americana*), and the Asian cockroach (*Blattella asahinai*), a recent arrival in North America and similar in appearance to the German cockroach. In addition to their annoying presence, cockroaches spread diseases, such as salmonella poisoning and parasitic toxoplasmosis, by contaminating food with their infected droppings. These pests proliferate in areas where food and water are available. Cockroaches prefer starchy foods but will feed on any human or pet food scraps, garbage, paper, or fabrics soiled with food. Unless infestations are heavy or their hiding places are disturbed, they are rarely seen in exposed locations during the day. These nocturnal insects seek out dark, protected areas in which to live and breed. Usually they congregate in kitchens and bathrooms. They may be found behind or under sinks, refrigerators, and water heaters, within the walls of household appliances, behind baseboards and molding, in wall voids, around pipes, in garbage cans, and in piles of cluttered paper or grocery bags. They may be present in cracks or crevices in cupboards, cabinets, desks, dressers, and closets. They may infest basements, crawl spaces, and sewers. Cockroaches move from one room to another through wall voids or through cracks in walls, floors, and ceilings and along pipes and conduits. If their living conditions become too crowded, they may migrate. Infestations usually begin when stray insects or egg cases are brought into the home with shipped items, secondhand furniture or appliances, grocery bags, or debris. They may also move into homes from sewers.

Solution: Eliminate cockroach food sources by keeping the kitchen and other areas of the home free of food scraps. Clean up the kitchen after each meal and store food in tightly sealed metal, glass, or heavy plastic containers. Empty household garbage and pet litter regularly. Do not leave pet food out overnight. Fix leaking faucets and pipes. Clean up water puddles or moist areas around the kitchen, basement, and other infested areas. Plug cracks around baseboards, shelves, cupboards, sinks, and pipes with a filling material, such as putty or caulk. Remove food and utensils, then apply ORTHO HomeDefense Indoor & Outdoor Insect Killer, ORTHO Ant-Stop Ant Killer Dust, or ORTHO HomeDefense Hi-Power Brand Roach, Ant & Spider Killer in cracks in cupboards, on surfaces underneath sinks, along molding, behind appliances, and in other areas where insects are likely to congregate. Allow the spray to dry, then reline the shelves with fresh paper before replacing food and utensils.

Flour beetles

Red flour beetle (6× life size).

Problem: Reddish to dark brown elongated beetles, 1/10 to 1/7 inch long, and yellowish white wiry grubs, 1/5 inch long, are feeding in flour, cereals, cake mixes, macaroni, and other flour and grain products.

Analysis: Several species of flour beetles (*Tribolium* and *Oryzaephilus* species), including the sawtoothed grain beetle, the red flour beetle, and the confused flour beetle, infest grain products in the pantry, grocery store, and packing plant. These pests feed on and reproduce in stored flour products. They can migrate to and infest nearby broken or poorly sealed containers and can also chew through and infest flimsy paper and cellophane packages. Even when infested packages are removed, the beetles can live on flour and cereals that sift into cracks in the cupboard.

Solution: Discard all infested food. Clean out cupboards thoroughly before restocking. If infestation is widespread, remove all food items from the pantry and treat cracks and area behind shelves with ORTHO HomeDefense Home & Garden Insect Killer, or fog the infested area with ORTHO HomeDefense Hi-Power Brand Indoor Insect Fogger. Let the treatment dry before replacing food. Keep foods in airtight glass, plastic, or metal containers. If you suspect food is contaminated, kill the beetles, grubs, and eggs by deep-freezing food for 4 days or heating it in a shallow pan in the oven at 150°F for half an hour. Keep the pantry clean, and avoid buying broken packages; they are more likely to be infested.

Grain weevils

Rice weevils (6× life size).

Problem: Reddish brown to black beetles 1/8 to 1/6 inch long with elongated snouts are feeding in stored whole-grain rice, corn, wheat, and beans. The beetles may be seen crawling around the pantry. Yellow-white grubs may be found inside infested kernels.

Analysis: Grain weevils (*Sitophilus* species), a pantry pest, include the granary weevil and the rice weevil. They usually damage whole grains but occasionally infest flour and other broken or processed grain products. The adult weevils lay eggs inside grain kernels. Larvae that hatch from the eggs mature inside the kernels, pupate, and emerge as adult weevils. The adults wander about and are often seen far from the site of infestation.

Solution: Discard all infested food. Clean out cupboards thoroughly before restocking. If the infestation is widespread, remove all food items and fog the kitchen with ORTHO HomeDefense Hi-Power Brand Indoor Insect Fogger; or spray shelves and cracks in pantry cupboards with ORTHO HomeDefense Home & Garden Insect Killer. Do not spray countertops or other food preparation areas. Reline shelves with paper and replace food after the spray has dried. Keep foods in airtight glass, plastic, or metal containers. Avoid buying broken packages, which are more likely to be infested.

Cigarette beetle and drugstore beetle

Cigarette beetle, larva, and pupa (4× life size).

Problem: Reddish or reddish brown oval-shaped beetles, 1/8 inch long, or yellowish white curved grubs are feeding in stored tobacco, cigars, or cigarettes. They may also be found in spices such as red pepper and paprika, coffee beans, and other stored foods derived from plants.

Analysis: Cigarette beetles (*Lasioderma serricorne*) and drugstore beetles (*Stegobium paniceum*) are native to tropical parts of the world and can survive in the United States only in warm buildings (65°F and up). These pests feed on and reproduce in foods and spices. They may also feed on wool, leather, paper, drugs, and other household items. When infested products are brought into the home, the beetles can invade nearby uncontaminated foods kept in unsealed, broken, or flimsy containers. They may also chew through sealed paper containers.

Solution: Remove and destroy all infested foods. Clean out cupboards thoroughly before restocking. If the infestation is widespread, remove all food items and spray shelves and cracks with an insecticide containing *malathion* or *rotenone*. Let the spray dry before replacing food. Kill insects in packaged products by freezing for a day. Keep foods and spices in airtight glass, metal, or plastic containers. Refrigerate food kept in paper packages. Do not purchase items in broken or unsealed packages; they are more likely to be infested.

Appendix

TABLE OF CONTENTS

In this appendix you will find a wealth of information packed into more than 100 lists, charts, and maps.

For More Information

ABOUT PLANT PROBLEMS

Animal and Plant Health Inspection Service (APHIS)
This USDA site describes problem pests, diseases, and weeds, and the steps being taken to combat them.
www.aphis.usda.gov

Factsheet Database
Searches universities, government agencies, and extension factsheets in all states. From Ohio State.
http://plantfacts.ohio-state.edu/

National IPM Network
Links to pest-control sites.
http://www.reeusda.gov/agsys/nipmn/index.htm

Digital Diagnostics
Oklahoma State University
www.ento.okstate.edu/ddd/ddd.html

INSECTS

Insects on WWW
Virginia Tech
A vast and well-organized collection of links to insect sites.
http://atum.isis.vt.edu/~fanjun/text

Insect Notes
North Carolina State University
www.ces.ncsu.edu/depts/ent/notes

PLANT DISEASES

American Phytopathological Society Plant Pathology Resource Center
Information and links about plant diseases.
www.scisoc.org/resource

New and Emerging Plant Diseases Project
North Carolina State University
Information about new plant diseases from all over the country.
http://www.ces.ncsu.edu/depts/ent/clinic/Emerging/index.htm

WEEDS

Weed Science Society of America
http://ext.agn.uiuc.edu/wssa/

USDA Noxious Weeds Home Page
http://www.aphis.usda.gov/ppq/weeds/
Weed Photo Gallery
University of California, Davis
http://www.ipm.ucdavis.edu/PMG/weeds_common.html

ANIMAL PESTS

Controlling Nuisance Birds & Wildlife
University of Wisconsin
cf.unex.edu/ics/infosource/birds.cfm

PESTICIDES

EXTOXNET Pesticide Information Profiles
Information about the toxicity of pesticides.
http://ace.orst.edu/info/extoxnet/pips/ghindex.html

CDMS Label and Material Safety Data Sheets
Full text of labels and Material Safety Data Sheets for most pesticides, provided by the manufacturers.
http://www.cdms.net/manuf/manuf.asp

Pesticide Management Education Program
Cornell University
http://pmep.cce.cornell.edu/

HOUSEPLANTS

Horticulture Solutions: Houseplants
University of Illinois Extension
http://www.ag.uiuc.edu/~robsond/solutions/horticulture/house.html

Diagnosing Problems of Indoor Plants
Ohio State University
http://www.ag.ohio-state.edu/~ohioline/hyg-fact/3000/3068.html

LAWNS

Lawn Challenge
University of Illinois
http://www.urbanext.uiuc.edu/lawnchallenge/index.html

Floridaturf
University of Florida
http://floridaturf.com/index.html

Lawn Diseases
North Dakota State University
http://www.ext.nodak.edu/extpubs/plantsci/landscap/pp950w.htm

Managing Lawn and Turf Insects
University of Minnesota
www.extension.umn.edu/distribution/horticulture/DG1008.html

ANNUALS, PERENNIALS, AND BULBS

Ohioline: Flowers
Ohio State University
http://ohioline.ag.ohio-state.edu/lines/flwrs.html

Flowers Outdoors
University of Wisconsin
cf.uwex.edu/infosource/flowers.cfm

TREE AND SHRUB PROBLEMS

Tree Injuries—Prevention and Care
University of Nebraska, Lincoln
http://www.ianr.unl.edu/pubs/forestry/g1035.htm

Insect Pests of Shrubs
North Carolina State University
http://ipmwww.ncsu.edu/AG189/html/index.html

FRUIT TREE PROBLEMS

Home Fruit Production—Citrus
Texas Agricultural Extension Service
http://aggie-horticulture.tamu.edu/extension/homefruit/citrus/citrus.html

Common Tree Fruit Insects
Michigan State University Extension
http://www.canr.msu.edu/vanburen/fruitbug.htm

Tree Fruit Diseases in the Pacific Northwest
University of Washington
http://fruit.wsu.edu/diseases.htm

VEGETABLE PROBLEMS

Vegetable Growing
Oregon State University
http://eesc.orst.edu/agcomwebfile/garden/vegetable/

Small Fruit in the Home Garden
Virginia Cooperative Extension
http://www.ext.vt.edu/pubs/envirohort/426-840/426-840.html#L4

Insect and Disease Control in the Home Vegetable Garden
University of Vermont Extension
http://ctr.uvm.edu/ctr/pubs/br1158.htm

Weeds in the Home Vegetable Garden
Virginia Tech
http://www.ext.vt.edu/pubs/envirohort/426-364/426-364.html

GENERIC AND TRADE NAMES OF COMMON CHEMICALS

This is a partial list of trade names to assist in locating pesticides. It contains only a few of the most widely available products. Ortho cannot guarantee the quality of any products except those sold by Ortho or Scotts, our parent company.

Generic Name	Action	Trade Names
2,4–D	Herbicide, selective for broadleaf weeds	ORTHO Weed-B-Gon Lawn Weed Killer
Acephate	Insecticide, systemic	ORTHO Orthene Systemic Insect Control
Acifluorfen	Herbicide	ORTHO Kleeraway Grass & Weed Killer
Allethrin	Insecticide	ORTHO HomeDefense Flying & Crawling Insect Killer
Atrazine	Herbicide, selective for broadleaf weeds	Atrazine
Benomyl	Fungicide, broad-spectrum	Benlate
Bensulide	Herbicide, preemergence	Betasan® Pre-San®
Captan	Fungicide, broad-spectrum	ORTHO Home Orchard Spray
Carbaryl	Insecticide	Sevin®
Chloroneb	Fungicide, seed-treatment	Terraneb®
Chlorothalonil	Fungicide	ORTHO Multi-Purpose Fungicide Daconil 2787® Plant Disease Control
Chlorpyrifos	Insecticide	ORTHO Borer & Leaf Miner Spray ORTHO Dursban® Lawn & Garden Insect Control
Cycloheximide	Fungicide	Acti-dione
DCPA	Herbicide, preemergence	Dacthal®
Diazinon	Insecticide	ORTHO Bug-B-Gon Insect Killer ORTHO Diazinon Granules ORTHO Diazinon Ultra Insect Spray
Dicamba	Herbicide	Banvel D
Dichlobenil	Herbicide, preemergence	ORTHO Casoron Granules
Dicofol	Acaricide	Kelthane®
Dimethoate	Insecticide	Cygon
Disulfoton	Insecticide, systemic	ORTHO RosePride Systemic Rose & Flower Care
Dithiopyr	Herbicide	Dimension®
DSMA	Herbicide, selective for broadleaf weeds	DSMA
Endosulfan	Insecticide	Thiodan
Eptam	Herbicide	Eptam®
Ethazole	Fungicide, soil	Terrazole® Truban®
Fenarimol	Fungicide	Dodine® Rubigan®
Ferbam	Fungicide	Ferbam®
Ferrous sulfate monohydrate	Moss killer	Scotts Moss Control Granules
Fluazifop-P-butyl	Herbicide, selective for grassy weeds	ORTHO Grass-B-Gon Grass Killer
Fosetyl-aluminum	Bactericide, fungicide	Aliette®
Glyphosate	Herbicide, systemic	Roundup®
Halofenozide	Insecticide, soil	ORTHO Grub-B-Gon
Hexakis	Acaricide	Vendex®
Imazapyr	Herbicide	ORTHO GroundClear Triox Total Vegetation Killer
Imidacloprid	Insecticide	Scotts Grubex®
Generic Name	Action	Trade Names
Lime-sulfur	Fungicide, insecticide, miticide	ORTHO Dormant Disease Control Lime-Sulfur Spray
Linuron	Herbicide	Lorox®

Generic Name	Action	Trade Names
Malathion	Insecticide	Malathion
Mancozeb	Fungicide	Dithane® M–45 Manzate®
Maneb	Fungicide	Maneb
MCPP	Herbicide	Mecoprop
Metaldehyde	Molluscicide	ORTHO Bug-Geta Snail & Slug Killer
Methanearsonate, calcium acid	Herbicide	ORTHO Crabgrass & Nutgrass Killer
Methoprene	Insecticide (insect growth regulator)	ORTHO Flea-B-Gon Total Flea Killer Indoor Fogger Precor®
Methoxychlor	Insecticide	ORTHO Home Orchard Spray
Myclobutanil	Fungicide	Eagle® Nova® Rally®
Napropamide	Herbicide	Devrinol®
Oryzalin	Herbicide, preemergence	Surflan®
Oxyfluorfen	Herbicide	ORTHO GroundClear Triox Total Vegetation Killer
PCNB	Fungicide, soil	Terraclor®
Pendimethalin	Herbicide	Scotts Halts® Crabgrass Preventer Scotts Turf Weedgrass Control
Permethrin	Insecticide	Ambush® ORTHO Flea-B-Gon Total Flea Killer Indoor Spray
Phosmet	Insecticide	Imidan®
Prometon	Herbicide, preemergence and postemergence	ORTHO GroundClear Triox Total Vegetation Killer
Propiconazole	Fungicide	Alamo® Banner®
Pyrethrins	Insecticide	ORTHO RosePride Rose & Flower Insect Killer ORTHO Tomato & Vegetable Insect Killer
Resmethrin	Insecticide	ORTHO RosePride Orthenex Insect & Disease Control
Rotenone	Insecticide	Rotenone
Siduron	Herbicide, preemergence	Siduron
Simazine	Herbicide	Princep®
Streptomycin	Bactericide	Streptomycin
Sumithrin	Insecticide	ORTHO HomeDefense Home & Garden Insect Killer
Tetramethrin	Insecticide	ORTHO HomeDefense Home & Garden Insect Killer
Thiophanate-methyl	Fungicide, systemic	Cleary 3336® Topsin®
Thiram	Fungicide	Thiram
Triadimefon	Fungicide	Bayleton® Scotts ProTurf® Fungicide
Triclopyr	Herbicide	ORTHO Brush-B-Gon Poison Ivy, Poison Oak & Brush Killer
Trifluralin	Herbicide	Treflan®
Triforine	Fungicide	ORTHO RosePride Funginex Rose & Shrub Disease Control
Zineb	Fungicide	Cuprothex®
Ziram	Fungicide	Ziram

THE USDA PLANT HARDINESS ZONE MAP OF NORTH AMERICA

Plants are classified according to the amount of cold weather they can handle. For example, a plant listed as hardy to Zone 6 will survive a winter in which the temperature drops to –10° F.

Warm weather also influences whether a plant will survive in your region. Although this map does not address heat hardiness, in general, if a range of hardiness zones is listed for a plant, the plant will survive winter in the coldest zone as well as tolerate the heat of the warmest zone.

To use this map, find the location of your community, then match the color band marking that area to the zone key at left.

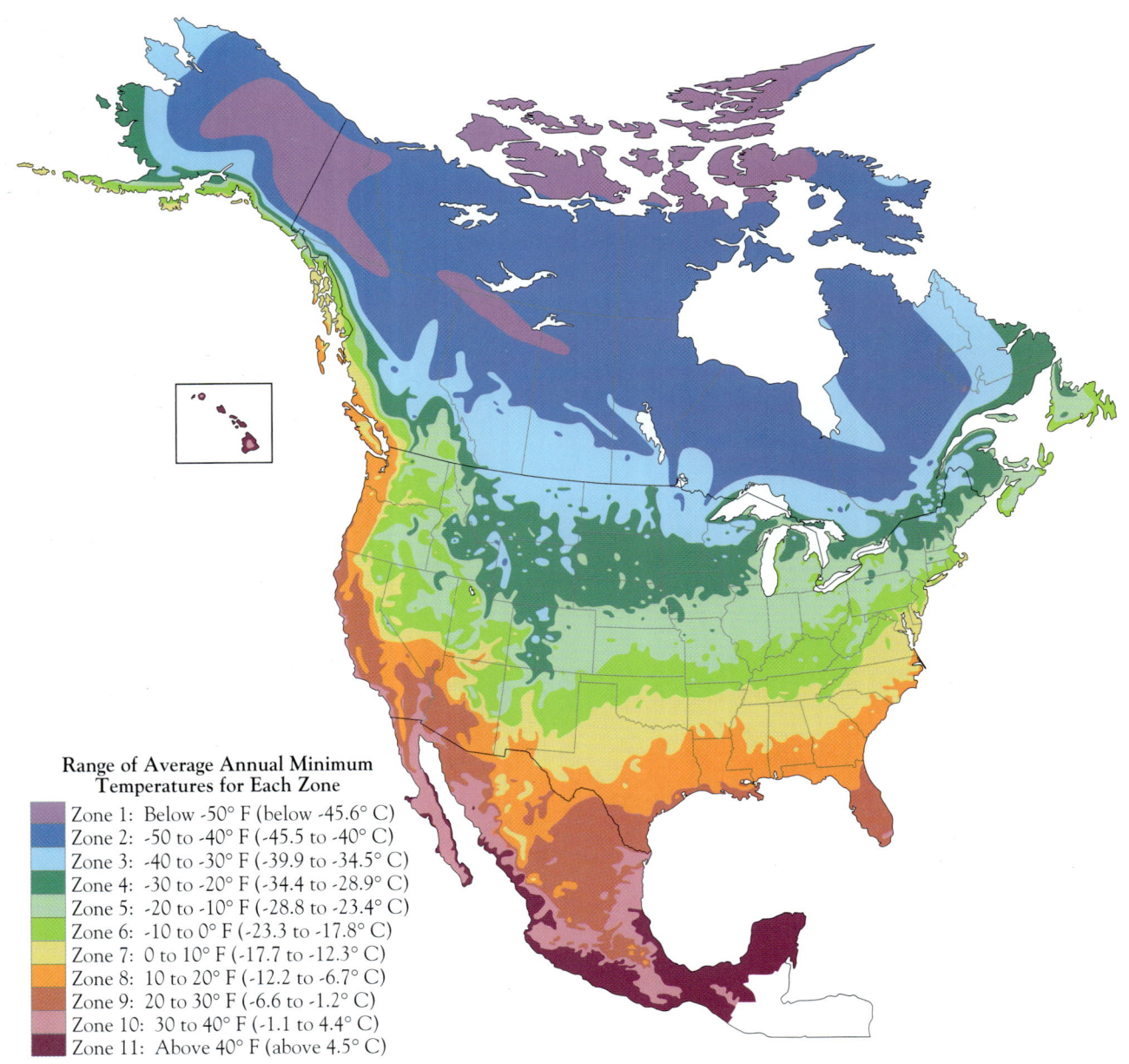

Range of Average Annual Minimum Temperatures for Each Zone

Zone 1: Below -50° F (below -45.6° C)
Zone 2: -50 to -40° F (-45.5 to -40° C)
Zone 3: -40 to -30° F (-39.9 to -34.5° C)
Zone 4: -30 to -20° F (-34.4 to -28.9° C)
Zone 5: -20 to -10° F (-28.8 to -23.4° C)
Zone 6: -10 to 0° F (-23.3 to -17.8° C)
Zone 7: 0 to 10° F (-17.7 to -12.3° C)
Zone 8: 10 to 20° F (-12.2 to -6.7° C)
Zone 9: 20 to 30° F (-6.6 to -1.2° C)
Zone 10: 30 to 40° F (-1.1 to 4.4° C)
Zone 11: Above 40° F (above 4.5° C)

METRIC CONVERSIONS

U.S. Units to Metric Equivalents			Metric Units to U.S. Equivalents		
To Convert From	Multiply By	To Get	To Convert From	Multiply By	To Get
Inches	25.4	Millimeters	Millimeters	0.0394	Inches
Inches	2.54	Centimeters	Centimeters	0.3937	Inches
Feet	30.48	Centimeters	Centimeters	0.0328	Feet
Feet	0.3048	Meters	Meters	3.2808	Feet
Yards	0.9144	Meters	Meters	1.0936	Yards
Square inches	6.4516	Square centimeters	Square centimeters	0.1550	Square inches
Square feet	0.0929	Square meters	Square meters	10.764	Square feet
Square yards	0.8361	Square meters	Square meters	1.1960	Square yards
Acres	0.4047	Hectares	Hectares	2.4711	Acres
Cubic inches	16.387	Cubic centimeters	Cubic centimeters	0.0610	Cubic inches
Cubic feet	0.0283	Cubic meters	Cubic meters	35.315	Cubic feet
Cubic feet	28.316	Liters	Liters	0.0353	Cubic feet
Cubic yards	0.7646	Cubic meters	Cubic meters	1.308	Cubic yards
Cubic yards	764.55	Liters	Liters	0.0013	Cubic yards

To convert from degrees Fahrenheit (F) to degrees Celsius (C), first subtract 32, then multiply by 5/9.

To convert from degrees Celsius to degrees Fahrenheit, multiply by 9/5, then add 32.

QUANTITIES OF GROUND LIMESTONE NEEDED TO RAISE pH TO 6.5

Present pH of soil	Pounds needed per 100 sq ft of sandy loam	Pounds needed per 100 sq ft of loam	Pounds needed per 100 sq ft of clay loam
4.0	11.5	16	23
4.5	9.5	13.5	19.5
5.0	8	10.5	15
5.5	6	8	10.5
6.0	3	4	5.5

Dolomitic limestone is recommended because it adds magnesium as well as calcium to the soil. The limestone should be cultivated into the soil. Adapted from Soil Acidity Needs of Plants, New York County Extension Service, publication D-2-25.

THE pH SCALE

Some familiar foods and materials		Soils
	ACID	
Grapefruit	3	Peat moss
Grape	4	
Bread	5	Best for rhododendron, azalea, and other acid-loving plants
Milk	6	
Pure water: Neutral	7	Average eastern soils
		Average southwestern soils
Baking soda	8	
Soap	9	
Milk of magnesia	10	Alkali soils
	11	
	ALKALINE	

PLANTS THAT WILL GROW IN ACID SOIL (pH OF 4.5 TO 5.5)

Common Name	Botanical Name
TREES AND SHRUBS	
Andromeda	Pieris
Broom	Cytisus
Camellia	Camellia
Crape myrtle	Lagerstroemia indica
Gardenia	Gardenia
Heath	Erica
Heather	Calluna
Hemlock	Tsuga
Holly	Ilex
Hydrangea	Hydrangea
Leucothoe	Leucothoe
Magnolia	Magnolia
Manzanita	Arctostaphylos
Mountain ash	Sorbus
Mountain laurel	Kalmia
Pin oak	Quercus palustris
Pine	Pinus
Quaking aspen	Populus tremuloides
Rhododendron, azalea	Rhododendron
Serviceberry	Amelanchier
Spruce	Picea
Weeping willow	Salix babylonica
FLOWERS	
Baby's breath	Gypsophila
Coreopsis	Coreopsis
Lily-of-the-valley	Convallaria
Lupine	Lupinus

PLANTS FOR SANDY SOIL

Common Name	Botanical Name
TREES	
American holly	Ilex opaca
Crabapple	Malus
Eastern red cedar	Juniperus virginiana
Eastern white pine	Pinus strobus
Jack pine	P. banksiana
Japanese black pine	P. thunbergii
Japanese pagoda tree	Sophora japonica
Jerusalem thorn	Parkinsonia aculeata
London plane tree	Platanus × acerifolia
Pin oak	Quercus palustris
Pitch pine	Pinus palustris
Post oak	Quercus stellata
Red pine	Pinus resinosa
Russian olive	Elaeagnus angustifolia
Short-leaf pine	Pinus echinata
Slash pine	Pinus elliottii
Sour gum	Nyssa sylvatica
Washington hawthorn	Crataegus phaenopyrum
White oak	Quercus alba
White poplar	Populus alba
White spruce	Picea glauca
SHRUBS	
Amur privet	Ligustrum amurense
Bayberry	Myrica pensylvanica
Beach plum	Prunus maritima
Bush clover	Lespedeza thunbergii
Butterfly bush	Buddleia davidii
Common buckthorn	Rhamnus catharticus
Firethorn	Pyracantha coccinea
Flowering quince	Chaenomeles speciosa
Highbush blueberry	Vaccinium corymbosum
Japanese barberry	Berberis thunbergii
Japanese kerria	Kerria japonica
Japanese spirea	Spiraea japonica

Common Name	Botanical Name
Mockorange	Philadelphus coronarius
Mountain laurel	Kalmia latifolia
Pfitzer juniper	Juniperus × media 'Pfitzerana'
Red chokeberry	Aronia arbutifolia
Rugosa rose	Rosa rugosa
Sheep laurel	Kalmia angustifolia
Shrubby cinquefoil	Potentilla fruticosa
Smooth sumac	Rhus glabra
Tamarisk	Tamarix parviflora
Tatarian honeysuckle	Lonicera tatarica
Wax myrtle	Myrica cerifera
Weigela	Weigela florida
VINES AND GROUND COVERS	
American bittersweet	Celastrus scandens
Bearberry	Arctostaphylos uva-ursi
Bracken	Pteridium aquilinum
Common thyme	Thymus vulgaris
Creeping juniper	Juniperus horizontalis
Fragrant sumac	Rhus aromatica
Grape	Vitis
Hall's Japanese honeysuckle	Lonicera japonica 'Halliana'
Hardy kiwi	Actinidia arguta
Japanese garden juniper	Juniperus procumbens
Lilyturf	Liriope spicata
Lippia	Phyla nodiflora
Memorial rose	Rosa wichuraiana
Shore juniper	Juniperus conferta
Stonecrop	Sedum acre
Trailing lantana	Lantana montevidensis
Trumpet creeper	Campsis radicans
Virginia creeper	Parthenocissus quinquefolia
Wedelia	Wedelia trilobata

PLANTS TOLERANT OF SALINE SOIL

Common Name	Botanical Name
Aleppo pine	Pinus halepensis
Blue dracaena	Cordyline indivisa
Bougainvillea	Bougainvillea
Brush cherry	Syzygium paniculatum
Capeweed	Arctotheca calendula
Coyote brush	Baccharis pilularis
European fan palm	Chamaerops humilis
Firethorn	Pyracantha
Gazania	Gazania
Karo	Pittosporum crassifolium
Lippia	Phyla nodiflora
Mirror plant	Coprosma repens
Natal plum	Carissa grandiflora
Norfolk Island pine	Araucaria heterophylla
Oleander	Nerium oleander
Pampas grass	Cortaderia selloana
Rosea ice plant	Drosanthemum
Rosemary	Rosmarinus officinalis 'Lockwood de Forest'
Spindle tree	Euonymus japonica
Trailing ice plant	Lampranthus
Weeping bottlebrush	Callistemon viminalis
White ice plant	Delosperma 'Alba'

PLANTS FOR GROWING IN AREAS WITH RESTRICTED ROOT SPACE

Common Name	Botanical Name
Amur maple	Acer tataricum ginnala
Black haw	Viburnum prunifolium
Chinese pistachio	Pistacia chinensis
Cinquefoil	Potentilla fruticosa
Cotoneaster	Cotoneaster
Crabapple	Malus
Dogwood	Cornus
Dwarf mugo pine	Pinus mugo var. mugo
Flowering cherry	Prunus
Golden rain tree	Koelreuteria paniculata
Hawthorn	Crataegus
Hedge maple	Acer campestre
Holly	Ilex
Hornbeam	Carpinus
Japanese maple	Acer palmatum
Juniper	Juniperus
Maidenhair tree	Ginkgo
Mountain mahogany	Cercocarpus
Peach	Prunus
Pinyon pine	Pinus edulis
Plum	Prunus
Redbud	Cercis
Rosemary	Rosmarinus
Russian olive	Elaeagnus angustifolia
Saucer magnolia	Magnolia × soulangiana
Siebold viburnum	Viburnum sieboldii
Silk tree	Albizia julibrissin
Snowbell	Styrax
Southern black haw	Viburnum rufidulum
Star magnolia	Magnolia stellata
Strawberry tree	Arbutus unedo

PLANTS FOR WET SOIL

Common Name	Botanical Name	Common Name	Botanical Name
TREES		Willow	Salix
Alder	Alnus	Winterberry	Ilex verticillata
American holly	Ilex opaca	**PERENNIALS**	
American larch	Larix laricina	Astilbe	Astilbe
Bald cypress	Taxodium distichum	Bloodroot	Sanguinaria canadensis
Dahoon	Ilex cassine	Bugbane	Cimicifuga racemosa
Horsetail tree	Casuarina equisetifolia	Calla	Zantedeschia
Kanooka tristania	Tristania laurina	Canada lily	Lilium canadense
Pin oak	Quercus palustris	Cardinal flower	Lobelia cardinalis
Plane tree, sycamore	Platanus	Common cattail	Typha latifolia
Poplar	Populus	Creeping Jenny	Lysimachia nummularia
Red maple	Acer rubrum	Elephant's ear	Colocasia esculenta
River birch	Betula nigra	Fern family	Polypodiaceae
Serviceberry	Amelanchier arborea	Forget-me-not	Myosotis scorpioides
Silver maple	Acer saccharinum	Globe flower	Trollius
Sour gum	Nyssa sylvatica	Golden-eyed grass	Sisyrinchium californicum
Swamp white oak	Quercus bicolor	Great lobelia	Lobelia siphilitica
Sweet bay magnolia	Magnolia virginiana	Japanese iris	Iris ensata
Sweet gum	Liquidambar styraciflua	Japanese primrose	Primula japonica
Willow	Salix	Joe Pye weed	Eupatorium maculatum
SHRUBS		Lance-leaved violet	Viola lanceolata
Arborvitae	Thuja	Marsh mallow	Althaea officinalis
Bayberry	Myrica pensylvanica	Marsh marigold	Caltha palustris
Buttonbush	Cephalanthus occidentalis	Mint	Mentha
Cranberry bush	Viburnum trilobum	Monkey flower	Mimulus
Fern-leaf bamboo	Bambusa disticha	Monkshood	Aconitum
Inkberry	Ilex glabra	New England aster	Aster novae-angliae
Pink-shell azalea	Rhododendron vaseyi	Piggyback plant	Tolmiea menziesii
Red chokecherry	Aronia arbutifolia	Ranunculus	Ranunculus
Red osier dogwood	Cornus sericea	Ribbon grass	Phalaris arundinacea 'Picta'
Spicebush	Lindera benzoin	Sedge	Carex
Swamp rose	Rosa palustris	Siberian iris	Iris sibirica
Sweet azalea	Rhododendron arborescens	Sweet flag	Acorus calamus
Sweet shrub	Calycanthus	Sweet white violet	Viola blanda
Water birch	Betula occidentalis	Virginia waterleaf	Hydrophyllum virginianum

PLANTS TOLERANT OF INDUSTRIAL POLLUTION (SO$_2$)

Common Name	Botanical Name
TREES AND SHRUBS	
American holly	Ilex opaca
Black locust	Robinia pseudoacacia
Cottonwood	Populus deltoides
Dwarf mugo pine	Pinus mugo var. mugo
Eastern arborvitae	Thuja occidentalis
Eastern sycamore	Platanus occidentalis
English holly	Ilex aquifolium
English oak	Quercus robur
European hornbeam	Carpinus betulus
Grapefruit	Citrus × paradisi
Green ash	Fraxinus pennsylvanica
Hedge maple	Acer campestre
Juniper	Juniperus
Lemon	Citrus limon
Lime	Citrus aurantiifolia
Linden	Tilia
London plane tree	Platanus × acerifolia
Maidenhair tree	Ginkgo biloba
Mountain maple	Acer spicatum
Privet	Ligustrum
Red oak	Quercus rubra
Sour gum	Nyssa sylvatica
Sourwood	Oxydendrum aboreum
Sweet orange	Citrus sinensis
Western red cedar	Thuja plicata
VEGETABLES AND FRUITS	
Cabbage	
Celery	
Corn	
Muskmelon	
Onion	

PLANTS TOLERANT OF SMOG (OZONE AND PAN)

Common Name	Botanical Name
TREES AND SHRUBS	
American arborvitae	Thuja occidentalis
Balsam fir	Abies balsamea
Chinese azalea	Rhododendron molle
English oak	Quercus robur
European white birch	Betula pendula
Gray dogwood	Cornus racemosa
Norway maple	Acer platanoides
Red pine	Pinus resinosa
Sugar maple	Acer saccharum
White fir	Abies concolor
Winged euonymus	Euonymus alatus
FLOWERS	
California poppy	Eschscholzia californica
Columbine	Aquilegia
Coral bells	Heuchera sanguinea
Daffodil	Narcissus
Iris	Iris
Lily	Lilium
Lily of the Nile	Agapanthus
Snapdragon	Antirrhinum
GROUND COVERS	
English ivy	Hedera helix
Periwinkle	Vinca minor
VEGETABLES AND FRUITS	
Strawberry	
Sweet potato	

PLANTS ATTRACTIVE TO BEES

Common Name	Botanical Name
TREES AND SHRUBS	
Abelia	Abelia
Acacia	Acacia
Barberry	Berberis
Bottlebrush	Callistemon
Broom	Cytisus
California lilac	Ceanothus
Cotoneaster	Cotoneaster
Escallonia	Escallonia
Firethorn	Pyracantha
Heath	Erica
Heather	Calluna
Honeylocust	Gleditsia
Honeysuckle	Lonicera
India hawthorn	Rhaphiolepis
Lantana	Lantana
Loquat	Eriobotrya
Manzanita	Arctostaphylos
Myrtle	Myrtus
Oleander	Nerium
Pittosporum	Pittosporum
Privet	Ligustrum
Rosemary	Rosmarinus
Star jasmine	Trachelospermum
Thyme	Thymus
Wisteria	Wisteria
FLOWERS	
Bellflower	Campanula
Flowering tobacco	Nicotiana
Forget-me-not	Myosotis
Lavender	Lavandula
Sage	Salvia
Sunflower	Helianthus
Sweet alyssum	Lobularia
Yarrow	Achillea

PLANTS RELATIVELY FREE OF INSECTS AND DISEASES

Common Name	Botanical Name	Common Name	Botanical Name
Anise magnolia	Magnolia salicifolia	Japanese kerria	Kerria
Bald cypress	Taxodium	Katsura tree	Cercidiphyllum japonicum
Bayberry	Myrica	Kentucky coffee tree	Gymnocladus dioica
Bottletree	Brachychiton		
Broom	Cytisus	Kobus magnolia	Magnolia kobus
Buckthorn	Rhamnus	Maidenhair tree	Ginkgo
Carob	Ceratonia siliqua	Mediterranean hackberry	Celtis australis
Castor-aralia	Kalopanax pictus		
Cedar	Cedrus	Myrtle	Myrtus
Chinese pistachio	Pistacia chinensis	Persian parrotia	Parrotia persica
Cinquefoil	Potentilla	Podocarpus	Podocarpus
Cork tree	Phellodendron	Shiny xylosma	Xylosma congestum
Cornelian cherry	Cornus mas		
Cucumber tree	Magnolia acuminata	Siebold viburnum	Viburnum sieboldii
Dawn redwood	Metasequoia	Silk oak	Grevillea robusta
Fig	Ficus	Smoke tree	Cotinus
Franklin tree	Franklinia	Snowbell	Styrax
Golden rain tree	Koelreuteria paniculata	Sour gum	Nyssa
		Star magnolia	Magnolia stellata
Goldenchain tree	Laburnum	Stewartia	Stewartia
Hardy rubber tree	Eucommia ulmoides	Tamarisk	Tamarix
		Tree of heaven	Ailanthus
Hop hornbeam	Ostrya	Turkish filbert	Corylus colurna
Hornbeam	Carpinus	Umbrella pine	Sciadopitys verticillata
Incense cedar	Calocedrus decurrens		
Japanese cornelian cherry	Cornus officinalis	Wax-leaf privet	Ligustrum lucidum
Japanese pagoda tree	Sophora japonica		

PLANTS SUSCEPTIBLE TO CEDAR-APPLE RUST (GYMNOSPORANGIUM JUNIPERI-VIRGINIANAE)

Common Name	Botanical Name
RESISTANT	
Chinese juniper 'Foemina' 'Keteleeri'	Juniperus chinensis
Common juniper 'Aureospica' 'Saxatilis' 'Suecica'	J. communis
Dwarf juniper	J. communis var. depressa
Eastern red cedar 'Tripartita'	J. virginiana
Sargent's juniper	J. sargentii
Savin juniper 'Broadmoor' 'Knap Hill' 'Skandia'	J. sabina
Singleseed juniper	J. squamata
SUSCEPTIBLE	
Apple, crabapple	Malus
Eastern red cedar	Juniperus virginiana
Rocky Mountain juniper	J. scopulorum
PARTICULARLY SUSCEPTIBLE APPLE VARIETIES	
Bechtel (crabapple)	
Imperial	
Jonathan	
Parkman (crabapple)	
Rome	
Wealthy	
York	

PLANTS SUSCEPTIBLE TO FIRE BLIGHT

Common Name	Botanical Name	Common Name	Botanical Name
TREES AND SHRUBS		Photinia	Photinia
Apple, crabapple	Malus	Rose	Rosa
Cotoneaster	Cotoneaster	Serviceberry	Amelanchier
Flowering almond, plum, and cherry	Prunus	Spirea	Spiraea
		FRUIT TREES AND BERRIES	
Flowering quince	Chaenomeles	Apple	
Hawthorn	Crataegus	Pear	
Loquat	Eriobotrya	Quince	
Mountain ash	Sorbus	Raspberry	

341

PLANTS SUSCEPTIBLE AND RESISTANT TO PHYTOPHTHORA CINNAMONI AND P. LATERALIS

Common Name	Botanical Name	Common Name	Botanical Name	Common Name	Botanical Name
SUSCEPTIBLE TREES AND SHRUBS		Juniper	Juniperus		obtusum
Abelia	Abelia	Larch	Larix	Meyer juniper	Juniperus squamata 'Meyeri'
Acacia	Acacia	Manzanita	Arctostaphylos		
Andromeda	Pieris	Myrtle	Myrtus	Pfitzer juniper	Juniperus × media 'Pfitzerana'
Aralia	Fatsia	Oak	Quercus		
Arborvitae	Thuja	Olive	Olea	Rock daphne	Daphne cneorum
Bald cypress	Taxodium	Pine	Pinus	Sasanqua camellia	Camellia sasanqua
Beefwood	Casuarina	Pittosporum	Pittosporum		
California lilac	Ceanothus	Plane tree, sycamore	Platanus	Savin juniper	Juniperus sabina
Camphor tree	Cinnamomum	Rhododendron, azalea	Rhododendron	Sawara cypress	Chamaecyparis pisifera
Cedar	Cedrus	St. Johnswort	Hypericum		
Chestnut	Castanea	Spruce	Picea	White cedar	C. thyoides
Coast redwood	Sequoia sempervirens	Sweet bay	Laurus	**SUSCEPTIBLE FRUITS AND BERRIES**	
		Viburnum	Viburnum	Apricot	
Common camellia	Camellia japonica	Walnut	Juglans	Avocado	
Cypress	Cupressus	Willow	Salix	Blueberry (highbush)	
Daphne	Daphne	Yew	Taxus	Cherry	
Dogwood	Cornus	**RESISTANT TREES AND SHRUBS**		Citrus	
Douglas fir	Pseudotsuga	Alaska cedar	Chamaecyparis nootkatensis	Peach	
Eucalyptus	Eucalyptus			Pear	
Fir	Abies	American arborvitae	Thuja occidentalis	**RESISTANT BERRY**	
Heath	Erica	Dwarf mugo pine	Pinus mugo var. mugo	Blueberry (rabbiteye)	
Heather	Calluna				
Hibiscus	Hibiscus	Hiryu azalea	Rhododendron		
Incense cedar	Calocedrus				

PLANTS SUSCEPTIBLE TO SOUTHERN BLIGHT* (SCLEROTIUM ROLFSII)

Common Name	Botanical Name	Common Name	Botanical Name	Common Name	Botanical Name
SHRUBS		Gladiolus	Gladiolus	Tulip	Tulipa
Daphne	Daphne	Hollyhock	Alcea	Viola	Viola
Hydrangea	Hydrangea	Iris	Iris	Violet	Viola
Pittosporum	Pittosporum	Lily	Lilium	Zinnia	Zinnia
Rose	Rosa	Lupine	Lupinus	**VEGETABLES AND FRUITS**	
FLOWERS		Marguerite	Chrysanthemum	Apple	Onion
Anemone	Anemone	Marigold	Tagetes	Artichoke	Pea
Bellflower	Campanula	Mum	Chrysanthemum	Avocado	Peanut
Black-eyed Susan	Rudbeckia	Pansy	Viola	Bean	Pepper
Canna	Canna	Phlox	Phlox	Beet	Potato
Carnation	Dianthus	Pincushion flower	Scabiosa	Cabbage	Rhubarb
China aster	Callistephus	Pink	Dianthus	Cantaloupe	Squash
Cosmos	Cosmos	Pot marigold	Calendula	Carrot	Strawberry
Daffodil	Narcissus	Stonecrop	Sedum	Cucumber	Tomato
Dahlia	Dahlia	Sweet pea	Lathyrus	Eggplant	Turnip
Delphinium	Delphinium	Sweet William	Dianthus	Lettuce	Watermelon
				Okra	

Southern blight has been reported on hundreds of plants. This is a partial list of plants that are frequently infected by this disease.

Powdery Mildews and Some of the Plants They Infect

Common Name	Botanical Name	Common Name	Botanical Name	Common Name	Botanical Name
ERYSIPHE CICHORACEARUM		Candytuft	Iberis	Rhododendron, azalea	Rhododendron
Aster	Aster	Carnation	Dianthus	Tulip tree	Liriodendron
Bachelor's button	Centaurea	Cherry	Prunus	Viburnum	Viburnum
Begonia	Begonia	Columbine	Aquilegia	Walnut	Juglans
Black-eyed Susan	Rudbeckia	Delphinium	Delphinium	California lilac	Ceanothus
Carpet bugle	Ajuga	Laurel	Prunus	**PHYLLACTINIA CORYLEA**	
Cineraria	Senecio	Pansy	Viola	Bramble	Rubus
Cosmos	Cosmos	Pea		Dogwood	Cornus
Coyote brush	Baccharis	Periwinkle	Vinca	Horsechestnut	Aesculus
Dahlia	Dahlia	Pink	Dianthus	Mockorange	Philadelphus
Daisy	Chrysanthemum	Radish		Oak	Quercus
Eucalyptus	Eucalyptus	Sweet alyssum	Lobularia	**PODOSPHAERA SPECIES**	
Forget-me-not	Myosotis	Sweet pea	Lathyrus	Apple, crabapple	Malus
Hebe	Hebe	Sweet William	Dianthus	Ash	Fraxinus
Hollyhock	Alcea	Viola	Viola	Firethorn	Pyracantha
Lettuce	Lactuca	Violet	Viola	Maple	Acer
Marguerite	Chrysanthemum	**MICROSPHAERA ALNI**		Pear	Pyrus
Mum	Chrysanthemum	Alder	Alnus	Photinia	Photinia
Painted tongue	Salpiglossis	Hazelnut	Corylus	Plum, peach, apricot	Prunus
Poppy	Papaver	Honeysuckle	Lonicera	Spirea	Spiraea
Pot marigold	Calendula	Lilac	Syringa	**SPHAEROTHECA FULIGINEA**	
Ranunculus	Ranunculus	Oak	Quercus	Cantaloupe	
Sage	Salvia	Plane tree, sycamore	Platanus	Cucumber	
Smoke tree	Cotinus	Snowberry	Symphoricarpos	Squash	
Snapdragon	Antirrhinum	**OTHER MICROSPHAERA SPECIES**		**SPHAEROTHECA SPECIES**	
Spiraea	Spiraea	Acacia	Acacia	Blanket flower	Gaillardia
Sumac	Rhus	Blueberry	Vaccinium	Cinquefoil	Potentilla
Sunflower	Helianthus	Catalpa	Catalpa	Coral bells	Heuchera
Transvaal daisy	Gerbera	Crape myrtle	Lagerstroemia	Cotoneaster	Cotoneaster
Verbena	Verbena	Euonymus	Euonymus	Currant, gooseberry	Ribes
Watermelon	Citrullus	Heath	Erica	Flowering tobacco	Nicotiana
Yarrow	Achillea	Honeysuckle	Lonicera	Hawthorn	Crataegus
Zinnia	Zinnia	Hydrangea	Hydrangea	Heath	Erica
ERYSIPHE POLYGONI		India hawthorn	Rhaphiolepis	Kalanchoe	Kalanchoe
Amaranth	Amaranthus	Locust	Robinia	Petunia	Petunia
Bean		Magnolia	Magnolia	Phlox	Phlox
Beet		Passion flower	Passiflora	Piggyback plant	Tolmiea
Begonia	Begonia	Plane tree, sycamore	Platanus	Rose	Rosa
Cabbage family		Poplar	Populus	Strawberry	Fragaria
California poppy	Eschscholzia	Privet	Ligustrum		

PLANTS SUSCEPTIBLE TO QUINCE RUST (GYMNOSPORANGIUM CLAVIPES OR G. LIBOCEDRI)

Common Name	Botanical Name
Chokeberry	Aronia
Common juniper	Juniperus communis
Eastern red cedar	J. virginiana
Hawthorn	Crataegus
Incense cedar	Calocedrus decurrens
Mountain ash	Sorbus
Serviceberry	Amelanchier

PLANTS THAT MAY BE INFESTED BY THE MEDITERRANEAN FRUIT FLY

Apple	Peach
Apricot	Pear
Avocado	Pepper
Cantaloupe	Persimmon
Cherry	Plum
Citrus	Pumpkin
Cucumber	Pyracantha
Fig	Quince
Grape	Squash (Hubbard)
Guava	Strawberry
Loquat	Tomato
Olive	Walnut

Adapted from California Agriculture, March–April 1981.

PLANTS SUSCEPTIBLE TO BOTRYOSPHAERIA RIBIS

Common Name	Botanical Name
TREES AND SHRUBS	
Chestnut	Castanea
Dogwood	Cornus
Firethorn	Pyracantha
Forsythia	Forsythia
Holly	Ilex
Maple	Acer
Poplar	Populus
Redbud	Cercis
Rhododendron, azalea	Rhododendron
Rose	Rosa
Sour gum	Nyssa
Sweet gum	Liquidambar
Willow	Salix
FRUIT AND NUT TREES AND SMALL FRUITS	
Apple	
Avocado	
Citrus	
Currant	
Fig	
Hickory	
Pear	
Pecan	

PLANTS SUSCEPTIBLE TO BACTERIAL BLIGHT (PSEUDOMONAS SYRINGAE)

Almond	Oleander
Apple	Pea
Avocado	Peach
Bean	Pear
Cherry	Plum
Citrus	Rose
Lilac	Stock

PLANTS RESISTANT TO COTTON ROOT ROT (PHYMATOTRICHUM OMNIVORUM)

Common Name	Botanical Name
TREES AND SHRUBS	
Deutzia	Deutzia
Ferns	
Hackberry	Celtis
Oak	Quercus
Palms	
Pomegranate	Punica
Weeping mulberry	Morus alba 'Pendula'
FLOWERS	
Amaranth	Amaranthus
Baby's breath	Gypsophila
California poppy	Eschscholzia
Calla	Zantedeschia
Candytuft	Iberis
Cyclamen	Cyclamen
Daffodil	Narcissus
Foxglove	Digitalis
Freesia	Freesia
Hyacinth	Hyacinthus
Iris	Iris
Moss rose	Portulaca
Nasturtium	Tropaeolum
Petunia	Petunia
Phlox	Phlox
Poppy	Papaver
Primrose	Primula
Slipper flower	Calceolaria
Snapdragon	Antirrhinum
Stock	Matthiola
Sweet alyssum	Lobularia
Zinnia	Zinnia
VEGETABLES AND FRUITS	
Asparagus	Grape
Cabbage family	Leek
Cantaloupe	Onion
Celery	Pumpkin
Cranberry	Spinach
Cucumber	Squash
Currant	Strawberry
Dewberry	Watermelon
Garlic	

PLANTS RESISTANT TO ARMILLARIA ROOT ROT

Common Name	Botanical Name	Common Name	Botanical Name	Common Name	Botanical Name
TREES AND SHRUBS		Japanese flowering crabapple	*Malus floribunda*	Southern catalpa	*Catalpa bignonioides*
American elder	*Sambucus canadensis*	Japanese maple	*Acer palmatum*	Southern magnolia	*Magnolia grandiflora*
Austrian pine	*Pinus nigra*	Japanese pagoda tree	*Sophora japonica*	Star acacia	*Acacia verticillata*
Bald cypress	*Taxodium distichum*	Judas tree	*Cercis siliquastrum*		
Barberry	*Berberis polyantha*	Lawson's cypress	*Chamaecyparis lawsoniana 'Ellwoodii'*	Swamp birch	*Betula pumila*
Bigleaf maple	*Acer macrophyllum*			Sweet gum	*Liquidambar styraciflua*
Boxleaf honeysuckle	*Lonicera nitida*	Leyland cypress	*Cupressocyparis leylandii*	Ternstroemia	*Ternstroemia*
Boxwood	*Buxus sempervirens*	Madrone	*Arbutus menziesii*	Torrey pine	*Pinus torreyana*
		Maidenhair tree	*Ginkgo biloba*	Tree heath	*Erica arborea*
Bush acacia	*Acacia longifolia*	Mayten tree	*Maytenus boaria*	Tree of heaven	*Ailanthus altissima*
Canary Island pine	*Pinus canariensis*	Modesto ash	*Fraxinus velutina 'Modesto'*		
Carob	*Ceratonia siliqua*			Tulip tree	*Liriodendron tulipifera*
Carrotwood	*Cupaniopsis anacardioides*	Monterey pine	*Pinus radiata*	Valley oak	*Quercus lobata*
		Mulberry	*Morus*	Western redbud	*Cercis occidentalis*
Catalina cherry	*Prunus lyonii*	Northern Bayberry	*Myrica pensylvanica*		
Chaste tree	*Vitex agnus-castus*			White fir	*Abies concolor*
Cherry laurel	*Prunus caroliniana*	Oregon grapeholly	*Mahonia aquifolium*	Yedda hawthorn	*Rhaphiolepis umbellata*
Chinese elm	*Ulmus parvifolia*				
Chinese pistachio	*Pistacia chinensis*	Oriental sweet gum	*Liquidambar orientalis*	**FRUITS, NUTS, AND BERRIES**	
Chinese wisteria	*Wisteria sinensis*			American chestnut	
Coast redwood	*Sequoia sempervirens*	Palms	*Palmae*	Apple	
		Plane tree, sycamore	*Platanus*	Avocado	
Dawn redwood	*Metasequoia glyptostroboides*	Queensland pittosporum	*Pittosporum rhombifolium*	Black cherry	
				Black walnut	
English holly	*Ilex aquifolium*	Red gum	*Eucalyptus camaldulensis*	Callery pear	
Eugenia	*Eugenia*			Cherry plum	
Evergreen ash	*Fraxinus uhdei*	Rose of Sharon	*Hibiscus syriacus*	American persimmon	*Diospyros virginiana*
Flowering quince	*Chaenomeles speciosa*	Russian olive	*Elaeagnus angustifolia*		
		St. Johnswort	*Hypericum patulum*	Japanese persimmon	*D. kaki*
Fragrant sumac	*Rhus aromatica*			Kadota fig	
Hackberry	*Celtis*	Scotch pine	*Pinus sylvestris*	Loganberry	
Heavenly bamboo	*Nandina domestica*	Shademaster locust	*Gleditsia triacanthos 'Shademaster'*	Mission fig	
Holly-leaved cherry	*Prunus ilicifolia*			Olallieberry	
Holly-leaved oak	*Quercus ilex*			Pear	
Incense cedar	*Calocedrus decurrens*	Silver-dollar gum	*Eucalyptus polyanthemos*	Pecan	
Jacaranda	*Jacaranda mimosifolia*	Smoke tree	*Cotinus coggygria*		
Japanese cedar	*Cryptomeria japonica*	Smooth Arizona cypress	*Cupressus arizonica var. glabra*		

PLANTS RESISTANT TO VERTICILLIUM WILT

Common Name	Botanical Name	Common Name	Botanical Name
TREES AND SHRUBS		Anemone	Anemone
American yellowwood	Cladrastis lutea	Baby-blue-eyes	Nemophila
Apple	Malus	Baby's breath	Gypsophila
Ash	Fraxinus	Balloon flower	Platycodon
Bamboo		Begonia	Begonia
Barberry	Berberis	Blanket flower	Gaillardia
Beech	Fagus	Bulbs	
Birch	Betula	Cacti	
Black locust	Robinia pseudoacacia	Candytuft	Iberis
Box elder	Acer negundo	Carnation	Dianthus
Boxwood	Buxus	Christmas rose	Helleborus
California lilac	Ceanothus	Cinquefoil	Potentilla
Conifers (cypress, fir, juniper, larch, pine, sequoia, spruce, and others)		Columbine	Aquilegia
		Coral bells	Heuchera
Crabapple	Malus	Ferns	
Currant	Ribes	Gladiolus	Gladiolus
Dogwood	Cornus	Hollyhock	Alcea
Eucalyptus	Eucalyptus	Impatiens	Impatiens
Firethorn	Pyracantha	Iris	Iris
Ginkgo	Ginkgo biloba	Lantana	Lantana
Hawthorn	Crataegus	Monkey flower	Mimulus
Holly	Ilex	Moss rose	Portulaca
Honeylocust	Gleditsia	Nasturtium	Tropaeolum
Hornbeam	Carpinus	Nemesia	Nemesia
Juniper	Juniperus	Orchids	
Katsura tree	Cercidiphyllum japonicum	Ornamental grasses	
		Pansy	Viola
Larch	Larix	Penstemon	Penstemon
Linden	Tilia	Periwinkle	Vinca
Manzanita	Arctostaphylos	Pincushion flower	Scabiosa
Maple	Acer	Pink	Dianthus
Mulberry	Morus	Pot marigold	Calendula
Oak	Quercus	Primrose	Primula
Oleander	Nerium	Ranunculus	Ranunculus
Palms		Sunrose	Helianthemum
Pear	Pyrus	Sweet William	Dianthus
Plane tree	Platanus	Verbena	Verbena
Redbud	Cercis canadensis	Viola	Viola
Russian olive	Elaeagnus	Violet	Viola
Sweet gum	Liquidambar	Zinnia	Zinnia
Sycamore	Platanus	**VEGETABLES**	
Viburnum	Viburnum	Asparagus	
Walnut	Juglans	Bean	
Willow	Salix	Carrot	
Yew	Taxus	Celery	
FLOWERS		Corn	
Ageratum	Ageratum	Lettuce	
Alyssum	Alyssum	Pea	
		Sweet potato	

Houseplants

HOUSEPLANTS THAT TOLERATE FULL LIGHT

Common Name	Botanical Name
Amaryllis	*Hippeastrum*
Asparagus fern	*Asparagus*
Cactus family	*Cactaceae*
Caladium	*Caladium*
Chrysanthemum	*Chrysanthemum*
Citrus	*Citrus*
Coffee plant	*Coffea*
Coleus	*Coleus*
Column flower	*Columnea*
Croton	*Codiaeum*
False aralia	*Dizygotheca*
Geranium	*Pelargonium*
Ivy	*Hedera*
Japanese aralia	*Fatsia*
Lipstick plant	*Aeschynanthus*
Miniature rose	*Rosa*
Ornamental fig, rubber plant	*Ficus*
Ornamental pepper	*Capsicum*
Parlor ivy, string-of-beads	*Senecio*
Passion flower	*Passiflora*
Poinsettia	*Euphorbia*
Sago palm	*Cycas*
Schefflera	*Brassaia*
Velvet plant	*Gynura*
Wax begonia	*Begonia*
Wax plant	*Hoya*

HOUSEPLANTS THAT TOLERATE LOW LIGHT

Common Name	Botanical Name
Cast-iron plant	*Aspidistra*
Chinese evergreen	*Aglaonema*
Dracaena	*Dracaena*
Dumb cane	*Dieffenbachia*
Grape ivy	*Cissus*
Mother-in-law's tongue	*Sansevieria*
Nephthytis	*Syngonium*
Palms	*Palmae*
Kentia palm	*Howea*
Lady palm	*Rhapis*
Parlor palm	*Chamaedorea*
Peace lily	*Spathiphyllum*
Philodendron	*Philodendron*
Pothos	*Scindapsus* and *Epipremnum*

HOUSEPLANTS THAT ROOT EASILY

Common Name	Botanical Name
African violet	*Saintpaulia*
Aluminum plant	*Pilea*
Begonia	*Begonia*
Chinese evergreen	*Aglaonema*
Chrysanthemum	*Chrysanthemum*
Coleus	*Coleus*
Geranium	*Pelargonium*
Grape ivy	*Cissus*
Impatiens	*Impatiens*
Ivy	*Hedera*
Jade plant	*Crassula*
Nephthytis	*Syngonium*
Nerve plant	*Fittonia*
Philodendron	*Philodendron*
Piggyback plant	*Tolmiea*
Pothos	*Scindapsus*
Prayer plant	*Maranta*
Spider plant	*Chlorophytum*
Swedish ivy	*Plectranthus*
Velvet plant	*Gynura*
Wandering Jew	*Tradescantia*

HOUSEPLANTS SENSITIVE TO SALTS IN THE SOIL

Common Name	Botanical Name
Avocado	*Persea*
Cast-iron plant	*Aspidistra*
Citrus	*Citrus*
Coffee plant	*Coffea*
Fern family	*Polypodiaceae*
Grape ivy	*Cissus*
Haworthia	*Haworthia*
Ivy	*Hedera*
Japanese aralia	*Fatsia*
Kentia palm	*Howea*
Ornamental fig, rubber plant	*Ficus*
Parlor palm	*Chamaedorea*
Peace lily	*Spathiphyllum*
Philodendron	*Philodendron*
Piggyback plant	*Tolmiea*
Prayer plant	*Maranta*
Sago palm	*Cycas*
Screw pine	*Pandanus*
Spider plant	*Chlorophytum*
Split-leaf philodendron	*Monstera*
Strawberry geranium	*Saxifraga*
Ti plant	*Cordyline*
Zebra plant	*Aphelandra* and *Calathea*

TURFGRASS CLIMATE ZONES

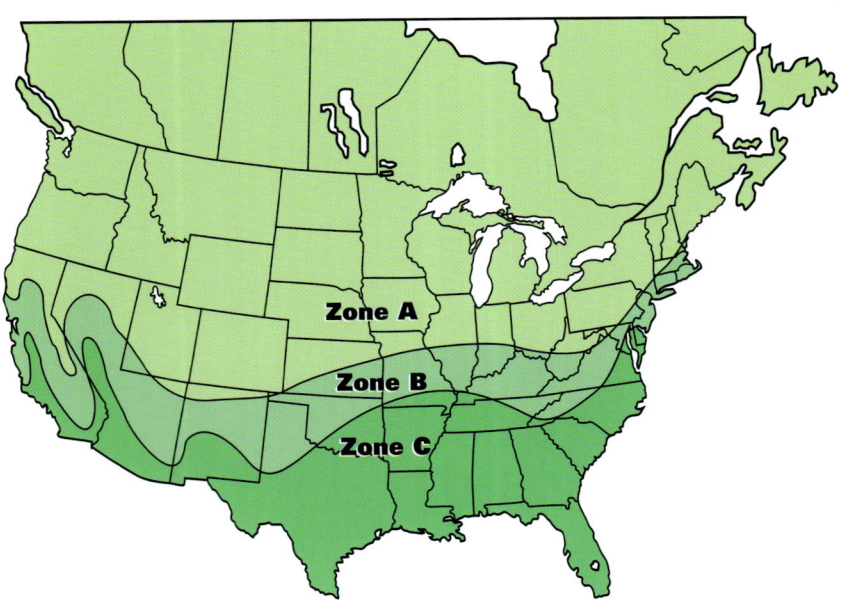

Zone A Cool-season grasses.
Zone C Warm-season grasses.
Zone B This is a transition zone in which both warm-season and cool-season grasses are grown. Because warm-season grasses have long dormant periods in this zone, cool-season grasses are usually preferred. Tall fescue does particularly well in this zone.

From *Turn Managers Handbook*, W. H. Daniel and R. P. Freeborg.

CHARACTERISTICS OF SOME TURFGRASSES

Grass	Zone[1]	Drought Resistant	Shade Tolerant	Days to Germinate	Low Maintenance[1,2]
Bahiagrass	C			21–28	
Bentgrass	A		•	5–12	
Bermudagrass, common	C	•		14–20	•
Bermudagrass, improved	C	•		Sprigs[3]	•
Carpetgrass	C			21	
Centipedegrass	C			14–20	
Fescue, red	A	•	•	5–12	•
Fescue, tall	A	•	•	5–12	•
Kentucky bluegrass	A			20–30	
St. Augustinegrass	C		•	Sprigs[3]	
Zoysiagrass	C	•	•	Sprigs[3]	•

[1] For information on Zones, see map above.
[2] Low-maintenance turfgrasses are those that tolerate irregular fertilizing, watering, and mowing.
[3] Usually planted as sprigs or plugs, rather than as seed.

TURFGRASSES RESISTANT TO DOLLAR SPOT

BLUEGRASS

A–20	Midnight
Adelphi	Parade
Ascot	Park
Bonnieblue	Preakness
Bristol	Princeton
Columbia	Rita
Eagleton	SR 2000
Eclipse	Touchdown
Kenblue	Unique
Liberty	Vantage
Majestic	Victa

FINE FESCUE

Chewings	**Creeping**
Brittany	Aruba
Jamestown	Dawson
Molina	Florentine
Sandpiper	Flyer II
Tiffany	Jasper
Treazure	Pennlawn
Victory	Shademaster II
Victory II	Silverlawn

Hard
Defiant
Ecostar
SR 3100

BENTGRASS

Allure	Egmont
Arlington	Tracenta
Bardot	

BLUEGRASS RESISTANT TO RUST

A–20	Glade
A–34	Haga
Alpine	Majestic
Baron	Miracle
Bartitia	Park
Bonnieblue	Pennstar
Classic	Rugby
Fylking	Suffolk
Georgetown	Washington

BLUEGRASS RESISTANT TO FUSARIUM BLIGHT

RESISTANT

A–20	Parade
Adelphi	Rugby
Bonnieblue	Sydsport
Columbia	Trenton
Enmundi	Vantage
Glade	Windsor

SUSCEPTIBLE

Arboretum	Modena
Belturf	Newport
Brunswick	Nugget
Campus	Park
Cougar	Pennstar
Delft	Plush
Enita	Ram #1
Fylking	South Dakota
Geronimo	
Merion	

BLUEGRASS RESISTANT TO FUSARIUM PATCH

Adelphi	Cynthia
Barcelona	Gnome
Bartitia	Liberty
Barzan	Merit
Birka	Miracle
Bonnieblue	Nublue
Bronco	Ram #1
Crest	Touchdown

BLUEGRASS RESISTANT TO HELMINTHOSPORIUM LEAF SPOT

RESISTANT

A–20	Eclipse
Able I	Limousine
Adelphi	Majestic
Alpine	Merion
Apex	Minstrel
Ascot	Noblesse
Barblue	Nugget
Barcelona	Parade
Birka	Pennstar
Blacksburg	Rita
Bonnieblue	Rugby
Bristol	SR 2000
Brunswick	Sydsport
Caliber	Touchdown
Cardiff	Vantage
Cobalt	Victa

SUSCEPTIBLE

Allure	Greenley
Barzan	Kenblue
Chelsea	Miracle
Cynthia	Nottingham
Delta	Park
Eagleton	Ram I
Fairfax	Raven
Geary	Ronde
Ginger	South Dakota
Glade	

BLUEGRASS RESISTANT TO RED THREAD

A–34	Eclipse
Able I	Fortuna
Adelphi	Gnome
Ampella	Indigo
Apex	Kelly
Ascot	Marquis
Aspen	Merion
Baron	Minstrel
Baronie	Miranda
Belmont	Nassau
Birka	Nublue
Blacksburg	Nustar
Bonnieblue	Princeton
Broadway	Rita
Cannon	SR 2000
Challenger	Touchdown
Conni	Trenton
Destiny	Viva

BLUEGRASS RESISTANT TO STRIPE SMUT

A–20	Cobalt
A–34	Conni
Adelphi	Dawn
Alpine	Destiny
Apex	Eclipse
Aquila	Enmundi
Aspen	Eva
Banff	Freedom
Barcelona	Georgetown
Baronie	Glade
Barzan	Indigo
Belmont	Julia
Birka	Newport
Blacksburg	Plush
Bonnieblue	Ram I
Brunswick	Sydsport
Caliber	Touchdown
Cardiff	Vantage
Classic	

Ground Covers

GROUND COVERS FOR SUNNY AREAS

Common Name	Botanical Name
Bearberry	Arctostaphylos uva-ursi
Bearberry cotoneaster	Cotoneaster dammeri
California lilac	Ceanothus griseus var. horizontalis
Chadwick yew	Taxus × media 'Chadwicki'
Creeping cotoneaster	Cotoneaster adpressus
Creeping thyme	Thymus
Dwarf coyote brush	Baccharis pilularis
Dwarf rosemary	Rosmarinus officinalis 'Prostratus'
Ice plant	Carpobrotus, Lampranthus, and other genera
Juniper	Juniperus
Lantana	Lantana
Lavender cotton	Santolina
Lippia	Phyla nodiflora
Low-bush blueberry	Vaccinium angustifolium
Moss phlox	Phlox subulata
Rockspray cotoneaster	Cotoneaster horizontalis
Rockcress	Arabis
St. Johnswort	Hypericum
Santa Cruz firethorn	Pyracantha koidzumii 'Santa Cruz'
Snow-in-summer	Cerastium tomentosum
Star jasmine	Trachelospermum jasminoides
Stonecrop	Sedum
Sun rose	Helianthemum nummularium
Woolly yarrow	Achillea tomentosa
Wormwood	Artemisia

GROUND COVERS FOR SHADY AREAS

Common Name	Botanical Name
Asparagus fern	Asparagus densiflorus
Baby's tears	Soleirolia soleirolii
Barrenwort	Epimedium
Carpet bugle	Ajuga
Dwarf Himalayan sweet box	Sarcococca hookerana var. humilis
Japanese holly fern	Cyrtomium falcatum
Ivy	Hedera
Japanese painted fern	Athyrium nipponicum 'Pictum'
Japanese spurge	Pachysandra terminalis
Lily turf	Liriope
Maidenhair fern	Adiantum pedatum
Mock strawberry	Duchesnea indica
Mondo grass	Ophiopogon japonicus
Periwinkle	Vinca
St. Johnswort	Hypericum
Sweet violet	Viola odorata
Sweet woodruff	Galium odoratum
Wild ginger	Asarum
Winter creeper	Euonymus fortunei
Wood fern	Dryopteris

DROUGHT-RESISTANT GROUND COVERS

Common Name	Botanical Name
Bearberry cotoneaster	Cotoneaster dammeri
Blue fescue	Festuca glauca
Creeping cotoneaster	C. adpressus
Crown vetch	Coronilla varia
Dwarf coyote bush	Baccharis pilularis
Dwarf rosemary	Rosmarinus officinalis 'Prostratus'
Goutweed	Aegopodium podagraria
Ice plant	Carpobrotus, Lampranthus, and other genera
Juniper	Juniperus
Lavender cotton	Santolina
Lippia	Phyla nodiflora
Manzanita	Arctostaphylos uva-ursi
Peruvian verbena	Verbena peruviana
Ribbon grass	Phalaris arundinacea var. picta
Rockspray cotoneaster	C. horizontalis
Rock rose	Cistus
St. Johnswort	Hypericum
Stonecrop	Sedum
Sun rose	Helianthemum nummularium
Thyme	Thymus
Wormwood	Artemisia

Annuals, Perennials, and Bulbs

PLANTING AND BLOOMING TIMES OF FLOWERS

Flower	Plant	Planting Season	Blooming Season	Ideal Soil Temperature for Germination (°F)	Days to Germination
Ageratum	A	Sp*	Sp–Su	60–65	10
Amaranth	A	Sp	Sp–Su	60–70	12
Anemone	Tu	F or Sp	Sp	**	**
Aster	P	Sp	Sp–Su–F	70	12–14
Astilbe	P	Sp	Sp–Su	60–70	14–21
Begonia	Tp	Sp	Sp–Su	70–75	15–20
Bellflower	P	Sp	Sp–Su–F	68–86	10–14
Black-eyed Susan	A, P	Sp	Su–F	70–75	5–10
Blanket flower	A, P	Sp*	Su–F	70	15–20
Candytuft	A, P	Sp*	Sp–Su	70–85	7–15
Canna	Rh	Sp	Su–F	**	**
Cape marigold	A	Sp	Su	70–85	15–20
Chrysanthemum	A, P	Sp	Su–F	70	7–10
Cineraria	A	Sp*	W–Sp	45–60	20
Cockscomb	A	Sp	Su	70–85	7–14
Columbine	P	F or Sp	Sp–Su	70–85	21–28
Coral bells	P	F or Sp	Sp–Su	70–85	5–20
Coreopsis	A, P	F or Sp	Su–F	70	15–20
Cosmos	A	Sp	Su–F	70–85	10–15
Crocus	Corm	F	Sp	**	**
Dahlia	Tu	Sp	Su–F	70–85 (seed)	15–20
Daylily	Tu	Sp*	Su	**	**
Delphinium	A, P	Sp*	Su	55–60	15–30
Flowering onion	Bu	F or Sp	Sp–Su	**	**
Foxglove	Bi, P	Sp	Sp–Su	70–85	15–20
Freesia	Corm	Sp*	Sp	**	**
Geranium	P	Sp	Sp–Su–F	68–85	15–65
Gladiolus	Corm	Sp	Su–F	**	**
Hollyhock	Bi	Sp–Su	Sp–Su	68	7–21
Hyacinth	Bu	F	Sp	**	**
Impatiens	A, Tp	Sp	Su	70	15–20
Iris	Rh, Bu	F	Sp–Su	**	**
Lantana	Tp	Sp	Sp–F	70	40–50
Lily	Bu	Sp	Su	**	**
Lily-of-the-valley	P	F (from pips)	Sp	**	**
Lobelia	A	Sp*	Su–F	70–85	15–20
Marigold	A	Sp	Su–F	70–75	5–7
Ornamental pepper	A	Sp	Su	70	15–20
Painted tongue	A	Sp	Su	70–75	15–20
Phlox	A, P	Sp*	Su–F	55–65	8–20
Pink	A, P	Sp*	Sp–Su	70	15–30
Pot marigold	A	Sp*	W–Sp	70–85	7–14
Primrose	A, P	Sp*	W–Sp	55–65	20
Ranunculus	Tu	F or Sp	Sp–Su	**	**
Sage	A, P	Sp	Su	70	12–15
Snapdragon	A	Sp*	Sp–Su	70	15
Solanum	A, P	Sp	Su–F	70	15–30
Spider flower	A	Sp*	Su	55–85	10–14
Stock	A	Sp	Sp–Su	55–90	15
Sunflower	A, P	Sp	Su–F	70–85	15–20
Sunrose	P	F or Sp	Su	70	15–20
Sweet alyssum	A, P	Sp	Sp–Su–F	70	7–15
Sweet pea	A	W–Sp	Sp	70	15
Transvaal daisy	Tp	Sp	Su	70	15
Yarrow	P	Sp	Su–F	70	5–15

PLANT KEY

A	Annual
Bi	Biennial
Bu	Bulb
P	Perennial
Rh	Rhizome
Tp	Tender Perennial, grown as annual in all but Zones 9 and 10
Tu	Tuber

PLANTING SEASON KEY

Sp	Spring
Su	Summer
F	Fall
W	Winter

Florida gardeners may find differences in planting times and soil temperatures. Check with your local county extension service.

* *Planted in the fall in Zones 9 and 10. See USDA Plant Hardiness Zone Map, page 336.*

** *Not usually planted from seed.*

FLOWERS TOLERANT OF PHYTOPHTHORA

Common Name	Botanical Name
RESISTANT OR VERY TOLERANT	
Begonia	Begonia
Cockscomb	Celosia
Flossflower	Ageratum
Flowering tobacco	Nicotiana
Geranium	Pelargonium
Marigold	Tagetes
Pincushion flower	Scabiosa
SOMEWHAT TOLERANT	
Coneflower	Rudbeckia
Impatiens	Impatiens
Petunia	Petunia
Zinnia	Zinnia

CHRYSANTHEMUM VARIETIES RESISTANT TO CHRYSANTHEMUM RUST

Achievement
Copper Bowl
Escapade
Helen Castle
Mandalay
Matador
Miss Atlanta
Orange Bowl
Powder Puff

Adapted from Chrysanthemum Cultivars Resistant to Verticillium Wilt and Rust, University of California County Extension leaflet 21057.

PLANTS SUSCEPTIBLE TO ASTER YELLOWS

Common Name	Botanical Name
FLOWERS	
Anemone	Anemone
Aster	Aster
Bachelor's button	Centaurea
Blanket flower	Gaillardia
Cape marigold	Dimorphotheca
Carnation, pink, sweet William	Dianthus
China aster	Callistephus
Cockscomb	Celosia
Coreopsis	Coreopsis
Cosmos	Cosmos
Delphinium	Delphinium
Gladiolus	Gladiolus
Lobelia	Lobelia
Marigold	Tagetes
Mum, daisy, marguerite	Chrysanthemum
Petunia	Petunia
Phlox	Phlox
Pincushion flower	Scabiosa
Pot marigold	Calendula
Snapdragon	Antirrhinum
Strawflower	Helichrysum
VEGETABLES	
Broccoli	Parsley
Cabbage	Parsnip
Carrot	Potato
Cauliflower	Pumpkin
Celery	Radish
Endive	Spinach
Lettuce	Squash
New Zealand spinach	Tomato
Onion	
WEEDS	
Dandelion	Taraxacum
Fleabane	Erigeron
Horseweed	Conyza
Plantain	Plantago
Ragweed	Ambrosia
Plum thistle	Cirsium
Wild carrot	Daucus

LILIES TOLERANT OF AND SUSCEPTIBLE TO VIRUSES

TOLERANT OR MODERATELY TOLERANT

SPECIES
Lilium amabile
L. bulbiferum
L. callosum
L. candidum
L. cernuum
L. concolor
L. dauricum
L. davidii
L. hansonii
L. henryi
L. humboldtii
L. leichtlinii var. maximowiczii
L. leucanthum var. centifolium
L. longiflorum
L. maculatum
L. martagon
L. monadelphum
L. pardalinum
L. parryi
L. pumilum
L. pyrenaicum
L. regale
L. speciosum
L. taliense

L. tsingtauense
L. wardii
L. wilsonii

HYBRIDS
Asiatics (1a)[1]
Connecticut King
Pollyanna
Yellow Blaze
Impact
Gran Paradiso
Redsong
Montreaux
Nepal
White Ballerina
Asiatics (1b, 1c)[1]
Ariadne
Aloft
Discovery
Citronella
George Slate
Iowa Rose
Red Velvet
Pixies
Buff Pixie
Butter Pixie
Lemon Pixie
Aurelians (5c)[1]
Gold Eagle
White Henryi

Orientals (7)[1]
Allegra
Casablanca
Journey's End
Orienpets (8)
Black Beauty
Leslie Woodriff
Scheherazade
Silk Road
Starburst Sensation
SUSCEPTIBLE
Lilium auratum
L. brownii var. australe
L. canadense
L. duchartrei
L. formosanum
L. lancifolium[2] (L. tigrinum)
L. lankongense
L. mackliniae
L. nepalense
L. philippinense
L. rubellum
L. sulphureum
L. superbum
L. wallichianum var. neilgherrense

[1] Horticultural classification.
[2] Easily infected, but not apparently injured; can become a carrier to other susceptible species.

Trees, Shrubs, and Vines

TREES AND SHRUBS THAT WILL GROW IN ALKALINE SOIL (pH OF 7.5 TO 8.4)

Common Name	Botanical Name
Arrowwood	Viburnum dentatum
Beefwood	Casuarina
Box elder	Acer negundo
Bridalwreath	Spiraea × vanhouttei
Bush cinquefoil	Potentilla fruticosa
Common jujube	Ziziphus jujuba
Date palm	Phoenix dactylifera
Deutzia	Deutzia
Forsythia	Forsythia
Fragrant honeysuckle	Lonicera fragrantissima
Fremont cottonwood	Populus fremontii
Hackberry	Celtis
Japanese barberry	Berberis thunbergii
Japanese pagoda tree	Sophora japonica
Japanese rose	Kerria japonica
Linden viburnum	Viburnum dilatatum
Locust	Robinia
Mockorange	Philadelphus
Mountain mahogany	Cercocarpus
Rose of Sharon	Hibiscus syriacus
Russian olive	Elaeagnus angustifolia
Sargent crabapple	Malus sargentii
Silk tree	Albizia
Arizona ash	Fraxinus velutina
Washington palm	Washingtonia

TREES SUSCEPTIBLE TO LIGHTNING INJURY

Common Name	Botanical Name
SUSCEPTIBLE	
Ash	Fraxinus
Elm	Ulmus
Oak	Quercus
Pine	Pinus
Poplar	Populus
Spruce	Picea
Tuliptree	Liriodendron
LESS SUSCEPTIBLE*	
Beech	Fagus
Birch	Betula
Horse chestnut	Aesculus

* No species is totally immune, and location and size of the tree also influence susceptibility.

Adapted from Tree Maintenance, P. P. Pirone. Copyright 1978 Oxford University Press. Reprinted by permission.

TREES SUSCEPTIBLE TO LANDFILL DAMAGE

Common Name	Botanical Name
MOST SUSCEPTIBLE	
Beech	Fagus
Dogwood	Cornus
Oak	Quercus
Pine	Pinus
Spruce	Picea
Sugar maple	Acer saccharum
Tuliptree	Liriodendron
MODERATELY SUSCEPTIBLE	
Birch	Betula
Hemlock	Tsuga
Hickory	Carya
LEAST SUSCEPTIBLE	
Elm	Ulmus
Locust	Robinia
Pin oak	Quercus palustris
Plane tree	Platanus
Poplar	Populus
Sycamore	Platanus
Willow	Salix

Adapted from Tree Maintenance, P. P. Pirone. Copyright 1978 Oxford University Press. Reprinted by permission.

TREES COMMONLY DAMAGED BY SAPSUCKERS

Common Name	Botanical Name
Acacia	Acacia
Apple, crabapple	Malus
Beech	Fagus
Beefwood	Casuarina
Birch	Betula
Douglas fir	Pseudotsuga menziesii
Fir	Abies
Hemlock	Tsuga
Larch	Larix
Loquat	Eriobotrya
Magnolia	Magnolia
Palms	Palmae
Pine	Pinus
Quaking aspen	Populus tremuloides
Red maple	Acer rubrum
Red spruce	Picea rubens
Silk oak	Grevillea
Sugar maple	Acer saccharum
Willow	Salix

TREES WITH WEAK FORKS AND BRITTLE WOOD

Common Name	Botanical Name
Acacia	Acacia
Bottlebrush	Callistemon citrinus
Chestnut oak	Quercus prinus
Chinaberry	Melia azedarach
Coast redwood	Sequoia sempervirens
Eucalyptus	Eucalyptus
Horse chestnut	Aesculus
Locust	Robinia
Melaleuca	Melaleuca
Modesto ash, Arizona ash	Fraxinus velutina 'Modesto'
Poplar	Populus
Sassafras	Sassafras
She oak	Casuarina stricta
Siberian elm	Ulmus pumila
Silver maple	Acer saccharinum
Southern magnolia	Magnolia grandiflora
Tree of heaven	Ailanthus
Tuliptree	Liriodendron
White mulberry	Morus alba
Willow	Salix

TREES AND SHRUBS RESISTANT TO CROWN GALL

Common Name	Botanical Name
Abelia	Abelia
Andromeda	Pieris
Barberry	Berberis
Beech	Fagus
Birch	Betula
Boxwood	Buxus
Catalpa	Catalpa
Cedar	Cedrus
Cryptomeria	Cryptomeria
Deutzia	Deutzia
Elderberry	Sambucus
Firethorn	Pyracantha
Goldenchain tree	Laburnum
Golden rain tree	Koelreuteria
Heather	Calluna
Hemlock	Tsuga
Holly	Ilex
Hornbeam	Carpinus
Kentucky coffee tree	Gymnocladus
Larch	Larix
Leucothoe	Leucothoe
Magnolia	Magnolia
Maidenhair tree	Ginkgo
Mountain laurel	Kalmia
Oregon grapeholly	Mahonia aquifolium
Redbud	Cercis
Sassafras	Sassafras
Serviceberry	Amelanchier
Silk tree	Albizia
Smoke tree	Cotinus
Sour gum	Nyssa
Spruce	Picea
Sumac	Rhus
Sweet gum	Liquidambar
Tree of heaven	Ailanthus
Tuliptree	Liriodendron
Yellowwood	Cladrastis
Zelkova	Zelkova

Adapted from Crown Gall, W. A. Sinclair and W. T. Johnson. Cornell University Tree Pest leaflet A-5.

TREES AND SHRUBS WITH SHALLOW ROOT SYSTEMS

Common Name	Botanical Name
Acacia	Acacia
Alder	Alnus
Black locust	Robinia
Elm	Ulmus
Eucalyptus	Eucalyptus
Evergreen ash	Fraxinus uhdei
Fig	Ficus
Honeylocust	Gleditsia
Mulberry	Morus
Pacific dogwood	Cornus nuttallii
Plane tree, sycamore	Platanus
Poplar	Populus
Silver maple	Acer saccharinum
Sumac	Rhus
Tree of heaven	Ailanthus altissima
Willow	Salix

TREES THAT TOLERATE COMPETITION FROM LAWN GRASS

Common Name	Botanical Name
Amur maple	Acer tataricum ginnala
Crape myrtle	Lagerstroemia
Dogwood	Cornus
Flowering cherry, peach, plum	Prunus
Golden rain tree	Koelreuteria
Hawthorn	Crataegus
Hedge maple	Acer campestre
Japanese maple	A. palmatum
Modesto ash	Fraxinus velutina 'Modesto'
Saucer magnolia	Magnolia × soulangiana
Silk tree	Albizia julibrissin
Smoke tree	Cotinus
Star magnolia	Magnolia stellata

LOW-GROWING TREES SUITABLE FOR PLANTING UNDER OVERHEAD WIRES

Common Name	Botanical Name
Amur maple	Acer tataricum ginnala
Crabapple	Malus
Crapemyrtle	Lagerstroemia indica
Flowering cherry, peach, plum	Prunus
Golden rain tree	Koelreuteria paniculata
Hedge maple	Acer campestre
Japanese maple	A. palmatum
Mountain maple	A. spicatum
Redbud	Cercis
Smoke tree	Cotinus
Snowbell	Styrax japonicus
Tatarian maple	Acer tataricum

SMALL TREES FOR AREAS WITH RESTRICTED ROOT SPACE

Common Name	Botanical Name
Amur maple	Acer tataricum ginnala
Black haw	Viburnum prunifolium
Chinese pistache	Pistacia chinensis
Crabapple	Malus
Dogwood	Cornus
Flowering cherry	Prunus
Golden rain tree	Koelreuteria
Hawthorn	Crataegus
Hedge maple	Acer campestre
Holly	Ilex
Hornbeam	Carpinus
Japanese maple	Acer palmatum
Peach	Prunus
Plum	Prunus
Redbud	Cercis
Russian olive	Elaeagnus angustifolia
Saucer magnolia	Magnolia × soulangiana
Siebold viburnum	Viburnum sieboldii
Silk tree	Albizia julibrissin
Snowbell	Styrax
Southern black haw	Viburnum rufidulum
Star magnolia	M. stellata

Adapted from Trees for American Gardens, Donald Wyman. Copyright 1951, 1965, MacMillan Publishing Co. Reprinted by permission.

TREES AND SHRUBS FOR SHADY AREAS

Common Name	Botanical Name
Andromeda	*Pieris*
Arborvitae	*Thuja*
Azalea	*Rhododendron*
Boxwood	*Buxus*
Camellia	*Camellia*
Daphne	*Daphne*
David viburnum	*Viburnum davidii*
Dogwood	*Cornus*
False cypress	*Chamaecyparis*
Heavenly bamboo	*Nandina*
Hemlock	*Tsuga*
Holly	*Ilex*
Hydrangea	*Hydrangea*
Japanese aucuba	*Aucuba japonica*
Japanese maple	*Acer palmatum*
Laurel	*Laurus nobilis*
Laurustinus	*Viburnum tinus*
Leucothoe	*Leucothoe*
Mountain laurel	*Kalmia*
Oregon grapeholly	*Mahonia aquifolium*
Pittosporum	*Pittosporum*
Privet	*Ligustrum*
Redbud	*Cercis*
Rhododendron	*Rhododendron*
Serviceberry	*Amelanchier*
Silverbell	*Halesia*
Sweet box	*Sarcococca*
Sweet shrub	*Calycanthus*
Vine maple	*Acer circinatum*
Witch hazel	*Hamamelis*
Yew	*Taxus*

CRAPE MYRTLE VARIETIES IMMUNE TO POWDERY MILDEW

'Acoma'	'Muskogee'
'Apalachee'	'Osage'
'Choctaw'	'Tonto'
'Fantasy'	

RESISTANT VARIETIES

'Bashams Party Pink'	'Pecos'
'Comanche'	'Regal Red'
'Glendora White'	'Sioux'
'Hopi'	'Tuscarora'
'Lipan'	'Tuskegee'
'Miami'	'Wichita'
'Near East'	'Yuma'

DISEASE-RESISTANT CRABAPPLES

Species or Variety	Scab	Fire Blight	Powdery Mildew	Description
'Adams'	VR	VR	VR	Flowers opening to pink; fruit vivid red, ¾" across.
'Beverly'	VR	SR		Flowers opening to white; fruit red, ½" to ¾" across.
'Centurion'	R			Flowers rose red; fruit glossy cherry red, ⅝" across.
'Christmas Holly'	VR			Flowers opening to white; fruit bright red and lasting, ⅛" across.
'David'	VR	R		Flowers opening to white; fruit scarlet, ½" across.
'Dolgo'	VR	VR		Flowers white; fruit red, 1¼" across.
'Donald Wyman'	VR	SR	VR	Flowers opening to white; fruit glossy red, ¾" across.
'Harvest Gold'	SR			Flowers white; fruit gold and lasting, ⅜" across.
'Henning'	VR			Flowers white; fruit orange red, ⅝" across.
'Jewelberry'	VR	VR		Flowers pink and white; fruit glossy red, ½" across; dwarf, shrubby tree.
'Liset'	VR	VR	R	Flowers rose red to light crimson; fruit glossy dark red, ⅝" across.
Malus baccata var. *jackii*	VR	SR	VR	Flowers white, fragrant; fruit glossy red, ½" across.
M. floribunda	VR	R	VR	Flowers opening to pink and white; fruit yellow and red, ⅜" across.
M. sargentii	VR	R		Flowers white, fragrant; fruit dark red, ¼" across; broad habit.
'Mary Potter'	R	R	R	Flowers opening to white; fruit red, ½" across.
'Ormiston Roy'	VR	VR		Flowers pink; fruit yellow, ⅝" across.
'Red Baron'	R	VR		Flowers very dark red; fruit glossy dark red, medium-sized; columnar habit.
'Red Jewel'	R	SR	VR	Flowers white; fruit cherry red, ⅜" across; broad habit.
'Selkirk'	R	VR	R	Flowers purplish pink; fruit glossy bright red.
'Sentinel'	R	VR		Flowers pale pink; fruit red, small.
'Silver Moon'	VR			Flowers white; fruit tiny, red, persistent.
'Sugartyme'	VR			Flowers white; fruit bright red, persistent, ¼" across.
'White Angel'	VR	R		Flowers opening to white; fruit red, ½" across.
'White Cascade'	VR	VR	VR	Flowers opening to white; fruit lime yellow; weeping habit.

VR: Very resistant
R: Resistant
SR: Somewhat resistant

ELM VARIETIES RESISTANT TO DUTCH ELM DISEASE

'Accolade'	'Pathfinder'
'Autumn Gold'	'Patriot'
'Cathedral'	'Pioneer'
'Dynasty'	'Princeton'
'Frontier'	'Prospector'
'Homestead'	'Regal'
'Independence'	'Sapporo'
'New Harmony'	'Urban'
'New Horizon'	'Valley Forge'
'Ohio'	

FUCHSIAS RESISTANT TO GALL MITE

'Baby Chang'	'Lena'
'Berg NIMF'	'Liebesfraud'
'Cara Mia'	'Machu Picchu'
'Chance Encounter'	'Mary'
'Chang'	'Mendocino Mini'
'Chickadee'	'Miniature Jewels'
'Cinnabarina'	'Ocean Mist'
'Curly Q'	'Perky'
'Encliandra' hybrids	'San Francisco'
'Fabian Franck'	'Scarlet Ribbons'
'Fanfare'	'Space Shuttle'
'First Success'	'Tangerine'
'Isis'	'Texas Longhorn'
'Jamboree'	'Wave of Life'

OAKS THAT NEED ADDED WATER DURING DROUGHTS

Common Name	Botanical Name
Bur oak	Quercus macrocarpa
English oak	Q. robur
Pin oak*	Q. palustris
Red oak	Q. rubra
Scarlet oak	Q. coccinea
Swamp white oak*	Q. bicolor
White oak	Q. alba
Willow oak	Q. phellos

Will tolerate wet soil.

SUSCEPTIBILITY OF JUNIPERS TO KABATINA TWIG BLIGHT

Common Name	Botanical Name
HIGHLY SUSCEPTIBLE JUNIPERS	
Chinese juniper 'Spartan'	Juniperus chinensis
Creeping juniper 'Bar Harbor' 'Blue Rug' 'Plumosa Compacta' 'Wiltonii'	J. horizontalis
Rocky Mountain juniper 'Skyrocket'	J. scopularum
Hollywood juniper 'Torulosa'	J. chinensis
RESISTANT JUNIPERS	
Chinese juniper 'Glauca Hetzii' 'Parsonii'	J. chinensis
Common juniper 'Hornibrookii'	J. communis
Creeping juniper 'Marcella'	J. horizontalis
Hybrid juniper 'Gold Coast' 'Pfitzeriana Aurea'	J. × media
Sargent's juniper 'Glauca' 'Viridis'	J. sargentii
Savin juniper 'Tamariscifolia'	J. sabina
Single-seed juniper 'Prostrata'	J. squamata

OAKS THAT NEED NO WATER AFTER THE FIRST TWO YEARS

Common Name	Botanical Name
Blue oak	Quercus douglasii
California black oak	Q. kelloggii
California scrub oak	Q. dumosa
Canyon oak	Q. chrysolepis
Coast live oak	Q. agrifolia
Cork oak	Q. suber
Holly oak	Q. ilex
Interior live oak	Q. wislizenii
Mesa oak	Q. engelmannii
Oregon white oak	Q. garryana
Valley oak	Q. lobata

JUNIPERS RESISTANT TO PHOMOPSIS TWIG BLIGHT

Common Name	Botanical Name
Chinese juniper 'Foemina' 'Keteleeri'	Juniperus chinensis
Common juniper 'Repanda' 'Suecica'	J. communis
Single-seed juniper 'Prostrata'	J. squamata
Creeping juniper	J. horizontalis
Dwarf juniper	J. communis var. depressa
Hybrid juniper 'Pfitzerana Aurea'	J. × media
Savin juniper 'Broadmoor' 'Fargesi' 'Pumila' 'Skandia'	J. sabina

SUSCEPTIBILITY OF PINES TO PINE WILT

Common Name	Botanical Name
SUSCEPTIBLE PINES	
Austrian pine	Pinus nigra
Japanese black pine	P. thunbergii
Japanese red pine	P. densiflora
Loblolly pine	P. taeda
Lodgepole pine	P. contorta ssp. latifolia
Maritime pine	P. pinaster
Monterey pine	P. radiata
Mugo pine	P. mugo
Scotch pine	P. sylvestris
Scrub pine	P. virginiana
Sugar pine	P. lambertiana
Western white pine	P. monticola
RESISTANT PINES	
Cuban pine	Pinus caribaea
Jack pine	P. banksiana
Jeffrey pine	P. jeffreyi
Longleaf pine	P. palustris
Pitch pine	P. rigida
Short-leaf pine	P. echinata
Slash pine	P. elliottii
Table mountain pine	P. pungens
White pine	P. strobus

PALMS FREQUENTLY ATTACKED BY PALM LEAF SKELETONIZER

Common Name	Botanical Name
Cabbage palm	Livistona australis
Chinese fan palm	Livistona chinensis
Coconut palm	Cocos nucifera
Date palm	Phoenix
Paurotis palm	Acoelorrhaphe wrightii
Pindo palm	Butia capitata
Washington palm	Washingtonia

PYRACANTHA AND HAWTHORN TOLERANT OF FIRE BLIGHT

Common Name	Botanical Name
PYRACANTHA	
Laland's firethorn	Pyracantha coccinea var. lalandei P. fortuneana
HAWTHORN	
Washington hawthorn	Crataegus phaenopyrum

These species are not immune to fire blight but are not damaged by it as severely as are other species.

PYRACANTHA RESISTANT TO SCAB

'Mohave' 'Shawnee'
'Orange Glow' 'Watereri'
'Rogersiana'

GYPSY MOTH FOOD PREFERENCES

Common Name	Botanical Name
MOST PREFERRED	
Apple, crabapple	Malus
Gray birch	Betula alleghaniensis
Hawthorn	Crataegus
Linden	Tilia
Mountain ash	Sorbus
Oak	Quercus
Paper birch	Betula papyrifera
Quaking aspen	Populus tremuloides
Rose	Rosa
Serviceberry	Amelanchier
Tamarisk	Tamarix
Willow	Salix
Witch hazel	Hamamelis
INTERMEDIATE	
Beech	Fagus
Cottonwood	Populus deltoides
Elm	Ulmus
Hackberry	Celtis
Hemlock	Tsuga
Hickory	Carya
Magnolia	Magnolia
Maple	Acer
Pine	Pinus
Redbud	Cercis
Sassafras	Sassafras
Sour gum	Nyssa
Sweet cherry	Prunus avium
Sweet gum	Liquidambar
LEAST PREFERRED	
Ash	Fraxinus
Black walnut	Juglans nigra
Catalpa	Catalpa
Dogwood	Cornus
Fir	Abies
Holly	Ilex
Horse chestnut	Aesculus
Locust	Robinia
Mountain laurel	Kalmia
Mulberry	Morus
Plane tree	Platanus
Sycamore	Platanus
Tuliptree	Liriodendron

ROSE VARIETIES RESISTANT TO BLACK SPOT, POWDERY MILDEW, AND RUST

HYBRID TEAS				
'Audie Murphy'	R		'Trade Winds'	R
'Aztec'	PM, R		'Tropicana'	BS
'Carousel'	PM		'White Bouquet'	R
'Charlotte Armstrong'	BS		**FLORIBUNDAS**	
'Chrysler Imperial'	BS		'Alain'	R
'Command Performance'	BS, R		'Burma'	R
'Coronado'	BS		'Donald Prior'	R
'Ernest H. Morse'	BS		'Etiole De Hollande'	R
'Fortyniner'	BS		'Fashionette'	R
'Fred Howard'	R		'Garden Party'	R
'Garden Party'	BS, R		'Gold Cup'	R
'Golden Rapture'	PM		'Red Gold'	BS, R
'Grand Opera'	BS		'Red Radiance'	R
'Jamaica'	PM		'Sarabande'	BS, PM, R
'John F. Kennedy'	BS, R		'Simplicity'	BS
'Lowell Thomas'	PM		'Summer Snow'	R
'Lucy Cramphorn'	BS		'Tiara'	BS
'Matterhorn'	PM		'Wildfire'	PM
'Miss All American Beauty'	PM		**CLIMBERS**	
'Pascali'	PM		Blaze	BS, PM
'Pink Favorite'	R		Bonfire	PM
'Queen Charlotte'	PM		Cecile Brunner	PM
'Radiant"	BS		Paul's Scarlet	BS, PM
'Sante Fe'	PM			
'Sierra Dawn'	PM, R		**DISEASE KEY**	
'Simon Bolivar'	PM, R		BS	Black Spot
'Sphinx'	BS		PM	Powdery Mildew
'Sutter's Gold'	BS, PM			
'Tiffany'	BS		R	Rust

APPLE DISEASE RESISTANCE AND ADAPTATION

Variety	Pollination	AS	BR	CAR	FB	PM	2	3	4	5	6	7	8	9	10[3]
Anna	D	S											•	•	•
Baldwin	A	S		VR	S	HS	•	•	•	•					
Braeburn	B	S	S	HS	HS	R									
Cortland	B	HS		S	S	HS		•	•	•	•				
Dorsett Golden	D	S										•	•	•	•
Ein Shemer	D	S											•	•	•
Empire	B	S	HS	R	MR	S		•	•	•	•				
Enterprise	B	VR		VR	R	MR				•	•	•	•		
Freedom	B	VR	S	VR	R	R		•	•	•					
Fuji	B	S		HS	HS	R					•	•	•	•	
Gala	B	S		HS	HS	R				•	•	•	•	•	
Golden Delicious	A	S		HS	S	S				•	•	•	•	•	
Goldrush	B	VR		R	MR	R				•	•	•	•		
Granny Smith	B	S		R	HS	HS							•	•	
Gravenstein	E	S		VR		HS					•	•	•	•	
Honeycrisp	B	R		S	HS			•	•	•	•	•			
Idared	E	S		S	HS	HS			•	•	•	•			
Jonagold	E	HS		S	HS	S				•	•	•	•		
Jonathan	B	S		HS	HS	HS			•	•	•	•			
Liberty	B	VR		VR	MR	S		•	•	•					
McIntosh	B	HS		VR	S	S		•	•	•	•				
Northern Spy	B	S		S	S	S	•	•	•	•					
Prima	C	VR		HS	MR	R				•	•	•	•		
Priscilla	C	VR		R	MR	R				•	•	•	•		
Pristine	B	VR		R	MR	R				•	•	•	•		
Red Delicious	B	S	R	VR	R	R				•	•	•	•		
Rhode Island Greening	E	S		S	HS	S		•	•	•					
Rome Beauty	A	HS		HS	HS	HS					•	•	•	•	
Sir Prize	C	VR		HS	MR	R				•	•	•	•		
Stayman Winesap	E	HS	R	S	S	HS				•	•	•	•		
Winesap	E	HS		R	S	R				•	•	•	•		
Winter Banana	B	S		HS	S						•	•	•	•	
Yellow Transparent	B	R		R	HS	R	•	•	•	•	•				

[1] Adapted from www.caf.wvu.edu/kearneysville/wvufarm8.html.
[2] Based on USDA Plant Climate Zone Map (see page 336).
[3] Florida only.

POLLINATION KEY

A Self-fruitful, but crop is improved with a pollinator
B Pollinate with any A or B
C Prima and Priscilla cross-pollinate well; Sir Prize needs either Prima or Priscilla for pollination but will not pollinate them
D Pollinate with another D
E Not a pollinator; pollinate with an A or a B

DISEASE KEY		RESISTANCE KEY	
AS	Apple Scab	HS	Highly Susceptible
BR	Brown Rot	S	Susceptible
CAR	Cedar-Apple Rust	MR	Moderately Resistant
FB	Fire Blight	R	Resistant
PM	Powdery Mildew	VR	Very Resistant

CITRUS COLD HARDINESS

Varieties are listed from most hardy to least hardy.

Kumquat
Orangequat
Sour orange
Meyer lemon
Rangpur lime
Mandarin orange (tangerine)
Sweet orange
Bearss lime
Tangelo
Lemon
Grapefruit
Limequat
Mexican lime

Home gardeners in Florida and California can plant a wide variety of citrus in the milder areas of their states. In the warmest areas, you will have success with the more tender citrus, such as grapefruit, lemons, and limes. South Texas gardeners can plant Meyer lemon, Satsuma mandarin, and 'Marrs Early' orange.

AVOCADO HARDINESS

Varieties are listed from most hardy to least hardy.

For California	For Florida
Bacon	Brogdon
Zutano	Tonnage
Fuerte	Choquette
Hass	Pollock

APRICOT DISEASE RESISTANCE AND ADAPTATION

Variety	Pollination	Brown Rot Resistance	Zone Adaptation[1]						
			4	5	6	7	8	9[2]	10
Blenheim	A	S				•	•	•	•
Goldcot	A				•	•	•	•	
Perfection	B				•	•	•		
Royal	A	S			•	•	•		
Stella	A				•	•	•	•	
Tilton	A	R				•	•	•	•

[1] Based on USDA Plant Climate Zone Map (see page 336).
[2] California only.

CHERRY DISEASE RESISTANCE AND ADAPTATION

Variety	Pollination	Disease Resistance		Zone Adaptation[1]								
		BC	CLS	2	3	4	5	6	7	8	9[2]	10
SWEET CHERRIES												
Bing	A	S					•	•	•	•	•	
Black Tartarian	B						•	•	•	•	•	
Corum	C	R					•	•	•			
Early Burlat	C	VR					•	•	•	•	•	
Lambert	A	S					•	•	•			
Lapins	D						•	•	•	•		
Royal Ann	A	S					•	•	•			
Sam	C	R					•	•	•			
Stella	D						•	•	•	•		
Sue	C	R					•	•	•			
Van	B	S					•	•	•	•	•	•
SOUR CHERRIES												
Balaton	D						•	•	•	•		
Meteor	D		R			•	•	•	•			
Montmorency	D						•	•	•			
North Star	D		R			•	•	•	•	•		

[1] Based on USDA Plant Climate Zone Map (see page 336).
[2] California only.

PEAR DISEASE RESISTANCE AND ADAPTATION

Variety	Pollination	Fire Blight Resistance	Zone Adaptation[1]						
			4	5	6	7	8	9[2]	10
Bartlett	B	HS		•	•	•		•	
Bosc	B	HS		•	•	•		•	
Clapp	B	HS		•	•	•			
Comice	B	S		•	•	•		•	
D'Anjou	B	HS		•	•	•			
Kieffer[3]	A	R		•	•	•	•	•	•
Moonglow	A	VR		•	•	•	•		
Orient[3]	A	VR		•	•	•	•	•	•
Seckel	C	R		•	•	•	•		

[1] Based on USDA Plant Climate Zone Map.
[2] California only, except for Orient.
[3] Grown in Florida.

PEACH DISEASE RESISTANCE AND ADAPTATION

Variety	Pollination	Resistance to Bacterial Leaf Spot	2	3	4	5	6	7	8	9	10[2]
					Zone Adaptation[1]						
Belle of Georgia	A	VR				•	•	•	•		
Desert Gold	A								•	•	•
Early Red Free	A	VR			•	•	•	•			
Elberta	A	S				•	•	•	•		
Flordasun	A				•	•	•	•			
J. H. Hale	B					•	•	•	•		
Harbrite	A	R				•	•	•	•		
Loring	A							•	•	•	
Madison[3]	A	R				•	•	•	•		
Newhaven	A	S					•	•	•	•	
Redhaven	A	R					•	•	•	•	
Redskin	A	R					•	•	•	•	
Reliance[3]	A				•	•	•	•			
Rio-Oso-Gem	A	S				•	•	•	•		
Sunhaven	A	R				•	•	•	•		

[1] Based on USDA Plant Climate Zone Map (see page 336).
[2] Florida only.
[3] Does well in colder areas of Zone 5.

POLLINATION KEY

A Self-fruitful; requires no pollinator
B Requires pollinator; use an A

RESISTANCE KEY

VR Very Resistant
R Resistant
S Susceptible

PLUM DISEASE RESISTANCE AND ADAPTATION

Variety	Type	Pollination	Resistance to Black Knot	4	5	6	7	8	9[2]
				Zone Adaptation[1]					
Blue Damson	E	B	S		•	•	•		
Burbank	J	D			•	•	•	•	•
Ember	J	E		•	•	•	•	•	
Greengage	E	B				•	•	•	
Italian Prune (Fellenberg)	EPP	F	R		•	•	•	•	
Methley	J	C	R		•	•	•	•	•
Ozark Premier	J	D			•	•	•	•	
President	E	A	VR		•	•	•	•	
Santa Rosa	J	C	R		•	•	•	•	•
Shiro	J	D	R	•	•	•	•	•	
Stanley	EPP	F	S		•	•	•	•	•
Underwood	J	E			•	•	•	•	•

[1] Based on USDA Plant Climate Zone Map (see page 336).
[2] California only.

TYPE KEY

E   European plum
EPP European prune plum
J   Japanese red plum

RESISTANCE KEY

VR Very Resistant
R Resistant
S Susceptible

POLLINATION KEY

A Not self-fruitful; pollinate with a B
B Self-fruitful; no pollinator necessary
C Self-fruitful, but crop is improved by a pollinator; use a D
D Not self-fruitful; pollinate with another D
E Cross-pollinates well with another E
F Self-fruitful, but crop is improved by a pollinator; use an F

VEGETABLE SEED INFORMATION

Vegetable	Optimum Germination Temperatures	Days to Germination
Asparagus	70–75°F	14–21
Bean, lima	70	7–10
Bean, snap	70	6–10
Beet	50–85	10–14
Broccoli	70–75	10–14
Brussels sprouts	70–75	10–14
Cabbage	70–75	10–14
Carrot	50–85	14–21
Cauliflower	70–75	8–10
Cucumber	70	7–10
Eggplant	70	10–15
Lettuce	65–70	7–10
Melon	75	5–7
Onion	70–75	10–14
Parsley	70–75	14–21
Pea	40–75	7–10
Pepper	75–80	10
Radish	45–85	4–6
Spinach	70	8–10
Squash	70–75	7–10
Sweet corn	70	5–7
Tomato	70–75	5–8
Turnip	60–85	7–10

EARLIEST DATES FOR SAFE SPRING PLANTING OF VEGETABLES

| Crop | Average Date of Last Spring Freeze | | | | | | | |
	Feb. 1	Feb. 15	Mar. 1	Mar. 15	Apr. 1	Apr. 15	May 1	May 15	June 1
Asparagus	—	—	—	2/1	2/15	3/15	3/15	4/15	5/1
Bean, lima	2/1	3/1	3/15	4/1	4/15	5/1	5/15	6/1	—
Bean, snap	2/1	3/1	3/15	3/15	4/1	4/15	5/1	5/15	6/1
Beet	1/1	1/15	2/15	2/15	3/1	3/15	4/1	4/15	5/1
Broccoli[1]	1/1	1/15	2/1	2/15	3/1	3/15	4/1	4/15	5/15
Brussels sprouts[1]	1/1	1/15	2/1	2/15	3/1	3/15	4/1	4/15	5/15
Cabbage[1]	1/1	1/1	1/15	2/1	2/15	3/1	3/15	4/15	5/15
Carrot	1/1	1/15	2/1	2/15	3/1	3/15	4/1	5/1	5/15
Cauliflower[1]	1/1	1/15	1/15	2/1	2/15	3/1	4/1	4/15	5/15
Cucumber	2/15	2/15	3/1	4/1	4/15	5/1	5/15	6/1	—
Eggplant[1]	2/1	2/15	3/15	4/1	4/15	5/1	5/15	6/1	—
Lettuce	1/1	1/1	1/1	2/1	2/15	3/15	4/1	4/15	5/15
Muskmelon	2/15	2/15	3/1	4/1	4/15	5/1	6/1	—	—
Onion	1/1	1/1	1/1	2/1	2/15	3/1	3/15	4/15	5/1
Parsley	1/1	1/1	1/15	2/1	2/15	3/15	4/1	4/15	5/15
Pea	1/1	1/1	1/15	2/1	2/15	3/1	3/15	4/15	5/1
Pepper[1]	2/1	3/1	3/15	4/1	4/15	5/1	5/15	6/1	6/1
Potato	1/1	1/15	1/15	2/1	3/1	3/15	4/1	4/15	5/1
Radish	1/1	2/2	1/1	1/15	2/15	3/1	3/15	4/1	5/1
Spinach	1/1	1/1	1/1	1/15	2/1	2/15	3/15	4/1	4/15
Squash	2/1	3/1	3/15	4/1	4/15	5/1	5/1	5/15	6/1
Tomato[1]	2/1	3/1	3/15	4/1	4/15	5/1	5/15	5/15	6/1
Turnip	1/1	1/15	2/1	2/1	2/15	3/1	3/15	4/1	5/1
Watermelon	2/15	2/15	3/1	3/15	4/15	5/1	5/15	6/1	—

[1] Seeds may be started indoors 4 to 6 weeks before planting date.

To find the average date of the last freeze, ask at your local nursery or call your county extension service.

VEGETABLE VARIETIES RESISTANT TO SOUTHERN ROOT KNOT NEMATODE

Bean	Corn	Pea	Pepper	Sweet Potato	Tomato
Bountiful	Carmel Cross	Burpeeana Early	All Big	Apache	All Round
Brittle Wax	Golden Beauty Hybrid	Wando	Bontoc Sweet Long	Carver	Anahu
Tender Pod	Golden Cross Bantam		World Beater	Hopi	Anahu-R
Wingard Wonder	Span Cross			Jasper	Atkinson
				Jewel	Auburn 76
				Nemagold	Beefeater
				Nugget	Beefmaster
				Ruby	Big Seven
				Sunnyside	Calmart
				White Bunch	Chicogrande
				White Triumph	Coldset
				Whitestar	Eurocross
					Extase
					Monte Carlo
					Nemared
					Nematex
					Patriot
					Peto 662 VFN
					Ponderosa
					VFN–8

VEGETABLES SUSCEPTIBLE TO FUSARIUM

Asparagus	Onion
Bean	Pea
Brussels sprouts	Pepper
Cabbage	Radish
Cauliflower	Spinach
Celery	Sweet potato
Cucumber	Tomato
Melon	Turnip
Okra	

BEAN VARIETIES RESISTANT TO RUST

Cape
Dade
Kentucky Wonder
Resisto

BEAN VARIETIES RESISTANT TO MOSAIC VIRUS

Aristocrop	Golden Rod
Arrow	Harvester
Astro	Improved Tendergreen
Bonanza Wax	M.R.
Bush Blue Lake	Peak
Bush Blue Lake 47 and 274	Provider
	Resistant Cherokee
Cape	Resisto
Cherokee	Roma II PVP
Contender	Romano
Dade (pole)	Spartan
Del Rey	Spurt
Eagle	Strike
Early Gallatin	Stringless Blue Lake
Early Harvest	Sungold
Flo	Tendercrop
FM–IK	Tenderlake
Gallatin 50	Topcrop
Gator Green 15	Win
Gold Crop	

VEGETABLES SUSCEPTIBLE TO VERTICILLIUM WILT

Artichoke	Pepper
Beet	Potato
Brussels sprouts	Pumpkin
Cabbage	Radish
Eggplant	Rhubarb
Melon[1]	Spinach
New Zealand spinach	Strawberry[2]
Okra	Tomato
Peanut	

[1] Watermelon, cantaloupe, and honeydew become infected but are not seriously damaged. Persian, casaba, and crenshaw melons are very susceptible.

[2] See page 366 for strawberry varieties resistant to verticillium.

Adapted from Plants Resistant or Susceptible to Verticillium Wilt, University of California County Extension leaflet 2703.

CORN VARIETIES TOLERANT OF DISEASES

Apache	S, B, Bl, M
Atlantic	Bl
Aztec	S, B
Bellringer	S, B
BiQueen	Bl
Calico	S, B
Calumet	S, B, M
Capitan	Bl
Cherokee	B, Bl, M
Comanche	S, B
Comet	S, B, Bl
Florida Staysweet	Bl
Gold Cup	S, B
Guardian	Bl
Merit	S, B, M
Mevak	S, B
Quicksilver	S, B, M
Seneca Sentry	B, M
Silver Queen	B
Sweet Sue	S
Wintergreen	S, B, Bl, M

DISEASE KEY

S   Smut

B   Bacterial Wilt

Bl Southern and Northern Leaf Blights

M   Maize Dwarf Mosaic

BEET VARIETIES THAT PRODUCE SMOOTH, ROUND ROOTS

Albino White Beet
Detroit Dark Red
Earlisweet Hybrid
Early Wonder, Green Top
Early Wonder, Tall Top
Garnet
Golden Beet
Perfected Detroit
Red Ace Hybrid
Red Ball
Ruby Queen

BLUEBERRY REGIONAL ADAPTATION

FOR THE SOUTH AND SOUTHERN CALIFORNIA

Rabbiteye (Vaccinium ashei)
Bluebelle
Bluegem
Climax
Tifblue
Woodward
Highbush (Vaccimium corymbosum)
Avonblue
Flordablue
Sharpblue

FOR THE VERY COLDEST REGIONS

Meader
Northland

WIDELY ADAPTED

Berkeley
Bluecrop
Blueray
Bluetta
Collins
Jersey
Patriot

DISEASE-RESISTANT CUCUMBERS

SLICING CUCUMBERS

A & C Hybrid Imp	L+, S+, A+, P+, D+, C+
A & C Hybrid 1810	L+, S+, A+, P+, D+, C+
Cherokee 7	L+, S+, A+, P, D, C+
Dasher	L, S, A, P, D, C
Dasher II	L, S+, A+, P+, D+, C+
Early Triumph	L, S, A, P, D, C
Gemini 7	L, S, A, P, D, C
Medalist	S+, P, D, C+
Poinsett	L+, A+, P+, D+
Poinsett 76	L+, S+, P+, D+
Roadside Fancy	L, S, A, P, D, C
Setter	L, S+, A, P+, D+
Shamrock	L, S, P, D, C
Slicemaster	L, S+, A, P, D, C
Slice-Mor	S+, A+, P, D+, C
Southernsett	L, S+, P, D, C+
Sprint 440	L, S, A, P, D
Sweet-Slice	L, S+, A, P, D, C
Sweet Success	S, P, D, C

PICKLING CUCUMBERS

Addis	L, A, P, D, C
Bounty	L, S, A, P, D, C
Calypso	L, S, A, P, D, C
Carolina	L, S, A, P, D, C
Chipper	L, A, P, D, C
County Fair	S, A, P, D, C
Explorer	L, P, D
Flurry	L, S, A, P, D, C
Liberty	L, S, D, C
Lucky Strike	S, A, P, C
Multipik	L, S, A, D, C
Panorama	L, S, A, P, D, C
Peto Triplemech	L, S, A, P, D, C
Picarow	L, S, A, P, D, C
Premier	L, S, A, P, D, C
Salty	S, P, D, C
Sampson	L, A, P, D, C
Score	L, S, A, D, C
Spear-It	L, S, A, P, D, C
Sumter	L, S, A, P, D, C
Tamor	L, S, A, P, D, C
Triple Crown	L, S, A, P, C
V.I.P.	L, S, A, P, D, C

DISEASE KEY

A variety's tolerance of a disease is indicated by the codes below. If a variety is resistant to the disease, the code is followed by a plus (+).

L Angular Leaf Spot
S Scab
A Anthracnose
P Powdery Mildew
D Downy Mildew
C Cucumber Mosaic Virus

DISEASE RESISTANCE OF BRAMBLES

	Disease Resistance	Widely Adapted	North	Pacific South	Pacific Northwest	California
ERECT BLACKBERRIES						
Black Satin	A, LS	•		•	•	•
Brazos				•		
Cherokee	A, OR	•				
Chester Thornless	C	•				
Darrow		•				
Ebony King	OR	•				
Eldorado	OR			•		
Flordagrand[1]	LS			•		
Hull Thornless		•				
Lawton	OR, V			•		
Navaho	A	•				
Shawnee	A, OR	•				
Triple Crown		•				
TRAILING BLACKBERRIES						
Black Butte				•	•	
Boysenberry	OR, V				•	•
Loganberry					•	•
Marionberry					•	
Olallieberry	M, V			•	•	
Youngberry	OR					
RED RASPBERRIES, SUMMER-BEARING						
Canby		•				
Chilliwack		•				
Latham		•				
Meeker	PM				•	
Sumner	PM	•			•	
Titan	M		•			
Willamette	PM				•	•
RED RASPBERRIES, EVERBEARING						
Amity		•				
August Red			•			
Autumn Bliss		•				
Heritage		•				
Redwing			•			
Southland	A, LS, PM			•		
Summit						•
Black Raspberries						
Bristol	PM	•				
Cumberland			•			
Munger	PM				•	

1 Needs cross-pollination.

DISEASE KEY

A Anthracnose
C Cane Blight
LS Leaf Spot
M Mosaic
OR Orange Rust
PM Powdery Mildew
V Verticillium Wilt

DISEASE RESISTANCE AND ADAPTATION OF GRAPES

| Variety | Disease Resistance | | | Adaptation | | | | |
	Black Rot	Downy Mildew	Powdery Mildew	Northeast	Midwest	Pacific Northwest	Southeast	California, Arizona
Aurora		R	R	•		•		
Beta	R			•	•	•		
Buffalo		R	R	•	•	•		
Campbells Early	R	R		•		•		
Canadice				•	•	•		
Concord		R		•	•	•		
Delaware	R			•		•		
Flame Seedless						•		•
Fredonia	R			•		•		
Glenora		R	R	•	•	•		
Himrod			R	•	•	•		
Interlaken			R	•	•	•		
Magnolia							•	
Niagara			R	•	•	•		•
Reliance		R	R	•	•	•		
Scuppernong[1]							•	
Suffolk Red			R	•	•	•		
Thompson Seedless								•
Tokay								•
Worden	R		R	•	•	•		

1 Muscadine grape; use Magnolia as pollinator.

RESISTANCE KEY

R   Resistant

REGIONAL ADAPTATION OF ONION VARIETIES

FOR THE SOUTH
Excel
Granex
Texas Grano
Tropicana Red
White Granex

FOR THE WEST
California Early Red
Early Yellow Globe
Southport Yellow Globe
Yellow Bermuda

FOR THE NORTH
Downing Yellow Globe
Early Yellow Globe
Empire
Nutmeg
Spartan Lines

LETTUCE VARIETIES TOLERANT OF TIP BURN

Calmar
Climax
Empire
Empress
   Fairton
Great Lakes 118,
   366, 659, 659–700,
   and 6238
Green Lake
Ithaca
Merit
Mesa 659
Minetto
Montello
Montemar
New York 515
   Improved
Oswego
Parris Island Cos
Pennlake
Salinas
Super 59
Vanguard
Vanguard 75
Vanmax

WATERMELON VARIETIES TOLERANT OF ANTHRACNOSE

Blackstone
Calhoun
Charleston Gray
Crimson Sweet
Dixielee
Family Fun
Graybelle
Imperial
Madera
Smokylee
Sweet Favorite
   Hybrid
Verona
You Sweet
   Thing Hybrid

ONION VARIETIES TOLERANT OF PINK ROOT

Autumn Spice
Beltsville Bunching
Brown Beauty
Buccaneer
Colossal
Copper Coast
Danvers
Early Supreme
El Capitan
Evergreen
   White Bunch
Fiesta
Granada
Granex Yellow
Henry's Special
Majesty
Red
   Commander
Rialto
Ringer
Spanish Main
White Granex
White Robust
Yellow Globe
Yellow Grano-
   New Mexico

Peppers Tolerant of Tobacco Mosaic Virus

Ace
Allbig
Annabelle
Argo
Beater
Bell Boy
Big Bertha
Burlington
Early Canada Bell
Early Niagara Giant
Early Wonder
Emerald Giant
Gatorbelle
Gypsy
Hybelle
Lady Bell
Liberty Bell
Ma Belle
Merced
Mercury
Midway
Miss Belle
New Ace
Pennwonder
Pimientol
Puerto Rico
  Perfection
Puerto Rico
  Wonder
Resistant
  Florida Giant
Rutgers World
  Beater
Shamrock
Skipper
Staddon's
  Select
Thick Walled
  World Beater
Titan
Valley Giant
Yolo Wonder

Potato Varieties Tolerant of Scab

Alamo
Cascade
Cherokee
La Rouge
Lemhi
Nooksack
Norchip
Norgold Russet
Norland
Ona
Onaway
Ontario
Plymouth
Pungo
Russet Burbank
Shurchip
Sioux
Superior
Targhee

Spinach Varieties Resistant to Downy Mildew

Aden
Badger Savoy
Basra
Bismark
Bouquet
Califlay
Chesapeake
Chinook
Dixie Market
Duet
Early Smooth
Grandstand
High Pack
Long Standing
  Savoy
Marathon
Melody
Nares
Salma
Savoy Supreme
Skookum
Vienna
Winter
  Bloomsdale

Regional Adaptation and Disease Resistance of Tomato Varieties

	Disease Resistance	Widely Adapted	South	North	West
EARLY SEASON					
Early Cascade	V, F	●			
Jetfire	V, F				●
New Yorker	V			●	
Porter Improved					●
Small Fry	V, F, N	●			
Spring Set	V, F	●			
MIDSEASON					
Ace 55	V, F				●
Atkinson	F, N		●		
Better Boy	V, F, N	●			
Big Girl Hybrid	V, F	●			
Big Set	V, F, N	●			
Bonus	V, F, N	●			
Burpees VF	V, F	●			
Columbia	V, F, CT	●			
Floradel	F		●		
Floramerica	V, F	●			
Heinz 1350	V, F	●			
Jet Star	V, F			●	
Marglobe	F	●			
Parks Whopper	V, F, N, T	●			
Roma VF	V, F	●			
Rowpac	V, F, CT				●
Roza	V, F, CT				●
Salad Master	V, F, CT				●
Supersonic	V, F			●	●
Terrific	V, F, N	●			
Tripi-Red	V, F		●		
LATE SEASON					
Beefeater	V, F, N	●			
Beefmaster	V, F, N	●			
Manalucie	F		●		
Ramapo	V, F			●	
Tropic	V, F, T		●		
Vineripe	V, F, N	●			
Wonder Boy	V, F	●			

Disease Key

V Verticillium Wilt
F Fusarium Wilt
N Nematodes
T Tobacco Mosaic Virus
CT Curly Top

DISEASE RESISTANCE OF STRAWBERRIES

| Variety | Disease Resistance | | | | | | Adaptation | | | |
	Leaf Spot	Powdery Mildew	Red Stele	Scorch	Verticillium Wilt	Virus	South	North	California	Pacific Northwest
Allstar		VR	VR	VR	VR			•		
Badgerbelle	VR							•		
Badgerglo	R							•		
Benton			R			R				•
Blakemore	R			VR	VR		•	•		
Cardinal	VR	VR		VR			•	•		
Catskill				VR	VR			•		
Cyclone	VR							•		
Darrow	R	R	VR	R	R			•		
Delite	VR		VR	VR	VR		•	•		
Earlibelle	VR			VR			•			
Earliglow		R	VR	VR	VR			•		
Florida 90						R	•			
Fort Laramie								•		
Guardian		VR	VR	VR	VR			•		
Honeoye	R			R				•		
Hood	R	VR	VR		VR					•
Jewel	VR			VR				•		
Lateglow		R	VR	R	VR			•		
Midway			R		VR			•		
Ogallala	VR						•	•		
Ozark Beauty	VR			VR			•	•		
Pocahontas	R			VR			•			
Redchief		VR	VR	VR	R			•		
Scarlet	VR			VR				•		
Seascape									•	•
Selva									•	
Sequoia									•	
Shuksan			VR			R				•
Sparkle	VR		VR					•		
Stoplight	R			R				•		
Surecrop	VR		VR	VR	VR			•		
Tilikum										•
Tioga	R					R	•		•	
Totem	R	R	R	R						•
Tribute	VR	VR	VR				•			•
Tristar			VR	R	VR		•	•		•
Trumpeter					VR			•		

RESISTANCE KEY

VR Very Resistant

R   Resistant

Photographers

In order to make *Ortho's Home Gardener's Problem Solver* as useful as possible, we have published a color photograph of every problem in the book. Gathering this many photographs was a massive task. We relied extensively on college professors and County Extension agents for photographic contributions. Many of the photographs were taken originally to help teach students or the public about plant problems. We wish to thank the following photographers for their invaluable assistance in supplying these photographs. After each photographer's name, we list the pages on which his or her photographs appear. The letter that follows the page number shows the position of the photograph on the page, from left to right. A lowercase letter i refers to an inset photograph.

Adkins, Ralph J.: 138L, 138C, 150L, 161L, 162R, 167L, 189L, 193R, 196L, 210L, 272L, 275R, 299C, 301R; Adkins, Scott T.: 271Li, 279R; Allen, W.: 220L, 236L; American Phytopathological Society: 204R, 216R, 281Ci; Aplin, William C.: 38, 65; Badgley, Max E.: 25L, 52C, 55R, 87R, 193L, 193C, 196R, 196Ri, 208R, 229C, 252R, 277L, 284R, 291C, 303C, 310C, 311L, 313C, 319C, 324C, 327R, 330R; Bingham, Ray R.: 284C; Boger, Allen: 49L, 52R, 66R, 85C, 85R, 132L, 143L, 144C, 146R, 151C, 155L, 155R, 163C, 168R, 175L, 179C, 206L, 207C, 210R, 251L, 251C, 253C, 255L, 255C, 255R, 283R, 285L, 286L, 290C, 301L; Brame, R. Harper: 234C, 256C, 265Ri, 277R, 302C; Bridges, Bartow H. Jr.: 132C, 144R, 152L, 154C, 155C, 161C, 179L; Butler, Jackie D.: 51L, 55L, 110Ci, 219C, 325L; Byther, Ralph S.: 83R, 86L, 113L, 121R, 122L, 137C, 156L, 156C, 157C, 162C, 165C, 169L, 177L, 177C, 179R, 183C, 183Ci, 191R, 216C, 222L, 227L, 227C, 235L, 256L, 265C, 269L, 269C, 282C, 289L, 289C, 295L, 297R, 298C; Caldwell, D., Davey Tree Expert Co.: 24Ci, 52L, 68C, 109C, 109Ri, 158C, 158R; Callan, Kristie: 27C, 28C, 71L, 80L, 80R, 81C, 82L, 82C, 84L, 84C, 84R, 85L, 86C, 86R, 88R, 90C, 90R, 91C, 92L, 92C, 93L, 93C, 93R, 94L, 94C, 95C, 95R, 96R, 97L, 98R, 100L, 100C, 100R, 101C, 101R, 102L, 103L, 103R, 104C, 104R, 106C, 107L, 107C, 107R, 109C, 110L, 111R, 112C, 112R, 115R, 117C, 117R, 118C, 119L, 120R, 121L, 121C, 123C, 213R, 286C, 286R; Cardillo, Rob: 63, 64; Childress, Clyde: 131; Clark, Jack K.: 68L, 210C, 224Li, 281C, 294Ci, 317C, 320L, 326R, 327C, 329L; Clemson University: 277C, 315C; Coartney, James S.: 24L; Coatsworth, Josephine: 199, 201L; Collman, Sharon J.: 56C, 92R, 105C, 135C, 136R, 138C, 139L, 139C, 143C, 143R, 144L, 146C, 148L, 148C, 152R, 165R, 166C, 166R, 181L, 182C, 184L, 184R, 189L, 191C, 194C, 205L, 300C; Coop. Ext. Assoc. of Nassau County, NY: 49C, 49Ci, 58R, 167C, 182R; Copeland, Alan: 26C, 76BL; Coppert, David M.: 254L; Cotner, Samuel: 272C, 274L, 274R, 280R, 281L, 283L,

293Ri; Cranshaw, Whitney, Colorado SU: 184Ri, 194C, 194Ci; Cross, David J.: 325R; Crozier, J. A.: 56Li, 58L, 69C, 69Ci, 81R, 116R, 137L, 169R, 190C, 190R, 192L, 204C, 259C, 270R, 271L, 295R, 305L; Cummings, Maynard W.: 205L, 205R, 225R, 316L, 317L, 318L, 318C; Daughtrey, Margery: 99R, 102C, 116L, 171L; Davey Tree Expert Co.: 149R, 185L, 185Li; Davis, Spencer H. Jr.: 22L, 105R, 122C, 148R, 166L, 172R, 177Ri, 178Ci; De Hertogh, A. A.: 111C; DeFilippis, Jim: 134R, 209L, 223R; Dill, James F.: 101L, 134L, 134C, 152C, 157L, 163L, 172C, 175R, 176R, 180C, 181L, 207L, 217L, 251R, 258R, 260L, 262R, 262Ri, 263R, 268R, 270C, 273R, 276L, 277Ri, 280C, 294L, 295C, 296C, 302R, 303L, 305R, 313L, 315R, 316C, 317R, 320R, 321L, 321C, 324L, 324R, 325C, 331C; Dirr, Michael A.: 119C, 133L, 164R, 183L, 188C, 197R; Ebeling, Walter: 314L, 319R, 320Ri, 321R, 328L, 329R; Elmore, C. L.: 41, 42L, 57C, 57R; Emerson, Barbara H.: 59L, 60L; Feucht, James: 110C, 114C, 114R, 150C, 153C, 157R, 180L, 184C, 197L, 211L; Fitch, Charles Marden: 24C, 108L, 186R, 260R; Foulds, Randolph: 74B, 77; Frese, Paul F.: 114L; Furniss, Malcomb: 178R; Gill, Raymond J.: 25C; Goldberg, David: 75, 76T, 78, 124, 128; Gray, Ken: 231L, 237L, 276R, 276Ri, 278R, 287R, 289R; Hafernik, John E. Jr.: 312R; Hale, Frank A.: 50C, 163R, 167R; Hall, Dennis H.: 262C, 267R, 282L, 299L, 303R; Hansen, Mary Ann: 186L; Harivandi, M. Ali: 49R, 51R, 54C, 57L; Hatch, Duane L.: 227Li, 250R, 256R, 273L; Hawkes, George R.: 211C, 211R, 224C, 250C; Hodgson, Larry, HortiCom Inc.: 24R; Holmes, G. J.: 304L; Holt, Saxon: 40, 43, 129, 307, 308, 309; Horst, R. K.: 29R, 97C, 110R, 116C, 252L; Howitt, Angus J.: 214R, 214Ri, 219L, 220R; Janne, E. E.: 318R; Johnson, Dr. Russell: 296L, 320C; Jones, Alan L.: 213L, 213C, 228C, 228R, 215C, 218R, 222R, 226R, 227R, 228R, 230R, 234R, 235C, 236R; Jones, Deborah: 238; Jones, R. K.: 115C; Joyner, Gene: 53L, 53Li, 61C, 170C, 225C, 285C; Kennedy, M. Keith: 50L, 54L, 67R, 70L 87L, 136L, 151L, 160C, 168L, 176L, 176C, 176Ci, 176R, 221L, 231R, 257C; Kinsey, Marvin G.: 311R; Koehler, C. S.: 187L, 298C; Korab, Balthazar: 72, 79; Kriner, R. R.: 51L, 181C, 212L, 215L, 215R, 216L, 226L, 228L, 229L, 264L, 268L, 268C, 271L, 273C, 284L, 297C, 299R, 314R, 328C, 328Ci; Laemmlen, F.: 91L, 280L, 291L; Landis, Michael: 32, 35, 36, 39, 44, 198, 200, 240L, 241BL, 241BR; LaRue, James H.: 233C; Liebman, David: 60R, 306BL, 310L, 310R, 326L; Lindtner, Peter: 89C, 195C, 206L, 207L, 219R; Lyons, Robert E.: 25R; MacDonald, J. D.: 70C, 70Ci; Mayer, Steven L.: 316R; McCay, Bryan: 34, 76BR, 125; McClurg, Charles A.: 278C; McFarland, Dennis: 328R; McGrew, J. R.: 285R; McKinley, Michael: 26R, 29L, 29C, 66L, 67L, 67C, 74T, 119L, 132L, 142L, 142C, 142R, 147L, 150R, 151R; Meritt, Richard W.: 326C, 327L; Miller, Jeffery C.:

157Ri; Miller, R. H.: 322C; Miller, Richard L.: 291R, 296R, 298L, 315L, 322L; Moore, Wayne S.: 122C, 133C, 168C, 214L, 214C; Nameth, Steve: 91L; Natter, Jean R.: 23R, 61L, 154R, 312C; Norton, R. A.: 204L, 208L, 235R; NY State Turfgrass Assoc: 56L, 58C; NYS Ag. Exper. Station, Harvo Tashiko: 298Li; Ohr, Howard D.: 55C; Orhcard Nursery, Lafayette, Ca.: 129; Ortho Photo Library: 37, 60C, 62, 201R, 202, 240R, 241T, 242, 243, 244, 245, 246, 247, 248, 249, 259C; OSU, Ken Gray Collection: 50Li, 51Ci, 52Ci, 113C, 133R, 182L; Parsons, Jerry M.: 261C, 261R, 288R, 304C; Pehrson, John: 223C; Peirce, Pamela K.: 61R; Perry, Sandra: 27R, 96L, 174C, 263L; Pindar, Mary C.: 73; Plant Photo Library: 137R; Pollet, D. K.: 173C, 232C, 232R, 233L, 236C; Potter, Daniel A.: 164L; Powell, C. C.: 31L, 66C, 69L, 69R, 70R, 96C, 99L, 104L, 111L, 159C, 159R, 160L, 174L, 174R, 178C, 179Ci, 189C; Quirino, Cecil B.: 26L, 89R, 278L, 311C, 314C, 330L, 330C, 331L; Raabe, Robert D.: 23L, 30C, 31C, 31R, 99L, 103C, 109L, 118R, 123R, 162L, 171L, 171R, 185C, 192C, 212R, 237C, 250L, 302L; Rhoads, Ann F.: 87C, 146L, 191L; Rice, Ken: 19; Rogers, Paul A.: 187C; Rothenberger, Barbara: 83L, 89L, 98L, 123L, 195R; Rutledge, Alvin D.: 269R; Sabarese, Anita: 50Ri, 82R, 88L, 89Ci, 135R, 140L, 141L, 141C, 141R, 169C, 208L, 253R, 254R, 257L, 293R, 321Ri; Salmon, Eric: 127; Sanders, Doug: 262L; Schoeneweiss, D. F.: 71R, 97R, 149C, 195L, 197C; Schuder, Donald L.: 164Li, 194R, 221C, 222C; Sherf, A. F.: 106L, 230C, 253L, 257R, 260C, 264C, 264R, 265L, 265R, 266L, 267C, 271R, 272C, 276C, 281R, 288L, 288C, 290R, 292R, 293L, 293C, 294C, 294R, 304R; Sibbett, G. Steve: 237R; Sikkema, James: 56R, 81L, 118L, 120C, 136C, 180R, 209C, 209R, 259R, 292L; Sites, Robert W.: 312L; Smiley, Richard W.: 48R, 48Ri; Smith, John J.: 42R; Smith, Michael D.: 23C, 48L, 48C, 53C, 59C, 80C, 135L, 139R, 140C, 140Ci, 145C, 145R, 147L, 147R, 149L, 153L, 153R, 154L, 156R, 188R, 218L, 218C, 221R, 224L, 224R, 233C, 254C, 258L, 258C, 266C, 282R, 323R; Sorensen, Kenneth, NCSU: 231C, 275C; Specker, Donald: 301C; Stacey, Leon: 266R, 267L, 305R; Stark, Arvil R.: 50R, 53R, 54R; Stein, Deni Weider: 108R, 192R; Struse, Steve: 18, 20R; Swezey, Lauren Bonar: 22C, 22R, 68R, 71L, 88L, 164C, 170C; The Studio Central: 20L, 21; Thomson, Sherman V.: 90L, 159Ri, 160Ci, 177R, 206R, 230L; UCB, Coop. Ext.: 159L; UCD, Coop. Ext.: 59R; University of California: 94R, 172L, 187R, 212L, 217C, 220L, 232L, 293Li, 331R; University of Illinois, Champaign-Urbana: 30R, 106C, 122Li; USDA Forest Service: 275L, 322L, 323L, 323C; Van Waters & Rogers: 319L, 329C; Weidhaas, John A. Jr.: 173L, 186C; West, Ron: 306TL, 306TR, 306BR, 313R; Wick, Robert L.: 161R; Wott, John A.: 83C, 140R; Zitter, Thomas: 274Li

367

Index